AF352817

Wild Crop Relatives and Associated Biocultural and Traditional Agronomic Practices for Food and Nutritional Security

Wild Crop Relatives and Associated Biocultural and Traditional Agronomic Practices for Food and Nutritional Security

Editors

Purushothaman Chirakkuzhyil Abhilash
Ajeet Singh
Rama Kant Dubey
Hailin Zhang
Othmane Merah

MDPI • Basel • Beijing • Wuhan • Barcelona • Belgrade • Manchester • Tokyo • Cluj • Tianjin

Editors
Purushothaman Chirakkuzhyil Abhilash
Banaras Hindu University
India

Ajeet Singh
Banaras Hindu
University India

Rama Kant Dubey
Research Unit
Germany

Hailin Zhang
China Agricultural University
China

Othmane Merah
Université de Toulouse
France

Editorial Office
MDPI
St. Alban-Anlage 66
4052 Basel, Switzerland

This is a reprint of articles from the Special Issue published online in the open access journal *Agronomy* (ISSN 2073-4395) (available at: https://www.mdpi.com/journal/agronomy/special_issues/wild_crop_biocultural_agronomic_practices_food_security).

For citation purposes, cite each article independently as indicated on the article page online and as indicated below:

LastName, A.A.; LastName, B.B.; LastName, C.C. Article Title. *Journal Name* **Year**, *Article Number*, Page Range.

ISBN 978-3-03943-400-8 (Hbk)
ISBN 978-3-03943-401-5 (PDF)

Cover image courtesy of Ajeet Singh.

Contents

About the Editors . . . . **vii**

Preface to "Wild Crop Relatives and Associated Biocultural and Traditional Agronomic Practices for Food and Nutritional Security" . . . . **ix**

Ajeet Singh, Rama Kant Dubey, Amit Kumar Bundela and Purushothaman C. Abhilash
The Trilogy of Wild Crops, Traditional Agronomic Practices, and UN-Sustainable Development Goals
Reprinted from: *Agronomy* **2020**, *10*, 648, doi:10.3390/agronomy10050648 . . . . **1**

Mariam Coulibaly, Chaldia O.A. Agossou, Félicien Akohoué, Mahamadou Sawadogo and Enoch G. Achigan-Dako
Farmers' Preferences for Genetic Resources of Kersting's Groundnut [*Macrotyloma geocarpum* (Harms) Maréchal and Baudet] in the Production Systems of Burkina Faso and Ghana
Reprinted from: *Agronomy* **2020**, *10*, 371, doi:10.3390/agronomy10030371 . . . . **11**

James Poornima Jency, Ravikesavan Rajasekaran, Roshan Kumar Singh, Raveendran Muthurajan, Jeyakumar Prabhakaran, Muthamilarasan Mehanathan, Manoj Prasad and Jeeva Ganesan
Induced Mutagenesis Enhances Lodging Resistance and Photosynthetic Efficiency of Kodomillet (*Paspalum Scrobiculatum*)
Reprinted from: *Agronomy* **2020**, *10*, 227, doi:10.3390/agronomy10020227 . . . . **31**

Difo Voukang Harouna, Pavithravani B. Venkataramana, Athanasia O. Matemu and Patrick Alois Ndakidemi
Agro-Morphological Exploration of Some Unexplored Wild *Vigna* Legumes for Domestication
Reprinted from: *Agronomy* **2020**, *10*, 111, doi:10.3390/agronomy10010111 . . . . **47**

Barthlomew Chataika, Levi Akundabweni, Enoch G. Achigan-Dako, Julia Sibiya, Kingdom Kwapata and Benisiu Thomas
Diversity and Domestication Status of Spider Plant (*Gynandropsis gynandra*, L.) amongst Sociolinguistic Groups of Northern Namibia
Reprinted from: *Agronomy* **2020**, *10*, 56, doi:10.3390/agronomy10010056 . . . . **73**

Sujith S. Ratnayake, Lalit Kumar and Champika S. Kariyawasam
Neglected and Underutilized Fruit Species in Sri Lanka: Prioritisation and Understanding the Potential Distribution under Climate Change
Reprinted from: *Agronomy* **2020**, *10*, 34, doi:10.3390/agronomy10010034 . . . . **87**

Anna Michalska-Ciechanowska, Aneta Wojdyło, Bożena Bogucka and Bogdan Dubis
Moderation of Inulin and Polyphenolics Contents in Three Cultivars of *Helianthus tuberosus* L. by Potassium Fertilization
Reprinted from: *Agronomy* **2019**, *9*, 884, doi:10.3390/agronomy9120884 . . . . **107**

Carla Guijarro-Real, Ana M. Adalid-Martínez, Katherine Aguirre, Jaime Prohens, Adrián Rodríguez-Burruezo and Ana Fita
Growing Conditions Affect the Phytochemical Composition of Edible Wall Rocket (*Diplotaxis erucoides*)
Reprinted from: *Agronomy* **2019**, *9*, 858, doi:10.3390/agronomy9120858 . . . . **119**

Amandine D.M. Akakpo and Enoch G. Achigan-Dako
Nutraceutical Uses of Traditional Leafy Vegetables and Transmission of Local Knowledge from
Parents to Children in Southern Benin
Reprinted from: *Agronomy* **2019**, *9*, 805, doi:10.3390/agronomy9120805 133

Juan B. Alvarez and Carlos Guzmán
Recovery of Wheat Heritage for Traditional Food: Genetic Variation for High Molecular Weight
Glutenin Subunits in Neglected/Underutilized Wheat
Reprinted from: *Agronomy* **2019**, *9*, 755, doi:10.3390/agronomy9110755 153

**Martin Brtnicky, Tereza Dokulilova, Jiri Holatko, Vaclav Pecina, Antonin Kintl,
Oldrich Latal, Tomas Vyhnanek, Jitka Prichystalova and Rahul Datta**
Long-Term Effects of Biochar-Based Organic Amendments on Soil Microbial Parameters
Reprinted from: *Agronomy* **2019**, *9*, 747, doi:10.3390/agronomy9110747 163

**Kanagaraj Muthu-Pandian Chanthini, Vethamonickam Stanley-Raja, Annamalai
Thanigaivel, Sengodan Karthi, Radhakrishnan Palanikani, Narayanan Shyam Sundar,
Haridoss Sivanesh, Ramaiah Soranam and Sengottayan Senthil-Nathan**
Sustainable Agronomic Strategies for Enhancing the Yield and Nutritional Quality of Wild
Tomato, *Solanum Lycopersicum* (l) Var *Cerasiforme* Mill
Reprinted from: *Agronomy* **2019**, *9*, 311, doi:10.3390/agronomy9060311 179

**Difo Voukang Harouna, Pavithravani B. Venkataramana, Athanasia O. Matemu
and Patrick Alois Ndakidemi**
Wild *Vigna* Legumes: Farmers' Perceptions, Preferences, and Prospective Uses for
Human Exploitation
Reprinted from: *Agronomy* **2019**, *9*, 284, doi:10.3390/agronomy9060284 201

**Teresa Borelli, Danny Hunter, Stefano Padulosi, Nadezda Amaya, Gennifer Meldrum,
Daniela Moura de Oliveira Beltrame, Gamini Samarasinghe, Victor W. Wasike, Birgül Güner,
Ayfer Tan, Yara Koreissi Dembélé, Gaia Lochetti, Amadou Sidibé and Florence Tartanac**
Local Solutions for Sustainable Food Systems: The Contribution of Orphan Crops and Wild
Edible Species
Reprinted from: *Agronomy* **2020**, *10*, 231, doi:10.3390/agronomy10020231 227

**Fernand S. Sohindji, Dêêdi E. O. Sogbohossou, Herbaud P. F. Zohoungbogbo,
Carlos A. Houdegbe and Enoch G. Achigan-Dako**
Understanding Molecular Mechanisms of Seed Dormancy for Improved Germination in
Traditional Leafy Vegetables: An Overview
Reprinted from: *Agronomy* **2020**, *10*, 57, doi:10.3390/agronomy10010057 253

Sangam Dwivedi, Irwin Goldman and Rodomiro Ortiz
Pursuing the Potential of Heirloom Cultivars to Improve Adaptation, Nutritional, and Culinary
Features of Food Crops
Reprinted from: *Agronomy* **2019**, *9*, 441, doi:10.3390/agronomy9080441 281

**Ajeet Singh, Pradeep Kumar Dubey, Rajan Chaurasia, Rama Kant Dubey, Krishna Kumar
Pandey, Gopal Shankar Singh and Purushothaman Chirakkuzhyil Abhilash**
Domesticating the Undomesticated for Global Food and Nutritional Security: Four Steps
Reprinted from: *Agronomy* **2019**, *9*, 491, doi:10.3390/agronomy9090491 303

About the Editors

Purushothaman Chirakkuzhyil Abhilash is a Senior Assistant Professor at the Institute of Environment and Sustainable Development, Banaras Hindu University in Varanasi, India, and Lead of the Agroecosystem Specialist Group of IUCN's Commission on Ecosystem Management (CEM). He is a fellow (FNAAS) of the National Academy of Agricultural Sciences, India. His research interest lies in restoring marginal and degraded lands for regaining ecosystem services and supporting a bio-based economy, land system management, sustainable utilization of agrobiodiversity, nature-based solutions and ecosystem-based adaptations for climate-resilient and planet-healthy food production, and sustainable agriscape management for food and nutritional security. He is particularly interested in sustainability analysis, system sustainability, sustainability indicators, circular economy principles, policy realignment, and the localization of UN-SDGs for sustainable development. He serves on the editorial board of prestigious journals in ecology, environment, and sustainability from leading international publishers and also serves as a subject expert for UN-IPBES, IRP-UNEP, UNDP-BES Network, IPCC, UNCCD, APN, GLP, and IUCN Commissions (CEM, CEC, CEESP, and SSC) for fostering global sustainability.

Ajeet Singh is a Research Fellow at the Institute of Environment and Sustainable Development, Banaras Hindu University, Varanasi. He earned his B.Sc. in Botany (Hons.), M.Sc. in Environmental Science (Banaras Hindu University), M.Phil. and Ph.D. in Environment and Sustainable Development from the Institute of Environmental and Sustainable Development, Banaras Hindu University, India. He is currently working on agrobiodiversity management, including the study of neglected, underutilized, and wild crops and associated biocultural, traditional, and indigenous ecological knowledge for enhancing the food and nutritional security of the growing human population. He has received the 'Jawaharlal Nehru Scholarship' and 'NASI-Springer Award' for young scientists. He was involved as a lead author of 'Land Restoration for Achieving the Sustainable Development Goals: An International Resource Panel Think Piece' by the International Resource Panel (IRP) of the United Nations Environment Programme (UNEP).

Rama Kant Dubey is a Postdoctoral Fellow at the National University of Singapore. He received his doctoral degree in Industrial Microbiology from the Institute of Environment and Sustainable Development, Banaras Hindu University, India, in 2019. During his doctoral research, he developed various climate-smart and resource conservation agro-biotechnological practices for improving the soil carbon stock, soil quality, yield, and nutritional quality of agricultural produce. In parallel to this, his newly developed agro-practices were able to reduce the microbial and soil respiration and improve key soil sustainability indicators under diverse agroecological zones of Uttar Pradesh, India. He also studied the consequences of changing climate conditions (elevated temperature) on microbial communities, trace gas emission, soil quality changes, and the subsequent effect on agricultural production. He is a 2017 Green Talent Awardee (BMBF, German Government), and he also worked as a Guest Researcher at Helmholtz Zentrum München, Germany, during 2018, where he explored the role of microbial communities inhabiting the rhizosphere, endosphere, and phyllosphere of Zea mays L. for maximizing the agro-environmental sustainability. His current research focuses on microbial ecology and aboveground–belowground multi-trophic interactions for regaining ecosystem services of Southeast Asian peatlands.

Hailin Zhang is a Professor of the College of Agronomy and Biotechnology (CAB) at China Agricultural University and Associate Dean of the CAB. He is a vice-chairman of the China Association of Farming System Research (CAFSR). His studies focus on conservation agriculture, climate change and cropping systems, soil carbon sequestration, cropping system management, and soil ecology. He linked conservation agriculture, soil carbon sequestration, and climate change and mitigation in his study, particularly focusing on changes in soil carbon, soil carbon fraction, and soil quality under conservation agriculture. He is also interested in the meta-analysis of management-induced changes in agroecosystems. He serves on the editorial boards of Agronomy and Green Reports and the advisory board of Sci. He was awarded by Publons as being in the top 1% of reviewers in the Agricultural Science, Environment and Ecology, and Cross-Field categories.

Othmane Merah is an Associate Professor at the University Paul Sabatier and in the Laboratory of Agroindustrial Chemistry, Toulouse, France. After obtaining a degree in Engineering in Agronomic Sciences from the National Agronomic School of Rennes (France) and the University of Blida (Algeria), he completed his PhD at the Institute of Plant Biotechnology (Paris XI University) and the National Institute for Agronomic Research (Montpellier). He completed his Habilitation Diploma by supervising doctoral research at the National Polytechnic Institute of Toulouse. He investigated the genetic, morphological, and physiological diversity in cereals, oilseed crops, and aromatic plants. The impact of genetics and plant management on drought tolerance in cereals and the accumulation of bioactives was also studied. He is a crop scientist and agronomist with wide experience in large-scale research programs and in leading multidisciplinary and international collaboration projects. He has participated in the development of three cultivars of brown mustard in Burgundy (France) for the famous "Moutarde de Dijon" condiment. He has published over 130 publications. He participated as a Guest Editor for several Special Issues in international journals. He is a member of the Editorial and Advisory Boards of several international journals. He has attended many international congresses and symposiums.

Preface to "Wild Crop Relatives and Associated Biocultural and Traditional Agronomic Practices for Food and Nutritional Security"

Ensuring food and nutritional security for a rapidly growing human population is one of the major sustainability challenges in this twenty-first century as well as one of the immediate global priorities for attaining Sustainable Development Goals (UN-SDGs) such as no poverty, zero hunger, and good health and well-being. However, the sole dependence of food production on limited crop species is a real challenge for humanity, and therefore the conservation and management of traditional and wild crop varieties and associated biocultural, traditional, and ecological agricultural knowledge are essential for dietary diversification and also for breeding the next generation of climate-smart crops for futuristic climatic conditions. Though many of the traditional and wild varieties have higher nutritional values and better adaptation traits than modern varieties, they are being neglected and underutilized throughout the world, with their real potential remaining unknown. Similarly, the associated traditional and biocultural knowledge is also gradually vanishing. Therefore, the sustainable management of agrobiodiversity, including traditional and wild varieties of crop plants as well as the associated biocultural and traditional knowledge regarding their conservation, propagation, and exploitation, is essential for the dietary diversification programs and also for ensuring that the food and nutritional demands of the growing population will be met. In this context, the present Special Issue, *"Wild Crop Relatives and Associated Biocultural and Traditional Agronomic Practices for Food and Nutritional Security"*, was dedicated to highlighting the potential traditional and wild crop varieties of nutritional significance and associated biocultural knowledge from diverse agroecological regions of the world for attaining food and nutritional security. The novel recommendations provided by this Special Issue can serve as a stepping-stone for utilizing wild and neglected crops and associated biocultural knowledge for eradicating hunger and malnutrition across the world.

We are grateful to Yanping Mou and the Agronomy editorial office for their continuous support and help with the successful production of this Special Issue. We extend our gratitude to all authors of this Special Issue for their highly interesting contributions. We must not forget to thank all the reviewers for their invaluable help and support in selecting suitable manuscripts and enabling the timely publication of this Special Issue.

Purushothaman Chirakkuzhyil Abhilash, Ajeet Singh, Rama Kant Dubey, Hailin Zhang,
Othmane Merah
Editors

Editorial

The Trilogy of Wild Crops, Traditional Agronomic Practices, and UN-Sustainable Development Goals

Ajeet Singh, Rama Kant Dubey, Amit Kumar Bundela and Purushothaman C. Abhilash *

Institute of Environment & Sustainable Development, Banaras Hindu University, Varanasi 221005, India;
ajeetbhu97943@gmail.com (A.S.); ramakant.sls@gmail.com (R.K.D.); amitbundela02@gmail.com (A.K.B.)
* Correspondence: pca.iesd@bhu.ac.in; Tel.: +91-9415644280

Received: 20 April 2020; Accepted: 28 April 2020; Published: 2 May 2020

Abstract: The world population is projected to become 10 billion by the end of this century. This growing population exerts tremendous pressure on our finite food resources. Unfortunately, the lion-share of the global calorie intake is reliant upon a handful of plant species like rice, wheat, maize, soybean and potato. Therefore, it is the need of the hour to expand our dietary reliance to nutritionally rich but neglected, underutilized and yet-to-be-used wild plants. Many wild plants are also having ethnomedicinal and biocultural significance. Owing to their ecosystem plasticity, they are adapted to diverse habitats including marginal, degraded and other disturbed soil systems. Due to these resilient attributes, they can be considered for large-scale cultivation. However, proper biotechnological interventions are important for (i) removing the negative traits (e.g., low yield, slow growth, antinutritional factors, etc.), (ii) improving the positive traits (e.g., nutritional quality, stress tolerance, etc.), as well as (iii) standardizing the mass multiplication and cultivation strategies of such species for various agro-climatic regions. Besides, learning the biocultural knowledge and traditional cultivation practices employed by the local people is also crucial for their exploitation. The Special Issue *"Wild Crop Relatives and Associated Biocultural and Traditional Agronomic Practices for Food and Nutritional Security"* was intended to showcase the potential wild crop varieties of nutritional significance and associated biocultural knowledge from the diverse agroecological regions of the world and also to formulate suitable policy frameworks for food and nutritional security. The novel recommendations brought by this Special Issue would serve as a stepping stone for utilizing wild and neglected crops as a supplemental food. Nevertheless, long-term cultivation trials under various agro-climatic conditions are utmost important for unlocking the real potential of these species.

Keywords: agrobiodiversity; biocultural knowledge; crop improvement; dietary diversification; field gene banks; food and nutritional security; planetary healthy diet; traditional agronomic practices

1. Introduction

The recently published "EAT-Lancet Commission Report" [1] pinpoints an astonishing fact that the human diet across the globe is not people and planet friendly, as it is far behind the reference level of essential commodities stipulated for a so called "planetary healthy diet" composed of whole grains, vegetables (red, green and orange), fruits and nuts, etc. (www.eatforum.org). The report again pointed that the current global food production itself is not planet friendly as it is produced in an unsustainable manner [2,3]. Therefore, providing a healthy and balanced diet for a growing population is one of the major sustainability challenges for government agencies and policy makers across the world [4–6]. Hence, it is the need of the hour to expand our dietary dependence from a handful of species (i.e., mainly rice, wheat, maize, soybean and potatoes) to locally available and nutritionally rich underutilized wild edibles such as wild leafy vegetables, wild tubers, wild legumes, wild fruits, wild nuts, etc., for a "planetary healthy diet" [5–9].

The literature provide evidences that ~7000 plant species have been reported from the various agro-climatic regions of the world with nutritional or medicinal importance [10,11]. Most of such wild and underutilized species were part of the staple diet of the hunter-gatherers and still many species are the inseparable part of the diet of local people and those who are living in close proximity to nature [10,12–14]. Though some of these species have been widely cultivated for human use and have become a part of the major cropping systems of present-day agriculture [15,16], a vast majority of them are yet-to-be used for large-scale cultivation [5,6] (see Figure 1). As suggested by Pardo-de-Santayana et al. [17], the major factors restricting the day-to-day use of such species are (i) a restricted harvest period, (ii) limited availability due to the lack of market chains, and also due to the (iii) social stigma attached to such species [7,8].

Figure 1. Plants for feeding the future. While there are thousands of plant species that are reported to have food, nutritional and medicinal significance, the current human diet is solely based on a handful of plant species. (**a**–**f**) Showing few examples of highly nutritious but not so commonly cultivated species. (**a**) Jaboticaba or Brazilian grapetree (*Plinia cauliflora*); (**b**) Asparagus tender shoot (*Asparagus officinalis*); (**c**) Brazilian Alcachofra (*Alcachofra* sp.); (**d**) Mulberry (*Morus* sp.); (**e**) Sweet potato (*Ipomea batatas*); and (**f**) Barbados cherry (*Malpighia cauliflora*). *Photo credit:* Prof. Leonardo Fernandes Fraceto, Sao Paulo State University, Sorocaba, Brazil.

Nevertheless, the role of such species for diet complementation is recorded from various parts of the world [5–10]. Moreover, the health and nutritional benefits of some of the species are also available in the literature and even some of the wild species are recognized as functional foods [18,19]. Since most of the wild species are bestowed with unusual colors and flavors, they can be used in food industry as a coloring and flavoring agent [10]. Like in the case of modern crop varieties, they also contain industrially important bioactive molecules such as ascorbic acid, tartaric acid, malic acid, citric acid, oxalic acid, succinic acid, etc. [10,20,21]. Despite these attributes, they are being neglected and disregarded and their multiple roles are yet to be understood [4]. Therefore, the present editorial was penned to highlight the implications of wild and underutilized plant species for food and nutritional security. The ensuing sections elucidate the trilogy, the content and coverage of the various articles published in this Special Issue as well as the various inventive measures for the sustainable use of wild crops for a good quality of life and human wellbeing.

2. The Trilogy: Wild Crops, Traditional Agronomic Practices & UN-SDGs

The triology of wild crops, traditional agronomic practices and UN-Sustainable Development Goals (UN-SDGs) is illustrated in Figure 2. From this depiction, it is apparent that there is an intricate, explicable and interconnected relationship between wild plants and SDGs.

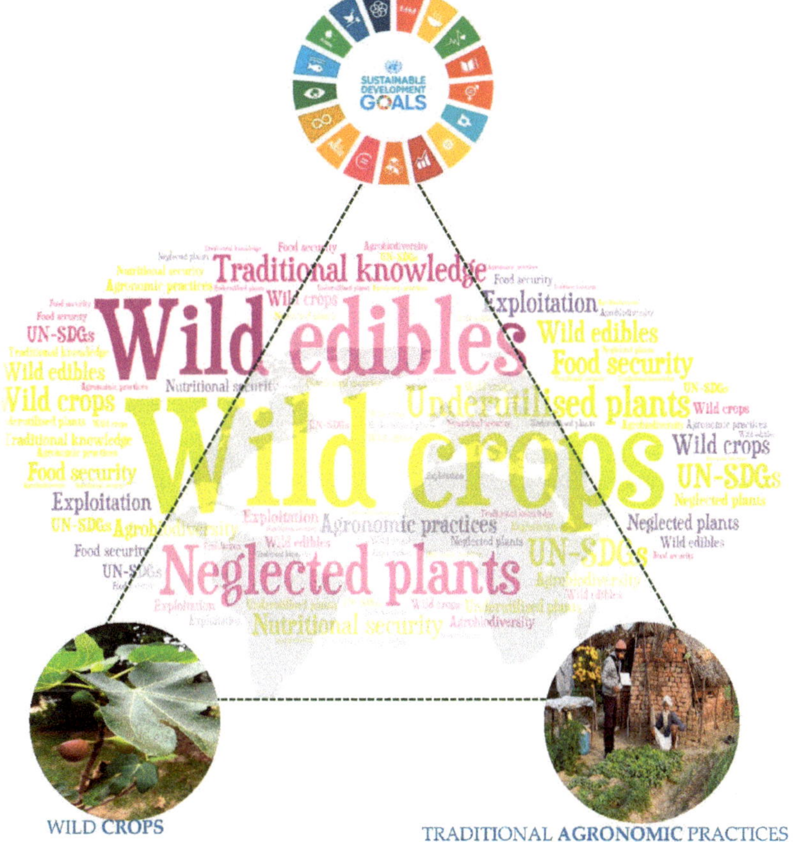

Figure 2. The trilogy of wild crops, traditional agronomic practices and UN-Sustainable Development Goals (UN-SDGs). Though there are many wild crops with food and nutritional significance, their largescale exploitation is mainly limited due to the lack of standard cultivation practices. Therefore, we cannot exploit the intended potential of such species for attaining UN-SDGs. Understanding the biocultural knowledge of the local people as well as learning the traditional agronomic practices employed by them is essential for standardizing the large-scale cultivation practices of such species. Moreover, an integrated understanding is also essential for exploiting their multiple utilities for global sustainability.

For example, the sustainable use of wild and neglected plants will directly or indirectly help in attaining several SDGs such as no poverty (SDG No. 1), zero hunger (SDG No. 2) and good health and wellbeing (SDG No.3) [4,7–9]. Moreover, the cultivation of such species in family and kitchen gardens will pave an opportunity for empowering women and so achieving gender equality (SDG No. 5). The attainment of food sovereignty through the wise use of wild crops will also reduce inequalities (SDG No. 10) while fostering responsible consumption and production (SDG No. 12). Furthermore, the large-scale adoption of low-input, climate resilient and hardy wild varieties in suitable cropping systems such as intercropping, mixed cropping, boarder cropping, etc., is a part of climate action (SDG No. 13) and encourages agrobiodiversity conservation and thereby the attainment of goal No. 15 (life on land) [7–9]. In a nutshell, the judicious use of wild crops in modern cultivation systems will foster sustainable development.

Despite their nutritional and ecological significances, many of these wild crops are still being neglected and underutilized as stakeholders (such as researchers, farmers and policy makers) are not aware of the agronomic practices required for such species [5,6]. Since modern crops are the result of the thousands of years of domestication, the key investments such as time, money and technology are utmost important for refining the cultivable traits of wild species [4,5]. Moreover, a detailed social and ecological impact assessment must be done to ascertain the socio-ecological complexities behind the introduction of such unknown species into modern agricultural frameworks [5]. So, in order to move further, there is an urgent need to document the traditional agronomic practices employed by the indigenous and local community and pass on these dying wisdoms to mainstream farmers for further refinement through field validation. Biotechnological interventions (including conventional and modern biotechnological approaches) are important for standardizing suitable agronomic practices for their large-scale cultivation in various agro-climatic regions of the world. Moreover, crop improvement is also essential for further enhancing the nutritional qualities and palatability of such species and also for removing the anti-nutritional factors if any [8,9]. Importantly, such crop improvement initiatives must be done in a participatory mode (i.e., with the involvement of farmers) to optimize crop-specific and site-specific agronomic practices of these lesser known edibles for human welfare. Without these initiatives, unlocking the real potential of wild species for food and nutritional security will be a mirage rather than a reality.

3. From Exploration to Crop Improvement: The Content and Coverage of the Special Issue

The current Special Issue is an assemblage of the diverse topics related to wild crops and their sustainable exploitation like agrobiodiversity, adaptive agronomic practices, biocultural knowledge, crop improvement programs, dietary diversification, field gene banks, food and nutritional security, genetic diversity, UN-Sustainable Development Goals, wild crop varieties and traditional agronomic practices, among others. The ensuing part provides a glimpse of the various articles published in this Special Issue.

Harouna et al. [22] investigated the preferences and perspectives of local farmers in the Arusha and Kilimanjaro regions in Tanzania regarding the use of wild Vigna legumes such as *Vigna racemosa*, *V. reticulata*, *V. vexillata* and *V. ambacensis*. While all of these four species have multiple utilities like food, feed and medicinal uses as well as other environmental significance such as a mulching agent and also for controlling soil erosion, very few farmers of the study regions (26%–28%) were actually aware of the significance of these species. The takeaways clearly ascertain the fact that exploration as well as popularization of such lesser known species is utmost important for their successful exploitation. Similarly, Coulibaly et al. [23] studied the farmers' preferences over the genetic resources of wild Kersting's groundnut (*Macrotyloma geocarpum*) in Burkina Faso and Ghana. The field results revealed that the famer's preferences were influenced by their sociocultural background and also depended upon the promising traits of the genetic resources such as the yield potential, drought tolerance and pest resistance.

In addition to wild pulses and legumes, neglected vegetables like the spider plant (*Gynandropsis gynandra* L. (Briq.) are also important species of food and nutritional relevance in sub-Saharan Africa. Chataika et al. [24] explored the diversity and domestication status of the spider plant in northern Namibia for identifying the client-preferred traits for a customized breeding program. Extensive field experiments for understanding the agro-morphological variations in wild crops are also important for selecting elite germplasms. In this way, Harouna and co-workers [25] conducted an extensive field trial of 160 accessions of wild Vigna in Tanzania and studied the variations in the quantitative as well as qualitative traits (15 each) for identifying the superior accessions for breeding and crop improvements. Identifying the most suitable habitats for the large-scale cultivation of wild crops as well as predicting their growth response under futuristic climatic condition is a changing paradigm in wild crop research. Interestingly, Ratnayake and co-authors [26] studied the potential range changes of four neglected and underutilized fruit species like *Aegle marmelos*, *Annona muricata*, *Limonia acidissima* and *Tamarindus indica* in Sri Lanka under different climate change scenarios and mapped their high-potential agro-ecological regions for large-scale cultivation. The study underpins the significance of predictive modelling as an effective tool for identifying the most suitable regions (Figure 3) of neglected species for large-scale exploitation.

Figure 3. Predictive modelling is essential for identifying the best suitable agro-ecological regions of various neglected and wild species for large-scale cultivation and popularization. Moreover, apart from the habitat requirement, the biocultural preferences are also considered for promoting a particular wild edible for a particular region. Some of the candidate species for habitat suitability mapping and large-scale cultivation: (**a**) Lablab or Hyacinth bean (*Lablab purpureus*); (**b**) wild ridge gourd (*Luffa acutangular*); (**c**) pseudo-cereal or red amaranth (*Amaranthus cruentus*); (**d**) wild sweet potato (*Ipomoea batatas*); (**e**) air potato (*Dioscorea bulbifera*); and (**f**) purple yam (*Dioscorea alata*).

The medicinal and nutraceutical uses of three traditional leafy vegetables, viz. *Crassocephalum crepidioides*, *Launaea taraxacifolia* and *Vernonia amygdalina* Del., in southern Benin and the legacy of the transmission of traditional and local knowledge from parents to children were studied by Akakpo and Achigan-Dako [27]. Since traditional leafy vegetables are an integral part of the African diet, the majority of these leafy vegetables (for example *Bidens pilosa, Brassica carinata, Gynandropsis gynandra, Corchorus* spp., *Launaea taraxacifolia, Talinum triangulare*, etc.) are not domesticated yet [27,28]. Some of the traditional leafy vegetables also exhibit seed dormancy. Therefore, understanding the molecular mechanisms that underpin the seed dormancy is essential for devising a suitable breeding program. In this context, Sohindji and co-workers [28] used *Gynandropsis gynandra* as a model plant to elucidate the mechanisms of seed dormancy in traditional leafy vegetables and proposed suitable strategies for the molecular breeding of wild leafy vegetables with improved germination.

Innovative breeding approaches are also essential for improving the quality and adaptability of wild crops. For example, Alvarez and Guzman [29] studied the genetic variation in the high molecular weight glutenin subunits (*Glu-A1, Glu-B1* and *Glu-D1*) in three neglected and underutilized wheat crops such as club wheat (*Triticum aestivum*), macha wheat (*T. aestivum* ssp. *macha*) and Indian dwarf wheat (*T. aestivum* L. ssp. *sphaerococcum*) for quality improvements. Similarly, Jency et al. [30] reported induced mutagenesis using ethyl methane sulphonate (EMS) and gamma radiation to develop photosynthetically efficient and lodging-resistant kodo millet (*Paspalum scrobiculatum*) for large-scale cultivation in peninsular India.

The standardization of agronomic practices is also important for improving the nutritional quality and other featured traits in wild crops (Figure 4). Moreover, previous studies have also reported that such innovative agronomic practices can also improve the soil quality and microbial activity [31–33]. Brtnicky et al. [34] proved that the long-term addition of biochar resulted in increases in the microbial biomass' carbon and microbial community abundance, and soil dehydrogenase activity in luvisols of arable land in the Czech Republic.

Figure 4. Standardization of suitable agronomic practices as well as the optimization of suitable cultivation models are essential for the large-scale cultivation and also for the successful introduction of such species into family and kitchen gardens. (**a**,**b**) The cultivation of traditional species in the backyard garden of Prof. Leonardo Fernandes Fraceto in Sorocaba, Sao Paulo, Brazil and (**c**,**d**) the wild species in the kitchen garden of local farmers in Rajgarh block, Mirzapur, Uttar Pradesh, India.

Chanthani et al. [35] demonstrated the positive effect of the seaweed extract from *Ulva flexuosa* on the seed priming of wild tomatoes for increasing the seed germination, growth and yield. The experimental results confirmed that the application of seaweed extract as a biostimulant has increased the nutritional and biochemical profile of the test plant by several folds, i.e., TSS (93%), phenol (92%), lycopene (12%) and ascorbic acid (86.8%). Similarly, Michalska-Ciechanowska and colleagues [36] moderated the inulin and polyphenolics contents in three cultivars of Jerusalem artichoke (*Helianthus tuberosus* L.) by optimizing the potassium (K) application. The study on the effect of the different growing conditions (i.e., greenhouse and open field) on the quality and phytochemical composition of edible wall rocket (*Diplotaxis erucoides*) revealed that the field conditions gave better results than the controlled conditions [37]. All these studies have given novel insights regarding the cultivation and management of lesser known edibles for human wellbeing.

In another work, Dwivedi et al. [38] reviewed the need of tapping unique traits in heirloom cultivars such as crop growth and yield governing attributes, biotic and abiotic stress tolerance traits as well as traits governing proximate composition, flavor, color, etc., into local elite cultivars through suitable breeding programs. Interestingly, the importance of creating inventive market channels, for example farmer–breeder–chef collaborations and seed-saver organizations for promoting and popularizing heirloom cultivars, are also suggested by them. Similarly, in the article on *Local Solutions for Sustainable Food Systems: The Contribution of Orphan Crops and Wild Edible Species*, Borelli and her colleagues [39] examined the "role of locally available; affordable and climate-resilient orphan crops, traditional varieties and wild edible species to support local food system transformation" in Brazil, Kenya, Guatemala, India, Mali, Sri Lanka and Turkey.

For the sustainable food system transition based on orphan crops and wild edibles, authors have basically employed a "three-pronged approach to (i) increase the evidence of the nutritional value and biocultural importance of these foods, (ii) better link the research to policy to ensure these foods are considered in national food and nutrition security strategies and actions, and (iii) improve consumer awareness of the desirability of these alternative foods so that they may more easily be incorporated in diets, food systems and markets" [39] (Figure 5).

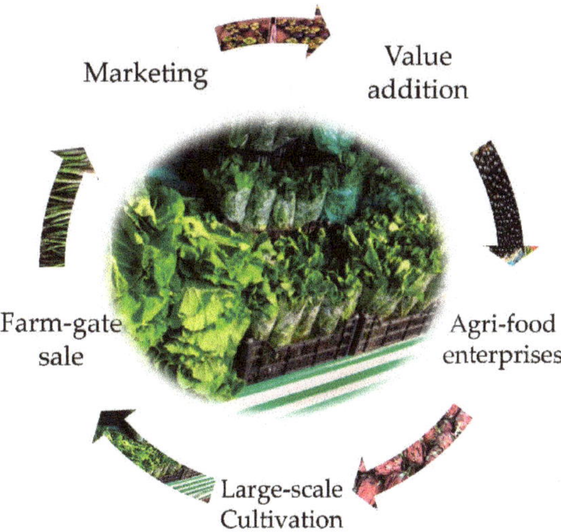

Figure 5. Creating marketing channels including the provisions for value-addition and also the creation of suitable agri-food enterprises based on wild, neglected and underutilized crops are essential for a sustainable food system transition. The availability and affordability of these wild edibles will gradually increase the adoption and acceptance of these lesser known edibles among the community.

The deliberations of the above two reviews lead to a common thread that apart from the exploration, documentation and large-scale exploitation of various wild and lesser known edibles for diet supplementation, the creation of market channels including supply-chain management and value additions are the key aspect governing the success of a transition towards a wild, neglected and underutilized edibles-based food system [38,39]. Community mobilization can be done for enhancing the sale of these items at the farm-gate level itself. Since a majority of these lesser-known edibles are seasonal, value-addition will provide an opportunity for establishing agri-food start-ups.

Last but not least, the crux of the overall issue was discussed by Singh et al. [40] and proposed four inventive strategies for the domestication of undomesticated crops for global food and nutritional security such as "(i) exploring the unexplored, (ii) refining the unrefined traits, (iii) cultivating the uncultivated, and (iv) popularizing the unpopular for the sustainable utilization". Furthermore, authors have also recommended the need of starting coordinated efforts at the national, regional and international level, especially under the aegis of UN organizations and internationally renowned agricultural institutions, policy makers, governments and voluntary organizations to conserve, promote and popularize wild, neglected and underutilized species for the wellbeing of both people and the planet [40].

4. Conclusions

In conclusion, neglected, wild, underutilized, traditional or orphan crops possess enormous nutritional, nutraceutical, industrial, ethnomedicinal and biocultural significance, and therefore their sustainable utilization will directly or indirectly lead to sustainable development. While multiple utilities of some of the species are known to local framers and the scientific community, the real potential of the vast majority of such species is unknown. Therefore, inventive measures are essential for the exploration, documentation and bioprospecting of such species for identifying their multipurpose benefits as well as the optimization of agronomic practices including suitable crop improvement programs that are essential for unlocking their real potential for human heath and wellbeing.

Author Contributions: Conceptualization, A.S., R.K.D. and P.C.A.; writing—original draft preparation, A.S., R.K.D, A.K.B. and P.C.A.; writing—review and editing, A.S., R.K.D., A.K.B. and P.C.A.; supervision, P.C.A. All authors have read and agreed to the published version of the manuscript.

Acknowledgments: Authors are grateful to Yanping Mou, Agronomy office for their continuous support and help for the successful edition of this Special Issue. We are also grateful to all authors of this Special Issue for their highly interesting contributions. We are also grateful to all the reviewers for their invaluable help and support for selecting suitable manuscripts and also the timely publication of this Special Issue.

Conflicts of Interest: The authors declare no conflict of interest.

References

1. Willett, W.; Rockström, J.; Loken, B.; Springmann, M.; Lang, T.; Vermeulen, S.; Garnett, T.; Tilman, D.; Wood, A.; Clark, M.; et al. Food in the Anthropocene: The EAT–Lancet Commission on Healthy Diets from Sustainable Food Systems. *Lancet* **2019**, *393*, 447–492. [CrossRef]
2. Dubey, P.K.; Abhilash, P.C. Agriculture in a changing climate. *J. Clean. Prod.* **2016**, *100*, 1046–1047. [CrossRef]
3. Dubey, P.K.; Singh, G.S.; Abhilash, P.C. *Adaptive Agricultural Practices: Building Resilience in a Changing Climate*; Springer: Berlin/Heidelberg, Germany, 2019; ISBN 978-3-030-15518-6. [CrossRef]
4. Singh, A.; Abhilash, P.C. Agricultural biodiversity for sustainable food production. *J. Clean. Prod.* **2018**, *172*, 1368–1369. [CrossRef]
5. Singh, A.K. Wild Relatives of Cultivated Plants in India. In *A Reservoir of Alternative Genetic Resources and More*; Springer: Berlin/Heidelberg, Germany, 2017; ISBN 9789811051166.
6. Hunter, D.; Guarino, L.; Spillane, C.; McKeown, P.C. (Eds.) *Routledge Handbook of Agricultural Biodiversity*; Routledge: Thames, UK, 2017; p. 692, Price 175, 00£; ISBN 9780415746922.
7. Singh, A.; Dubey, P.K.; Chaurasiya, R.; Mathur, N.; Kumar, G.; Bharati, S.; Abhilash, P.C. Indian spinach: An underutilized perennial leafy vegetable for nutritional security in developing world. *Energ. Ecol. Environ.* **2018**, *3*, 195–205. [CrossRef]

8. Singh, A.; Dubey, P.K.; Abhilash, P.C. Food for thought: Putting wild edibles back on the table for combating hidden hunger in developing countries. *Curr. Sci.* **2018**, *115*, 611–613. [CrossRef]

9. Singh, A.; Abhilash, P.C. Varietal dataset of nutritionally important *Lablab purpureus* (L.) sweet from Eastern Uttar Pradesh, India. *Data Brief* **2019**, *24*, 103935. [CrossRef]

10. Sánchez-Mata, M.C.; Cabrera Loera, R.D.; Morales, P.; Fernández-Ruiz, V.; Cámara, M.; Marqués, C.D.; de Santayana, M.P.; Tardío, J. Wild vegetables of the Mediterranean area as valuable sources of bioactive compounds. *Genet. Resour. Crop Evol.* **2012**, *59*, 431–443. [CrossRef]

11. Ghane, S.G.; Lokhande, V.H.; Ahire, M.L.; Nikam, T.D. *Indigofera glandulosa* Wendl. (Barbada) a potential source of nutritious food: Underutilized and neglected legume in India. *Genet. Resour. Crop Evol.* **2010**, *57*, 147–153. [CrossRef]

12. Termote, C.; Van Damme, P.; Dhed'a Djailo, B. Eating from the wild: Turumbu, Mbole and Bali traditional knowledge on non-cultivated edible plants, District Tsh- opo, DRCongo. *Genet. Resour. Crop Evol.* **2011**, *58*, 585–618. [CrossRef]

13. Vazquez-Garcia, V. Gender, ethnicity, and economic status in plant management: Uncultivated edible plants among the Nahuas and Popolucas of Veracruz, Mexico. *Agric. Hum. Values* **2008**, *25*, 65–77. [CrossRef]

14. Price, L.L. Farm womens rights and roles in wild plant food gathering and management in North-East Thailand. In *Women and Plants. Gender Relations in Biodiversity Management and Conservation*; Howard, P.L., Ed.; Zed Books and IDRC: London, UK, 2003; pp. 101–114.

15. Bharucha, Z.; Pretty, J. The roles and values of wild foods in agricultural systems. *Philos. Trans. R. Soc. B* **2000**, *365*, 2913–2926. [CrossRef] [PubMed]

16. Hammer, K.; Knüpffer, H.; Laghetti, G.; Perrino, P. Seeds from the past. In *A Catalogue of Crop Germplasm in Central and North Italy*; IG of CNR: Bari, Italy, 1999.

17. Pardo-de-Santayana, M.; Tardío, J.; Blanco, E.; Carvalho, A.M.; Lastra, J.J.; San Miguel, E.; Morales, R. Traditional knowledge on wild edible plants in the northwest of the Iberian Peninsula (Spain and Portugal): A comparative study. *J. Ethnobiol. Ethnomed.* **2007**, *3*, 27. [CrossRef] [PubMed]

18. Ansari, N.M.; Houlihan, L.; Hussain, B.; Pieroni, A. Anti- oxidant activity of five vegetables traditionally consumed by South-Asian Migrants in Bradford, Yorkshire, UK. *Phytother. Res.* **2005**, *19*, 907–911. [CrossRef] [PubMed]

19. Salvatore, S.; Pellegrini, N.; Brenna, O.V.; Del Rio, D.; Frasca, G.; Brighenti, F.; Tumino, R. Antioxidant characterization of some Sicilian edible wild greens. *J. Agric. Food Chem.* **2005**, *53*, 9465–9471. [CrossRef]

20. Phillips, K.M.; Tarrago-Trani, M.T.; Gebhardt, S.E.; Exler, J.; Patterson, K.Y.; Haytowitz, D.B.; Pehrsson, P.R.; Holden, J.M. Stability of vitamin C in frozen raw fruit and vegetable homogenates. *J. Food Compost. Anal.* **2010**, *23*, 253–259. [CrossRef]

21. Oliveira, A.P.; Pereira, J.A.; Andrade, P.B.; Valentão, P.; Seabra, R.M.; Silva, B.M. Organic acids composition of *Cydonia oblonga* Miller leaf. *Food Chem.* **2008**, *111*, 393–399. [CrossRef]

22. Harouna, D.V.; Venkataramana, P.B.; Matemu, A.O.; Ndakidemi, P.A. Wild Vigna legumes: Farmers' perceptions, preferences, and prospective uses for human exploitation. *Agronomy* **2019**, *9*, 284. [CrossRef]

23. Coulibaly, M.; Agossou, C.O.; Akohoué, F.; Sawadogo, M.; Achigan-Dako, E.G. Farmers' Preferences for Genetic Resources of Kersting's Groundnut *Macrotyloma geocarpum* (Harms) Maréchal and Baudet] in the Production Systems of Burkina Faso and Ghana. *Agronomy* **2020**, *10*, 371. [CrossRef]

24. Chataika, B.; Akundabweni, L.; Achigan-Dako, E.G.; Sibiya, J.; Thomas, B. Diversity and Domestication Status of Spider Plant (*Gynandropsis gynandra*, L.) amongst sociolinguistic groups of Northern Namibia. *Agronomy* **2020**, *10*, 56. [CrossRef]

25. Harouna, D.V.; Venkataramana, P.B.; Matemu, A.O.; Ndakidemi, P.A. Agro-morphological exploration of some unexplored wild vigna legumes for domestication. *Agronomy* **2020**, *10*, 111. [CrossRef]

26. Ratnayake, S.S.; Kumar, L.; Kariyawasam, C.S. Neglected and Underutilized Fruit Species in Sri Lanka: Prioritisation and understanding the potential distribution under climate change. *Agronomy* **2020**, *10*, 34. [CrossRef]

27. Akakpo, A.D.; Achigan-Dako, E.G. Nutraceutical uses of traditional leafy vegetables and transmission of local knowledge from parents to children in Southern Benin. *Agronomy* **2019**, *9*, 805. [CrossRef]

28. Sohindji, F.S.; Sogbohossou, D.E.; Zohoungbogbo, H.P.; Houdegbe, C.A.; Achigan-Dako, E.G. Understanding molecular mechanisms of seed dormancy for improved germination in traditional leafy vegetables: An overview. *Agronomy* **2020**, *10*, 57. [CrossRef]

29. Alvarez, J.B.; Guzmán, C. Recovery of wheat heritage for traditional food: Genetic variation for high molecular weight glutenin subunits in neglected/underutilized wheat. *Agronomy* **2019**, *9*, 755. [CrossRef]

30. Jency, J.P.; Rajasekaran, R.; Singh, R.K.; Muthurajan, R.; Prabhakaran, J.; Mehanathan, M.; Prasad, M.; Ganesan, J. Induced mutagenesis enhances lodging resistance and photosynthetic efficiency of Kodomillet (*Paspalum Scrobiculatum*). *Agronomy* **2020**, *10*, 227. [CrossRef]

31. Dubey, R.K.; Dubey, P.K.; Chaurasia, R.; Singh, H.B.; Abhilash, P.C. Sustainable agronomic practices for enhancing the soil quality and yield of *Cicer arietinum* L. under diverse agroecosystems. *J. Environ. Manag.* **2020**, *262*, 110284. [CrossRef]

32. Dubey, R.K.; Dubey, P.K.; Abhilash, P.C. Sustainable soil amendments for improving the soil quality, yield and nutrient content of *Brassica juncea* (L.) grown in different agroecological zones of eastern Uttar Pradesh. *Soil Tillage Res.* **2019**, *195*, 104418. [CrossRef]

33. Dubey, R.K.; Tripathi, V.; Prabha, R.; Chaurasia, R.; Singh, D.P.; Rao, C.H.; El-Keblawy, A.; Abhilash, P.C. *Unravelling the Soil Microbiomes: Perspectives for Environmental Sustainability*; Springer-Nature: Berlin/Heidelberg, Germany, 2020.

34. Brtnicky, M.; Dokulilova, T.; Holatko, J.; Pecina, V.; Kintl, A.; Latal, O.; Vyhnanek, T.; Prichystalova, J.; Datta, R. Long-term effects of biochar-based organic amendments on soil microbial parameters. *Agronomy* **2019**, *9*, 747. [CrossRef]

35. Chanthini, K.M.P.; Stanley-Raja, V.; Thanigaivel, A.; Karthi, S.; Palanikani, R.; Shyam Sundar, N.; Sivanesh, H.; Soranam, R.; Senthil-Nathan, S. Sustainable agronomic strategies for enhancing the yield and nutritional quality of wild tomato, *Solanum Lycopersicum* (l) var Cerasiforme Mill. *Agronomy* **2019**, *9*, 311. [CrossRef]

36. Michalska-Ciechanowska, A.; Wojdyło, A.; Bogucka, B.; Dubis, B. Moderation of inulin and polyphenolics contents in three cultivars of *Helianthus tuberosus* L. by potassium fertilization. *Agronomy* **2019**, *9*, 884. [CrossRef]

37. Guijarro-Real, C.; Adalid-Martínez, A.M.; Aguirre, K.; Prohens, J.; Rodríguez-Burruezo, A.; Fita, A. Growing conditions affect the phytochemical composition of edible Wall Rocket (*Diplotaxis erucoides*). *Agronomy* **2019**, *9*, 858. [CrossRef]

38. Dwivedi, S.; Goldman, I.; Ortiz, R. Pursuing the potential of heirloom cultivars to improve adaptation, nutritional, and culinary features of food crops. *Agronomy* **2019**, *9*, 441. [CrossRef]

39. Borelli, T.; Hunter, D.; Padulosi, S.; Amaya, N.; Meldrum, G.; Beltrame, D.M.D.O.; Samarasinghe, G.; Wasike, V.W.; Güner, B.; Tan, A.; et al. Local solutions for sustainable food systems: The contribution of orphan crops and wild edible species. *Agronomy* **2020**, *10*, 231. [CrossRef]

40. Singh, A.; Dubey, P.K.; Chaurasia, R.; Dubey, R.K.; Pandey, K.K.; Singh, G.S.; Abhilash, P.C. Domesticating the undomesticated for global food and nutritional security: Four steps. *Agronomy* **2019**, *9*, 491. [CrossRef]

Article

Farmers' Preferences for Genetic Resources of Kersting's Groundnut [*Macrotyloma geocarpum* (Harms) Maréchal and Baudet] in the Production Systems of Burkina Faso and Ghana

Mariam Coulibaly [1,2], Chaldia O.A. Agossou [1], Félicien Akohoué [1], Mahamadou Sawadogo [2] and Enoch G. Achigan-Dako [1,*]

[1] Laboratory of Genetics, Horticulture and Seed Science, Faculty of Agronomic Sciences, University of Abomey-Calavi, 01 BP 526 Abomey-Calavi, Republic of Benin; mary.coulibaly6@gmail.com (M.C.); achaldia.aboegnonhou@gmail.com (C.O.A.A.); akohoue.f@gmail.com (F.A.)

[2] Laboratory of Biosciences, Faculty of Earth and Life Science, University of Ouaga I Pr. Joseph Ki-Zerbo, 03 BP 7021 Ouagadougou, Burkina Faso; sawadogomahamadou@yahoo.fr

[*] Correspondence: enoch.achigandako@uac.bj

Received: 19 February 2020; Accepted: 4 March 2020; Published: 8 March 2020

Abstract: Pulses play important roles in providing proteins and essential amino-acids, and contribute to soils' nutrients cycling in most smallholder farming systems in Sub-Saharan Africa (SSA). These crops can be promoted to meet food and nutrition security goals in low-income countries. Here, we investigated the status of Kersting's groundnut (*Macrotyloma geocarpum*, Fabaceae), a neglected pulse in West Africa. We explored its diversity, the production systems, the production constraints and farmers' preferences in Burkina Faso and Ghana. Focus groups and semi-structured interviews were conducted in 39 villages with 86 respondents grouped in five sociolinguistic groups. Our results indicated that *Macrotyloma geocarpum* was produced in three cultivation systems: in the first system, farmers grew Kersting's groundnut in fields, mostly on mounds or on ridges; in the second system, farmers grew it as field border; and in the third system, no clear tillage practice was identified. The main constraints of those farming systems included: difficulty to harvest, the lack of manpower and the damage due to high soil humidity at the reproductive stage. A total of 62 samples were collected and clustered in six landraces based on seed coat colors including cream, white mottled with black eye, white mottled with greyed orange eye, black, brown mottled, and brown. All six groups were found in the southern-Sudanian zone whereas only white mottled with black eye and black colors were found in the northern-Sudanian zone. The white mottled with black eye landrace was commonly known and widely grown by farmers. Farmers' preferences were, however, influenced by sociolinguistic membership and the most preferred traits included high yielding, drought tolerance, and resistance against beetles. These findings offer an avenue to develop a relevant breeding research agenda for promoting Kersting's groundnut in Burkina Faso and Ghana.

Keywords: breeding; *Macrotyloma geocarpum*; farmers' preferences; cropping systems; constraints; cultivar development; landraces; conservation; sociolinguistic groups

---

1. Introduction

To feed the growing population of sub-Saharan Africa (SSA), agricultural productivity needs to be increased significantly and the existing crops diversity strengthened. The crop yield per unit area in SSA is projected to decrease [1], against an estimated increase in demand for cereals of 335% between 2010 and 2050 [2]. Similarly, the demand for legumes is also expected to increase as consumers' income increases with a likely shift in preferences from cereal grains to more nutrient-dense foods [3]. Moreover, the increasing world population will result in a substantial demand for additional proteins. This growing gap between demand and supply of food and nutrients in SSA will require a major re-focusing on grain legumes, with intensive research and development to identify climate-resilient species and cultivars with improved grain qualities [4]. While 19,500 species of legume were reported [5], only five are grown most widely in sub-Saharan Africa (SSA), namely common bean (*Phaseolus vulgaris* L.), chickpea (*Cicer arietinum* L.), cowpea [*Vigna unguiculata* (L.) Walp.], groundnut (*Arachis hypogaea* L.), and pigeonpea [*Cajanus cajan* (L.) Huth] [6]. Significant research and development work has been done in the past decade on those grain legumes through collaborative bilateral and multilateral projects as well as the CGIAR Research Program on Grain Legumes (CRP-GL) [7]. However, the bulk of edible legume species are mostly overlooked by research and development initiatives. Thus, they are often referred to as orphan legume crops. Key orphan legume crops in sub-Saharian Africa include Bambara groundnut [*Vigna subterranean* L. (Verdc.)], Yam bean [*Sphenostylis stenocarpa* (Hoechst ex. A. Rich.) Harms.)], Faba bean (*Vicia faba* L.) and Kersting's groundnut (*Macrotyloma geocarpum*).

Kersting's groundnut is an example of an orphan legume crop of West-Africa [8] with a wide spectrum of importance, for both human and animal health. The crop is valuable to human nutrition, particularly in areas where access to animal protein is limited. Its seeds have a high protein content (21.3%) [9], carbohydrate (61.53%–73.3%) and crude fibre (6.2%) [10]. *M. geocarpum* is considered as healthy because its grains contain a very low crude fat (1.0%) and a high concentration in arginine, an amino acid for pediatric growth; the seeds are also a good source of mineral elements such as iron, zinc, calcium and magnesium [9]. This crop can be used for complementary food formulation for children to combat malnutrition [11]. Moreover, it provides substantial incomes for rural population in Benin, Togo, Ghana, and Burkina Faso; its price can hike from CFA 1000 (USD 2) per kg in abundance period to CFA 4000–5000 (USD 8–10) per kg in scarcity period [12,13]. The crop also exhibits several medicinal properties and is used in traditional healthcare by local communities [13–15]. The root system of Kersting's groundnut fixes atmospheric nitrogen and improves the quality and structure of soils [16]. With climate variability and the occurrence of prolonged drought, *M. geocarpum* is a candidate resilient crop for sustainable cropping systems.

Kersting's groundnut production systems, and landraces diversity were documented in some West-African countries including Benin, Togo, Ghana and Nigeria [12,13,17–19]. These studies showed that its cropping systems varied across its production areas, and revealed the existence of different groups of landraces basing on the seed coat colors. However, in Burkina Faso and Ghana, research on the species has been limited to utilization [14] and the farming systems [17,19]. In both countries, *M. geocarpum* production systems, germplasm collection and breeding objectives have received little attention. In addition, the diversity revealed in those countries—white mottled with black eye and black in Burkina Faso [14] and white mottled with black eye, black, and brown mottled in Ghana [16,17]—is less than the diversity reported in other countries such as Benin and Togo, where authors found five landraces: white, white mottled with black eye, white mottled with yellow eye, black, and red [12]. This knowledge discrepancy/complementarity needs to be highlighted when we seek to promote Kersting's groundnut production and its genetic improvement in those countries. In other words, the identification of farmers' preferred traits is useful for breeders in developing varieties and for extension agents in appropriate choice of varieties to be popularized [20]. Furthermore, knowledge on production systems and the rationale driving farmers' practices are key requirements for guiding successful variety introduction [21]. In the production systems of Burkina Faso and Ghana, we inquire about how Kersting's groundnut production system organized, and the factors constraining or influencing its production. How does the crop diversity vary across agroecological zones and how is it managed by sociolinguistic groups? Furthermore, what are the preferred traits of farmers and how are they correlated to social factors? In this study, we seek to understand the current traditional cropping systems and analyze factors limiting *M. geocarpum* production for its promotion in Burkina Faso and Ghana. We assumed that: i) on-farm practices, constraints and farmers' preferences are influenced by sociolinguistic group membership; ii) the genetic diversity of Kersting's groundnut in Burkina

Faso and Ghana is higher than findings reported by previous studies; iii) the distribution of the crop landraces varies with agro-ecological zones.

2. Materials and Methods

2.1. Study Area

The study was carried out in Burkina Faso and Ghana in West-Africa (Figure 1). Burkina Faso is a landlocked country located between latitudes 9° and 15° N, and longitudes 6° W and 3° E bordered to the north and west by Mali, to the east by Niger and to the south by Benin, Togo, Ghana and Côte d'Ivoire. It has an area of 274,200 km^2. Ghana is located on the southern coast of West Africa, between latitudes 4°44′ N and 11°11′ N and longitudes 3°11′ W and 1°11′ E with an area of 238,540 km^2 and limited by Burkina to the north. The study was conducted in two agroecological zones of Burkina Faso and Ghana: the northern-Sudanian (or Sudano-Sahelian) and southern-Sudanian (Sudanian). The two agroecological zones included are both characterized by the unimodal rainfall season (Table 1).

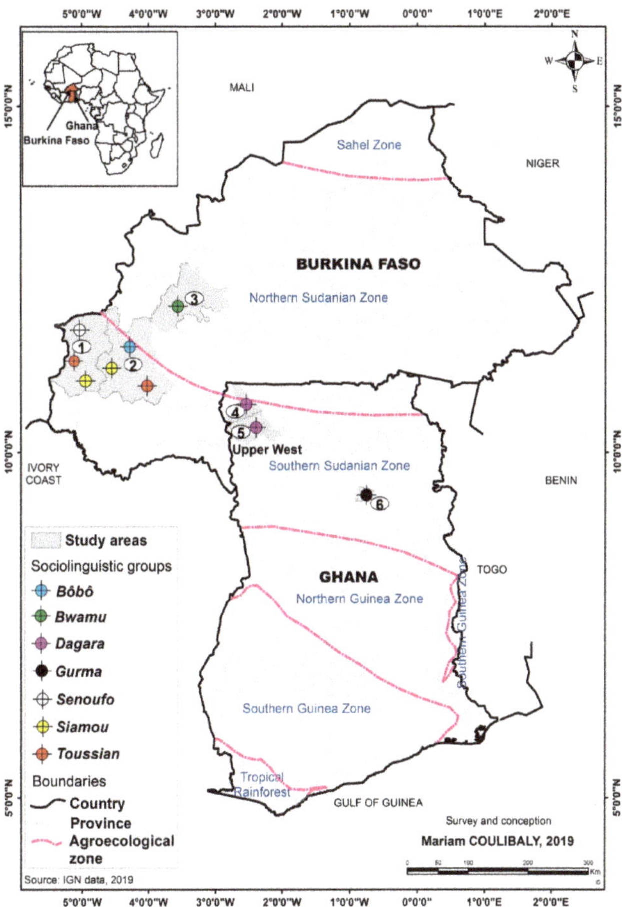

Figure 1. Study areas with agroecological zones in Burkina Faso and Ghana. 1: Kénédougou, 2: Houet, 3: Mouhoun, 4: Jirapa Lambussie, 5: Nadowli, and 6: Tamale Metropolis.

Table 1. Characteristics of southern-Sudanian and northern-Sudanian agroecological zones of Burkina Faso and Ghana.

Country	Agroecological Zone	Latitude	Climatic Condition				Soil Condition		Vegetation	Cropping System
			Growing Seasons	Rainfall mm/year	Temperature (°C)	Humidity (%)	Texture	Type		
Burkina Faso	Northern-Sudanian	11°30′–14°00′ N	150 days	700–900	28	75	Gravel	Ferruginous	Savanna with trees or shrubs	Cotton, sorghum, millet, cowpea, groundnut
	Southern-Sudanian	9°00′–11°30′ N	150–180 days	900–1200	27	85	Sand-clayey	Ferriferous	Semi-deciduous forest	Cotton, yams, cassava, sorghum, millet, maize, rice, cowpea and mangos, citrus, cashew
Ghana	Southern-Sudanian	8°00′–11°30′ N	180–200 days	800–1200	28.1	61	Loamy sand granular	Ferruginous	Grassland with few trees	

The northern-Sudanian zone is approximately located between latitudes 14°00′ and 11°30′ N, and is characterized by an annual precipitation of 600–900 mm. The vegetation in the zone is a savanna with trees or shrubs. This zone is mainly a cotton (*Gossypium hirsutum* L.) production area. Other crops include sorghum [*Sorghum bicolor* L. (Moensch)], millet [*Pennisetum glaucum* (L.) R.Br.], groundnut (*Arachis hypogea*), and cowpea (*Vigna unguiculata*). In the northern-Sudanian zone, Kersting's groundnut production was limited to a few villages represented mainly by the Bwaba linguistic group of the Gur family, and representing 2.1% of the population in Burkina Faso [22].

Moving southwards, the rainfall increases to 900–1200 mm in southern-Sudanian zone located between latitudes 11°30′ and 9°00′ N in Burkina Faso and between 8°00′ and 11°30′ N in Ghana. The vegetation is characterized by a savanna with trees or shrubs, sparse forests in Burkina Faso and is made up of grassland with few trees in Ghana. Agriculture in the southern-Sudanian zone is characterized by perennial crops such as mango (*Mangifera indica* L.), citrus (*Citrus* spp.), cashew (*Anacardium occidentale* L.), etc. Farmers in this zone also grow cotton, cowpea, yams (*Dioscorea* spp.), cassava (*Manihot* spp.), sorghum, millet, maize (*Zea mays* L.), and rice (*Oryza sativa* L.). The Bôbô, Senufo, Siamou and Toussian of Burkina Faso and the Gurma and Dagara of Ghana are the main sociolinguistic groups producing and managing Kersting's groundnut in the southern-Sudanian zone. Senufo and Toussian groups are related to the Gur family [23] and represent respectively, 1.4% and less than 0.1% of the Burkina Faso population [22]. The Bôbô group belongs to the Mande language family and represents 1.4% of the population. The Siamou linguistic group was previously declared to belong to the Kru language family (from Ivory Coast and Liberia) [24]. The Siamou group represents 0.1% of the population of Burkina Faso [22]. These people live in the department of Orodara located in Kénédougou province. The Dagaaba and Gurma linguistic groups in Ghana belong to the Gur language family [25]. Linguistically, the Dagara are part of the Mole-Dagbane group which represents 16.6% of Ghanaians while the Gurma people represent 5.7% of the Ghanaian population.

2.2. Sampling Strategy and Respondents' Consent

Field trips and collection of ethnobotanical data were carried out from February to May 2018. Semi-structured interviews [26] were conducted with 86 respondents in Burkina Faso and Ghana. In each country, the investigations started by market visits to identify the sources/suppliers of Kersting's groundnut seed sellers and to get growers' contacts. The surveyed localities were identified with the help of extension agents in the regions and communities. We shared the pictures of *M. geocarpum* seeds, pods and plant with the focal points who identified and confirmed the growing areas of the crop in their localities. The chain-referral sampling technique [27], a non-probability sampling technique, was used in both countries to select the respondents. In this sampling, the first contact with the community was selected as a well-known expert; in a subsequent phase, the expert indicated another expert until all the informants in the community were covered. The informant was a grower who knew more about the crop, who used it and grew or used to grow it. In each surveyed locality, we explained the objectives of the study to local authorities and farmers and requested prior informed consent from all participants, and ensured participants' confidentiality by anonymizing their identities in databases and publications.

2.3. Data Collection

One focus group was conducted in each village. Participants included male and female farmers selected based on their willingness to participate in the study. However, in some areas, to eliminate gender dominance in discussions, separate discussions were held with men and women farmers. The focus group discussions provided an insight into the different types of landraces grown, the constraints to Kersting's groundnut production and farmers' major landrace preference criteria. The intent of this study was not to rank farmers' preference criteria but to identify the most important traits farmers seek in a variety at village level. Different seed coat colors were shown to farmers and the questionnaire was used to interview farmers who recognized and grew or used to grow the crop.

In Ghana, the Agriculture technical services and the Savanna Agricultural Research Institute (SARI) agents also assisted the surveyors to identify the respondents in the villages by using the crop pictures and its local names.

The questionnaire was drawn up using Sphinx Plus version 4.5 [28]. For each informant, we recorded personal information about age, gender, sociolinguistic group, education level, occupation, cropland areas and the area allocated for Kersting's groundnut production. During the interviews, the respondents were requested to indicate vernacular names of the plant and the landraces they knew and the landraces they grew. Information regarding the farmers' practices and the constraints to the crop production were also recorded. Other questions were related to the criteria to choose to grow a specific landrace, and the traits of interest for the crop improvement. All interviews were conducted in the respondent's native language to ensure that questions were well understood. Local measurement units named 'tomato box', 'tine' and 'Yoruba plat' were estimated/converted in kg to determine farmers' yield of Kersting's groundnut.

In addition to interviews, seeds collection was organized with farmers who still possessed seeds of *M. geocarpum*. A color-chart was used to better identify the seed coat color and classify samples collected into different phenotypic groups.

2.4. Data Analysis

Statistical analyses of both quantitative and qualitative data were performed using R software version 3.6.1 [29]. Descriptive statistics were used to estimate frequencies, proportions and means for citations, socio-economic variables and Kersting's groundnut diversity data. The study areas and the diversity of Kersting's groundnut across the agroecological zones were analyzed by incorporating the germplasm collection data into a geographic information system framework to map the distribution of Kersting's groundnut diversity using ArcMap software version 10.5.

The varietal diversity analysis was performed using the farmers' knowledge, the landraces they grow and the landraces collected in the study areas. The proportion of citations of each Kersting's groundnut landrace known and landrace grown were estimated and used to calculate ratios for each sociolinguistic group.

The total number of samples per landrace collected in the two agroecological zones was calculated. In addition, ecological models were employed to analyse the level of Kersting's groundnut diversity. Following Magurran [30], we defined landrace diversity as the number of landraces collected in the study area. Based on a simple count of landraces cultivated per farmer, Margalef's and Simpson's diversity indices were computed [30]. Landraces' richness (locally adapted inter-varietal diversity) among the two agroecological zones was compared using Margalef's index (D_{Mg}) as follows:

$$D_{Mg} = (L - 1)/\ln (S)$$

where D_{Mg} stands for the Kersting's groundnut diversity maintained in an agroecological zone by farmers; L is the total number of Kersting's groundnut phenotypic groups or landraces collected from farmers and S the total number of samples collected in an agroecological zone and ln is the natural logarithm.

The Simpson's index (D) was measured to take into account the number of landraces as well as the relative abundance of each landrace. The following formula was used following Tadesse [31]:

$$D = \sum_{i=1}^{n} (pi)^2$$

where pi, the proportional abundance of the ith landrace = (n_i/N), n_i is the number of samples for the landrace i and N the total number of samples. As D increases, the diversity decreases; consequently, the sum of the squared proportions was subtracted from 1, (1—D), to express the abundance of each landrace. The value of this index ranges between 0 and 1; the greater the value, the greater the phenotypic group diversity.

Furthermore, Factorial Analysis of Correspondence (FAC) was performed to assess the relationships between farmers' preferences and sociolinguistic groups. To better understand Kersting's groundnut cultivation systems, the Factorial Analysis of Mixed Data (FAMD) was used to cluster farmers' cropping practices into farming systems. Both analyses were performed using the packages FactoMineR version 2.0 [32] and factoextra version 1.0.6 [33]. Graphs were generated using ggplot2 version 3.2.1 package [34].

3. Results

3.1. Socio-Demographical Characteristics of Respondents

60.47% of the respondents in this study were women. The respondents' age was between 27 and 103 years old with a mean value of 54.31 ± 1.42. Most of the respondents were illiterate (74.42%) with farming as the main activity (75.58%). Seven sociolinguistic groups were surveyed across the agroecological zones: Bôbô, Senufo, Siamou, Toussian, Gurma, Dagara in the southern-Sudanian zone and the Bwamu in the northern-Sudanian zone. The Dagara and Toussian were the most represented in the study with a proportion of 26.74% each. They were followed by the Bwamu (16.28%) and the Bôbô (12.79%) groups. Other sociolinguistic groups, with less than 10 respondents, included the Senufo (8), Siamou (5) and Gurma (2). They represented 17.44% of the total samples. The socio-demographical characteristics of the respondents are presented in the Table 2.

Table 2. Socio-demographical characteristics of respondents in southern-Sudanian and northern-Sudanian zones of Burkina Faso and Ghana.

Variables	Modalities	Southern-Sudanian	Northern-Sudanian	Total
Number of villages		28	11	39
Number of respondents		72	14	86
Gender (%)	Male	36.11	57.14	39.53
	Female	63.89	42.86	60.47
Age (*Years* ± SE)		54.43 ± 1.53	53.71 ± 3.84	54.31 ± 1.42
	<50 years	38.89	28.57	37.21
	≥50 years	61.11	71.43	62.79
Marital status (%)	Married	81.94	85.71	82.56
	Widow	18.06	14.29	17.44
Study degree (%)	Illiterate	80.56	42.86	74.42
	Primary school	13.89	21.43	15.12
	High school	5.56	35.71	10.47
Socio-linguistic groups (%)	Bôbô	15.28	0.00	12.79
	Bwamu	0.00	100	16.28
	Dagara	31.94	0.00	26.74
	Toussian	31.94	0.00	26.74
	Others	20.83	0.00	17.44
Socio-professional groups (%)	Farmer	75.58	85.71	76.74
	Other	24.42.	14.29	23.26

3.2. Kersting's Groundnut Diversity and Folk Description

The vernacular names used by famers to identify *M. geocarpum* were relatively different across sociolinguistic groups (Table 3). The folk nomenclature of the crop landraces depended mainly on the farmers' own criteria of describing it. Therefore, the names given by farmers had different meanings to the generic names of the crops. The criterion 'seed coat color' was widely used by farmers (84.88%) to identify Kersting's groundnut different landraces. The identification of the crop was based also on the agronomic properties (33.72%), the seed size (8.14%), the religious utilizations of the crop (5.81%), and the growth habits (3.47%). Farmers' descriptions of *M. geocarpum* referred sometimes to other legume crops, such as Bambara groundnut, because of the underground nature of both crops, and cowpea, because of their similarity in seed coat color (white mottled with black eye) and taste.

Different local names can be assigned to the crop by the same sociolinguistic group. Furthermore, the same name can be used by different groups. For instance, the Toussian group referred to Kersting's groundnut mainly using three different names, such as Gwandjessi, Sidiin and Soonbia. The names Soonbia and Sidiin were also used by Bôbô and Siamou groups, respectively. In addition to these names, the Siamou called the crop Sissi, and the Bôbô used Zaka or Bôn to refer to the crop. The Dagara sociolinguistic group used names such as Sonsuonii and Sonsuolii, while the Gurma group referred to the crop as Susuonu. The Bwamu and Senufo groups, respectively, used Watié and Dougouvouguê as names for the crop.

Table 3. Local names of Kersting's groundnut landraces, farmers' description criteria and their characteristics across sociolinguistic groups in Burkina Faso and Ghana.

Description Criteria and Percentages of Respondents (%)	Vernacular Names	Local Languages	Description
Seed coat color (84.88)	Zaka dimi	Bôbô	Black seed coat color
	Doforo fiman, Dougouma sôssô fiman	Dioula *	
	Sissi, Sidiin	Siamou	
	Dougouvouguê, Davouguê	Senufo	
	Watié	Bwamu	
	Sonsuolii, Sonsuonii	Dagara	
	Zaka flô	Bôbô	White mottled seed color or cowpea seed color
	Doforo yingueni, Dougouma sôssô gueman	Dioula	
	Sissi	Siamou	
	Watié	Bwamu	
	Soonbia, Sindiin, Dîî, Sôbia, S'daï	Toussian	
	Dougouvouguê, Davouguê	Senufo	
	Watié	Bwamu	
	Sonsuolii,	Dagara	
	Sonsuonii	Dagara	Mixed colors seeds
	Susuonu	Gurma	
	Bôn	Bôbô	White seed color
	Doforo gueman	Dioula	
Agronomic properties (33.72)	Zaka, Soonbia	Bôbô	Tolerant to drought
	Bôn	Bôbô	Late matured cowpea
	Soonbia, Gwandjessi, Sindiin	Toussian	Tolerant to drought cowpea, late matured crop
	Djîn	Toussian	Early matured Bambara groundnut
Seed size (8.14)	Djîn	Toussian	Small seed size
	Sonsuonii, Sonsuolii	Dagara	Small Bambara groundnut seeds
Religious utilisations (5.81)	Zaka	Bôbô	Crop hunting the evil eye
	Djîn, Dîî	Toussian	Mystical crop
	Watié	Bwamu	Crop with multiple functions
	Dougouvouguê, Davouguê	Senufo	Crop for family protection and richness
Growth habit (3.47)	Bôn	Bôbô	Underground cowpea
	Dougouma sôssô, Solomia	Dioula	

Dioula *: a common language spoken mainly in the Western part of Burkina Faso.

3.3. Farmers' Knowledge and Extent of Kersting's Groundnut Diversity in Burkina Faso and Ghana

According to the frequency of citations for each landrace known and grown by farmers and the ratio values, the six landraces (Figure 2) recorded across sociolinguistic groups could be arranged into four groups (Table 4). Group 1 included well-known (>50% of respondents) and frequently grown (ratio > 50%) landraces; Group 2 was made up of well-known (>50%) but less-grown (<50%) landraces; Group 3 was represented by less-known (<50%) but frequently grown landraces; Group 4 included less-known and less-grown landraces. The two landraces—White mottled with black eye and

Black—were known by all the linguistic groups; they made the Group 1 and Group 2, respectively. The White mottled with black eye was mainly adopted by the Toussian (ratio = 95.65%) while it was less cultivated by the Bôbô (ratio = 14.29%). Apart from the Dagara, all the linguistic groups grew the Black landrace which was widely grown by the Bwamu people. Group 3 was composed of the Brown mottled with greyed orange eye, the Brown and the white mottled with greyed orange eye landraces mostly produced by the Dagara. In fact, the landraces White mottled with greyed orange eye and the Brown were cited only by the Dagara group. The Brown mottled with greyed orange eye was mentioned by Gurma and Dagara groups in Ghana. The White landrace was the least known and the least cultivated. It is produced by some Bôbô communities in Burkina Faso, while the Toussian and other groups knew it but did not produce it. The knowledge of Kersting's groundnut landraces varied greatly across sociolinguistic groups. The Dagara were richer in terms of knowledge on *M. geocarpum* folk diversity while the Bwamu knew little about the wide diversity of this crop.

Figure 2. Kersting's groundnut seed coat colors. (**a**) Black, (**b**) White mottled with greyed orange eye, (**c**) Brown mottled with greyed orange eye, (**d**) White, (**e**) White mottled with black eye, (**f**) Brown.

Table 4. Kersting's groundnut diversity known (%) and grown (%) and grown/known ratios (%) across sociolinguistic groups in Burkina Faso and Ghana. Ratio (%) = 100 × frequency of grown/frequency of known.

Sociolinguistic Group	Bôbô			Bwamu			Dagara		
Landrace seed colors	Known	Grown	Ratio	Known	Grown	Ratio	Known	Grown	Ratio
White	18.18	9.09	50	-	-		-	-	
White mottled with black eye	63.64	9.09	14.29	35.71	21.43	60	17.39	4.35	25
White mottled with greyed orange eye	-	-		-	-		4.35	4.35	100
Black	90.91	81.82	90	100	100	100	73.91	-	0
Brown	-	-		-	-		69.57	60.87	87.50
Brown mottled with greyed orange eye	-	-		-	-		95.65	91.30	95.45

Sociolinguistic Group	Toussian			Others			Total		
Landrace seed colors	Known	Grown	Ratio	Known	Grown	Ratio	Known	Grown	Ration
White	8.70	-	0	13.33	-	0	6.98	1.16	16.62
White mottled with black eye	100	95.65	95.65	46.67	40	85.71	53.49	38.37	71.73
White mottled with greyed orange eye	-	-		-	-		1.16	1.16	100
Black	52.17	4.35	8.33	86.67	53.33	61.54	76.74	37.21	48.49
Brown	-	-		-	-		18.60	16.28	87.5
Brown mottled with greyed orange eye	-	-		13.33	13.33	100	27.91	26.74	95.81

3.4. Distribution of Kersting's Groundnut Landraces and Diversity Estimation

Landraces are the genetic bases for crop improvement. A total of 62 Kersting's groundnut samples were collected in the study districts and varied morphologically in seed coat color. The samples were clustered in six different landraces according to the seed coat color: the cream (known as White), White mottled with black eye, White mottled with grey orange eye, Black, Brown and Brown mottled with grey orange eye (Figure 2). The distribution of these groups varied across the agroecological zones. All the six landraces were found in the southern-Sudanian zone while two (i.e., White mottled with black eye and Black) were found in the northern-Sudanian zone. There were four landraces specific to the southern-Sudanian zone, while no landrace was specific to the northern-Sudanian zone. The most common and widely grown landraces listed by farmers across the two agroecological zones were the White mottled with black eye and the Black. In the southern-Sudanian zone, the six landraces comprised in total 51 samples collected, and in the northern-Sudanian zone, the two landraces found comprised 11 samples. The distribution of landraces is represented in Figure 3 and summarized in Table 5. The landraces were collected in 25 villages out of the 39 villages surveyed. In the southern-Sudanian zone samples were found in 16 villages located in six departments namely: Bôbô Dioulasso, Péni, Orodara, Kourinion, Morolaba and Jirapa. In the northern-Sudanian zone, samples were found in seven villages located in the departments of Ouarkoye and Bondokuy. Furthermore, some landraces were specific to each country. In fact, the White one was found only in Burkina Faso while the Brown, Brown mottled with greyed orange eye, and the White mottled with greyed orange eye ones were specific to Ghana.

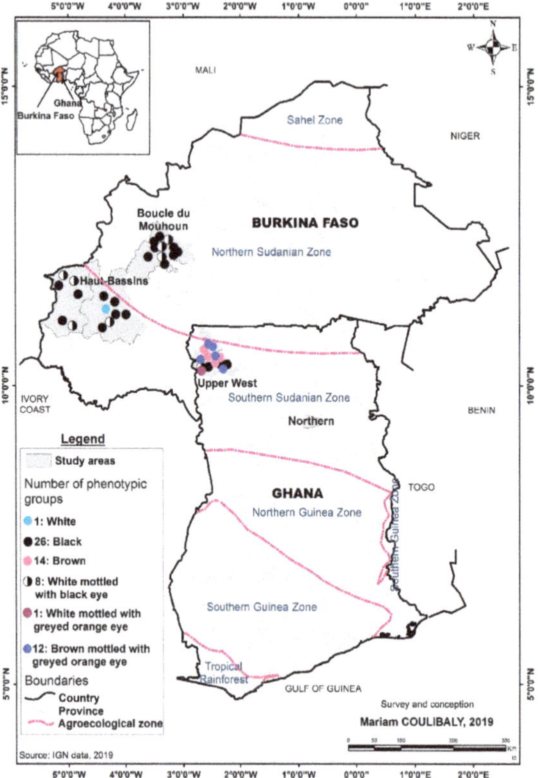

Figure 3. Distribution of Kersting's groundnut landraces collected across agroecological zones of Burkina Faso and Ghana.

Table 5. Kersting's groundnut landraces collected with diversity indices across agroecological zones of Burkina Faso and Ghana.

Landrace by Seed Coat Color	Southern-Sudanian	Northern-Sudanian
White mottled with black eye	5	3
Black	18	8
White	1	
White mottled with greyed orange eye	1	
Brown	14	
Brown mottled with greyed orange eye	12	
Diversity indices		
Number of landraces	6	2
Number of samples	51	11
Margalef's index	1.27	0.42
Simpson's index (1-D)	0.73	0.40

Diversity estimates (Table 5), based on the number of landraces (richness) and samples collected (abundance), revealed that the southern-Sudanian zone exhibited higher richness (Margalef = 1.27) and relative abundance (Simpson = 0.73) in terms of number of landraces and samples collected. The northern-Sudanian zone was found to be less diverse (Margalef = 0.42; Simpson = 0.40).

3.5. Criteria for Choosing Kersting's Groundnut Landraces to Grow

Farmers had many criteria in selecting *M. geocarpum* landraces (Table 6). These criteria concerned the crop characteristics and the use categories. The three main criteria were the seed coat color, the medicinal purposes and the organoleptic qualities. Preference for a specific landrace was dependent on the area and/or the farmer. The Black landrace was mainly grown by farmers for medicinal and socio-cultural purposes, while the White mottled with black eye landrace was grown for consumption and was preferred by farmers because of its organoleptic qualities. In other villages, the Brown and Brown mottled landraces were grown mainly because of the eating habits of consumers.

Table 6. Criteria used by farmers for growing Kersting's groundnut landraces in Burkina Faso and Ghana.

Criteria	Proportions of Citations (%)
Seed coat	37.21
Medicinal purposes	33.72
Organoleptic qualities	26.74
Eating habit	24.42
Socio-cultural purposes	23.26
Resistance to drought	10.59
Early maturity	4.71

3.6. Production Systems and Management across Sociolinguistic Groups in Burkina Faso and Ghana

In Burkina Faso and Ghana, Kersting's groundnut was produced on an average cropping area of 0.079 ± 0.0162 ha with an average yield of 546.10 ± 64.30 Kg/ha. Farmers in these countries grew *M. geocarpum* in rotation with cereals or legumes or in pure stand. However, few farmers (4.65%) intercropped it with cereals, roselle or cassava. Another cultivation practice used by farmers was the utilization of the crop around the fields of cereals or legumes as field border. Most of the farmers adopted to grow the crop on ridges or on flat sowing, while some farmers adopted the mounds as tillage practice. Field management included weeding (21–30 days after planting) and often earthing up at the reproduction phase (45–75 days after planting) in order to facilitate the pods development. Moreover, farmers planted Kersting's groundnut between May and August during the raining season and harvested after 4–5 months. Farmers stored the crop as dehulled seed and/or as pods in jars (61.63% of respondents), in garrets (26.74%), in plastic bottles (20.93%) and in bags (15.12%). Other material such as gourds, bottles and barrels were also used by a few farmers to store Kersting's groundnut. Farmers mixed Kersting's groundnut seeds with sand, ash, extracts of plants or in some rare cases with chemical products (1.16% of respondents) to reduce beetles' attacks.

The Factorial Analysis of Mixed Data (FAMD) of the cultivation practices revealed three farming systems in Burkina Faso and Ghana (Table 7). In the first system, all farmers grew Kersting's groundnut on mounds and 92.59% of them grew it on ridges; in this system, seeds were mainly stored in garrets as dehulled seed mixed with sand. In the second system, cropping Kersting's groundnut as field border and earthing up practices were used by most of the farmers (90.91% and 81.82%, respectively); in this system *M. geocarpum* was planted on flat sowing tillage and seeds were stored with plastic bottles mostly mixed with ash. In the third farming system there is no specific tillage practices; farmers in this system stored the crop as pod mixed with ash in jar.

Table 7. Results of Factorial Analysis of Mixed Data (FAMD) to cluster the cropping practices of Kersting's groundnut in Burkina Faso and Ghana.

Characteristics of Systems	Cla/Mod (%)	Mod/Cla (%)	Global	*p*-Value	v-Test
System 1					
Cultivation practices = field border	9.09	2.38	12.79	<0.001	−2.81
Tillage = ridges	92.59	59.52	31.40	<0.001	5.65
Tillage = mounds	100.00	19.05	9.30	<0.001	3.06
Tillage = flat sowing	23.26	23.81	50.00	<0.001	−4.75
Earthing up = yes	4.55	2.38	25.58	<0.001	−5.02
Storage form − as dehulled seed	52.50	100.00	93.02	<0.001	2.43
Storage material = garret	73.91	40.48	26.74	<0.001	2.76
Storage material = plastic bottle	22.22	9.52	20.93	<0.001	−2.50
Storage material = bag	15.38	4.76	15.12	<0.001	−2.58
Storage product = sand	88.46	54.76	30.23	<0.001	4.91
Storage product = ash	21.05	19.05	44.19	<0.001	−4.59
Seeds sources = family	15.79	7.14	22.09	<0.001	−3.26
System 2					
Cultivation practices = field border	90.91	25.64	12.79	<0.001	3.20
Cultivation practices = pure stand	31.71	33.33	47.67	<0.001	−2.38
Tillage = flat sowing	69.77	76.92	50.00	<0.001	4.54
Tillage = ridges	3.70	2.56	31.40	<0.001	−5.53
Earthing up = yes	81.82	46.15	25.58	<0.001	3.95
Storage material = plastic bottle	66.67	30.77	20.93	<0.001	1.98
Storage material = garret	17.39	10.26	26.74	<0.001	−3.15
Storage product = ash	78.95	76.92	44.19	<0.001	5.61
Storage product = sand	11.54	7.69	30.23	<0.001	−4.23
Seeds sources = family	73.68	35.90	22.09	<0.001	2.74
System 3					
Storage form = pod	100.00	100.00	5.81	<0.001	5.55
Storage material = jar	15.15	100.00	38.37	<0.001	2.71
Storage product = ash	10.42	100.00	55.81	<0.001	1.97

Cla/Mod: percentage of all surveyed farmers who use a specific farming practice and belong to a cluster (system). Mod/Cla: percentage of all farmers in a specific cluster (system) that also use a farming practice; *p*-value: level of analysis significance, v-test: measures the association between variables and clusters. It reveals which variables are positively or negatively associated with the clusters.

Kersting's groundnut growers got seeds from their family as gift or heritage. Seeds used by farmers came also from their own production but also used for consumption and other utilizations.

Macrotyloma geocarpum farming systems varied significantly (*p* < 0.035) across sociolinguistic groups in Burkina Faso and Ghana (Figure 4). The Dagara (Dagari) and Toussian adopted the first and the second farming systems. The Bôbô practiced the second and third farming systems. The Bwamu practiced all three farming systems. In general, the third system was less represented in the surveyed regions while the second farming system was the most practiced by all sociolinguistic groups.

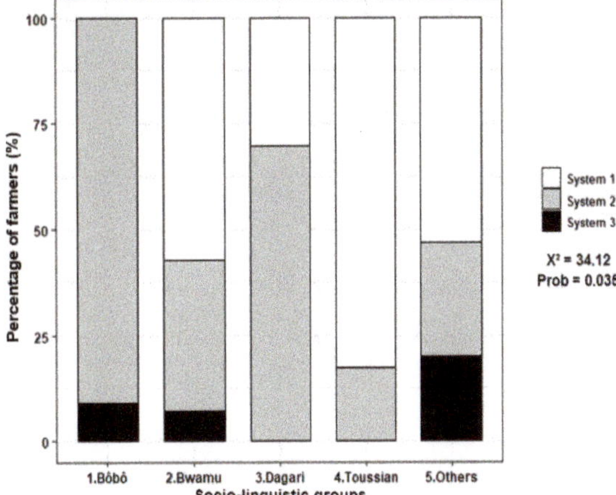

Figure 4. Frequency of Kersting's groundnut farming systems across socio-linguistic groups in Burkina Faso and Ghana.

3.7. Bottlenecks to Kersting's Groundnut Production

Farmers identified several factors constraining Kersting's groundnut production. Out of the fourteen cited, the five main constraints facing farmers included the difficulty to harvest the pods (66.28% of respondents), the lack of manpower (27.91%), the high soil moisture at the reproduction phase (24.42%), post-harvest pests damage (23.53%), and drought (22.09%) (Figure 5). The non-availability of seeds and the need for intensive soil preparation were also cited by farmers as factors limiting crop production. According to them, although present, disease pressures were not a big challenge in their production system. Pests disturbing and causing damages to the crop in the field included domestic herds, rodents foraging fresh pods, and termites. Constraints varied significantly among sociolinguistic groups (Table 8). For instance, apart from the Bôbô, all the groups cited the difficulty to harvest as the main constraint they encountered. The high soil moisture during the reproductive phase was also cited by all groups apart from the Dagara. For the Bôbô, Dagara and the Toussian, drought was an important factor constraining Kersting's groundnut production. Another factor cited by producers was the non-availability of seeds limiting the production in the farming systems of Bwamu and Toussian. The post-harvest pests were ranked by the Dagara group as important constraints in Kersting's groundnut production.

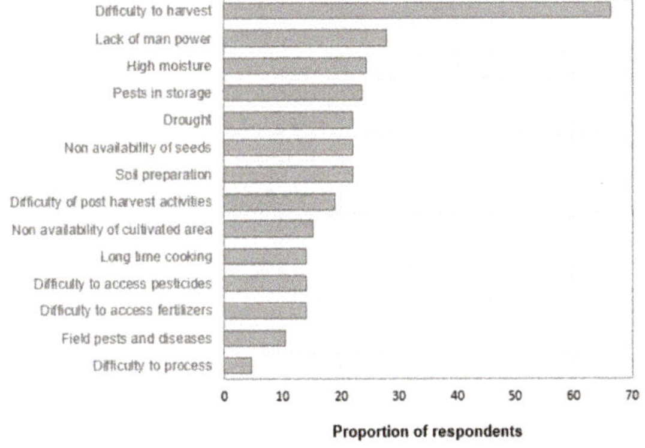

Figure 5. Constraints to Kersting's groundnut production in Burkina Faso and Ghana.

Table 8. Proportion of citations of Kersting's groundnut production bottlenecks across sociolinguistic groups in Burkina Faso and Ghana.

Constraints	Bôbô	Bwamu	Dagara	Toussian	Others	Kruskal Test
Difficulty to harvest	54.55	71.43	78.26	65.22	53.33	*
Non availability of seeds	0.00	28.57	0.00	39.13	40.00	***
Soil preparation	9.09	7.14	26.09	30.43	26.67	**
High soil moisture	27.27	28.57	13.04	34.78	20.00	*
Lack of labour	63.64	14.29	26.09	30.43	13.33	**
Pests in storage	9.09	21.43	43.48	17.39	13.33	***
Difficulty to access fertilizers	18.18	21.43	8.70	13.04	13.33	**
Difficulty to access pesticides	18.18	21.43	8.70	13.04	13.33	**
Field pests and diseases	27.27	14.29	0.00	8.70	13.33	***
Drought	45.45	7.14	26.09	26.09	6.67	***
Non availability of cultivated area	9.09	28.57	8.70	21.74	6.67	***
Difficulty of post harvesting activities	18.18	21.43	30.43	13.04	6.67	***
Long-time cooking	27.27	14.29	13.04	17.39	0.00	***
Difficulty to process	18.18	7.14	4.35	0.00	0.00	***

Kruskal Wallis test with frequencies of citations. *** p value < 0.001, ** p value < 0.01 and * p value < 0.05.

3.8. Preferred Traits for Kersting's Groundnut Improvement

Thirteen traits were listed by producers as important for crop improvement (Table 9). The top five preferred traits included early maturing, high yielding, resistance to bruchids, tolerance to drought, and tolerance to high soil moisture. Farmers were also interested in some specific seeds' attributes, as the nutritional content and time taken to cook on available cooking facilities. White mottled with black eye and White seeded landraces were preferred by farmers and can be used as background landraces.

Table 9. Traits identified by farmers for Kersting's groundnut new varieties in Burkina Faso and Ghana.

Criteria	Traits	Percentage of Respondents
Agronomic	Early maturity	34.88
	High yield	33.72
Genetic resistance	Tolerance to drought	26.74
	Resistance to bruchids	26.74
	Tolerance to high soil moisture	22.09
Seeds coat colors	White seed coat	25.58
	White mottled seed coat	25.58
	Black seed coat	18.60
	Red seed coat	10.47
	Brown seed coat	8.14
Seeds attributes	Good seed quality	15.12
	Facility to cook	10.47
	Big seed size	9.30

The Factorial Analysis of Correspondence revealed variation of farmers' preferred traits across sociolinguistic groups (Figure 6). The Bôbô group preferred the black seed coat color with resistance to high soil moisture. The Bwamu were interested in short seeds cooking time. The Dagara and Toussian groups preferred the White mottled with black eye landrace with high yielding, early maturing, and big seed size attributes.

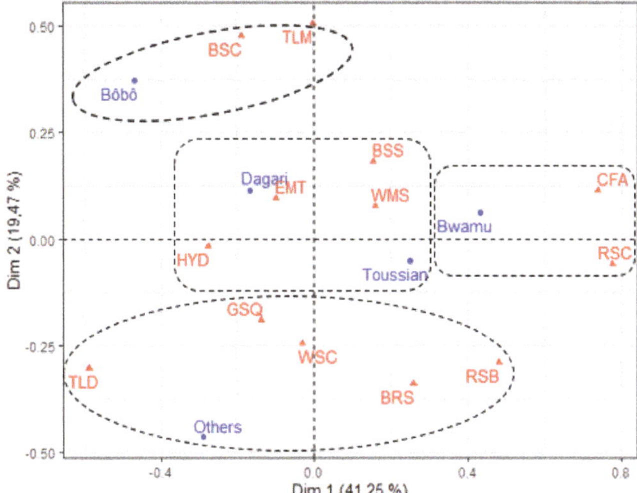

Figure 6. Traits identified by farmers for Kersting's groundnut improvement according to socio-linguistic groups. WSC: White Seed coat color; WMS: White mottled Seed coat color; RSC: Red Seed coat color; BSC: Black Seed coat color; BRS: Brown Seed coat color; TLD: Tolerance to drought; TLM: Tolerance to moisture; RSB: Resistance to bruchids; BSS: Big seed size; GSQ: Good seed quality; CFA: Facility to cook; HYD: High yield; EMT: Early maturity.

4. Discussion

4.1. Knowledge, Existence and Distribution of Kersting's Groundnut Diversity in Burkina Faso and Ghana

Sixty-two samples grouped in six different landraces were collected across both countries instead of the five landraces reported in previous studies [12,13,17]. Tamini [14] reported the White mottled with black eye and the Black landraces in Burkina Faso; whereas in this study, a White landrace was additionally recorded and collected. Similarly, the Brown and the White mottled with greyed orange eye were found only in Ghana and were not yet reported by previous investigations [17,35]. In this study, we therefore found additional diversity in the southern-Sudanian zone of Burkina Faso and Ghana. Those findings support the assumption that southern-Sudanian zone can be considered as the primary center of origin of Kersting's groundnut [12]. Previous reports indicated that the crop originated from central Benin and Northern Togo [36]. Amujoyegbe [18] mentioned Nigeria as the primary center of origin of the crop. Therefore, extended investigations followed by a wide diversity resource collection is expected across other West-African countries where the cultivation of the crop was mentioned. Future explorations could involve Nigeria, Mali and Ivory Coast in order to have a clear idea of the genetic diversity of this crop and the center of origin of that diversity. This large germplasm collection should be followed by the genetic characterization in different environments in order to better understand the genetic background and the stability of Kersting's groundnut specific traits. In addition, the association studies involving molecular tools could make more accurate the genetic characterization and can be useful in the identification of QTLs of functional and economic traits. Furthermore, ex-situ conservation strategies through National and International genebanks supported by the local conservation systems must be well organized to ensure the sustainable maintenance of the existing diversity. In Benin and Togo, the Black landrace was well known by farmers but less cultivated while the White one was widely grown and well appreciated by farmers and consumers [12,13]. As noticed by Akohoué [12] in Benin and Togo, farmers in Burkina Faso and Ghana also grew the Black landrace mainly because of medicinal and socio-cultural attributes. The White and the White mottled with black eye landraces were grown mainly for consumption and their organoleptic qualities. The preference of farmers for these landraces represents a good opportunity for their sustainable conservation on-farm and across generations. On-farm management of agricultural biodiversity implies that smallholder farmers select and develop the species they need to match their diets, culture, markets and environment. Hence, in-situ conservation could be more organized through the valorization of Community Seed Banks (CSB). CSBs are local-level institutions that contribute to seed conservation, in particular of farmer varieties, countering erosion of crop diversity or its loss following natural disasters [37]. CSBs have been established recently in Burkina Faso and Ghana, but are limited to few regions [38]. This platform should be extended to other regions in order to limit the loss of genetic resources including Kersting's groundnut.

Landraces are considered to be an integral influence on farmers' decisions on maintenance, management, and exchange [39]. Farmers had a good knowledge of Kersting's groundnut varieties across sociolinguistic groups in terms of classification, identification, description and uses. Our results revealed that farmers go beyond just naming, by grouping landraces together based on common characters. Their indigenous knowledge would be extremely useful to agronomists, geneticists and breeders. The criteria used by farmers to describe different landraces included the seeds characteristics, agronomic and religious properties, and the growth habits of the crop. Similar results were found in other studies including on the common bean [40], fonio [41], millet [42] and Kersting's groundnut [12]. These characteristics used to identify different Kersting's groundnut landraces reflect the consistence of folk taxonomy. The seed color used by farmers as the main folk descriptor of this legume was considered as a true discriminant trait by growers as well as by scientists who performed the agro-morphological characterization of the species [13,15,43].

4.2. Kersting's Groundnut Production Systems in Burkina Faso and Ghana

Most legumes including Kersting's groundnut have been overlooked by research and development agencies and are generally considered as women's crops in some areas since they are often produced by women producers on small land areas meant primarily for home consumption [44], whereas men tend to dominate in the production and marketing of cereals in the food value chain [45]. This was true for Kersting's groundnut in the farming systems of Burkina Faso and Ghana. However, in other communities, men were mostly involved in the crop production as the crop certainly exhibits a high economical value in those communities [12,17]. The economical and the nutritional importance of this crop should be clearly highlighted and promoted across its production areas in order to better use the genetic resources available for food and nutrition security as well as for income generation.

Farmers generally grew *M. geocarpum* in rotation. Others intercropped or used it as field border (mainly the Bôbô group). Farmers in Nigeria [18], Benin and Togo [12] also intercropped the crop. According to farmers, Kersting's groundnut helps improve the fertility of the soil, thereby reducing the cost of fertilizer inputs in crop farming. This is certainly due to the nodulation property of the crop for symbiotic nitrogen fixation [16,46] leading to the positive contribution of legume crops in the yield improvement of associated crops [47–49].

Kersting's groundnut was grown generally on ridges or on mound in the farming systems Burkina Faso and Ghana. The same types of tillage were used in the production systems of the crop in Ghana [19], Benin and Togo [12,13,50]. The type of tillage was often different across sociolinguistic groups. In Benin, Kouelo [50] showed that the type of tillage practice can contribute significantly to improve *M. geocarpum* yield and was specific to the soil conditions. Therefore, investigations should be done in the different growing areas of Kersting's groundnut in order to propose to farmers the suitable tillage practice adapted to their soils' characteristics.

In Burkina Faso and Ghana, farmers stored Kersting's groundnut as pods. This practice, also reported in other countries [12,13,17], contributes to limiting pulse beetles' damages and permits a longer storage life. Badii [51] found that the Black and Brown landraces were less susceptible to the bruchids (*Callosobruchus maculatus* Fabricius) attacks than the White mottled with black eye. The genetic determinism of this resistance to *C. maculatus* in Brown and Black landraces needs be deepened. Consequently, the possibility for introgression of bruchids-resistant gene from Brown and Black landraces into other landraces (White and White mottled with black eye) can be explored.

There was a number of barriers to *M. geocarpum* production, including the hard work required for harvesting while there is an increasing a lack of manpower for cultivation. In addition, the crop was grown mainly by elderly people (above 50 years old) while young people were focused on cereals and cash crops—generally ignoring legumes. This situation was also reflected by many governments' policy and research institution priorities [52]. The difficulty to harvest appeared as a major challenge for growers in Benin and Nigeria as well [13,18].

Furthermore, the non-availability of seeds to meet farmers' needs is another constraint limiting Kersting's groundnut production. In fact, in most of the growing areas where the crop production was decreasing or disappeared, farmers lost their seeds and did not have the possibility of a new supply. Similarly, in Benin, the lack of market for seed supply was one of the reasons for the low production of Kersting's groundnut. The possibility for *M. geocarpum* commercial seeds production is still inexistent; farmers saved their own seeds for planting. At the same time, the easy access to quality seed can be achieved and guaranteed only if there is a viable seed supply system to multiply and distribute seeds that have been produced or preserved by farmers [53]. According to the same authors, the most effective alternative is to create smallholder seed enterprises, located in farming communities, with lower capital investment needs. These enterprises should be closed to smallholder farmers and be able to distribute quality seeds of improved and local varieties such as Kersting's groundnut.

4.3. Farmers' Preferences for Kersting's Groundnut and Implication for its Genetic Improvement

Kersting's groundnut production has been decreasing drastically because of its low yield, the small seed size and the late maturing time [12,13,18]. This explains the need expressed by farmers to have improved cultivars with high yield, drought tolerance, and early maturing attributes. Authors revealed that the crop exhibits variations in flowering time, seed size and grains yield [13,43]. This offers room for genetic improvement and selection of desired cultivars. However, the scarce knowledge of the reproductive biology and the narrow genetic basis [24,54], require a lot of investment to develop crosses and broaden the genetic background. Failing to establish the clear nature of Kersting's groundnut floral biology may hinder the success of its varietal improvement through the cross pollination of good landraces. Although there is room for mutagenesis, no studies have so far embraced this pathway. Mutation breeding—using both physical and chemical mutagens—has been extensively and successfully used in the development of improved cultivars of legumes such as Bambara groundnut [55,56] and groundnut [57].

5. Conclusions

This study revealed the diversity of Kersting's groundnut landraces grown in the southern-Sudanian and northern-Sudanian zones of Burkina Faso and Ghana. Results showed that farmers grew six different landraces. All those landraces were grown in the southern-Sudanian zone, while only two were grown in the northern-Sudanian zone. Farmers preferred their varieties to be high yielding and tolerant to drought, resistant to bruchids and early maturing. This finding is relevant and important for the definition of the Kersting's groundnut breeding objectives. The study also analyzed the cropping systems of the crop, the main constraints to its production and examined farmers' preferences for its improvement. The study revealed that farming practices across the surveyed areas were diverse and strongly influenced by sociolinguistic membership. In Ghana and Burkina Faso, farmers' knowledge about the crop and its production practices remains vital for the future promotion of the resource.

Author Contributions: Conceptualization by M.C. and E.G.A.-D.; investigation by M.C. and C.O.A.A.; writing and editing M.C.; validation by E.G.A.-D., formal analysis by M.C. and F.A., review by E.G.A.-D. and M.S. All authors have read and agreed to the published version of the manuscript.

Funding: Mariam Coulibaly, from the University of Ouaga I Prof Joseph Ki-Zerbo, Burkina Faso, was a scholar of the "Intra-Africa Academic Mobility Scheme" under the project grant number 2016-2988 on "Enhancing training and research mobility for novel crops breeding in Africa (MoBreed)" funded by the Education, Audiovisual and Culture Executive Agency (EACEA) of the European Commission. The project provided a scholarship for academic training and research mobility and a research grant to the first Author to complete a PhD degree at the University of Abomey-Calavi (Republic of Benin).

Acknowledgments: Authors wish to thank MoBreed program for the financial support. We are thankful to all farmers and agriculture technical services in Burkina Faso and Ghana. We are grateful to local authorities in all surveyed villages. We thank also Xavier Matro Comlan, Ayenan Mathieu, Dao Abdalla for all assistance during the surveys.

Conflicts of Interest: The authors declared no conflict of interest.

References

1. Alexandratos, N.; Bruinsma, J. *World Agriculture Towards 2030/2050: The 2012 Revision*; Food and Agriculture Organization (FAO): Rome, Italy, 2012; pp. 1–160.

2. van Ittersum, M.K.; van Bussel, L.G.J.; Wolf, J.; Grassini, P.; van Wart, J.; Guilpart, N.; Claessens, L.; de Groot, H.; Wiebe, K.; Mason-D'Croz, D.; et al. Can sub-Saharan Africa feed itself? *Proc. Natl. Acad. Sci. USA* **2016**, *113*, 14964–14969. [CrossRef] [PubMed]

3. Syngenta. The Future of Farming. 2017. Available online: http://www.syngenta-us.com/thrive/research/future-offarming.html (accessed on 25 March 2018).

4. Considine, M.J.; Siddique, K.H.M.; Foyer, C.H. Nature's pulse power: Legumes, food security and climate change. *J. Exp. Bot.* **2017**, *68*, 1815–1818. [CrossRef] [PubMed]

5. Legume Phylogeny Working Group (LPWG). Legume phylogeny and classification in the 21st century: Progress, prospects and lessons for other species-rich clades. *Taxon* **2013**, *62*, 217–248. [CrossRef]

6. Snapp, S.; Rahmanian, M.; Batello, C. *Pulse Crops for Sustainable Farms in Sub-Saharan Africa*; Food and Agriculture Organization (FAO): Rome, Italy, 2018; pp. 1–60.

7. Ojiewo, C.; Monyo, E.; Desmae, H.; Boukar, O.; Mukankusi-Mugisha, C.; Thudi, M.; Pandey, M.K.; Saxena, R.K.; Gaur, P.M.; Chaturvedi, S.K.; et al. Genomics, genetics and breeding of tropical legumes for better livelihoods of smallholder farmers. *Plant Breed.* **2018**, *138*, 487–499. [CrossRef]

8. Mergeai, G. Influence des facteurs sociologiques sur la conservation des ressources phytogenetiques: Le cas de la lentille de terre [*Macrotyloma geocarpum* (Harms) Marechal et Baudet] au Togo. *Bull. Rech. Agron.* **1993**, *28*, 487–500.

9. Ajayi, O.B.; Oyetayo, F.L. Potentials of *Kerstingiella geocarpa* as a health food. *J. Med. Food* **2009**, *12*, 184–187. [CrossRef]

10. Aremu, M.O.; Osinfade, B.G.; Basu, S.K.; Ablaku, B.E. Development and nutritional quality evaluation of Kersting's groundnut-ogi for African weaning diet. *Am. J. Food Technol.* **2011**, *6*, 1021–1033. [CrossRef]

11. Dansi, A.; Vodouhè, R.; Azokpota, P.; Yedomonhan, H.; Assogba, P.; Adjatin, A.; Loko, Y.L.; Dossou-Aminon, I.; Akpagana, K. Diversity of the neglected and underutilized crop species of importance in Benin. *Sci. World J.* **2012**, *2012*, 19. [CrossRef]

12. Akohoué, F.; Sibiya, J.; Achigan-Dako, E.G. On-Farm practices, mapping, and uses of genetic resources of Kersting's groundnut [*Macrotyloma geocarpum* (Harms) Maréchal et Baudet] across ecological zones in Benin and Togo. *Genet. Resour. Crop Evol.* **2019**, *66*, 195–214. [CrossRef]

13. Assogba, P.; Ewedje, E.-E.B.K.; Dansi, A.; Loko, Y.L.E.; Adjatin, A.; Dansi, M.; Sanni, A. Indigenous knowledge and agro-morphological evaluation of the minor crop Kersting's groundnut [*Macrotyloma geocarpum* (Harms) Maréchal et Baudet] cultivars of Benin. *Genet. Resour. Crop Evol.* **2015**, *63*, 513–529. [CrossRef]

14. Tamini, Z. Étude ethnobotanique de la Lentille de Terre [*Macrotyloma geocarpum* (Harms) Maréchal & Baudet] au Burkina Faso. *J. Agric. Tradit. Bot. Appliquée* **1995**, *37*, 187–199.

15. Adu-Gyamfi, R.; Dzomeku, I.K.; Lardi, J. Evaluation of growth and yield potential of genotypes of kersting's groundnut (*Macrotyloma geocarpum* Harms) in Northern Ghana. *Int. Res. J. Agric. Sci. Soil Sci.* **2012**, *2*, 509–515.

16. Mohammed, M.; Jaiswal, S.K.; Sowley, E.N.K.; Ahiabor, B.D.K.; Dakora, F.D. Symbiotic N2 fixation and grain yield of endangered Kersting's groundnut landraces in response to soil and plant associated bradyrhizobium inoculation to promote ecological resource-use efficiency. *Front. Microbiol.* **2018**, *9*, 14. [CrossRef] [PubMed]

17. Adu-Gyamfi, R.; Fearon, J.; Bayorbor, T.B.; Dzomeku, I.; Avornyo, V. The status of Kersting's groundnut [*Macrotyloma Geocarpum* (Harms) Marechal and Baudet]. *Outlook Agric.* **2011**, *40*, 259–262. [CrossRef]

18. Amujoyegbe, B.; Obisesan, I.O.; Ajayi, A.O.; Aderanti, F.A. Disappearance of Kersting's groundnut (*Macrotyloma geocarpum* (Harms) Marechal and Baudet) in South-Western Nigeria: An indicator of genetic erosion. *Plant Genet. Resour. Newsl.* **2007**, *152*, 45–50.

19. Bampuori, A.H. Effect of traditional farming practices on the yield of indigenous Kersting's groundnut (*Macrotyloma geocarpum* Harms) Crop in the Upper West Region of Ghana. *J. Dev. Sustain. Agric.* **2007**, *2*, 128–144.

20. Ayenan, M.A.T.; Ofori, K.; Ahoton, L.E.; Danquah, A. Pigeonpea [(*Cajanus cajan* (L.) Millsp.)] production system, farmers' preferred traits and implications for variety development and introduction in Benin. *Agric. Food Secur.* **2017**, *6*, 11. [CrossRef]

21. Achigan-Dako, E.G.; Adjé, C.A.; N'Danikou, S.; Hotegni, N.V.F.; Agbangla, C.; Ahanchédé, A. Drivers of conservation and utilization of pineapple genetic resources in Benin. *SpringerPlus* **2014**, *3*, 1–11. [CrossRef]

22. Ouedraogo, M. *Etat et Structure de la Population: Résultats Définitifs du Recensement Général de la Population et de L'habitation (RGPH) de 2006–2009*; Institut National de la Statistique et de la Démographie: Ouagadougou, Burkina Faso, 2008; pp. 1–181.

23. Traoré, D. Le senar (langue Senufo du Burkina Faso): Éléments de description et d'influence du jula véhiculaire dans un contexte de contact de langues. In *Identité des Langues Gur du Burkina Faso*; Delplanque, A., Ed.; Cuvillier Verlag: Göttingen, Germany, 2015; pp. 1–21.

24. Boyd, R.; Fournier, A.; Nignan, S. Une base de données informatisée transdisciplinaire de la flore chez les Seme du Burkina Faso: Un outil pour l'étude du lien nature-société. In *Regards Scientifiques Croisés Sur le Changement Global et le Développement: Langue, Environnement, Culture*; Delplanque, A., Ed.; The Research Institute for Development: Marseille, France, 2012; pp. 1–239.

25. van der Geest, K.A.M. *The Dagara Farmer at Home and Away: Migration, Environment and Development in Ghana*; Center, L.A.S., Ed.; University of Amsterdam: Amsterdam, The Netherlands, 2011; pp. 1–19.

26. Albuquerque, U.P.; de Lucena, R.F.P.; de Freitas Lins Neto, M.E. Methods and techniques used to collect ethnobiological data. In *Methods and Techniques in Ethnobiology and Ethnoecology*; Albuquerque, U.P., de Lucena, R.F.P., de Freitas Lins Neto, M.E., Eds.; Springer Protocols Handbooks: New York, NY, USA, 2014; pp. 1–13.

27. Albuquerque, U.P.; Cruz da Cunha, L.V.F.; Lucena, R.F.P.; Alves, R.R.N. Selection of research participants. In *Methods and Techniques in Ethnobiology and Ethnoecology*; Albuquerque, U.P., Cruz da Cunha, L.V.F., Lucena, R.F.P., Alves, R.R.N., Eds.; Springer Protocols Handbooks: New York, NY, USA, 2014; pp. 15–37.

28. Sphinx Développement. *Initiation au Logiciel Sphinx: De la Conception du Questionnaire à la Communication des Résultats en Passant par la Collecte des Réponses et L'analyse des Données*; Sphinx Développement: Chavanod, France, 2007.

29. Team, R.C. *R: A Language and Environment for Statistical Computing*; R Foundation for Statistical Computing: Vienna, Austria, 2019.

30. Magurran, E. *Measuring Biological Diversity*; Blackwell Science Ltd.: Hoboken, NJ, USA, 2004.

31. Tadesse, D.; Asres, T. On farm diversity of barley landraces in North Western Ethiopia. *Int. J. Biodivers. Conserv.* **2019**, *11*, 1–7.

32. Le, S.; Josse, J.; Husson, F. FactoMineR: A package for multivariate analysis. *J. Stat. Softw.* **2008**, *25*, 1–18. [CrossRef]

33. Kassambara, A.; Mundt, F. Factoextra: Extract and Visualize the Results of Multivariate Data Analyses; 2019. Available online: http://www.sthda.com/english/rpkgs/factoextra (accessed on 6 March 2020).

34. Wickham, H. *Ggplot2: Elegant Graphics for Data Analysis*; Springer: New York, NY, USA, 2016.

35. Bayorbor, T.B.; Dzomeku, I.; Avornyo, V.; Opoku-Agyeman, M. Morphological variation in Kersting's groundnut (*Kerstigiella geocarpa* Harms) landraces from northern Ghana. *Agric. Biol. J. N. Am.* **2010**, *1*, 290–295. [CrossRef]

36. Achigan-Dako, E.G.; Vodouhè, S.R. *Macrotyloma geocarpum* (Harms) Maréchal & Baudet. In *Plant Resources of Tropical Africa 1: Cereals and Pulses*; Brink, M.B.G., Ed.; Backhuys Publishers CTA, PROTA: Wageningen, The Netherlands, 2006; pp. 111–114.

37. Vernooy, R.; Shrestha, P.; Sthapit, B. The rich but little known chronicles of community seed banks. In *Community Seed Banks: Origins, Evolution and Prospects*; Vernooy, R., Shrestha, P., Sthapit, B., Eds.; Routeledge Taylor and Francis Group: New York, NY, USA, 2015; pp. 2–7.

38. Vernooy, R.; Shrestha, P.; Sthapit, B. Comparative analysis of key aspects of community seed banks: Origins and evolution. In *Community Seed Banks: Origins, Evolution and Prospects*; Vernooy, R., Shrestha, P., Sthapit, B., Eds.; Routeledge Taylor and Francis Group: New York, NY, USA, 2015; pp. 11–19.

39. Olango, T.M.; Tesfaye, B.; Catellani, M.; Pè, M.E. Indigenous knowledge, use and on-farm management of enset (*Ensete ventricosum* (Welw.) Cheesman) diversity in Wolaita, Southern Ethiopia. *J. Ethnobiol. Ethnomed.* **2014**, *10*, 1–18. [CrossRef] [PubMed]

40. Loko, L.E.Y.; Toffa, J.; Adjatin, A.; Akpo, A.J.; Orobiyi, A.; Dansi, A. Folk taxonomy and traditional uses of common bean (*Phaseolus vulgaris* L.) landraces by the sociolinguistic groups in the central region of the Republic of Benin. *J. Ethnobiol. Ethnomed.* **2018**, *14*, 52. [CrossRef] [PubMed]

41. Ballogou, V.Y.; Mohamed, S.; Fatiou, T.; Hounhouigan, D.J. Indigenous knowledge on landraces and fonio-based food in Benin. *Ecol. Food Nutr.* **2014**, *53*, 390–409. [CrossRef]

42. Rengalakshmi, R. Folk biological classification of minor millet species in Kolli Hills, India. *J. Ethnobiol.* **2005**, *25*, 59–70. [CrossRef]

43. Akohoue, F.; Achigan-Dako, E.G.; Coulibaly, M.; Sibiya, J. Correlations, path coefficient analysis and phenotypic diversity of a West African germplasm of Kersting's groundnut [*Macrotyloma geocarpum* (Harms) Maréchal & Baudet]. *Genet. Resour.Crop Evol.* **2019**, *66*, 1825–1842.

44. Mapfumo, P.; Campbell, B.M.; Mpepereki, S. Legumes in soil fertility management: The case of Pigeonpea in smallholder farming systems of Zimbabwe. *Afr. Crop Sci. J.* **2001**, *9*, 629–644. [CrossRef]

45. Bationo, A.; Waswa, B.; Okeyo, J.; Maina, F.; Kihara, J.; Mokwunye, U. *Fighting Poverty in Sub-Saharan Africa: The Multiple Roles of Legumes in Integrated Soil Fertility Management*; Springer: Dordrecht, The Netherlands, 2011; p. 60.

46. Mohammed, M.; Sowley, E.N.K.; Dakora, F.D. Symbiotic N 2 fixation, C assimilation and water-use efficiency (WUE) of five Rhizobium-inoculated Kersting's groundnut (*Macrotyloma geocarpum*) landraces measured using 15 N and 13 C isotopic techniques. *S. Afr. J. Bot.* **2015**, *98*, 192–210. [CrossRef]

47. Sanou, J.; Bationo, B.A.; Barry, S.; Nabie, L.D.; Bayala, J.; Zougmore, R. Combining soil fertilization, cropping systems and improved varieties to minimize climate risks on farming productivity in northern region of Burkina Faso. *Agric. Food Secur.* **2016**, *5*, 12. [CrossRef]

48. Kalonga, J.; Mbega, E.R.; Mtei, K. Potential of legume diversification in soil fertility management and food security for resource poor farmers in Sub-Saharan Africa. *J. Biodivers. Environ. Sci. JBES* **2017**, *11*, 192–205.

49. Lupwayi, N.Z.; Kennedy, A.C.; Chirwa, R.M. Grain legume impacts on soil biological processes in sub-Saharan Africa. *Afr. J. Plant Sci.* **2011**, *5*, 1–7.

50. Kouelo, K.A.F.; Badou, A.; Houngnandan, P.; Francisco Merinosy, M.F.; Gnimassoun, C.J.-B.; Sochime, D.J. Impact du travail du sol et de la fertilisation minérale sur la productivité de *Macrotyloma geocarpum* (Harms) Maréchal et Baudet au center du Bénin. *J. Biosci.* **2012**, *51*, 3625–3632.

51. Badii, K.B.; Asante, S.K.; Bayorbor, T.B. Susceptibility of some Kersting's groundnut landrace cultivars to infestation by *Callosobruchus maculatus* (Fab.) [Coleoptera: Bruchidae]. *J. Sci. Technol.* **2011**, *31*, 10. [CrossRef]

52. Isaacs, K.B.; Snapp, S.S.; Kelly, J.D.; Chung, K.R. Farmer knowledge identifies a competitive bean ideotype for maize–bean intercrop systems in Rwanda. *Agric. Food Secur.* **2016**, *5*, 15. [CrossRef]

53. Neate, P.J.H.; Guei, R.G. *Promoting the Growth and Development of Smallholder Seed Enterprises for Food Security Crops: Best Practices and Options for Decision Making*; Food and Agriculture Organization (FAO): Rome, Italy, 2010; p. 28.

54. Pasquet, R.S.; Mergeai, G.; Baudoin, J.-P. Genetic diversity of the African geocarpic legume Kersting's groundnut, *Macrotyloma geocarpum* (Tribe Phaseoleae: Fabaceae). *Biochem. Syst. Ecol.* **2002**, *30*, 943–952. [CrossRef]

55. Adebola, M.I.; Esson, A.E. Fast neutrons induced genetic variability on Bambara nut (*Vigna subterranean* (L.) Verdc.). *Horticult. Biotechnol. Res.* **2017**, *3*, 10–12.

56. Chitti, B. *Genetic improvement of Bambara Groundnut (Vigna subterranea (L.) Verdc) through Mutation Breeding, in Genetics and Plant Breeding*; University of Agricultural Sciences GKVK: Bengaluru, India, 2015; p. 121.

57. Janila, P.; Nigam, S.N.; Pandey, M.K.; Nagesh, P.; Varshney, R.K. Groundnut improvement: Use of genetic and genomic tools. *Front. Plant Sci.* **2013**, *4*, 23. [CrossRef]

 *agronomy*

Article

Induced Mutagenesis Enhances Lodging Resistance and Photosynthetic Efficiency of Kodomillet (*Paspalum Scrobiculatum*)

James Poornima Jency [1], Ravikesavan Rajasekaran [1,*], Roshan Kumar Singh [2], Raveendran Muthurajan [3], Jeyakumar Prabhakaran [4], Muthamilarasan Mehanathan [2], Manoj Prasad [2] and Jeeva Ganesan [1]

[1] Department of Millets, Centre for Plant Breeding and Genetics, Tamil Nadu Agricultural University, Coimbatore 641003, Coimbatore, India; poornimajames@gmail.com (J.P.J.); jeevagss@gmail.com (J.G.)

[2] National Institute of Plant Genome Research, Aruna Asaf Ali Marg, New Delhi 110067, Delhi, India; sroshan@nipgr.ac.in (R.K.S.); muthamilarasan@nipgr.ac.in (M.M.); manoj_prasad@nipgr.ac.in (M.P.)

[3] Centre for Plant Molecular Biology and Biotechnology, Tamil Nadu Agricultural University, Coimbatore 641003, Coimbatore, India; raveendrantnau@gmail.com

[4] Department of Crop Physiology, Tamil Nadu Agricultural University, Coimbatore 641003, Coimbatore, India; jeyakumar@tnau.ac.in

* Correspondence: chithuragul@gmail.com; Tel.: +62-910422-2450507

Received: 26 December 2019; Accepted: 27 January 2020; Published: 4 February 2020

Abstract: The present research was focused in the development of photosynthetically efficient (PhE) and non-lodging mutants by utilizing ethyl methane sulphonate (EMS) and gamma radiation in the kodomillet variety CO 3, prone to lodging. Striking variations in a number of anatomical characteristics of leaf anatomy for PhE and culm thickness for lodging resistance was recorded in M_2 (second mutant) generation. The identified mutants were subjected to transcriptomic studies to understand their molecular basis. Expression profiling was undertaken for pyruvate phosphate dikinase (PPDK), *Nicotinamide Adenine Dinucleotide Phosphate Hydrogen*—(NADPH) and NADP-dependent malate dehydrogenase (NADP-MDH) in the mutants CO 3-100-7-12 (photosynthetically efficient) and in CO 3-200-13-4 (less efficient). For lodging trait, two mutants CO 3-100-18-22 (lodged) and CO 3-300-7-4 (non-lodged) were selected for expression profiling using genes *GA2ox6* and *Rht-B*. The studies confirmed the expression of PPDK increased 30-fold, NADP-ME2 ~1-fold and NADP-MDH10 was also highly expressed in the mutant CO 3-100-7-12. These expression profiles suggest that kodomillet uses an NADP-malic enzyme subtype C_4 photosynthetic system. The expression of *Rht-B* was significantly up regulated in CO 3-300-7-4. The study highlights the differential expression patterns of the same gene in different lines at different time points of stress as well as non-stress conditions. This infers that the mutation has some effect on their expression; otherwise the expression levels will be unaltered. Enhancement in grain yield could be best achieved by developing a phenotype with high PhE and culm with thick sclerenchyma cells.

Keywords: induced mutation; lodging resistance; photosynthetic efficiency; transcriptomics

---

1. Introduction

India is rich in agro-biodiversity and a large number of crop species are being cultivated. The growing population to a great extent depend on two crops, rice and wheat, for sustenance. In this context, the potentials of indigenous crop plants including millets are being gradually revisited for their genetic constitution which is becoming increasingly relevant in the changing agricultural scenario. Climatic and edaphic opponents are the present challenges in agricultural production. As a consequence, any innate crop species with comparative advantages under these challenging environments need to be targeted [1]. One category that is expected to support the decline in agricultural production includes small millets that contribute for the food security in dry and marginal lands where the major cereal crops like rice and wheat are unsuccessful pertaining to yield.

The small millet group is represented by six crops, namely, finger millet, kodomillet, foxtail millet, little millet, prosomillet and barnyard millet. Among the six crops, kodomillet (*Paspalum scrobiculatum* L.) is indigenous to India [2]. Being a C_4 plant, it is gaining attention due to its suitability to changing agro-climatic conditions as most of the arable land (69%) in India are arid and dry.

Lodging is a limitation in most of the small millet crops causing considerable losses in grain yield. Cultivating lodging-resistant plants is the most productive way to shrink losses due to lodging. Also, stem lodging disturbs the photosynthetic effectiveness of the canopy by affecting the grain filling which was reported in rice [3]. By reducing rice canopy photosynthesis, rice grain yield and quality was reduced by 60%–80% since these traits had negative associations with lodging. With a view to increasing the yield potential, identification of the superior sources, assessment of variability for the different morphological traits, increased photosynthetic efficiency and lodging resistance lines will add more value to the crop.

Nowadays, as opposed to the main cereals, there is an impetus to develop millets because of the predominant nutritional level. It is been said that kodomillet has more free radicals than other millets [4]. In addition, it provides low cost protein, minerals and vitamins in the form of healthy food [5].

The present investigation is an assessment of the extent of genetic variability induced for two traits viz., photosynthetic efficiency and lodging resistance, through physical (gamma rays) and chemical (ethyl methane sulphonate, EMS) mutagens and to isolate productive mutants in M_2 (mutant second) generation. The mutants of the desirable types for both the traits were identified and subjected to transcriptomic studies to understand their molecular basis. For future genetic and genomic studies the transcriptomal data provided in this crop will help improve the less studied, but nutritionally rich and sustainable crop.

2. Materials and Methods

2.1. Plant Material

The genetic variability was induced in Kodomillet variety CO 3 using different treatments of physical mutagen (gamma rays) and chemical mutagen (EMS). The accessions were provided by the Department of Millets, Centre for Plant Breeding and Genetics, Tamil Nadu Agricultural University, Coimbatore.

2.2. Methodology for Generating Mutant Population and Identifying the Desirable Mutants

Mature and viable seeds (moisture 12.0%) of the kodomillet variety CO 3 were irradiated at 100, 200, 300 and 400 Gy with a radioisotope $_{60}Co$ (Cobalt-60) which served as a source at the gamma chamber installed at Bhabha atomic research centre, Kalpakkam. The percentage of moisture content was determined based on the difference between fresh weight and dry weight of the seeds following the International Seed Testing Association guideline [6]. About 500 seeds were packed in butter paper covers and placed in the gamma chamber at different time intervals for each dose based on the half-life of the source. Non-irradiated dry seeds were taken as control. For chemical treatments, presoaked (8 h) seeds were mutaginized with different doses of EMS, viz., (0.2%), (0.3%), (0.4%) and (0.5%) at a room temperature of 28 ± 1 °C. A sample of seeds soaked in distilled water for the respective duration was utilized as control. Initially, the doses for chemical and physical treatments were determined based on lethal dose (LD_{50}) values based on the probit analysis [7,8] from the germination and survival test [9]. 1000 seeds from each treatment were grown in the raised nursery beds established at the Department of Millets with respective controls. The spacing was maintained at 45 cm (between the rows) and 10 cm (between the seeds). A total set of 138 (58 gamma radiation and 80 EMS) self-pollinated fertile M_1 plants were harvested individually and healthy seeds from each harvested plant were sown in plant progeny rows for growing M_2 generation (Figure 1). A set of 102 mutants (43 gamma radiation and 59 EMS) selected in M_2 generation were evaluated and validated.

Figure 1. Field view of the M_2 (mutant second) generation.

The mutagenized population was screened for the deviations in phenotypic characters in comparison with the control plants for two characters 1) photosynthetic efficiency (PhE), and 2) lodging resistance. For PhE, traits like flag leaf length (cm), flag leaf breadth (cm), chlorophyll index, measured using the soil plant analysis development (SPAD) meter (SPAD-502, Minolta Co., Osaka, Japan) and stomatal distribution calculated from epidermal impressions of mature, fully expanded leaf abaxial surface [10] were recorded. The stomata length and width were measured at 40× magnification to the nearest micrometer. Stomatal density was measured by counting the number of stomata per field of view at a magnification of 40×. The following observations were made for lodging resistance, culm thickness (millimeter), culm strength by measuring the pushing resistance of the stem using a handy force-gauge meter, [11] (Figure 2). In microtome, cross sections of intermodal region from the unique mutants were studied for their difference in thickness (micrometres) of the culm and of the width in the sclerenchymatous cells. These quantitative phenotypic trait values were utilized for assessing the promising mutant line pertaining to PhE and lodging resistance which was further utilized for the transcriptome analysis.

Figure 2. Push-pull gauge used to measure the culm strength in the mutant population.

2.3. Stress Treatment, RNA Isolation, cDNA Synthesis and Quantitative RealTime—Polymerase Chain Reaction (qRT-PCR) Analysis

The seeds of four extreme mutants (CO3-100-7-12-photosynthetically efficient; CO 3-200-13-4-less efficient; CO 3-300-7-4-non-lodged and CO 3-100-18-22-lodged) along with the wild type were grown in a plant growth chamber (PGC-6L; Percival Scientific Inc., Perry, USA) facility at the National Institute of Plant Genome Research, New Delhi, under the following conditions: 28 ± 1 °C day/23 ± 1 °C night/$70 \pm 5\%$ relative humidity with a photoperiod of 14 h and a photosynthetic photon flux density of 500 μmol m^{-2} s^{-1}. The plants were watered daily with one-third strength Hoagland's solution. For abiotic stress treatments, 21-day-old seedlings were exposed to 20% polyethylene glycol (PEG) 6000 (dehydration). Whole seedlings were collected at 0 h, 6 h and 12 h post-treatments. Zero-hour samples were used as controls. All the tissues were immediately frozen in liquid nitrogen after harvesting and stored at −80°C until RNA isolation.

For all the samples i.e., control, photosynthetically efficient and non-efficient mutants, lodging resistant and lodged mutants, total RNA was prepared from the leaf samples using Trizol reagent [12] and treated with RNase-free DNase I (50 U/μL; Fermentas, Hanover, MD, USA). Quality and purity of isolated RNA was checked using a Nano Drop 1000 Spectrophotometer (Thermo Scientific, Wilmington, DE, USA) [OD260:OD280 nm absorption ratio (1.8–2.0)] and the integrity was ascertained by resolving on 1.5% agarose gel containing 18% formaldehyde. One microgram (1 μg) of total RNA was reverse transcribed to first strand cDNA by anchored oligodT priming and random priming using a Thermo Scientific Verso cDNA synthesis kit following manufacturer's instructions. Quantitative realtime polymerase chain reaction (qRT-PCR) analysis was performed in StepOne™ Real-Time PCR Systems (Applied Biosystems, Foster City, USA) [13] using the primers mentioned in Table 1. The experiment was performed in three technical replicates for each biological duplicate. The amount of transcripts accumulated for each gene normalized to the internal control *Act2* was analysed using the 2-ΔΔCt method [13]. The PCR efficiency was calculated as: Efficiency = 10(−1/slope) − 1 by the default software (Applied Biosystems, Foster City, USA).

Table 1. List of primers used for quantitative realtime-polymerase chain reaction (qRT-PCR) analysis.

Primer Name	Primer Sequence	Number of Bases
SiPPDK2	F: GGTCGCAAAGCATGGCCTAA	20
	R: GAAGGCTCCCCACCATGTT	19
SiNADP-ME2	F: TGAGCGCTGTGGTGCAAA	18
	R:GGCAAAGTCCTCAAACTGAATGA	23
SiNADP-ME9	F: AGATTGGGCCCTTCTTATTGGT	22
	R: GTAACGCAGCTCGCTCCATT	20
SiNADP-MDH1	F: GGCGTGACCACCCTAGATGTT	21
	R: TACATTGGCCTTCCCAGCAT	20
SiNADP-MDH4	F: GCAGCAGTACGAGCGATTCA	20
	R: GCCCCGCGTGTTGTTCT	17
SiNADP-MDH10	F: GTGGGAGAGGTTCTTGGACTTG	22
	R: AGCATGCCCACCAATGACA	19
SiNADP-MDH11	F: GGAATGGAGCGAGCTGACTT	20
	R: CCCCTGTTCCGCAAAAATC	19
Rht-B	F: ATGAAGCGSGAGTACCAGGA	20
	R:TCTGCGCCACGTCCGCCATGTC	22
SiGA2ox6	F: CGCCCTCATCGTCAA	15
	R: ACGCTCTTGTATCTGTTGTTG	21

3. Results and Discussion

3.1. Determination of Lethal Dose (LD_{50}) of Mutagens

Induced mutation is essential to increase the rate of genetic variability as spontaneous mutation is at a slow rate that hampers the breeders to utilize them in the crop breeding programs. The major gain in induced mutation is that multiple trait mutants can be isolated. Before the beginning of the mutation-breeding program, information on the relative effectiveness of the mutagens is essential, [14] to determine the correct dose/concentration of the mutagens. Probit analysis was carried out using seed germination values for both the mutagens (gamma rays and EMS) to determine the LD_{50}. The expected LD_{50} value for the seeds treated with gamma rays was 300 Gy, in concordance with a previous study on radiation mutation [15,16] in kodo millet and for EMS it was observed to be 34.83 mM (0.4%). To comprehend, higher doses caused injury to the cell, which may be vital, and inhibited may cellular activities, eventually causing death of the cells. It had been noticed that, due to these chaos of the mutagens, seeds treated at high doses (500 Gy and 0.5% EMS) did not germinate, or their seedlings could not survive beyond a few days [16]. A wide range of variations was noticed among the M_2 generations for novel altered phenotypes in the traits viz., chlorophyll index, flag leaf length and breadth, stomatal number and length, culm thickness and culm strength.

3.2. Identification of Photosynthetically Efficient Mutant Lines

In order for plants to operate with efficiency, they need to balance the gaseous exchange from inside and outside the leaf to maximize carbon dioxide uptake for carbon assimilation and to attenuate water loss through transpiration. A total of seven mutants were selected for increased photosynthetic rate in the M_2 generation (Table 2) out of which the mutant CO 3-100-7-12 had a flag leaf length of 28.47 ± 0.50 cm which was comparatively less than the wild type, but there was an increase in the flag leaf breadth (1.33 ± 0.02 cm) over the wild type (1.17 ± 0.01cm). It showed high chlorophyll index (47.97 ± 0.91) which possessed 79 ± 1.46 stomatas per unit leaf area accompanied by an increased stomatal length ($10.27 \pm 0.17\mu m$) which was higher than the wild type. Stomatal number and length is related to specific stomatal conductance responsible for increased photosynthesis leading to higher yield. There were significant differences in number of stomata and length of stomatal apparatus among the selected mutants (Figure 3).

Table 2. Mutants screened for photosynthetic efficiency in M_2 generation.

S.No.	Mutants	Stomatal Number	Stomatal Length (μm)	Chlorophyll Index	Flag Leaf Length (cm)	Flag Leaf Breadth (cm)
1	CO 3-100-1-5	45.00 ± 1.46	9.03 ± 0.17	38.97 ± 0.91	22.77 ± 0.50	1.30 ± 0.02
2	CO 3-100-7-12 (high efficient)	79.00 ± 1.46	10.27 ± 0.17	47.97 ± 0.91	28.47 ± 0.50	1.33 ± 0.02
3	CO 3-200-1-3	58.00 ± 1.46	7.99 ± 0.12	41.37 ± 0.70	18.10 ± 0.52	1.20 ± 0.01
4	CO 3-200-4-1	71.00 ± 1.46	9.16 ± 0.12	35.20 ± 0.70	19.57 ± 0.52	1.17 ± 0.01
5	CO 3-200-14-1	56.00 ± 1.46	8.85 ± 0.12	37.00 ± 0.70	20.27 ± 0.52	1.30 ± 0.01
6	CO3-40.25-12-4	58.00 ± 1.33	9.90 ± 0.13	37.57 ± 0.60	32.33 ± 0.62	1.30 ± 0.03
7	CO 3-40.25-30-2	61.00 ± 1.33	9.46 ± 0.13	41.97 ± 0.60	29.10 ± 0.62	1.57 ± 0.03
	CO 3 200-13-4 (low efficient)	27.00 ± 1.46	9.18 ± 0.12	34.10 ± 0.70	28.17 ± 0.52	1.00 ± 0.01
	Wild type	55.00 ± 2.03	9.34 ± 0.49	35.62 ± 1.38	32.38 ± 0.40	1.17 ± 0.01

Stomata are the "doorkeepers" that regulate the volume of CO_2 infiltrating the intercellular air spaces of the leaf for photosynthesis. Earlier studies suggest that the increase in stomatal density translates to an increase in stomatal conductance and 30% greater photosynthetic rate under high light conditions in *Arabidopsis* [17]. Increased stomatal density enhanced leaf photosynthetic capacity in *Arabidopsis* [18]. The mutant CO 3-200-13-4 had the least stomatal count (27 ± 1.46) per unit leaf area. Therefore, stomatal density may be a target trait for plant engineering to improve photosynthetic capacity.

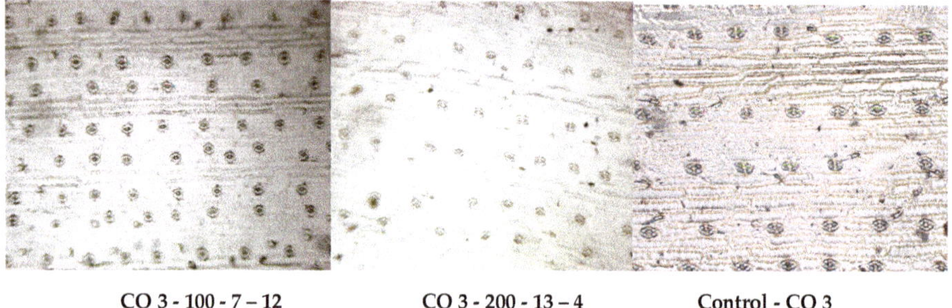

Figure 3. Stomatal distribution on the leaf epidermis of the identified extreme mutants along with control for photosynthetic efficient trait.

3.3. Identification of Non-Lodging Mutant Lines

Lodging is usually referred to as a condition in which the stem of a crop bends at or near the surface of the ground, which could lead to the collapse of the canopy. Nine mutants showed superior lodging resistance in the M_2 generation until the time of harvest. The mutant CO 3-300-7-4 exhibited the thickest culm diameter of 4 ± 0.24 mm over the wild type (2.53 ± 0.12 mm). The bending of the crop stem at the ground surface was measured using the handy force gauge meter, and the strength of the highly resistant mutant (CO 3-300-7-4) was 39.73 ± 1.75 newtons, while the culm of the wild type exerted a strength of 24.53 ± 1.50 newtons (Table 3). The association of pulling force with other characters indicated that the mutant lines with high culm strength were taller than the wild and possessed thick culm, more productive tillers, and increased yield [19]. Not much work has been published on kodomillet relating to culm strength and lodging resistance. Nevertheless, rice has been used as a model crop for understanding the mechanical properties of culm in which the phenotypical traits such as pushing resistance [20,21] and stem diameter [22] were involved in evoking culm strength and for lodging resistance.

Table 3. Mutants screened for lodging trait in M_2 generation.

S.No.	Mutants	Culm Thickness (mm)	Culm Strength (newtons)
1	CO 3-100-7-3	3.30 ± 0.07	34.73 ± 1.62
2	CO 3-100-10-5	3.47 ± 0.07	22.08 ± 1.62
3	CO 3-200-16-3	3.53 ± 0.05	29.73 ± 1.80
4	CO 3-200-17-2	3.53 ± 0.05	24.18 ± 1.80
5	CO 3-200-19-4	3.90 ± 0.05	39.63 ± 1.80
6	CO 3-300-2-5	2.90 ± 0.24	43.65 ± 1.75
7	CO 3-300-7-4 (highly stable)	4.00 ± 0.24	39.73 ± 1.75
8	CO 3-40.25-13-2	3.13 ± 0.05	30.27 ± 1.63
9	CO 3-40.25-18-5	3.10 ± 0.05	34.49 ± 1.63
	CO3-100-18-22 (unstable)	0.77 ± 0.07	6.88 ± 1.62
	Wild type	2.53 ± 0.12	24.53 ± 1.50

Based on culm thickness and culm strength, nine mutants (Table 3) were selected for non- lodging trait. The cross sectional anatomy for the identified mutants was analyzed using scanning electron microscopy (SEM, Figure 4).

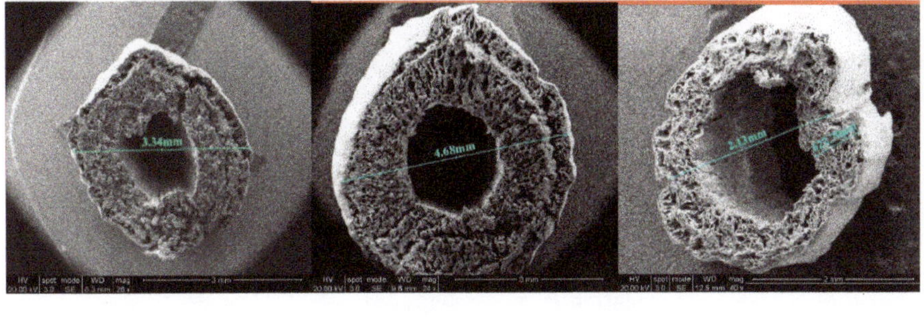

Control - CO 3   CO 3 - 300 - 7 – 4   CO 3 - 100 - 18 - 22.

Figure 4. Section of the internodal regions in the wild (CO 3) and extreme mutants studied through scanning electron microscopy (SEM).

Thickness of sclerenchymatous cell (TSC) showed distinct variations that provided rigidity and culm strength. TSC of the mutants ranged from 8.61 µm to 16.01 µm, whereas the thickness of the wild type (CO 3) was 11.09 µm. The mutant CO 3-300-7-4 had the thickest TSC (16.01 µm). The mutant CO 3-100-18-22 showed the thinnest TSC (8.61 µm) (Figure 5). The information that culm strength is positively correlated with the thickness of sclerenchyma cell walls explained the better lodging resistance in barley [23].

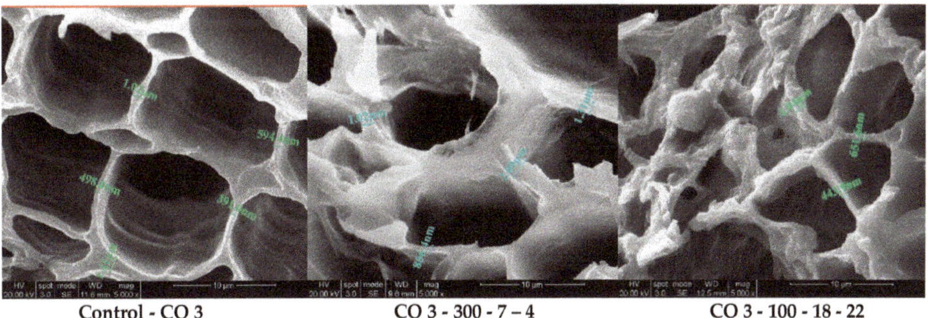

Control - CO 3   CO 3 - 300 - 7 – 4   CO 3 - 100 - 18 - 22

Figure 5. Transverse section of the sclerenchyma cells in the wild (CO 3) and in the extreme mutants.

Reduction in the mechanical strength of a plant may reflect alterations in cell wall structure, composition and fiber length. Therefore, cell-wall shape examined under SEM for the selected mutants revealed that the mutant's (CO 3-300-7-4) sclerenchyma cell wall was thickened (Figure 6), in striking contrast with the wild-type sclerenchyma's cell wall. The sclerenchyma of mutant CO 3-100-18-22 had the thinnest cell walls. Not much published evidence is available to substantiate this phenomenon in this crop.

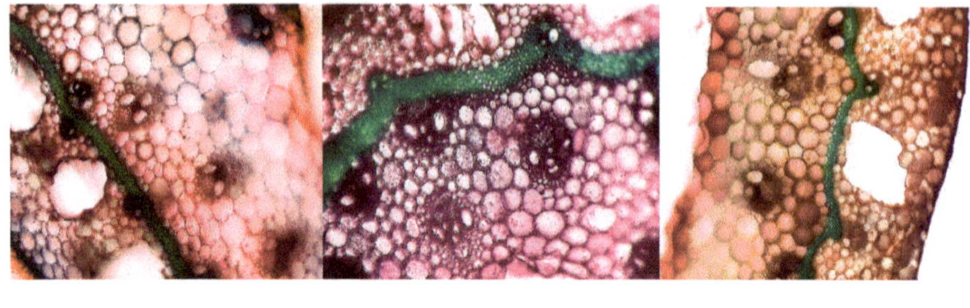

Figure 6. Cross-section of the extreme mutants depicting the thickness of the sclerenchyma cell along with the wild type (CO 3).

3.4. Molecular Characterization of Mutants through Transcriptomics

Characterization of the extreme mutants identified for non-lodging and photosynthetically efficient lines were done through transcriptomics. Comparative transcript profiling has always been a key to analyzing the dynamics of gene expression in two contrasting mutants. This provides preliminary evidence of transcript abundance of a given gene in two mutants that might be differentially regulated. Differential regulation of the same gene in two contrasting varieties of the same crop has been reported in several species. For example, a small heat shock protein-encoding gene of foxtail millet, *sHSP27*, has shown a 30-fold upregulation in the leaves of tolerant cultivar (cv. IC-4) at the early stages of heat stress as compared to susceptible cultivar (cv. IC-41) [24]. This demonstrates that the same gene could be differentially expressed in different mutants at a given time, which could be due to the difference in the regulation of gene expression at transcript-level. Keeping this in mind, the expression profiles of seven genes for photosynthetic efficiency and two candidate genes for lodging were chosen for expression profiling in the following mutant lines.

CO 3	:	Control
CO 3-100-7-12	:	Mutant line with better photosynthetic efficiency
CO 3-200-13-4	:	Mutant line with low photosynthetic efficiency
CO 3-100-18-22	:	Lodged mutant type
CO 3-300-7-4	:	Non-lodged mutant type

3.4.1. Expression Profiling of C_4 Photosynthetic Genes

Millets belong to the C_4 photosynthetic group of Poaceae where carbon assimilation is performed through three key steps [25]. First is the conversion of CO2 to $HCO3^-$ by carbonic anhydrase (CaH) which is then fixed to oxaloacetate, a C_4 acid, by phosphoenol pyruvate carboxylase (PEPC) in the mesophyll cell. This C_4 acid intermediate is converted to malate by Nicotinamide adenine dinucleotide phosphate-dependent malate dehydrogenase (NADP-MDH). Secondly, the malate diffuses to bundle sheath where an NADP-malic enzyme (NADP-ME) performs decarboxylation to release CO_2 and pyruvate. The refixation of CO_2 is attained by ribulose-1, 5-bisphosphate carboxylase (Rubisco) in the Calvin cycle. Thirdly, the pyruvate, orthophosphate dikinase (PPDK) converts pyruvate to phosphoenolpyruvate (PEP) in the mesophyll [26]. Thus, in contrast to C_3 cycle, photosynthetic efficiency of C_4 crops is higher (~50 folds) which confers several climate resilient adaptabilities including survival in high temperatures, high light intensities, radiations and drought conditions [27,28].

Recently, the complete C_4 enzyme repertoire of sequenced Poaceae genomes with emphasis on foxtail millet has been identified [29] (Figure 7).

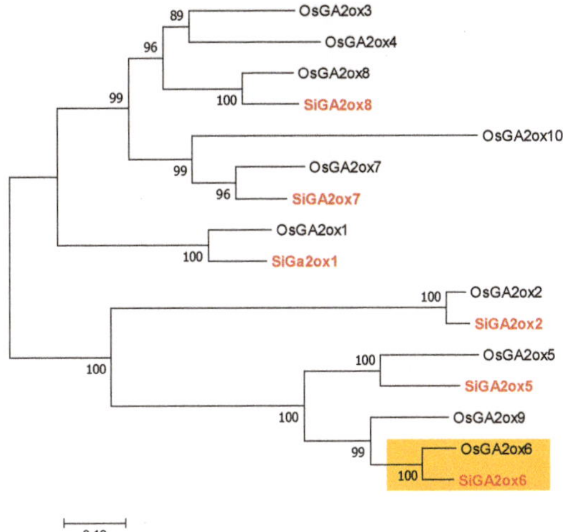

Figure 7. Evolutionary relationships of *OsGA2ox* and *SiGA2ox*. The candidate gene chosen for expression analysis is shown within the box.

3.4.1.1. Pyruvate, Orthophosphate Dikinase (PPDK)

The expression of PPDK in Control CO 3, CO 3-100-7-12 and CO 3-200-13-4 at 0 hr. was unaltered; however, a significant upregulation was observed in CO 3 and CO 3-100-7-12 at 6 h post-dehydration treatment (more than 10-fold). At 12 h, the expression was down regulated in CO 3 and upregulated in CO 3-100-7-12 and CO 3-200-13-4. Approximately 30-fold higher expression of PPDK was observed at 12 h post-treatment in CO 3-100-7-12 (Figure 8) which suggests that the expression may confer higher enzyme activity to ensure the effective functioning of C_4 photosynthesis. Similar observations were reported [27] where the expression of PPDK kept increasing in a bioenergy feedstock grass *Miscanthus* × *giganteus* during chilling stress. However, a decline in expression was observed in maize [30].'Over expression of maize PPDK in *Arabidopsis* showed a 4-fold upregulation in the expression, and the photosynthetic rates of transgenic plants had also increased [28].

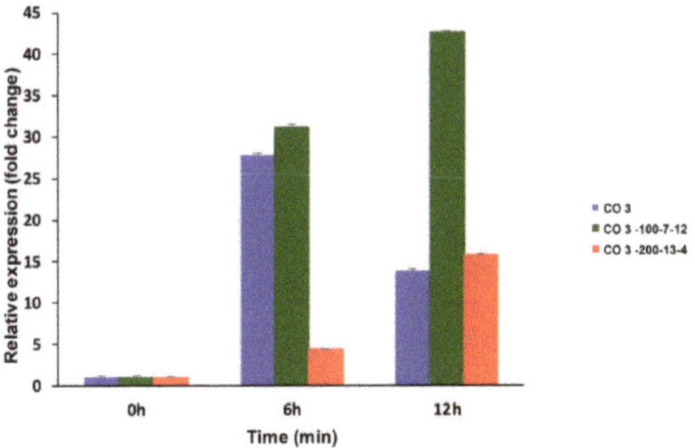

Figure 8. Relative expression profile of pyruvate, orthophosphate dikinase (PPDK) gene analyzed using qRT-PCR under dehydration stress for 0 h (control), 6 h (early) and 2 h (late) in 21 day-old seedlings of CO 3, CO 3-100-7-12, and CO 3-200-13-4. The relative expression ratio of each gene was calculated relative to its expression in control sample (0 h). *SiAct2* was used as an internal control to normalize the data. Error bars representing standard deviation were calculated based on three technical replicates for each biological duplicate.

3.4.1.2. Nicotinamide Adenine Dinucleotide Phosphate Hydrogen. (NADPH)

C_4 species are divided into subtypes, named for the primary decarboxylating enzyme that is localized to the bundle sheath. Maize, sorghum, and sugarcane use an NADP-dependent malic enzyme (NADP-ME subtype), whereas switch grass and teff (*Eragrostis tef*) use an NAD-malic enzyme (NAD-ME subtype) to generate a CO_2 pump in the bundle sheath cells. Most C_4 lineages are of the NADP-ME subtype [31].

In the present study, at 6 hr post-stress, NADP-ME2 was down regulated in CO 3-100-7-12 whereas NADP-ME9 did not show any difference as compared to control. At 12 h, NADP-ME2 was up regulated (~1-fold) in the better performing line, CO 3-100-7-12, whereas an incremental down regulation was observed in NADP-ME9. Both CO 3 and CO 3-200-13-4 showed increased levels of NADP-ME2, where CO 3-200-13-4 exhibited a 5-fold over expression in the 12 h sample. Similar elevated levels were observed for NADP-ME9; however, the relative expression levels were similar at both 6 h and 12 h samples in CO 3 and CO 3-200-13-4 (Figure 9). These expression profiles suggest that kodomillet uses an NADP-malic enzyme subtype C_4 photosynthetic system to fix carbon and, therefore, is a potentially powerful model system for dissecting C_4 photosynthesis.

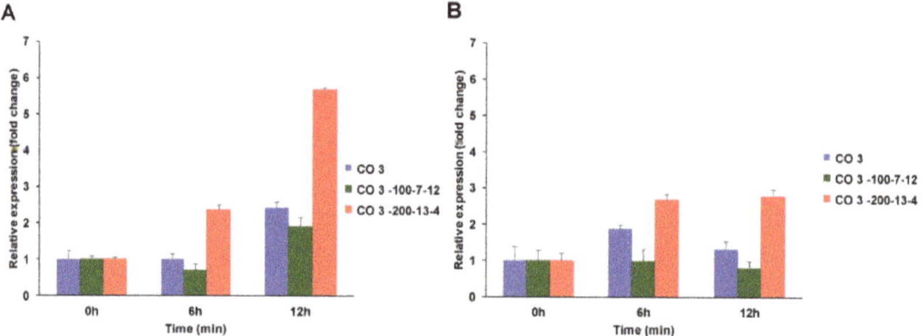

Figure 9. Relative expression profile of (**A**) *Nicotinamide Adenine Dinucleotide Phosphate-Malic Enzyme2 (NADP-ME2)* and (**B**) *Nicotinamide Adenine Dinucleotide Phosphate-Malic Enzyme9 (NADP-ME9)* genes analyzed using qRT-PCR under dehydration stress for 0 h (control), 6 h (early) and 2 h (late) in 21-day-old seedlings of CO 3, CO 3-100-7-12, and CO 3-200-13-4.

3.4.1.3. Nicotinamide Adenine Dinucleotide Phosphate –Malate dehydrogenase (NADP-MDH)

In case of NADP-MDH1, there was no much difference in the expression levels across the lines, but a 2- to 3-fold up regulation was observed in CO 3 at 6 and 12 h samples. The expression was unaltered in CO 3-100-7-12, whereas a significant down regulation was observed in CO 3-200-13-4 at the 12 h sample. NADP-MDH4 exhibited a similar expression profile in CO 3, where an interesting observation of ~4-fold up regulation was noted in 6 h and 12 h samples. At 6 h post-stress, the level of NADP-MDH4 was almost similar in CO 3-100-7-12 and CO 3-200-13-4, but at 12 h, CO 3-200-13-4 showed a 1-fold up regulation. By contrast, NADP-MDH10 was highly expressed in CO 3-100-7-12 and CO 3-200-13-4 during stress as compared to control, and in case of non-stress condition (0 h), elevated expression was observed in CO 3-200-13-4 followed by CO 3-100-7-12 as compared to CO 3. These results indicate regulation of this enzyme is sensitive to dehydration in mutated plants. The gene *NADP-MDH11* showed an initial up regulation at 6 h post-stress in CO 3-100-7-12 but down regulated at 12 h. In contrast, the transcript levels were increased with time in CO 3 and CO 3-200-13-4 (Figure 10) Similar differential expression of MDH genes was reported in wheat [32], *Arabidopsis* [33], maize [34], apple [35] and cotton [36].

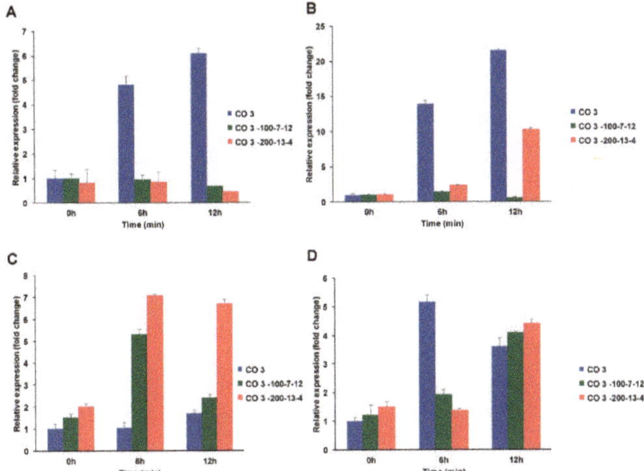

Figure 10. Relative expression profile of (**A**) *NADP-MDH1*, (**B**) *NADP-MDH4*, (**C**) *NADP-MDH10*, and (**D**) *NADP-MDH11* genes analyzed using qRT-PCR under dehydration stress for 0 h (control), 6 h (early) and 2 h (late) in 21-day-old seedlings of CO 3, CO 3-100-7-12, and CO 3-200-13-4.

Attempts have been made to overexpress maize NAPD-ME in rice through a transgenic approach, which showed a 20- to 70-fold increase in enzyme activity; however, the transgenic lines had aberrant chloroplast [30]. Over expressing sorghum NADP-ME in rice showed 1- to 7-fold increase in the enzyme activity in transgenic lines with null-positive effect on photosynthesis [37]. All the over expression studies have demonstrated enhanced photo inhibition of photosynthesis due to an increase in the level of NADPH inside the chloroplast by the action of the C_4-NADP-ME enzyme. These suggest that the cellular circuitry associated with NADP-ME is very complex in C_4 plants and thus over expressing a single gene from C_4 to C_3 will show detrimental effects in the transgenic lines.

3.4.2. Expression Profiling of Lodging-Related Genes

Two genes were reported to be responsible for lodging resistance in rice and wheat, and interestingly, both were related to the plant growth hormone, gibberellic acid (GA). GAs control a variety of growth and developmental processes during the entire life cycle of plants. Several loss-of-function mutants in GA biosynthesis showed typical GA-deficient phenotypes, such as dwarfism, small dark green leaves, prolonged germination dormancy, inhibited root growth, defective flowering, reduced seed production, and male sterility [38]. A major catabolic pathway for GAs is initiated by a 2β-hydroxylation reaction catalyzed by GA2 oxidases (*GA2ox*). In rice, 10 different classes of *GA2ox* were identified that were named *GA2ox1* to *GA2ox10*, and in the present study, the *GA2ox* orthologs of foxtail millet were first identified. However, foxtail millet possesses only seven *GA2ox* genes where the orthologs of *GA2ox4*, *GA2ox9* and *GA2ox10* were absent. A phylogenetic tree constructed to derive the evolutionary relationships between rice and foxtail millet *GA2ox* proteins showed that two distinct classes of these proteins exist in both the crops [38]. The proteins *GA2ox2*, *GA2ox5*, *GA2ox9* (ortholog absent in foxtail millet) and *GA2ox6* formed a separate clade, and as shown by [39], these proteins contain the three unique and conserved motifs. At 12 h post-dehydration treatment, the expression of *GA2ox6* was significantly reduced in the mutant CO 3-100-18-22, whereas an incremental decrease in gene expression was noticed in CO 3-300-7-4 (Figure 11). This could be due to the effect of mutation on the regulatory regions of the *GA2ox6* or other gene(s) that could have a negative effect on *GA2ox6* expression at the late-stage of dehydration stress.

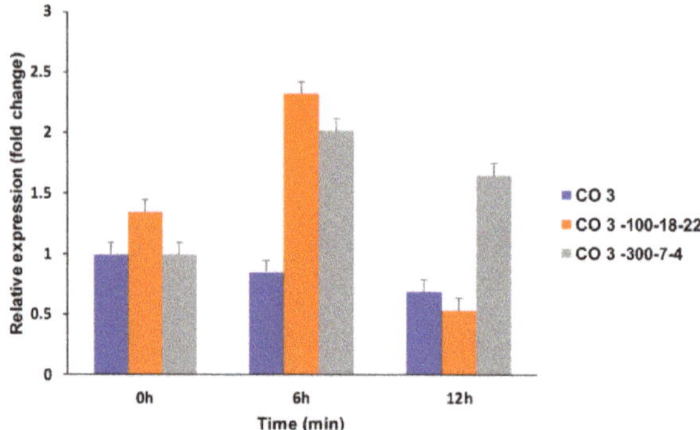

Figure 11. Relative expression profile of *GA2ox6* gene analyzed using qRT-PCR under dehydration stress for 0 h (control), 6 h (early) and 2 h (late) in 21-day-old seedlings of CO 3, CO 3-100-18-22, and CO 3-300-7-4. The relative expression ratio of each gene was calculated relative to its expression in control sample (0 h). *SiAct2* was used as an internal control to normalize the data. Error bars representing standard deviation were calculated based on three technical replicates for each biological duplicate.

The transcript abundance of the *Rht-B* gene in all the three lines (Control CO 3, CO 3-100-18-22, CO 3-300-7-4) suggests interesting evidence of GA-regulated stress responsive machinery that might operate in these lines and also, the mutation would have occurred in such a way as to impede the average expression of *Rht-B* in CO 3-100-18-22 (Figure 12). In wheat, the mutant lines of *Rht-B1b* and *Rht-D1b* were suggested to confer dwarfism by producing more active forms of growth repressors [40].

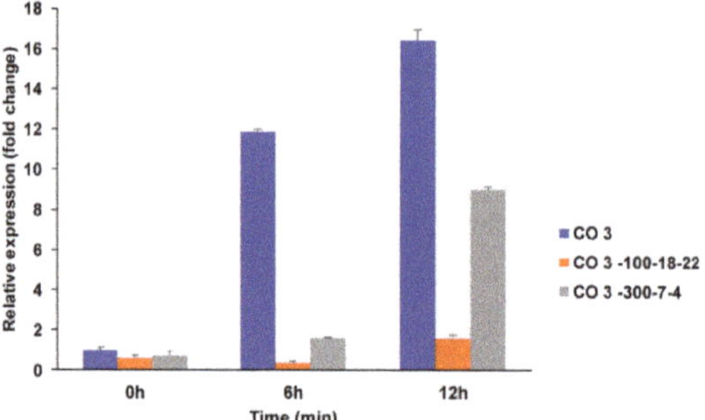

Figure 12. Relative expression profile of *Rht-B* gene analyzed using qRT-PCR under dehydration stress for 0 h (control), 6 h (early) and 2 h (late) in 21-day-old seedlings of CO 3, CO 3-100-18-22, and CO 3-300-7-4.

4. Conclusions

The outcome of this study have shown striking variations for PhE and lodging resistance in a number of anatomical characteristics of flag leaf anatomy and culm thickness. The study highlights differential expression patterns of the same gene in different lines at different time points of stress as well as non-stress conditions. This infers that the mutation has some effect on their expression; otherwise the expression levels will be unaltered. Mutation in regulatory elements of each gene might have altered the motifs for recognition and binding of transcription factors, and this could have led to a differed transcriptional regulation of gene expression. Cloning and sequencing of the genes as well as their promoter regions can provide a clue about the variations that have been inserted through mutagenesis. However, the present data is very preliminary for drawing a solid conclusion, and further functional characterization is required to delineate the precise role of each gene in stress response and the impact of the mutation on the functioning of that particular gene. This also suggests that an improvement in grain yield could be best attained by a developing a phenotype with high PhE and culm with thick sclerenchyma cells.

Author Contributions: Conceptualization, project administration R.R.; supervision R.M.; J.P.J., R.K.S. and J.G. performed the experiments. J.P.J. analyzed the data and wrote the manuscript. M.M.; M.P. data curation, validation, resources and supervision; J.P. formal analysis. All authors have read and agreed to the published version of the manuscript.

Funding: This research was funded by UGC: Maulana Azad National Fellowship for Minority Students, F1-17.1/2016-17/MANF-2015-17-TAM-55427.

Acknowledgments: The authors profoundly acknowledge the Department of Millets, Tamil Nadu Agricultural University, Lab members of 103, NIPGR and the scientists at BARC, Kalpakam for the support rendered.

Conflicts of Interest: The authors declare no conflict of interest.

References

1. Padulosi, S.; Mal, B.; Ravi, S.B.; Gowda, J.; Gowda, K.T.K.; Shanthakumar, G.; Yenagi, N.; Dutta, N. Food security and climate change: Role of plant genetic resources of minor millets. *Ind. J. Plant Genet. Res.* **2009**, *22*, 1–16.

2. De Wet, J.M.J.; Prasada Rao, K.E.; Mengesha, M.H.; Brink, D.E. Diversity in kodo millet, Paspalum scrobiculatum. *Econ. Bot.* **1983**, *37*, 159–163. [CrossRef]

3. Weng, F.; Zhang, W.; Wu, X.; Xu, X.; Ding, Y.; Li, G.; Liu, Z.; Wang, S. Impact of low-temperature, over cast and rainy weather during the reproductive growth stage on lodging resistance of rice. *Sci. Rep.* **2017**, *7*, 46596. [CrossRef]

4. Hegde, P.S.; Chandra, T.S. ESR spectroscopic study reveals higher free radical quenching potential in kodo millet (*Paspalum scrobiculatum*) compared to other millets. *Food Chem.* **2005**, *92*, 177–182. [CrossRef]

5. Yadava, H.S.; Jain, A.K. Advances in kodo millet research. Directorate of Information and Publications of Agriculture. *Indian Counc. Agric. Res.* **2006**.

6. ISTA. International rules for seed testing. *Seed Sci. Technol.* **1985**, *13*, 299–355.

7. Finney, D.J. *Probit Analysis*, 10th ed.; Cambridge University Press: Cambridge, UK, 1971.

8. Finney, D.J. *Statistical Method in Biological Assay*; Charles Griffin & Co.: London, UK, 1978.

9. PoornimaJency, J.; Ravikesavan, R.; Sumathi, P.; Raveendran, M. Effect of chemical mutagen on germination percentage and seedling parameters in Kodomillet variety CO 3. *Int. J. Chem. Std.* **2017**, *5*, 166–169.

10. Sampson, J. A method of replicating dry or moist surfaces for examination by light microscopy. *Nature* **1961**, *191*, 932. [CrossRef]

11. Kashiwagi, T.; Ishimaru, K. Identification and functional analysis of a locus for improvement of lodging resistance in rice. *Plant Physiol.* **2004**, *134*, 676–683. [CrossRef]

12. Longeman, J.; Schell, J.; Willmitzer, L. Improved method for the isolation of RNA from plant tissues. *Anal. Biochem.* **1987**, *163*, 16–20. [CrossRef]

13. Kumar, K.; Muthamilarasan, M.; Manoj, P. Reference genes for quantitative real-time PCR analysis in the model plant foxtail millet (*Setariaitalica* L.) subjected to abiotic stress conditions. *Plant Cell Tiss. Organ. Cult.* **2013**, *115*, 13–22. [CrossRef]

14. Smith, H.H. Comparative genetic effects of different physical mutagens in higher plants. In *Induced Mutations and Plant Breeding Improvement*; IAEA: Vienna, Austria, 1972; pp. 75–93.

15. PoornimaJency, J.; Ravikesavan, R.; Sumathi, P.; Raveendran, M. Determination of lethal dose and effect of physical mutagen on germination percentage and seedling parameters in kodomillet variety CO 3. *Electron. J. Plant Breed.* **2016**, *7*, 1122–1126.

16. Subramanian, A.; Nirmalakumari, A.; Veerabadhiran, P. Mutagenic Efficiency and Effectiveness in Kodomillet (Paspalum scrobiculatum L.). *Madras Agric. J.* **2011**, *98*, 1–3.

17. Schluter, U.; Muschak, M.; Berger, D.; Altmann, T. Photosynthetic performance of an Arabidopsis mutant with elevated stomatal density (sdd1-1) under different light regimes. *J. Exp. Bot.* **2003**, *54*, 867–874. [CrossRef]

18. Tanaka, Y.; Sugano1, S.S.; Shimada, T.; Nishimura, I.H. Enhancement of leaf photosynthetic capacity through increased stomatal density in Arabidopsis. *New Phytol.* **2013**, *198*, 757–764. [CrossRef]

19. Sreeja, R.; Subramanian, A.; Nirmalakumari, A.; Kannanbapu, J.R. Selection criteria for culm strength in Kodo millet (Paspalum scrobiculatum L.) to suit mechanical harvesting. *Electron. J. Plant Breed.* **2014**, *5*, 459–466.

20. Terashima, K.; Akita, S.; Sakai, N. Eco-physiological characteristics related with lodging tolerance of rice in direct sowing cultivation. *Jpn. J. Crop Sci.* **1992**, *61*, 380–387. [CrossRef]

21. Won, J.G.; Hirahara, Y.; Yoshida, T.; Imabayashi, S. Selection of rice lines using SPGP seedling method for direct seeding. *Plant Prod. Sci.* **1998**, *1*, 280–288. [CrossRef]

22. Zuber, U.; Winzeler, H.; Messmer, M.M.; Keller, B.; Schmid, J.E.; Stamp, P. Morphological traits associated with lodging resistance of spring wheat (*Triticumaestivum* L.). *J. Agron. Crop Sci.* **1999**, *182*, 17–24. [CrossRef]

23. Cenci, C.A.; Grando, S.; Ceccarelli, S. Culm anatomy in barley (*Hordeum vulgare*). *Can. J. Bot.* **1984**, *62*, 2023–2027. [CrossRef]

24. Roshan, K.S.; Jaishankar, J.; Muthamilarasan, M.; Shweta, S.; Dangi, A.; Manoj, P. Proteins in C_4 model, foxtail millet identifies potential candidates for crop improvement under abiotic stress. *Sci. Rep.* **2016**, *6*, 3264.

25. Pinto, H.; Sharwood, R.E.; Tissue, D.T.; Ghannoum, O. Photosynthesis of C3, C3-C4, and C4 grasses at glacial CO2. *J. Exp. Bot.* **2014**, *65*, 3669–3681. [CrossRef] [PubMed]

26. Furbank, R.T.; Chitty, J.A.; Jenkins, C.L.D.; Taylor, W.C.; Trevanion, J.S.; Caemmerer, S.V.; Ashton, A.R. Genetic manipulation of key photosynthetic enzymes in the C_4 plant *Flaveria bidentis*. *Aust. J. Plant. Physiol.* **1997**, *24*, 477–485. [CrossRef]

27. Wang, D.A.R.P.; Stephen, J.; Moose, P.; Long, P.S. Cool C_4 Photosynthesis: Pyruvate Pi Dikinase Expression and Activity Corresponds to the Exceptional Cold Tolerance of Carbon Assimilation in Miscanthus × giganteus. *Plant Physiol.* **2008**, *148*, 557–567. [CrossRef] [PubMed]

28. Wang, Y.M.; Xu, W.G.; Hu, L.; Zhang, L.; Li, Y.; Du, X.H. Expression of Maize Gene Encoding C4-Pyruvate Orthophosphate Dikinase(PPDK) and C4-Phosphoenolpyruvate Carboxylase (PEPC) in Transgenic Arabidopsis. *Plant Mol. Biol. Rep.* **2012**, *30*, 1367. [CrossRef]

29. Muthamilarasan, M.; Roshan, K.S.; Singh, N.K.; Prasad, M. C_4 photosynthetic enzymes of foxtail millet. *Sci. Reps.* 2018.

30. Takeuchi, Y.; Akagi, H.; Kamasawa, N.; Osumi, M.; Honda, H. Aberrant chloroplasts in transgenic rice plants expressing a high level of maize NADP-dependent malic enzyme. *Planta* **2000**, *211*, 265–274. [CrossRef]

31. Christin, P.A.; Salamin, N.; Kellogg, E.A.; Vicentini, A.; Besnard, G. Integrating phylogeny into studies of C_4 variation in the grasses. *Plant Physiol.* **2009**, *149*, 82–87. [CrossRef]

32. Ding, Y.; Ma, Q.H. Characterization of a cytosolic malate dehydrogenase cDNA which encodes an isozyme toward oxaloacetate reduction in wheat. *Biochimie* **2004**, *86*, 509–518. [CrossRef]

33. Beeler, S.; Liu, H.C.; Stadler, M.; Schreier, T.; Eicke, S.; Lue, W.L.; Truernit, E.S.; Zeeman, S.C.; Chen, J.; Kotting, O. Plastidial NAD-Dependent Malate Dehydrogenase is Critical for Embryo Development and Heterotrophic Metabolism in Arabidopsis. *Plant Physiol.* **2014**, *164*, 1175–1190. [CrossRef]

34. Menckhoff, L.; Ehret, N.M.; Buck, F.; Vuleti, M.; Luthje, S. Plasma membrane-associated malate dehydrogenase of maize (Zea mays L.) roots: Native versus recombinant protein. *J. Prot.* **2013**, *80*, 66–77. [CrossRef] [PubMed]

35. Yao, J.; Ma, H.; Zhang, P.; Ren, L.; Yang, X.; Yao, G.; Zhang, P.; Zhou, M. Inheritance of stem strength and its correlations with culm morphological traits in wheat (*Triticum aestivum* L.). *Can. J. Plant Sci.* **2011**, *91*, 1065–1070. [CrossRef]

36. Wang, Z.A.; Li, Q.; Ge, X.Y.; Yang, C.L.; Luo, X.L.; Zhang, A.H.; Xiao, J.L.; Tian, Y.C.; Xia, G.X.; Chen, X.Y.; et al. The mitochondrial malate dehydrogenase 1 gene GhmMDH1 is involved in plant and root growth under phosphorus deficiency deficiency conditions in cotton. *Sci. Rep.* **2015**, *5*, 10343. [CrossRef] [PubMed]

37. Wei, C.; Song, Z.J.; Fang, Z.; Hu, W.U.N. Photosynthetic Features of Transgenic Rice Expressing Sorghum C_4 Type NADP-ME. *Acta Bot. Sinica* **2004**, *46*, 873–882.

38. Lo, S.F.; Yang, S.Y.; Chen, K.T.; Hsing, Y.I.; Zeevaart, J.A.; Chen, L.J.; Yu, S.M. A Novel Class of Gibberellin 2-Oxidases Control Semi dwarfism, Tillering, and Root Development in Rice. *Plant Cell* **2008**, *20*, 2603–2618. [CrossRef]

39. Lee, D.J.; Zeevaart, A.D.J. Molecular Cloning of GA 2-Oxidase3 from Spinach and Its Ectopic Expression in *Nicotiana sylvestris*. *Plant Physiol.* **2005**, *138*, 243–254. [CrossRef]

40. Pearce, S.; Saville, R.; Vaughan, S.P.; Chandler, P.M.; Wilhelm, E.P.; Sparks, C.A.; Kaff, N.A.; Korolev, A.; Boulton, M.I.; Phillips, A.L.; et al. Molecular Characterization of Rht-1 Dwarfing Genes in Hexaploid Wheat. *Plant Physiol.* **2011**, *157*, 1820–1831. [CrossRef]

Article

Agro-Morphological Exploration of Some Unexplored Wild *Vigna* Legumes for Domestication

Difo Voukang Harouna [1,2,*], Pavithravani B. Venkataramana [2,3], Athanasia O. Matemu [1] and Patrick Alois Ndakidemi [2,3]

[1] Department of Food Biotechnology and Nutritional Sciences, Nelson Mandela African Institution of Science and Technology (NM-AIST), P.O. Box 447, Arusha 23311, Tanzania; athanasia.matemu@nm-aist.ac.tz

[2] Centre for Research, Agricultural Advancement, Teaching Excellence and Sustainability in Food and Nutrition Security (CREATES-FNS), Nelson Mandela African Institution of Science and Technology, P.O. Box 447, Arusha 23311, Tanzania; pavithravani.venkataramana@nm-aist.ac.tz (P.B.V.); Patrick.ndakidemi@nm-aist.ac.tz (P.A.N.)

[3] Department of Sustainable Agriculture, Biodiversity and Ecosystems Management, Nelson Mandela African Institution of Science and Technology, P.O. Box 447, Arusha 23311, Tanzania

* Correspondence: harounad@nm-aist.ac.tz or difovoukang@yahoo.fr

Received: 13 December 2019; Accepted: 26 December 2019; Published: 13 January 2020

Abstract: The domestication of novel or hitherto wild food crops is quickly becoming one of the most popular approaches in tackling the challenges associated with sustainable food crop production, especially in this era, where producing more food with fewer resources is the need of the hour. The crop breeding community is not yet completely unanimous regarding the importance of crop neo-domestication. However, exploring the unexplored, refining unrefined traits, cultivating the uncultivated, and popularizing the unpopular remain the most adequate steps proposed by most researchers to achieve the domestication of the undomesticated for food and nutrition security. Therefore, in the same line of thought, this paper explores the agro-morphological characteristics of some wild *Vigna* legumes from an inquisitive perspective to contribute to their domestication. One hundred and sixty accessions of wild *Vigna* legumes, obtained from gene banks, were planted, following the augmented block design layout of two agro-ecological zones of Tanzania, during the 2018 and 2019 main cropping seasons for agro-morphological investigations. The generalized linear model procedure (GLM PROC), two-way analysis of variance (two-way ANOVA), agglomerative hierarchical clustering (AHC) and principal component analysis (PCA) were used to analyze the accession, block and block vs. accession effects, as well as the accession × site and accession × season interaction grouping variations among accessions. The results showed that the wild species (*Vigna racemosa*; *Vigna ambacensis*; *Vigna reticulata*; and *Vigna vexillata*) present a considerable variety of qualitative traits that singularly exist in the three studied checks (cowpea, rice bean, and a landrace of *Vigna vexillata*). Of the 15 examined quantitative traits, only the days to flowering, pods per plant, hundred seed weight and yield were affected by the growing environment (accession × site effect), while only the number of flowers per raceme and the pods per plant were affected by the cropping season (accession × season effect). All the quantitative traits showed significant differences among accessions for each site and each season. The same result was observed among the checks, except for the seed size trait. The study finally revealed three groups, in a cluster analysis and 59.61% of the best variations among the traits and accessions in PCA. Indications as to the candidate accessions favorable for domestication were also revealed. Such key preliminary information could be of the utmost importance for the domestication, breeding, and improvement of these species, since it also determines their future existence—that is, so long as biodiversity conservation continues to be a challenging concern for humanity.

Keywords: undomesticated legumes; *Vigna racemosa*; *Vigna ambacensis*; *Vigna reticulata*; *Vigna vexillata*; wild food legumes; legumes; *Vigna* species; domestication; unexplored legumes

---

1. Introduction

It has been reported that only 12 crops contribute most to the current global food production, with only three of them (rice, wheat and maize) providing more than 50% of the world's calories [1–3]. Consequently, the Food and Agricultural Organization of the United Nations has predicted that the world is in need of about 70% more food to adequately feed the ever-growing population [4]. Therefore, a detailed screening and exploration of hitherto wild and novel species from various agro-climatic regions of the world could help to mitigate the need. Although the crop breeding community is not yet unanimous on the importance of the crop neo-domestication concept, it is increasingly becoming popular in research on this topic that domesticating the undomesticated is an ideal method, which could aid in mitigating the global food insecurity challenges [1,5]. The method will not only promote the successful utilization of hitherto wild and non-domesticated food crops in dietary diversification programs, but also help with biodiversity conservation.

Legumes (family: Fabaceae) constitute the third largest family of flowering plants. [6]. However, only ten species have been domesticated and recognized as human food [5]. Those few domesticated ones have incontestably proven to be of crucial nutritional value for both humans and animals due to their protein content, causing them to be recognized as the second most valuable plant source of nutrients [6]. Despite their positive impact on global food and nutrition security, it has also been reported that their production rate remains unsatisfying, as compared with their consumption rate, due to biotic and abiotic challenges [7]. Therefore, there is a need to look for sustainable alternative strategies to improve and diversify their production. A systematic exploration of the hitherto wild undomesticated and wild relatives of the domesticated species within the commonly known and the little-known genera of legumes might be a hopeful strategy.

The *Vigna* genus is a large group of legumes consisting of more than 200 species, of which some are of agronomic, economic, and environmental importance [8]. The most common domesticated ones include the mung bean [*V. radiata* (L.) Wilczek], urd bean [*V. mungo* (L.) Hepper], cowpea [*V. unguiculata* (L.) Walp], azuki bean [*V. angularis* (Willd.) Ohwi and Ohashi], bambara groundnut [*V. subterranea* (L.) Verdn.], moth bean [*V. aconitifolia* (Jacq.)], and rice bean [*V. umbellata* (Thunb.) Ohwi and Ohashi]. They have recognized usages, ranging from forage, green manure, and cover crops, in addition to their high-protein grains. However, they are very limited in number, as compared with the existing wild ones.

The *Vigna* genus legumes also comprise more than 100 wild species, which do not have common names, apart from their scientific appellation [9]. They are given different denotations, such as the under-exploited wild *Vigna* species, undomesticated *Vigna* species, wild *Vigna* or alien species, depending on the scientist [5,8,10]. These constitute the main subject of interest in this research, as very little information about them, including their agro-morphological characteristics, has been reported. For the purpose of this study, accessions of *Vigna racemose*, *Vigna ambacensis*; *Vigna reticulata*, and *Vigna vexillata* were first considered to carry out preliminary investigations based on the very little information gathered about them and their availability in the nearest gene bank.

Archeological reports suggested that the crop domestication processes were preceded by a period of pre-domestication, where humans first began to purposefully plant wild plants that had favorable traits [11]. This later on led to a purposive and intentional selection of crops with a group of traits, generated through human preferences for ease of harvest and growth advantages under human propagation, known as the domestication syndrome [11]. It is then important to screen the existing traits of the wild crop in order to examine and select crop accessions with traits that humans prefer for an effective domestication. It is from that line of thought that this paper aimed to explore the agro-morphological characteristics of some wild *Vigna* legumes in order to disseminate preliminary information that could lead to their future domestication.

2. Materials and Methods

2.1. Sample Collection and Preparation

One hundred and sixty (160) accessions of wild *Vigna* species of legumes were obtained from gene banks, as presented in Table S1. Approximately 20–100 seeds of each accession were supplied by the gene banks and planted in pots, which were placed in screen houses at the Nelson Mandela African Institution of Science and Technology, Arusha, Tanzania during the period of November 2017–March 2018. The pot experiment only allowed for seed multiplication and the preliminary observation of the growth behavior of the wild legumes, prior to experimentation in the field. In the field, all the accessions were planted in an experimental plot, following the augmented block design arrangement [12], and allowed to grow until full maturity. In addition to the wild accessions, three domesticated *Vigna* legumes—that is, cowpea (*V. unguiculata*), rice bean (*V. umbellata*), and a semi-domesticated landrace (*V. vexillata*)—were used as checks. The checks were obtained from the Genetic Resource Center (GRC-IITA), Nigeria (cowpea), the National Bureau of Plant Genetic Resources (NBPGR), India (rice bean), and the Australian Grain Gene bank (AGG), Australia (semi-domesticated landrace *V. vexillata*).

2.2. Study Sites and Meteorological Considerations

The study was conducted in two agro-ecological zones, located at two research stations in Tanzania, during two main cropping seasons (March–September 2018 and March–September 2019).

The first research station (site A) was at the Tanzania Agricultural Research Institute (TARI), Selian in the Arusha region, located in the northern part of Tanzania. TARI-Selian lies at a latitude of 3°21′50.08″ N and longitude of 36°38′06.29″ E at an elevation of 1390 m above sea level (a.s.l.).

The second site (site B) was at the Tanzania Coffee Research Institute (TaCRI), located in the Hai district, Moshi, Kilimanjaro region (latitude 3°13′59.59″ S, longitude 37°14′54″ E). The site is at an elevation of 1681 m above sea level.

The meteorological characteristics (monthly rainfall and temperature dynamics) of the two study sites for the two study cropping seasons were obtained from the Tanzania Meteorological Agency and are summarized in Figure 1 below.

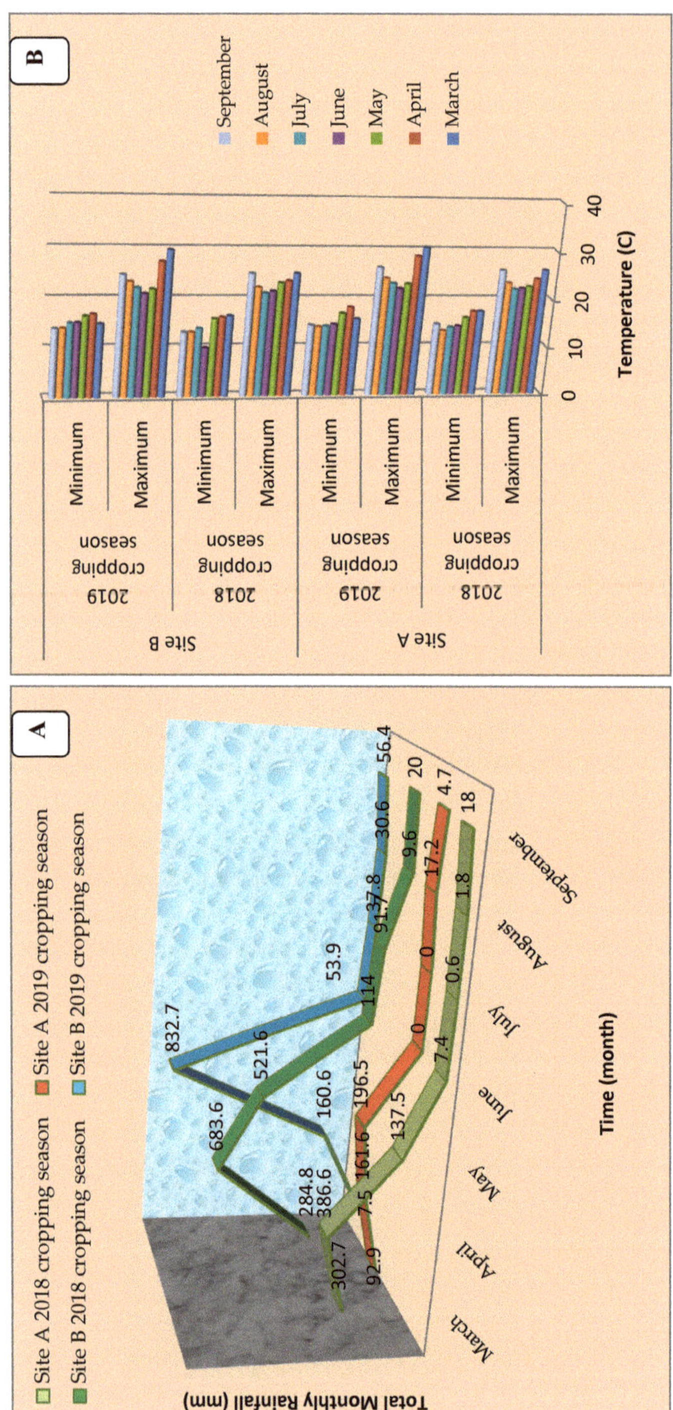

Figure 1. Meteorological characteristics of the two study sites. (**A**) rainfall dynamics of the study sites; (**B**) temperature dynamics of the study sites.

2.3. Experimental Design and Planting Process

The 160 accessions of wild *Vigna* legumes were planted in an augmented block design field layout, following the randomization generated by the statistical tool on the website developed by the Indian Agricultural Research Institute [13] for 160 accessions, with three checks. The software generated a total of 208 experimental plots, with 8 blocks each containing 26 experimental plots. Each plot represents a line of different wild accession or a check. Each check was repeated two times within a block. Ten seeds from each accession were planted in a line of 5 m in length, with a distance of 50 cm between each seed. The distances between the accessions (lines) within a block, as well as the distance between the blocks, were 1 m each. Data were then collected from five randomly selected plants in each line using the wild *Vigna* descriptors [12].

2.4. Data Collection and Analysis

Data on thirty (30) characteristics (both qualitative and quantitative) were recorded using both IPGRI (International Plant Genetic Resource Institute) and NBPGR (National Bureau of Plant Genetic Resources) descriptors [12]. Fifteen (15) qualitative and fifteen (15) quantitative characters were recorded. The descriptors used for the characterization of the wild *Vigna* in this study are found in Table 1 below. Data on the quantitative traits were recorded for five randomly selected individuals per accession.

The generalized linear model procedure (GLM PROC) of the SAS software (SAS University Edition, SAS Institute Inc., North Carolina State, USA) was used to analyze the accession and the block and block vs. accession effects, while the two-way analysis of variance (two-way ANOVA), agglomerative hierarchical clustering (AHC) and principle component analysis (PCA) of XLSTAT were used to analyze the accession × site and accession × season interactions, as well as the clustering and variations among accessions. The SAS University Edition version and the XLSTAT-Base version 21.1.57988.0 were used.

Table 1. Important descriptors for the characterization of the wild *Vigna* species germplasm [12].

	Parameters	Descriptors
	Qualitative Traits	
1	Seed germination habit	1. Epigeal, 2. Hypogeal
2	Attachment of primary leaves (at two-leaf stage)	1. Sessile, 2. Sub-sessile, 3. Petiolate
3	Growth habit (recorded at first pod maturity)	1. Erect, 2. Semi-erect, 3. Spreading, 4. Semi-prostrate, 5. Prostrate, 6. Climbing
4	Leafiness (at 50% flowering)	1. Sparse, 2. Intermediate, 3. Abundant
5	Leaf pubescence	1. Glabrous, 2. Very sparsely pubescent, 3. Sparsely pubescent, 4. Moderately Pubescent, 5. Densely pubescent
6	Petiole pubescence	1. Glabrous, 2. Pubescent, 3. Moderately pubescent, 4. Densely pubescent
7	Lobing of terminal leaflet (at first pod maturity)	1. Unlobed, 2. Shallow, 3. Intermediate, 4. Deep, 5. Very deep
8	Terminal leaflet shape	1. Lanceolate, 2. Broadly ovate, 3. Ovate, 4. Rhombic, 5. Others
9	Stipule size	1. Small, 2. Medium, 3. Large
10	Hypocotyl color	1. Green; 2. Purple, 3. Others
11	Stem pubescence	1. Glabrous, 2. Sparsely pubescent, 3. Moderately pubescent, 4. Highly pubescent
12	Pod attachment to peduncle	1. Erect, 2. Horizontal, 3. Horizontal-pendent 4. Pendent, 5. Others
13	Pod pubescence	1. Glabrous, 2. Sparsely pubescent, 3. Moderately pubescent, 4. Densely pubescent

Table 1. *Cont.*

	Parameters	Descriptors
14	Pod curvature	1. Straight, 2. Slightly curved, 3. Curved (sickle shaped)
15	Constriction of pod between seeds	1. Absent, 2. Slight, 3. Pronounced
	Pod cross section	1. Semi flat, 2. Round, 3. Others
	Quantitative Traits	
	1	Germination time (days)
	2	Terminal leaflet length (cm)
	3	Terminal leaflet width (cm)
	4	Petiole length (cm)
	5	Days to flowering
	6	Flower bud size (cm)
	7	Number of flowers per raceme
	8	Peduncle length (cm)
	9	Pods per peduncle
	10	Pod length (cm)
	11	Pods per plant
	12	Seeds per pod
	13	Seed size (mm^2)
	14	100-Seed weight (g)
	15	Yield (Kg/ha)

3. Results

3.1. Qualitative Traits Exploration of the Wild Unexplored Vigna Species

Figure 2 below gives a pictorial description of some distinguishing morphological characteristics of the wild *Vigna* legumes, studied based on the physical phenotypic observations during the pot experimental phase. Other qualitative characteristics were studied in the field at different stages of the plants' growth, i.e., the germination, vegetative, podding, and maturity stages.

Figure 2. Some qualitative morphological characteristics of the studied wild Vigna.

3.1.1. Germination Stage

The hypocotyl color, primary leaf attachment (at the two-leaf stage), and germination habit were the traits monitored at this growth stage. Figure 3 shows the percentage of the distribution of accessions for each trait variation. All the checks showed homogenous phenotypic characteristics, while the variations among accessions of the same species were observed for the wild species. *V. ambacensis* accessions showed a higher percentage of purple hypocotyl color, which they share with rice bean (*V. umbellata*). On the other hand, *V. vexillata*, *V. reticulata*, and *V. racemosa* showed higher percentages of green hypocotyl color, similar to cowpea and the landrace of *V. vexillata*.

The *V. vexillata*, *V. ambacensis*, and *V. racemosa* accessions showed a resemblance in the primary leaf attachment trait (sub-sessile) to cowpea, presenting a high percentage of accessions for the trait, while the *V. reticulata* accessions shared the sessile phenotype with the landrace of *V. vexillata*.

Both cowpea and the landrace of *V. vexillata* presented an epigeal germination habit, which they have in common with most accessions of *V. reticulata* and *V. racemosa*, while most accessions of *V. ambacensis* and *V. vexillata* shared a common phenotype (hypogeal) with rice bean.

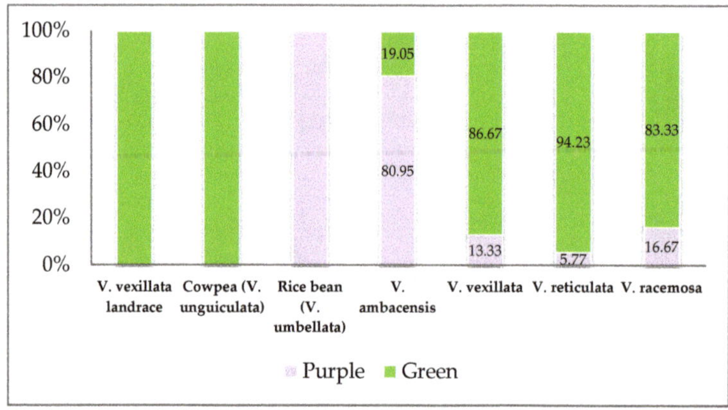

(**a**) Hypocotyl Color.

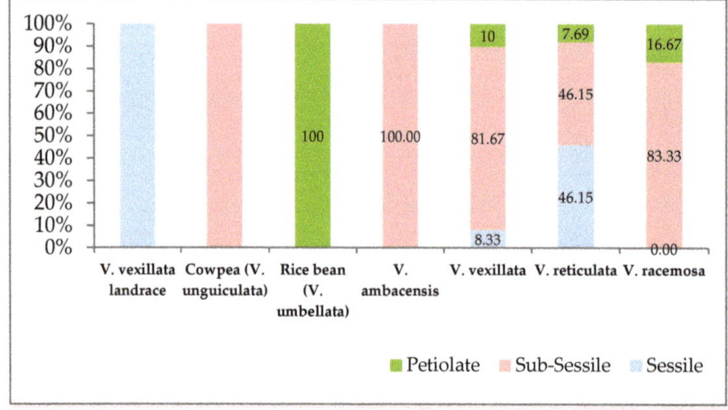

(**b**) Attachment of primary leaves (at the two-leaf stage).

Figure 3. *Cont.*

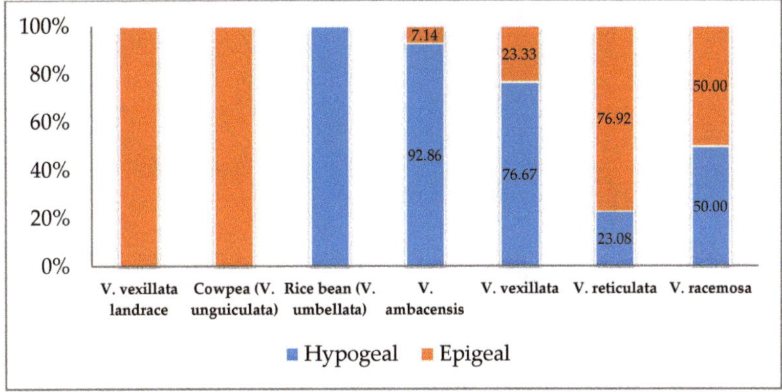

(**c**) Germination Habit.

Figure 3. Variations of some selected qualitative traits, evaluated at the germination stage. (**a**) hypocotyl color; (**b**) attachment of primary leaves (at the two-leaf stage); (**c**) germination habit.

3.1.2. Vegetative Stage

The frequency distributions of variations for the qualitative traits examined at the vegetative stage are presented in Figure 4. The leafiness, leaf pubescence, petiole pubescence, lobing of terminal leaflet, terminal leaflet shape, stipule size, and stem pubescence traits were monitored to characterize the wild accessions of Vigna legumes at this stage.

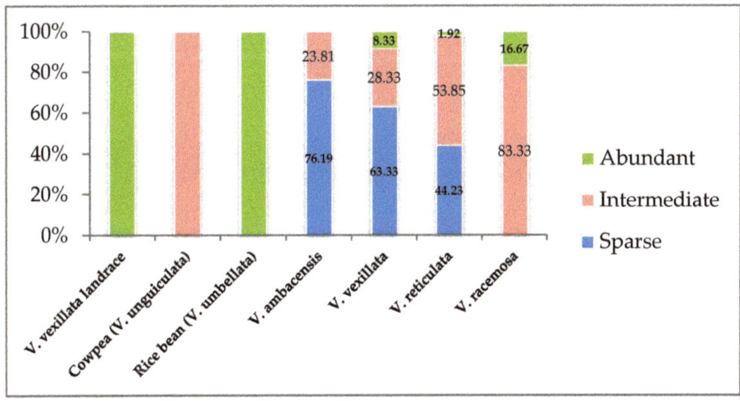

(**a**) Leafiness (at 50% flowering).

Figure 4. *Cont.*

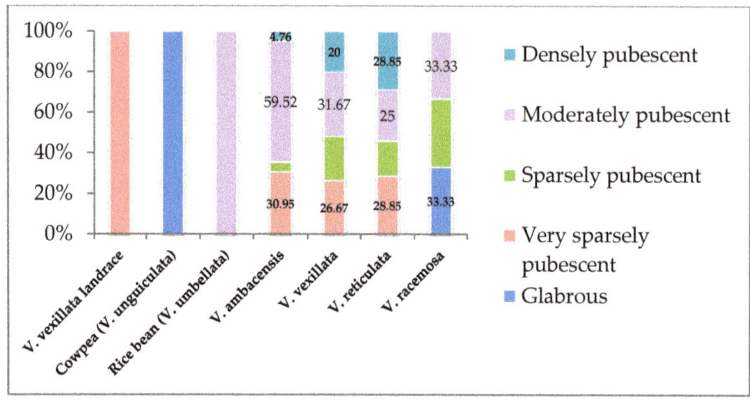

(**b**) Leaf pubescence.

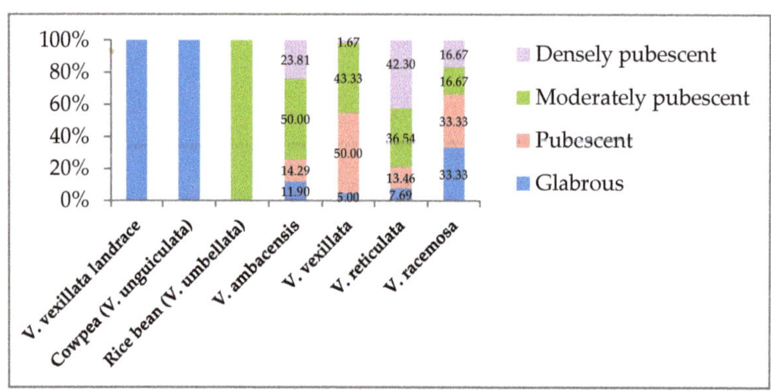

(**c**) Petiole pubescence.

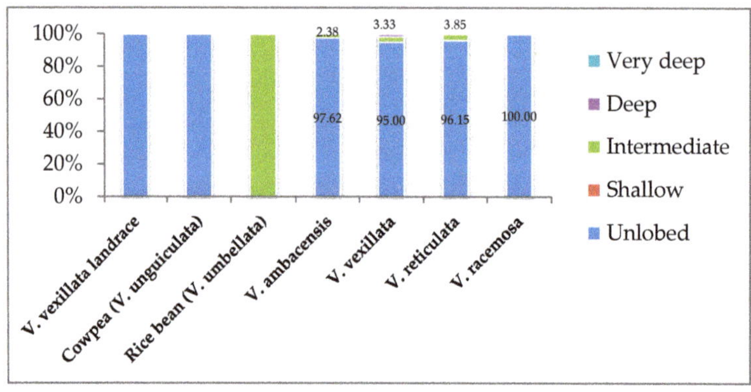

(**d**) Lobing of the terminal leaflet (at first pod maturity).

Figure 4. *Cont.*

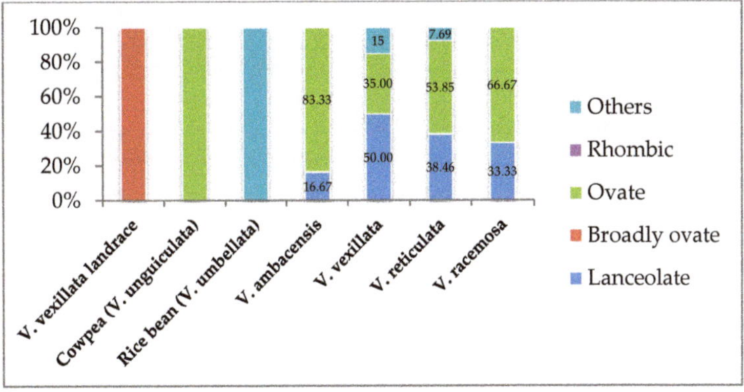

(**e**) Terminal leaflet shape.

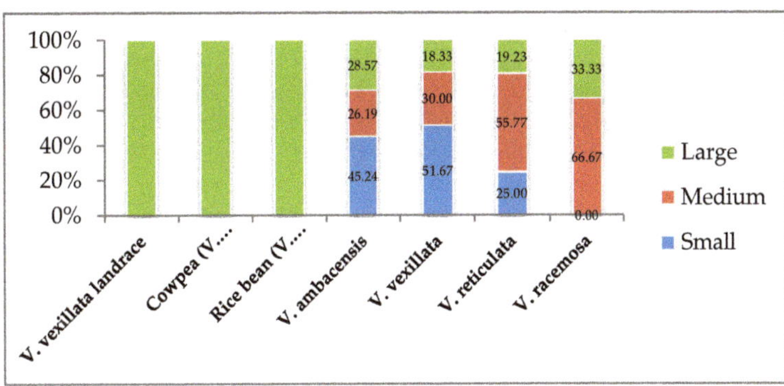

(**f**) Stipule size.

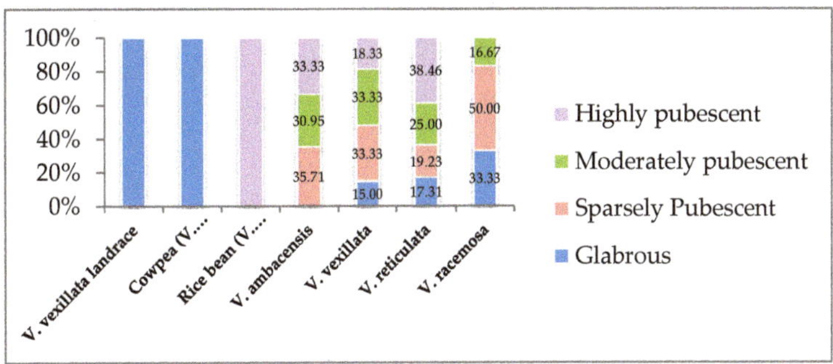

(**g**) Stem pubescence.

Figure 4. Variations in some selected qualitative traits, evaluated at the vegetative stage. (**a**) leafiness (at 50% flowering); (**b**) leaf pubescence; (**c**) petiole pubescence; (**d**) lobing of the terminal leaflet (at first pod maturity); (**e**) terminal leaflet shape, (**f**) stipule size; (**g**) stem pubescence.

Most *V. ambacensis* (76%) and *V. vexillata* (63%) accessions presented a sparse leafy character, which was not the case with any of the checks (Figure 4a). Most *V. racemosa* and *V. reticulata* accession shared a common feature of intermediate leafiness with cowpea. Rice bean and the landrace of *V. vexillata* presented an abundant leafiness, which they had in common with few *V. reticulata*, *V. racemosa*, and *V. vexillata*.

High variations were observed for the leaf pubescence traits among the wild accessions of all the species. A higher percentage (59%) of *V ambacensis* presented moderate leaf pubescence, as found in rice bean, while the other species presented less than 50% accession for the same trait variation. Cowpea had a glabrous leaf pubescence, while the landrace of *V. vexillata* was very sparsely pubescent (Figure 4b).

Considerable variations were also observed in the petiole pubescence trait (Figure 4c). Only the *V. racemosa* accession significantly (33%) shared the common feature of globrous petiole pubescence with cowpea and *V. vexillata* landrace, and 50% of the *V. ambacensis* accession showed a moderately pubescent characteristic, which is similar to that found in rice bean. On the other hand, 50% of the *V. vexillata* accessions were pubescent and did not share the trait intensity with any of the checks. The *V. reticulata* accessions presented the highest percentage (42%) of the densely pubescent variant within the trait and a considerable percentage (36%) of the moderately pubescent variant.

Lobing of the terminal leaflet trait varied little among the studied wild accessions (Figure 4d). The majority (more than 90%) of all the accessions of the studied wild species presented an unlobed variant of the trait, like cowpea and *V. vexillata* landrace (Figure 4d).

Most *V. racemosa*, *V. ambacensis*, and *V. reticulata* (66%, 83% and 53%, respectively) presented an ovate variant of the terminal leaflet shape trait, which is the same variant found in cowpea (Figure 4e). The broadly ovate variant of this trait was only found in the *V. vexillata* landrace, while the rice bean presented an irregular variant (with lobes), which was found in few accessions of *V. reticulata* and *V. vexillata*. The lanceolate variant of the trait was also found in *V. racemosa* (33%), *V. ambacensis* (17%), *V. vexillata* (50%), and *V. reticulata* (38%).

All three of the checks showed a large variant of the stipule size trait, while 67% of *V. racemosa* had the medium stitpule size variant, as well as 56% of the *V. reticulata* accessions (Figure 4f). The small size variant was observed in 45% and 52% of *V. ambacensis* and *V. vexillata*, respectively.

The stem pubescence trait also varied significantly among the wild accessions (Figure 4e). The *V. vexillata* landrace presented the glabrous variant of the trait, which matched with 15.00%, 17.31%, and 33% of *V. vexillata*, *V. reticulata*, and *V. racemosa*, respectively. The stem of rice bean presented the highly pubescent variant of the stem pubescent trait, as found in 33%, 18%, and 38% of *V. ambacensis*, *V. vexillata*, and *V. reticulata*, respectively. A moderately pubescent variant was found in 31%, 33%, 25%, and 17% of *V. ambacensis*, *V. vexillata*, *V. reticulata*, and *V. racemosa*, respectively. Finally, 50% of the *V. racemosa* accessions had the sparsely pubescent variant, while only 36%, 33%, and 19% of *V. ambacensis*, *V. vexillata*, and *V. reticulata* were found to have the same variant.

3.1.3. Pod Formation and Maturity Stage

At this stage, the following traits were observed and recorded for wild accessions and checks under study: pod attachment to peducle, pod pubescence, pod curvature, constriction of the pod between seeds, and pod cross section (Figure 5).

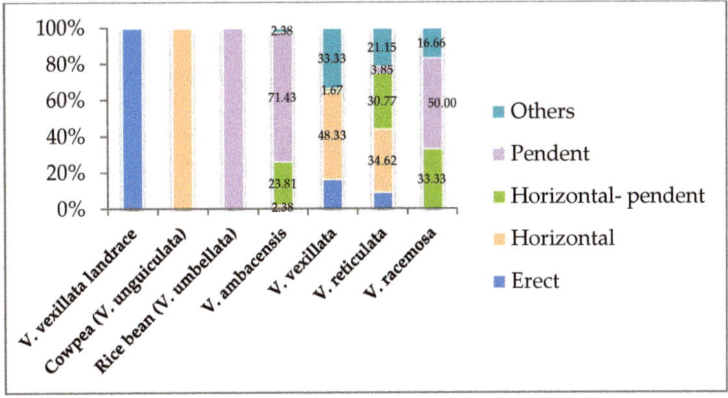

(**a**) Pod attachment to peduncle.

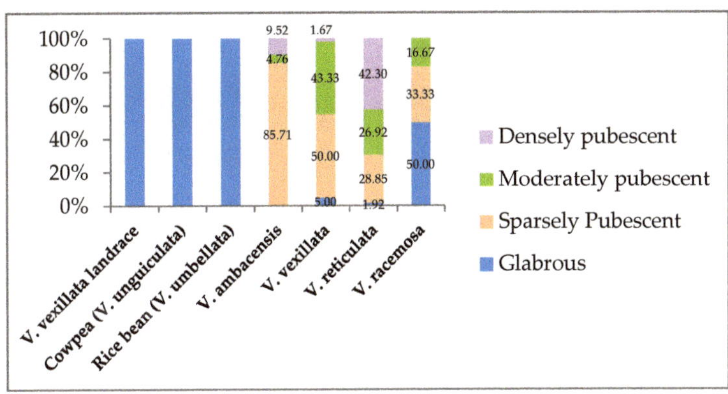

(**b**) Pod pubescence.

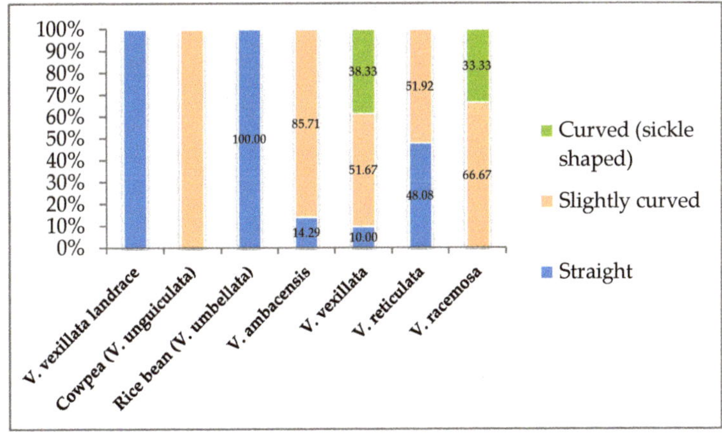

(**c**) Pod curvature.

Figure 5. *Cont.*

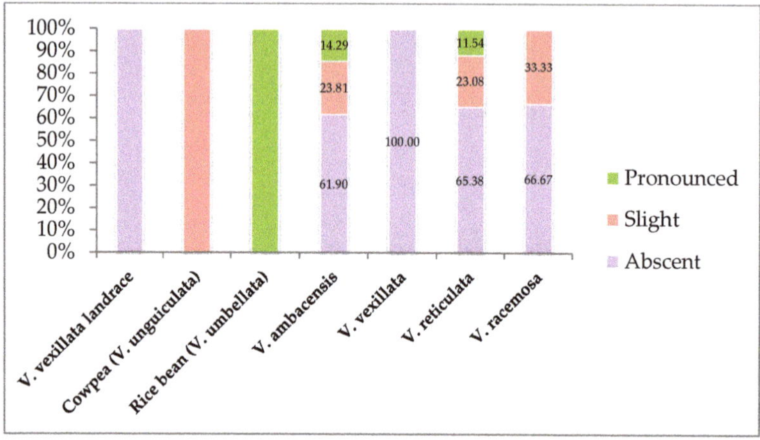

(**d**) Constriction of the pod between seeds.

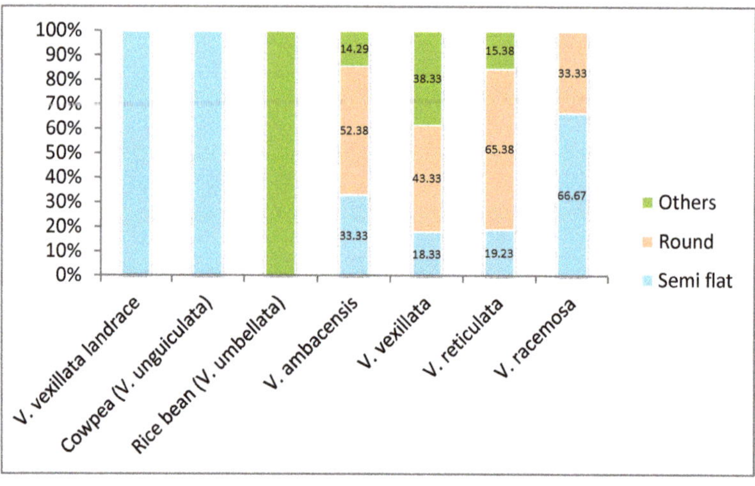

(**e**) Pod cross section.

Figure 5. Variations in some selected qualitative traits, evaluated at the maturity and podding stages. (**a**) pod attachment to peduncle; (**b**) pod pubescence; (**c**) pod curvature; (**d**) constriction of the pod between seeds; (**e**) pod cross section.

For the pod attachment to peducle trait, most accessions of *V. ambacensis* (71%) and *V. racemosa* (50%) were similar to rice bean for the "pendent" variant of the trait, and 48% of *V. vexillata* and 35% of *V. reticulata* accessions presented the "horizontal" variant of the trait, which is similar to cowpea. The "erect" variant of the trait was found in *V. vexillata* landrace and a few accessions of *V. vexillata* and *V. reticulata* (Figure 5a). The horizontal-pendent form of the trait was commonly found in all the wild accessions, with 24%, 2%, 31%, and 33% in *V. ambacensis*, *V. vexillata*, *V. reticulata*, and *V. racemosa*, respectively.

All the checks had a glabrous form for the pod pubescence trait (Figure 5b). Only *V. racemosa* (50%) accessions were found to be similar to the checks for this trait. Moreover, 5%, 43%, 27%, and 17% of *V. ambacensis*, *V. vexillata*, *V. reticulata* and *V. racemosa*, respectively, were moderately pubescent, while 86% of *V. ambacensis*, 50% of *V. vexillata*, 29% of *V. reticulata*, and 33% of *V. racemosa* were sparsely pubescent (Figure 5b). Only 42% of the accessions of *V. reticulata* and 10% of those of the *V. vexillata* pods were densely pubescent.

More than 50% of the studied wild accessions presented the "slightly curved" form of the pod curvature trait, which was similar to cowpea (Figure 5c). Rice bean and the *V. vexillata* landrace commonly shared the "straight" form of the trait with 14% of *V. ambacensis*, 10% of *V. vexillata* and and 48% of *V. reticulata* accessions. On the other hand, only *V. vexillata* (38%) and *V. racemosa* (33%) accessions showed the "curved" form of the pod curvature trait (Figure 5c).

Most of the studied wild Vigna accessions (more than 50%) had no constriction of the pod between seeds (variant: "absent"), as found in *V. vexillata* landrace (Figure 5d). The trait was found in a "pronounced" form only in rice bean and 15% of *V. ambacensis*, as well as 12% of *V. reticulata* accessions. A "slight" constriction of the pod between seeds was the form found in cowpea and 24% of *V. ambacensis*, 23% of *V. reticulata*, and 33% of *V. racemosa* accessions (Figure 5d).

Cowpea and *V. vexillata* landrace presented a "semi-flat" form of the pod cross section trait, together with 33% of *V. ambacensis*, 18% of *V. vexillata*, 19% of *V. reticulata*, and 67% of *V. racemosa* accessions (Figure 5e). Rice bean presented a flat ("other") form of the trait, together with 14% of *V. ambacensis*, 38% of *V. vexillata*, and 15% of *V. reticulata* accessions. The "round" variant of the trait was observed in 52% of *V. ambacensis*, 43% of *V. vexillata*, 65% of *V. reticulata*, and 33% of *V. racemosa* accessions (Figure 5e).

3.2. Quantitative Traits Exploration of the Wild Unexplored Vigna Species

Table S2 summarizes the means, ranges, and coefficients of variation for the selected quantitative traits studied at site A and B during the two cropping seasons. Furthermore, the adjusted mean values for the studied traits are summarized per species in Table S3. The two tables show the results for only one season (the 2018 cropping season) for site A (Tables S2 and S3).

To understand the variations of the means for the various traits studied within the cropping sites and seasons, the generalized linear model procedure (glm proc) of the SAS University Editions was run, and the results are summarized in Tables 2–5. One-way analysis of variance (ANOVA) and type III Sum of Squares Analysis, as well as the analysis of differences, helped to indicate the accession effect, block effect, and the differences among the accessions, checks, and check vs. accession.

The results from the site A study during the 2018 cropping season show that there was a significant difference ($p < 0.05$) between the checks and the wild accessions for all the analyzed traits (Table 2a). Accession effects were also found for all the traits, except for the number of flowers per raceme trait (trait 7) (Table 2a). Block effects were only found for the terminal leaflet width (trait 3), days to flowering (trait 5), number of flowers per raceme (trait 7), pods per peduncle (trait 9), pod length (trait 10), and seed size (trait 13) traits (Table 2a). Significant differences among the accessions, checks, and check vs. accession were observed, as shown in Table 2b for all the traits. Exceptionally, the seed size trait showed no significant difference among the checks ($p > 0.05$) (Table 2b).

Similarly, the results from site B during the 2018 cropping season show that there was a significant difference ($p < 0.05$) between the checks and the wild accessions for all the analyzed traits (Table 3a). Accession effects were also found for all the traits, with no exception, like in the case of trait 7 at site A (Table 3a). Block effects were only found for the terminal leaflet width (trait 3), days to flowering (trait 5), number of flowers per raceme (trait 7), pods per peduncle (trait 9), pod length (trait 10), seeds per pod (trait 12), and seed size (trait 13) traits (Table 3a). Significant differences among the accessions, checks, and check vs. accession were observed, as shown in Table 3b for all the traits. Exceptionally, the seed size trait showed no significant difference among the checks ($p > 0.05$) (Table 3b).

A similar pattern of results was observed in the site B study area during the second cropping season (2019 cropping season) (Table 4). It was found that there is a significant difference ($p < 0.05$) between the checks and the wild accessions for all the analyzed traits (Table 4a). Accession effects were also found for all the traits, with no exception, like in the case of trait 7 at site A (Table 4a). Block effects were only found for the terminal leaflet width (trait 3), days to flowering (trait 5), number of flowers per raceme (trait 7), pods per peduncle (trait 9), pod length (trait 10), seeds per pod (trait 12), and seed size (trait 13) traits (Table 4a). Significant differences among the accessions, checks, and check vs. accession were observed, as shown in Table 4b for all the traits. Exceptionally, the seed size trait showed no significant difference among the checks ($p > 0.05$) (Table 4b).

In order to establish the relationship (interactions) between the accessions and the cropping site and cropping seasons, a two-way analysis of variance was run, and the significant p-values are summarized in Table 5. Of the 15 quantitative traits examined, only the days to flowering, pods per plant, hundred seed weight, and the yield were affected by their growing environment (accession x site effect), while only the number of flowers per raceme and the pods per plant were affected by the cropping season (accession x season effect) (Table 5). All the quantitative traits showed significant differences among the accessions for each site and each season. The same result was observed among the checks, except for the seed size trait.

To determine whether some of the wild *Vigna* accessions share common quantitative traits and can be grouped together, an agglomerative hierarchical clustering analysis was performed, and a dendrogram of three clusters was obtained based on 138 accessions out of the 160 planted due to the exclusion of 22 accessions which did not germinate or did not perform well (Figure 6). The various accessions forming each cluster are presented in Table S4. Cluster I, which is made up of the majority of wild accessions also included two checks, the *Vigna vexillata* landrace and cowpea (*V. unguiculata*). Cluster II was made up of only check 3 (rice bean, *V. umbellate*), while cluster III contained 50 accessions of the wild *Vigna* species.

Furthermore, to examine the relationship that could exist between the quantitative traits and the accessions, as well as the relationship between the accessions themselves, a principal component analysis (XLSTAT) was performed using the adjusted means values, obtained earlier. A correlation circle, combined with an observation chart, was obtained, as shown in Figure 7. The analysis showed that the first (F1 = 45.39%) and second (F2 = 14.22%) PCA dimensions represent 59.61% of the initial information, which is the best combination and explains the variation among the accessions and traits. It was found that there is a positive correlation between the traits, except for the "days to flowering" trait, which is due to the angles between their vectors (Figure 7). It was also noted that all the checks, together with a set of wild accessions, are found on the right side of the F1 axis, forming a group of accessions with higher values for the examined quantitative traits, except for the days to flowering trait. Those accessions could share common features with the checks. A second group, made up of only wild accessions, was found on the left side of the F1 axis, representing the accessions with lower values for the evaluated traits. These accessions also present lower values for the "days to flowering" trait on the F2 axis (Figure 7).

Table 2. (**a**) Analysis of variance (ANOVA) and type III sum of squares analysis for the selected quantitative traits* at the Tanzania Agricultural Research Institute (TARI, Arusha region) during the 2018 cropping season; (**b**) analysis of the differences and interactions between accessions and checks for the selected traits (all species) at the Tanzania Agricultural Research Institute (TARI, Arusha region) during the 2018 cropping season.

(a)

| Traits | | ANOVA | | | | | Type III Sum of Squares Analysis | | | | | | | | |
| | | Model | | | | | Block Effect | | | | | Accession Effect | | | |
	DF	Sum of Squares	Mean Squares	F	Pr > F	DF	Sum of Squares	Mean Squares	F	p	DF	Sum of Squares	Mean Squares	F	p
1	172	230.50	1.34	0.00	<0.0001	7	0.00	0.00	-	-	165	230.19	1.40	0.00	<0.0001
2	172	2513.26	14.61	11.86	<0.0001	7	36.22	5.17	4.20	0.0019	165	1512.31	9.17	7.44	<0.0001
3	172	685.22	3.98	2.99	0.0001	7	15.18	2.17	1.63	0.1600	165	675.48	4.09	3.07	0.0001
4	172	1999.94	11.63	6.57	<0.0001	7	72.68	10.38	5.86	0.0001	165	1908.16	11.56	6.53	<0.0001
5	172	41,888.19	243.54	269.88	<0.0001	7	2.67	0.38	0.42	0.8818	165	41,253.80	250.02	277.07	<0.0001
6	172	92.42	0.54	21.64	<0.0001	7	0.63	0.09	3.65	0.0046	165	75.72	0.46	18.48	<0.0001
7	172	5907.56	34.35	14.23	<0.0001	7	24.98	3.57	1.48	0.2069	165	165	4000.12	24.24	10.05
8	172	16,033.67	93.22	68.36	<0.0001	7	25.65	3.66	2.69	0.0245	165	12,837.61	77.80	57.05	<0.0001
9	172	759.23	4.41	4.96	<0.0001	7	10.55	1.51	1.69	0.1427	165	539.13	3.27	3.67	<0.0001
10	172	4990.46	29.01	27.94	<0.0001	7	13.24	1.89	1.82	0.1139	165	3894.89	23.61	22.73	<0.0001
11	172	298,644.75	1736.31	365.40	<0.0001	7	81.81	11.69	2.46	0.0367	165	221,943.38	1345.11	283.07	<0.0001
12	172	3427.67	19.93	Infini	<0.0001	7	0.00	0.00			165	2597.62	15.74	Infini	<0.0001
13	172	26,079.22	151.62	29.99	<0.0001	7	49.43	7.06	1.40	0.2377	165	23,910.77	144.91	28.66	<0.0001
14	172	6155.01	35.78	14.06	<0.0001	7	60.79	8.68	3.41	0.0070	165	5923.05	35.90	14.11	<0.0001
15	172	225,200,114.2	1,309,303.0	14.99	<0.0001	7	2,007,022.1	286,717.4	3.28	0.0087	165	218,001,052.5	1,321,218.5	15.13	<0.0001

(b)

| Traits | | Contrast (Differences) | | | | | | | | | | | | | |
| | | Among Accessions | | | | | Among Checks | | | | | Check vs. Accession | | | |
	DF	Sum of Squares	Mean Squares	F	Pr > F	DF	Sum of Squares	Mean Squares	F	p	DF	Sum of Squares	Mean Squares	F	p
1	53	36.59	0.69	Infini	<0.0001	3	16.00	5.33	Infini	<0.0001	1	49.51	49.51	Infini	<0.0001
2	53	397.53	7.50	6.09	<0.0001	3	80.92	26.97	21.89	<0.0001	1	49.72	49.72	40.35	<0.0001
3	53	181.78	3.43	2.58	0.0019	3	28.15	9.38	7.05	0.0008	1	130.73	130.73	98.16	<0.0001
4	53	779.54	14.71	8.31	<0.0001	3	132.98	44.33	25.03	<0.0001	1	343.41	343.41	193.94	<0.0001
5	53	17,494.29	330.08	365.79	<0.0001	3	55.09	18.36	20.35	<0.0001	1	1095.79	1095.79	1214.34	<0.0001
6	53	17.56	0.33	13.34	<0.0001	3	0.29	0.10	3.84	0.0178	1	0.29	0.29	11.70	0.0016
7	53	984.25	18.57	7.70	<0.0001	3	137.32	45.77	18.97	<0.0001	1	159.39	159.39	66.05	<0.0001

Table 2. *Cont.*

8	53	1790.74	33.79	24.78	<0.0001	3	240.59	80.20	58.81	<0.0001	1	1751.07	1751.07	1284.07	<0.0001
9	53	141.29	2.67	3.00	0.0004	3	27.70	9.23	10.38	<0.0001	1	10.15	10.15	11.41	0.0018
10	53	822.94	15.53	14.95	<0.0001	3	516.15	172.049	165.66	<0.0001	1	91.85	91.85	88.45	<0.0001
11	53	98,712.64	1862.50	391.96	<0.0001	3	1092.09	364.03	76.61	<0.0001	1	6392.80	6392.80	1345.35	<0.0001
12	53	377.84	7.13	Infini	<0.0001	3	288.00	96.00	Infini	<0.0001	1	82.51	82.51	Infini	<0.0001
13	53	8853.77	167.05	33.04	<0.0001	3	5.64	1.88	0.37	0.7736	1	1092.99	1092.99	216.20	<0.0001
14	53	2595.73	48.98	19.24	<0.0001	3	59.56	19.85	7.80	0.0004	1	379.86	379.86	149.26	<0.0001
15	53	103,443,899.9	1,951,771.7	22.35	<0.0001	3	13,494,835.8	4,498,278.6	51.52	<0.0001	1	5,333,366.9	5,333,366.9	61.08	<0.0001

* **1:** Germination time; **2:** Terminal leaflet length; **3:** Terminal leaflet width; **4:** Petiole length; **5:** Days to flowering; **6:** Flower bud size; **7:** Number of flowers per raceme; **8:** Peduncle length; **9:** Pods per peduncle; **10:** Pod length; **11:** Pods per plant; **12:** Seeds per pod; **13:** Seed size; **14:** 100-Seed weight; **15:** Yield.

Table 3. (**a**) Analysis of variance (ANOVA) and type III sum of squares analysis for the selected quantitative traits* at the Tanzania Coffee Research Institute (TaCRI, Kilimanjaro region) during the 2018 cropping season; (**b**) analysis of the differences and interactions between accessions and checks for the selected traits (all species) at the Tanzania Coffee Research Institute (TaCRI, Kilimanjaro region) during the 2018 cropping season.

(a)

| | ANOVA | | | | | Type III Sum of Square Analysis | | | | | | | | |
| | Model | | | | | Block Effect | | | | | Accession Effect | | | |
Traits	DF	Sum of Squares	Mean Squares	F	Pr > F	DF	Sum of Squares	Mean Squares	F	p	DF	Sum of Squares	Mean Squares	F	p
1	172	2564.76	14.91	Infini	<0.0001	7	0.00	0.00			165	1217.73	7.38	Infini	<0.0001
2	172	2986.01	17.36	11.86	<0.0001	7	43.03	6.15	4.20	0.0019	165	1796.78	10.89	7.44	<0.0001
3	172	829.12	4.82	2.99	0.0001	7	18.36	2.62	1.53	0.1600	165	817.33	4.95	3.07	0.0001
4	172	2121.73	12.33	6.57	<0.0001	7	77.10	11.01	5.36	0.0001	165	2024.37	12.27	6.53	<0.0001
5	172	45,758.01	266.03	233.27	<0.0001	7	4.33	0.62	0.54	0.80	165	44,869.16	271.93	238.44	<0.0001
6	172	133.09	0.77	21.64	<0.0001	7	0.91	0.13	3.65	0.0046	165	109.03	0.66	18.48	<0.0001
7	172	4330.26	25.18	8.55	<0.0001	7	22.38	3.20	1.09	0.3929	165	3090.50	18.73	6.36	<0.0001
8	172	17,010.12	98.90	68.36	<0.0001	7	27.21	3.89	2.69	0.0245	165	13,619.42	82.54	57.05	<0.0001
9	172	748.83	4.35	4.82	<0.0001	7	10.10	1.57	1.74	0.1314	165	538.18	3.26	3.61	<0.0001
10	172	4792.84	27.87	27.94	<0.0001	7	12.72	1.817	1.82	0.1139	165	3740.65	22.67	22.73	<0.0001
11	172	60,475.56	351.60	365.40	<0.0001	7	16.57	2.37	2.46	0.0367	165	44,943.53	272.39	283.07	<0.0001
12	172	3387.31	19.69	236.32	<0.0001	7	0.58	0.08	1.00	0.4478	165	2581.24	15.64	187.73	<0.0001
13	172	23,536.50	136.84	29.99	<0.0001	7	44.61	6.37	1.40	0.2377	165	21,579.47	130.78	28.66	<0.0001
14	172	4993.97	29.03	14.40	<0.0001	7	48.15	6.88	3.41	0.0070	165	4836.06	29.31	14.54	<0.0001
15	172	182,274,678.9	1,059,736.5	15.32	<0.0001	7	1,589,762.2	227,108.9	3.28	0.0087	165	177,020,077.70	1,072,849.00	15.51	<0.0001

Table 3. *Cont.*

(b)

| | | Contrast (Differences) | | | | | | | | | | | | | | |
| | | Among Accessions | | | | | Among Checks | | | | | Check vs. Accession | | | |
Traits	DF	Sum of Squares	Mean Squares	F	Pr > F	DF	Sum of Squares	Mean Squares	F	p	DF	Sum of Squares	Mean Squares	F	p
1	53	67.20	1.27	Infini	<0.0001	3	16.00	5.33	Infini	<0.0001	1	118.26	118.26	Infini	<0.0001
2	53	472.30	8.91	6.09	<0.0001	3	96.14	32.04	21.89	<0.0001	1	59.07	59.07	40.35	<0.0001
3	53	219.95	4.15	2.58	0.0019	3	34.07	11.36	7.05	0.0008	1	158.18	158.18	98.16	<0.0001
4	53	827.02	15.60	8.31	<0.0001	3	141.08	47.03	25.03	<0.0001	1	364.33	364.33	193.94	<0.0001
5	53	17,480.49	329.82	289.20	<0.0001	3	58.84375	19.61	17.20	<0.0001	1	872.40	872.40	764.95	<0.0001
6	53	25.28	0.47	13.34	<0.0001	3	0.41	0.14	3.84	0.0178	1	0.42	0.42	11.70	0.0016
7	53	805.51	15.20	5.16	<0.0001	3	125.78	41.93	14.25	<0.0001	1	169.17	169.17	57.48	<0.0001
8	53	1899.80	35.84	24.78	<0.0001	3	255.25	85.08	58.81	<0.0001	1	1857.71	1857.71	1284.07	<0.0001
9	53	138.13	2.61	2.89	0.0006	3	29.55375	9.85	10.92	<0.0001	1	12.05	12.05	13.35	0.0008
10	53	790.35	14.91	14.95	<0.0001	3	495.71	165.24	165.66	<0.0001	1	88.22	88.22	88.45	<0.0001
11	53	19,989.31	377.16	391.96	<0.0001	3	221.15	73.72	76.61	<0.0001	1	1294.54	1294.54	1345.35	<0.0001
12	53	377.20	7.12	85.40	<0.0001	3	276.38	92.13	1105.50	<0.0001	1	78.07	78.070	936.85	<0.0001
13	53	7990.53	150.76	33.04	<0.0001	3	5.093673	1.70	0.37	0.7736	1	986.42	986.43	216.20	<0.0001
14	53	2134.87	40.28	19.98	<0.0001	3	47.18	15.73	7.80	0.0004	1	324.85	324.85	161.15	<0.0001
15	53	83,831,969.77	1,581,735.28	22.87	<0.0001	3	10,689,259.48	3,563,086.49	51.52	<0.0001	1	4,556,054.50	4,556,054.50	65.87	<0.0001

* **1**: Germination time; **2**: Terminal leaflet length; **3**: Terminal leaflet width; **4**: Petiole length; **5**: Days to flowering; **6**: Flower bud size; **7**: Number of flowers per raceme; **8**: Peduncle length; **9**: Pods per peduncle; **10**: Pod length; **11**: Pods per plant; **12**: Seeds per pod; **13**: Seed size; **14**: 100-Seed weight; **15**: Yield.

Table 4. (**a**) Analysis of variance (ANOVA) and type III sum of squares analysis for the selected quantitative traits * at the Tanzania Coffee Research Institute (TaCRI, Kilimanjaro region) during the 2019 cropping season; (**b**) analysis of the differences and interactions between accessions and checks for the selected traits (all species) at the Tanzania Coffee Research Institute (TaCRI, Kilimanjaro region) during the 2019 cropping season.

(a)

| | | ANOVA | | | | | Type III Sum of Square Analysis | | | | | | | | |
| | | Model | | | | | Block Effect | | | | | AccessionEffect | | | |
Traits	DF	Sum of Squares	Mean Squares	F	Pr > F	DF	Sum of Squares	Mean Squares	F	p	DF	Sum of Squares	Mean Squares	F	p
1	172	1886.23	10.97	Infini	<0.0001	7	0.00	0.00			165	1236.69	7.50	Infini	<0.0001
2	172	2312.36	13.44	11.86	<0.0001	7	33.32	4.76	4.20	0.0019	165	1391.43	8.43	7.44	<0.0001

Table 4. *Cont.*

Traits	DF	Sum of Squares	Mean Squares	F	p	DF	Sum of Squares	Mean Squares	F	p	DF	Sum of Squares	Mean Squares	F	p
3	172	627.56	3.65	2.99	0.0001	7	13.90	1.99	1.63	0.1600	165	618.64	3.75	3.07	0.0001
4	172	2185.86	12.71	6.57	<0.0001	7	79.43	11.35	5.86	0.0001	165	2085.55	12.64	6.53	<0.0001
5	172	45,758.01	266.03	233.27	<0.0001	7	4.33	0.62	0.54	0.7960	165	44,869.16	271.93	238.44	<0.0001
6	172	155.23	0.90	21.64	<0.0001	7	1.07	0.15	3.65	0.0046	165	127.18	0.77	18.48	<0.0001
7	172	108,256.43	629.40	8.55	<0.0001	7	559.500	79.93	1.09	0.3929	165	77,262.23	468.26	6.36	<0.0001
8	172	18,046.03	104.92	68.36	<0.0001	7	28.86	4.12	2.69	0.0245	165	14,448.84	87.57	57.05	<0.0001
9	172	1643.58	9.56	4.62	<0.0001	7	26.73	3.82	1.84	0.1094	165	1181.19	7.16	3.46	<0.0001
10	172	4918.26	28.59	27.94	<0.0001	7	13.05	1.86	1.82	0.1139	165	3838.54	23.26	22.73	<0.0001
11	172	216,024.75	1255.96	365.40	<0.0001	7	59.18	8.45	2.46	0.0367	165	160,542.80	972.99	283.07	<0.0001
12	172	4479.72	26.04	236.32	<0.0001	7	0.77	0.11	1.00	0.4478	165	3413.69	20.69	187.73	<0.0001
13	172	24,009.58	139.59	29.99	<0.0001	7	45.51	6.50	1.40	0.2377	165	22,013.22	133.41	28.66	<0.0001
14	172	5554.90	32.30	14.06	<0.0001	7	54.86	7.84	3.41	0.0070	165	5345.55	32.40	14.11	<0.0001
15	172	203,243,103.1	1,181,645.9	14.99	<0.0001	7	1,811,337.5	258,762.5	3.28	0.0087	165	196,745,949.9	1,192,399.7	15.13	<0.0001

(b)

Contrast (Differences)

Traits		Among Accessions					Among Checks					Check vs. Accession			
	DF	Sum of Squares	Mean Squares	F	Pr > F	DF	Sum of Squares	Mean Squares	F	p	DF	Sum of Squares	Mean Squares	F	p
1	53	44.13	0.83	Infini	<0.0001	3	16.00	5.33	Infini	<0.0001	1	103.45	103.45	Infini	<0.0001
2	53	365.75	6.90	6.09	<0.0001	3	74.45	24.82	21.89	<0.0001	1	45.74	45.74	40.35	<0.0001
3	53	166.48	3.14	2.58	0.0019	3	25.78	8.59	7.05	0.0008	1	119.73	119.73	98.16	<0.0001
4	53	852.01	16.08	8.31	<0.0001	3	145.34	48.45	25.03	<0.0001	1	375.34	375.34	193.94	<0.0001
5	53	17,480.49	329.82	289.20	<0.0001	3	58.84	19.61	17.20	<0.0001	1	872.40	872.40	764.95	<0.0001
6	53	29.49	0.56	13.34	<0.0001	3	0.48	0.16	3.84	0.0178	1	0.49	0.49	11.70	0.0016
7	53	20,137.78	379.96	5.16	<0.0001	3	3144.59	1048.20	14.25	<0.0001	1	4229.18	4229.18	57.48	<0.0001
8	53	2015.50	38.03	24.78	<0.0001	3	270.79	90.26	58.31	<0.0001	1	1970.84	1970.84	1284.07	<0.0001
9	53	295.70	5.58	2.70	0.0013	3	63.23	21.08	10.18	<0.0001	1	25.27	25.27	12.21	0.0013
10	53	811.04	15.30	14.95	<0.0001	3	508.68	169.56	165.66	<0.0001	1	90.53	90.53	88.45	<0.0001
11	53	71,403.81	1347.24	391.96	<0.0001	3	789.97	263.32	76.61	<0.0001	1	4624.23	4624.23	1345.35	<0.0001
12	53	498.84	9.41	85.40	<0.0001	3	365.51	121.84	1105.50	<0.0001	1	103.25	103.25	936.85	<0.0001
13	53	8151.14	153.80	33.04	<0.0001	3	5.20	1.70	0.37	0.7736	1	1006.25	1006.25	216.20	<0.0001
14	53	2342.65	44.20	19.24	<0.0001	3	53.76	17.92	7.80	0.0004	1	342.83	342.83	149.26	<0.0001
15	53	93,358,119.63	1,761,473.96	22.35	<0.0001	3	12,179,089.35	4,059,696.45	51.52	<0.0001	1	4,813,363.60	4,813,363.60	61.08	<0.0001

* **1**: Germination time; **2**: Terminal leaflet length; **3**: Terminal leaflet width; **4**: Petiole width; **5**: Days to flowering; **6**: Flower bud size; **7**: Number of flowers per raceme; **8**: Peduncle length; **9**: Pods per peduncle; **10**: Pod length; **11**: Pods per plant; **12**: Seeds per pod; **13**: Seed size; **14**: 100-Seed weight; **15**: Yield.

Table 5. Two-way analysis of variance for the interactions due to the site and season for the studied quantitative traits.

S/N	Traits	*p*-Values for Site Effects			*p*-Values for Season Effects		
		Site	Accession	Site x Accession	Season	Accession	Accession x Season
1	Germination time (days)	<0.0001	0.000	0.153	0.097	0.000	0.979
2	Terminal leaflet length (cm)	<0.0001	<0.0001	1.000	<0.0001	<0.0001	0.961
3	Terminal leaflet width (cm)	0.000	<0.0001	1.000	<0.0001	<0.0001	0.998
4	Petiole length (cm)	0.000	<0.0001	1.000	0.009	<0.0001	1.000
5	Days to flowering	<0.0001	<0.0001	0.032	<0.0001	<0.0001	1.000
6	Flower bud size (cm)	<0.0001	<0.0001	0.078	<0.0001	<0.0001	0.899
7	Number of flowers per raceme	<0.0001	<0.0001	0.995	<0.0001	0.000	0.003
8	Peduncle length (cm)	0.003	<0.0001	1.000	0.003	<0.0001	1.000
9	Pods per peduncle	0.742	<0.0001	0.973	<0.0001	<0.0001	0.054
10	Pod length (cm)	0.194	<0.0001	1.000	0.371	<0.0001	1.000
11	Pods per plant	<0.0001	<0.0001	<0.0001	<0.0001	<0.0001	<0.0001
12	Seeds per pod	0.894	<0.0001	1.000	<0.0001	<0.0001	0.712
13	Seed size (mm^2)	<0.0001	<0.0001	0.052	0.013	<0.0001	0.506
14	100-Seed weight (g)	<0.0001	<0.0001	0.037	<0.0001	<0.0001	0.068
15	Yield (Kg/ha)	<0.0001	<0.0001	0.032	<0.0001	<0.0001	0.055

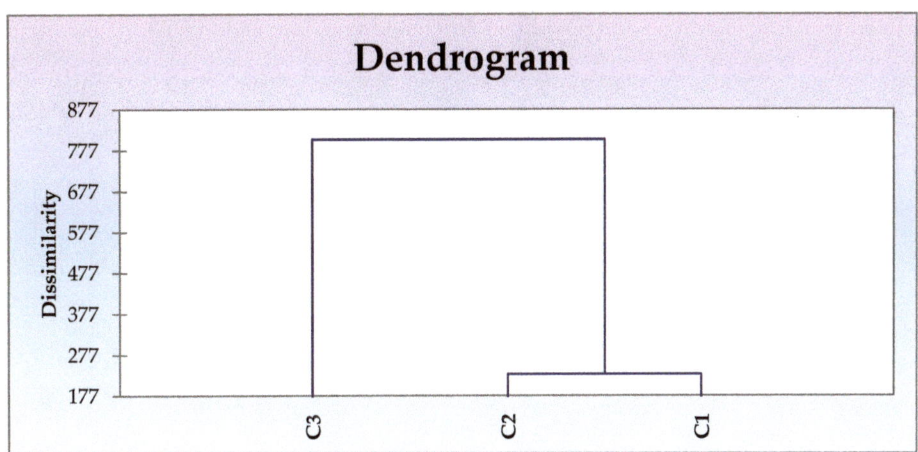

Figure 6. Dendrogram depicting the studied clusters of wild *Vigna* accessions for the 15 quantitative traits.

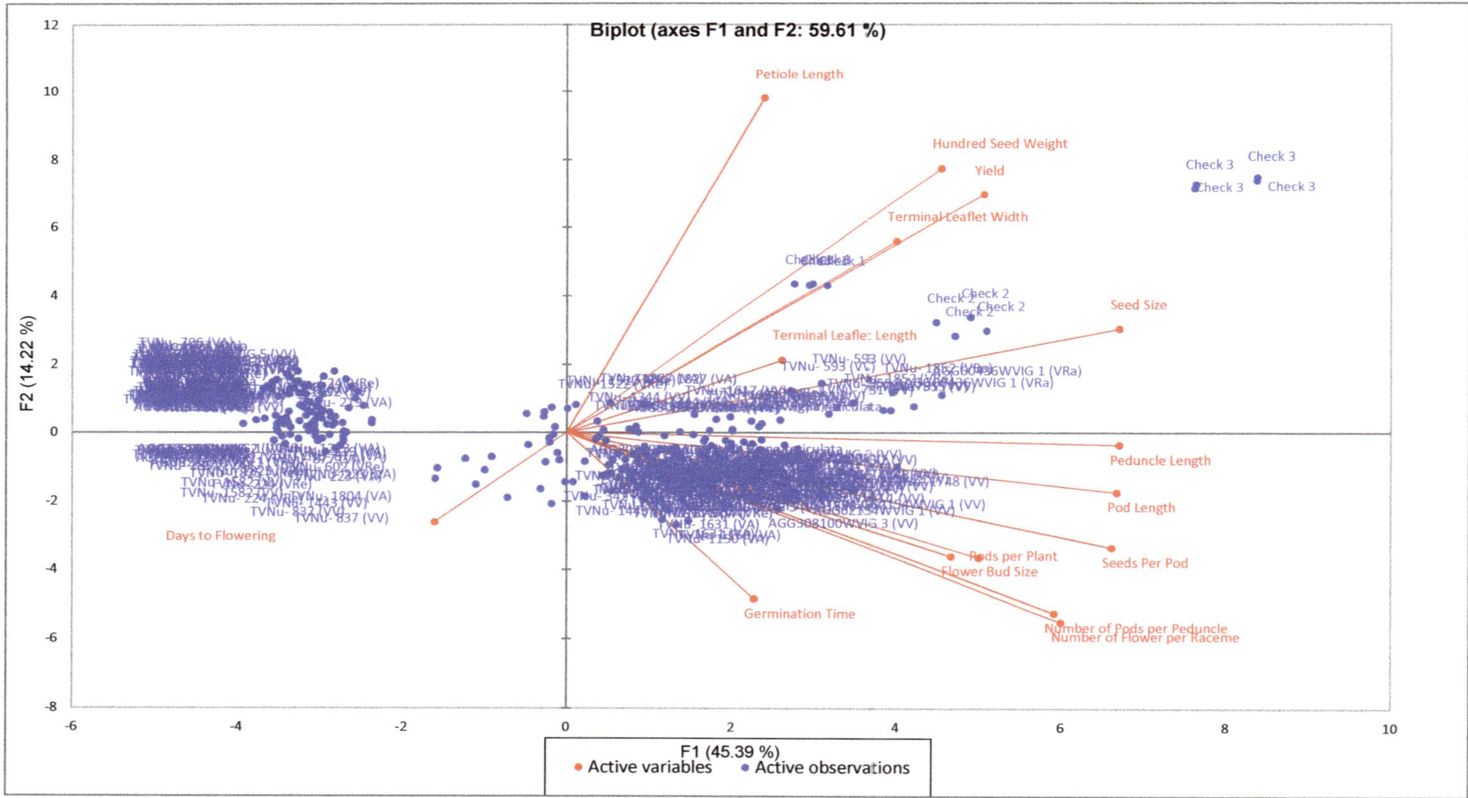

Figure 7. Principal component analysis result, showing the relationship between the traits and the accessions.

4. Discussion

The qualitative exploration of the wild *Vigna* species showed that there are variations in their characteristics for the same trait within the same species (Section 3.1), while all the checks expressed the same form of a particular trait throughout the experiments. Some of these qualitative characters are expressions of a genetic variation within the genome of the plant. A recent taxonomic differentiation was established between two wild Vigna species (*V. stipulacea* and *V. trilobata*) based on their morphological characteristics, such as germination habit, primary leaf attachment, etc. [14]. Therefore, the variations observed could be due to the heterogenous nature of the wild accessions, which have been homogenized in the checks through selection and breeding processes. It is a common opinion of many researchers that wild crop populations are much older and more diverse than domesticated crops, having undergone millennia of recombination, genetic drift, and natural selection [11]. Some of the trait forms found in the wild accessions might have only disappeared from the domesticated one during the domestication process. Some of the unique traits of the wild accessions, such as their leaf, stem, and petiole pubescence, which are not found in the checks, might have existed in those checks but disappeared with time during the domestication process. They could have a potential use, if they are domesticated, since they are thought to be responsible for some beneficial traits, such as the resistance to diseases and pests [15,16]. Therefore, it might be time to start examining some of the traits from the wild species that have disappeared in order to domesticate new species. The qualitative characteristics of the wild *Vigna* accessions found in this study were in line with most of the characteristics found in earlier works, carried out on other wild *Vigna* species [12,14].

Regarding the studied quantitative traits, Tables S2 and S3 summarized the means, ranges, and coefficients of variation at site A for only one season (the 2018 cropping season). This is due to the fact that, during the 2019 cropping season at site A, the rainfall was not enough (as per the pattern shown in Figure 1) to allow for germination and the growth of certain accessions. Most seeds did not germinate during that season, and those that did germinate (mainly checks) could not resist the harsh conditions. Figure 1 shows that the rainfall started at a very low rate (92.9 mm), then achieved its peak value (196.6 mm) and stopped. This amount of rainfall might have not been sufficient for the soil to allow the germination of the wild accessions. It is also known that the seed structure could influence the germination of seeds [17,18]. However, the characteristics of the seed structure of wild legumes are still yet to be reported.

The first cropping season at site A showed significant differences ($p < 0.05$) between the checks and the wild accessions for all the analyzed traits (Table 2a). Accession effects were found for all the traits, except for the number of flowers per raceme trait (trait 7) (Table 2a). This shows that the different accessions and species involved in the study possess different phenotypic and probably genetic characteristics. The number of flowers per raceme seemed to have no significant difference among the accessions and the checks. This might have been influenced by other agro-climatic conditions of the environment at that moment, which could affect some accessions, but probably not all. It has been reported that simple shading can affect the number of flowers per raceme [19]. The block effect observed at that site could be due to some particular factors of the field, ranging from agro-climatic to soil characteristics. The most probable explanation could be that the soil was heterogeneous in the same field and differently affected the checks. The ability of a plant to respond to soil characteristics can affect some of its physiological and phenotypic characteristics [20].

A similar pattern of results was observed during the two cropping seasons at site B. The observed phenomenon could have the same explanation as in the case of site A, mentioned above.

Based on the result shown in Table 5, only the days to flowering, pods per plant, hundred seed weight, and the yield were affected by their growing environment (accession × site effect), while only the number of flowers per raceme and the pods per plant were affected by the cropping season. These effects might be explained by the agro-climatic characteristics of each cropping site and season. As shown in Figure 1, site A has lower and shorter rainfall characteristics, which can affect the days to flowering. This is in line with earlier reports that predicted that the changes in the flowering time are associated with a reduction in precipitation [21]. The effect of the yield and yield parameters, such as the pods per plant and hundred seed weight traits, has been reported before in relation to other legumes, and these reports do not contradict the present findings [22,23]. Therefore, it should be recommended that these traits be taken into consideration during any attempt to domesticate or improve wild legumes. The number of flowers per raceme and the pods per plant were the only traits affected by the cropping season. These two traits are closely related, as confirmed by the positive correlation that exists between the two, shown in Figure 7. They could also be directly or indirected affected by the variations in temperature and rainfall, as per the earlier explanation [20,21]. The significant effect of the season on the number of flowers per raceme and pods per plant has also been reported in relation to the landraces of *Phaseolus vulgaris* and cowpea (*Vigna unguiculata*) [24,25]. These two traits also need to be considered in any attempt at domestication.

Figure 6 revealed that the wild *Vigna* accessions could be grouped into three clusters, with one larger cluster (cluster I), including two checks (Table S4). This shows that some of the wild accessions share common features and probably genetic characteristics. Cluster I, containing the checks, could offer a clear orientation for the selection of candidates for domestication. Cluster 1 could also offer recommendations pertaining to the cooking time and water absorption capacity traits as reported earlier [26]. These are clear indications that these wild legumes could be domesticated and made useful, as the preliminary finding showed that farmers would be interested in utilizing them for various purposes [27]. In fact, it has recently been reported that Vigna stipulacea, another wild legume species with biotic resistance traits is domesticable [28] However, it is also necessary to note that domestication process could also affect the nutritional and health characteristics of the domesticated product as alerted by some researchers [29]. Therefore, the choice of *V. vexillata*, *V. reticulata*, *V. ambacensis*, and *V. racemosa* species in this study was first based on their availability in genebanks and from the little preliminary information obtained from the authors earlier investigations [5,26,27].

Figure 7 provides further indications relating to the domestication of these wild legumes by grouping them based on their quantitative agro-morphological traits. It was shown that most of the quantitative traits are positively correlated, and there is a degree of commonality between the checks and a group of some wild species.

5. Conclusions

This study revealed that the wild *Vigna* species possesses a large variation range of qualitative and quantitative traits, which could be exploited in the improvement of domesticated species or guide their domestication. Specifically, it was found that only the days to flowering, pods per plant, hundred seed weight and the yield were affected by their growing environment (accession x site effect), while only the number of flowers per raceme and the pods per plant were affected by the cropping season (accession x season effect). All the quantitative traits showed significant differences among accessions for each site and each season. The study further provides indications relating to *the* candidate accessions favorable for domestication, based on the quantitative and qualitative traits. However, further characterization, focusing on the biochemical content of the wild species, may be of great value for the extension of domestication information.

Supplementary Materials: The following are available online at http://www.mdpi.com/2073-4395/10/1/111/s1, Table S1: Wild Vigna legumes accessions used in the study, Table S2: Means, ranges and coefficients of variation for the selected quantitative traitsa, analyzed at site A* and B* during the two cropping seasons, Table S3: Adjusted mean values for selected quantitative traits per species at various sites, Table S4: Distribution of the accessions, according to the clusters generated from the agglomerative hierarchical analysis (AHC).

Author Contributions: P.A.N. and P.B.V. conceived and designed the experiments; D.V.H. performed the experiments, collected data, analyzed the data, and made the first draft of the manuscript; P.B.V. and A.O.M. supervised the research and internally reviewed the manuscript; and P.A.N. made the final internal review and revised the final draft of the manuscript. All authors have read and agreed to the published version of the manuscript.

Funding: This research was partially funded by the Centre for Research, Agricultural Advancement, Teaching Excellence and Sustainability in Food and Nutrition Security (CREATES-FNS) through the Nelson Mandela African Institution of Science and Technology (NM-AIST) and the World Bank. The research also received funding support from the International Foundation for Science (IFS) through Grant No. I-3-B-6203-1.

Acknowledgments: The authors are grateful to the Centre for Research, Agricultural Advancement, Teaching Excellence and Sustainability in Food and Nutrition Security (CREATES-FNS), as well as its partners, the Nelson Mandela African Institution of Science and Technology (NM-AIST), and the World Bank. The authors also acknowledge the additional funding support from the International Foundation for Science (IFS) through Grant No. I-3-B-6203-1. The authors are also grateful to the Genetic Resources Center, International Institute of Tropical Agriculture (IITA), Ibadan-Nigeria, as well as the Australian Grains Gene bank (AGG) for providing supporting information and seed materials for research.

Conflicts of Interest: The authors declare no conflict of interest.

References

1. Singh, A.; Dubey, P.K.; Chaurasia, R.; Dubey, R.K.; Pandey, K.K.; Singh, G.S.; Abhilash, P.C. Domesticating the Undomesticated for Global Food and Nutritional Security: Four Steps. *Agronomy* **2019**, *9*, 491. [CrossRef]
2. Kew, R.B.G. *The State of the World's Plants Report*; Kew:Royal Botanic Gardens: London, UK, 2016.
3. Padulosi, S.; Thompson, J.; Rudebjer, P. *Fighting Poverty, Hunger and Malnutrition with Neglected and Underutilized Species (NUS): Needs, Challenges and the Way Forward*; Bioversity International: Rome, Italy, 2013; ISBN 9789290439417.
4. Diouf, J. FAO's Director-general on how to feed the world in 2050. *Popul. Dev. Rev.* **2009**, *35*, 837–839.
5. Harouna, D.V.; Venkataramana, P.B.; Ndakidemi, P.A.; Matemu, A.O. Under-exploited wild *Vigna* species potentials in human and animal nutrition: A review. *Glob. Food Sec.* **2018**, *18*, 1–11. [CrossRef]
6. Bhat, R.; Karim, A.A. Exploring the nutritional potential of wild and underutilized legumes. *Compr. Rev. Food Sci. Food Saf.* **2009**, *8*, 305–331. [CrossRef]
7. Ojiewo, C.; Monyo, E.; Desmae, H.; Boukar, O.; Mukankusi-Mugisha, C.; Thudi, M.; Pandey, M.K.; Saxena, R.K.; Gaur, P.M.; Chaturvedi, S.K.; et al. Genomics, genetics and breeding of tropical legumes for better livelihoods of smallholder farmers. *Plant Breed.* **2018**, 1–13. [CrossRef]
8. Pratap, A.; Malviya, N.; Tomar, R.; Gupta, D.S.; Jitendra, K. Vigna. In *Alien Gene Transfer in Crop Plants, Volume 2: Achievements and Impacts*; Pratap, A., Kumar, J., Eds.; Springer Science+Business Media, LLC: Berlin/Heidelberg, Germany, 2014; pp. 163–189. ISBN 9781461495727.
9. Tomooka, N.; Naito, K.; Kaga, A.; Sakai, H.; Isemura, T.; Ogiso-Tanaka, E.; Iseki, K.; Takahashi, Y. Evolution, domestication and neo-domestication of the genus *Vigna*. *Plant Genet. Resour. Characterisation Util.* **2014**, *12*, S168–S171. [CrossRef]
10. Tomooka, N.; Kaga, A.; Isemura, T.; Vaughan, D.; Srinives, P.; Somta, P.; Thadavong, S.; Bounphanousay, C.; Kanyavong, K.; Inthapanya, P. Vigna Genetic Resources. In Proceedings of the 14th NIAS International Workshop on Genetic Resources: Genetic Resources and Comparative Economics of Legumes (Glycine and Vigna), Tsukuba, Japan, 14 September 2009; pp. 11–21.
11. Smýkal, P.; Nelson, M.N.; Berger, J.D.; Von Wettberg, E.J.B. The impact of genetic changes during crop domestication. *Agronomy* **2018**, *8*, 119. [CrossRef]
12. Bisht, I.S.; Bhat, K.V.; Lakhanpaul, S.; Latha, M.; Jayan, P.K.; Biswas, B.K.; Singh, A.K. Diversity and genetic resources of wild *Vigna* species in India. *Genet. Resour. Crop Evol.* **2005**, *52*, 53–68. [CrossRef]
13. IASRI Augmented Block Designs. Available online: http://www.iasri.res.in/design/AugmentedDesigns/home.htm (accessed on 18 January 2019).
14. Gore, P.G.; Tripathi, K.; Pratap, A.; Bhat, K.V.; Umdale, S.D.; Gupta, V.; Pandey, A. Delineating taxonomic identity of two closely related *Vigna* species of section *Aconitifoliae*: *V. trilobata* (L.) Verdc. and *V. stipulacea* (Lam.) Kuntz in India. *Genet. Resour. Crop Evol.* **2019**, *66*, 1155–1165. [CrossRef]
15. Oyatomi, O.; Fatokun, C.; Boukar, O.; Abberton, M.; Ilori, C. Screening Wild *Vigna* Species and Cowpea (*Vigna unguiculata*) Landraces for Sources of Resistance to Striga gesnerioides. In *Enhancing Crop Genepool Use: Capturing Wild Relatives and Landrace Diversity for Crop Improvement*; CAB International: Wallingford, UK, 2016; pp. 27–31.

16. Popoola, J.; Adebambo, A.; Ejoh, S.; Agre, P.; Adegbite, A.; Omonhinmin, C. Morphological Diversity and Cytological Studies in Some Accessions of *Vigna vexillata* (L.) A. Richard. *Annu. Res. Rev. Biol.* **2017**, *19*, 1–12. [CrossRef]

17. Swanson, B.G.; Hughes, J.S.; Rasmussen, P.H. Seed Microstructure: Review of Water Imbibition in Legumes. *Food Struct.* **1985**, *4*, 115–124.

18. Smykal, P.; Vernoud, V.; Blair, M.W.; Soukup, A.; Thompson, R.D. The role of the testa during development and in establishment of dormancy of the legume seed. *Front. Plant Sci.* **2014**, *5*, 351. [PubMed]

19. Jiang, H.; Egli, D.B. Shade Induced Changes in Flower and Pod Number and Flower and Fruit Abscission in Soybean. *Agron. J.* **1993**, *85*, 221–225. [CrossRef]

20. Morgan, J.B.; Connolly, E.L. Plant-Soil Interactions: Nutrient Uptake. *Nat. Educ. Knowl.* **2013**, *4*, 2.

21. Kigel, J.; Konsens, I.; Rosen, N.; Rotem, G.; Kon, A.; Fragman-Sapir, O. Relationships between flowering time and rainfall gradients across Mediterranean-desert transects. *Isr. J. Ecol. Evol.* **2011**, *57*, 91–109. [CrossRef]

22. Satish, D.; Rc, J.; Patil, R.K.; Masuthi, D. Genotype x environment interaction and stability analysis in Recombinant inbred lines of French bean for growth and yield components. *J. Pharmacogn. Phytochem.* **2017**, *6*, 216–219.

23. Sabaghnia, N.; Dehghani, H.; Sabaghpourb, S.H. Graphic Analysis of Genotype by Environment Interaction for Lentil Yield in Iran. *Agron. J.* **2008**, *100*, 760–764. [CrossRef]

24. Arteaga, S.; Yabor, L.; Torres, J.; Solbes, E.; Muñoz, E.; Vicente, O.; Boscaiu, M. Morphological and agronomic characterization of Spanish landraces of *Phaseolus vulgaris* L. *Agriculture* **2019**, *9*, 149. [CrossRef]

25. Adewale, B.D.; Okonji, C.; Oyekanmi, A.A.; Akintobi, D.A.C.; Aremu, C.O. Genotypic variability and stability of some grain yield components of Cowpea. *Afr. J. Agric. Res.* **2010**, *5*, 874–880.

26. Harouna, D.V.; Venkataramana, P.B.; Matemu, A.O.; Ndakidemi, P.A. Assessment of Water Absorption Capacity and Cooking Time of Wild Under-Exploited *Vigna* Species towards their Domestication. *Agronomy* **2019**, *9*, 509. [CrossRef]

27. Harouna, D.V.; Venkataramana, P.B.; Matemu, A.O.; Ndakidemi, P.A. Wild *Vigna* Legumes: Farmers' Perceptions, Preferences, and Prospective Uses for Human Exploitation. *Agronomy* **2019**, *9*, 284. [CrossRef]

28. Henry, R. Domesticating *Vigna* Stipulacea: A Potential Legume Crop With Broad Resistance to Biotic Stresses. *Front. Plant Sci.* **2019**, *10*, 1–12.

29. Smýkal, P.; Nelson, M.N.; Berger, J.D.; Von Wettberg, E.J.B. The impact of genetic changes during crop domestication on healthy food development. *Agronomy* **2018**, *8*, 26. [CrossRef]

 *agronomy*

Article

Diversity and Domestication Status of Spider Plant (*Gynandropsis gynandra*, L.) amongst Sociolinguistic Groups of Northern Namibia

Barthlomew Chataika [1,*], Levi Akundabweni [1], Enoch G. Achigan-Dako [2], Julia Sibiya [3], Kingdom Kwapata [4] and Benisiu Thomas [1]

1   Crop Science Department, Faculty of Agriculture and Natural Resources, University of Namibia, Ogongo Campus, Private Bag 5520, Oshakati, Namibia; lakundabweni@unam.na (L.A.); bthomas@unam.na (B.T.)
2   Laboratory of Genetics, Horticulture and Seed Science, Faculty of Agronomic Sciences, University of Abomey-Calavi, Cotonou 999105, Benin; enoch.achigandako@uac.bj
3   School of Agricultural, Earth and Environmental Sciences, University of KwaZulu-Natal, P.Bag X01, Scottsville, Pietermaritzburg 3209, South Africa; sibiyaj@ukzn.ac.za
4   Department of Horticulture, Lilongwe University of Agriculture and Natural Resources, P.O. Box 219, Lilongwe, Malawi; kwapata@yahoo.com
*   Correspondence: barthchataika@gmail.com

Received: 28 October 2019; Accepted: 9 December 2019; Published: 1 January 2020

Abstract: Knowledge on the diversity and domestication levels of the spider plant (*Gynandropsis gynandra*) has the potential to affect pre-breeding for client-preferred traits, yet information is scarce in Namibia due to limited research. We investigated indigenous knowledge on the species diversity and domestication levels in the regions of Kavango West, Ohangwena, Omusati, Oshana, and Oshikoto of northern Namibia. Semi-structured interviews involving 100 randomly selected farming households, four key informant interviews, and a focus group discussion were conducted. Descriptive and chi-square tests were conducted using IBM SPSS version 20. Out of the possible four morphotypes, the results suggested that only one with green stem and green petiole existed and was associated with soils rich in organic manure. Spider plant abundance was reported to be on the decline, due to declining soil fertility. On a scale of 0 (wild species) to 6 (highest level of domestication), an index of 1.56 was found and this implied very low domestication levels. Furthermore, the study found significant differences in the trends of domestication across the sociolinguistic groups (χ^2 (12, N = 98) = 46.9, $p < 0.001$) and regions studied (χ^2 (12, N = 100) = 47.8, $p < 0.001$), suggesting cultural and geographical influences. In conclusion, the findings constituted an important precedent for guiding subsequent pre-breeding efforts.

Keywords: pre-breeding; morphotypes; domestication index; indigenous knowledge; sociolinguistic groups; client-preferred traits

---

1. Introduction

In many parts of sub-Saharan Africa (SSA), the spider plant (*G. gynandra* L. (Briq.) is an important indigenous vegetable that is neglected and underutilized but plays a crucial role in food and nutrition security and income generation of the rural poor [1–4]. It grows as a volunteer weedy crop in farmers' fields and in the wild during the rainy season [5–7]. Depending on the knowledge of the farmers, the vegetable is either removed as a weed or spared so that it can be harvested for use as a vegetable / relish or for sale in local markets.

No studies, however, have been reported on genetic diversity and domestication trends of the species, despite the emerging shift aimed at integrating indigenous and neglected vegetables in smallholder farming systems. The lack of attention means that the potential value of the spider plant remains underestimated and underexploited. This article reports on a study conducted in northern Namibia to assess the potential for domestication of the spider plant and promotion of its cultivation. The plant is well known in northern Namibia amongst different sociolinguistic groups. Farmers harvest the green leaves when they are abundant in the rainy season between November and March. The harvested leaves are either cooked as a relish, sold in the market as green or processed vegetable. Processing is done to preserve the vegetable for use in the dry season. Farmers preserve spider plants either by sun drying or bleaching followed by drying or forming of pellets which are used as a relish, sold on the open markets [6] and used during some of the traditional practices such as "Olufuko" amongst the Oshiwambo sociolinguistic groups [8]. (Source of information???). The cultural and economic importance of spider plant amongst sociolinguistic groups of northern Namibia calls for the need to promote its production which can be achieved by utilizing its genetic diversity.

Plant domestication and genetic improvement can be enhanced through the utilization of genetic diversity [9]. Diversity refers to the number of morphotypes or accessions that are found and used in a particular region [10] resulting from culture, traditional knowledge, the introduction of new species, domestication and crop improvement by farmers.

In addition, our interest in plant domestication was based on whether, in northern Namibia, there was a process of plant population evolution [11], leading to genetic change emanating from exploitation, selection, cultivation of the selected wild plants, and adaption to the agroecosystems and the human needs [12]. Such diversity and domestication, if found, would provide an opportunity to identify the 'elite' species with desirable utilization attributes such as nutritional and medicinal traits for further propagation to serve the rural residents.

These research interests were based on previous recommendations of this nature amongst ethnobotanists in Africa. For example, Dansi et al. [1] and Sogbohossou et al. [13] recommended an ethnobotanical investigation to evaluate, identify, document and prioritize interventions for reducing production constraints, improving agricultural practices and assessing the species contribution to household income.

2. Materials and Methods

2.1. Research Sites Sampling Design and Research Tools

The study was conducted in Kavango West, Ohangwena Omusati, Oshana, and Oshikoto regions of Northern Namibia between June and July, 2018. Namibia is located in south western part of Africa, surrounded by Angola to the north, Botswana to the east, South Africa to the south and Atlantic Ocean to the west (Figure 1).

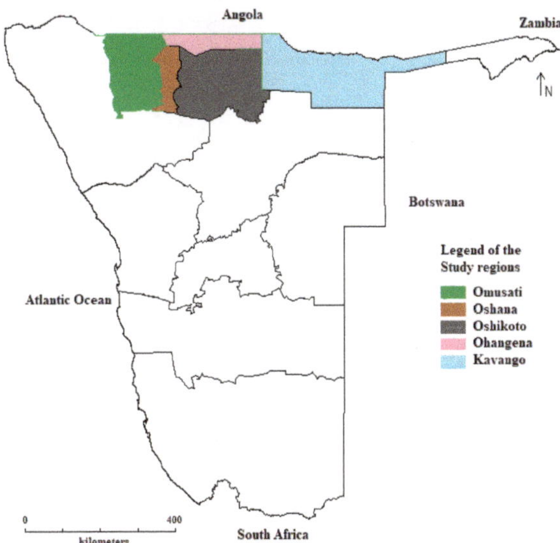

Figure 1. Map of Namibia showing the five study regions.

Previous studies have shown that the spider plant grows naturally and mostly in north central and north eastern Namibia [14]. The plant is known by several names as it is a familiar plant to several different sociolinguistic groups in that region as: Cat's whiskers or Spider flower (English), Ombidi (Oshikwanyama), Omboga (Oshindonga), Ombowa yozongombe or Ombowayozondu (Herero) and Gomabeb (Damara) [14].

The study used three tools to capture the diversity and domestication levels of spider plant in Northern Namibia. These tools were semi structured interviews targeting farming households; focus group discussion (FGD) which involved a farmer group and key informant interviews (KII) targeting agricultural extension staff in the Ministry of Agriculture.

Sampling was done at three stages. Firstly, purposeful selection of the four regions was done based on the outcome of the preliminary field survey and literature review. The second stage was random selection of constituencies in each of the regions. Constituencies are local administrative areas that form a region. One constituency was selected in each of the five study regions. Finally, a sample of 100 farming households was randomly drawn from the five regions (20 farming households per sampled constituency in each of the regions) for semi-structured interviews. Cochran's sample size formula ($n0 = \left(z^2 pq\right)/e^2$.) was used to determine the minimum sample size [15] based on 8% margin of error (e) and 95% confidence interval (z-value = 1.96). An estimate of the population which had knowledge of spider plant (p) was set at 80% based on the preliminary assessment. The choice of the margin of error took into consideration the resource constraints, while ensuring that it fell in between 10% and 5%. Using the formula, a sample size of 96 farming households was found and this was adjusted upwards to 100 households. In order to account for non-responses, the sample size was adjusted upwards by 10% per constituency and only the first 20 households were interviewed (Table 1). The interviews targeted the head of the household or the spouse.

Table 1. Study constituencies and number of sampled farming households per region.

Region	Constituency	Number of Households	* Percentage of Farming Households	Farming Households Sampled
Omusati	Anamulenge	2500	53	20
Oshana	Okatana	2600	12	20
Ohangwena	Engela	4900	36	20
Oshikoto	Omuntele	3300	33	20
Kavango West	Kapako	4200	31	20
Overall		17500	33	100

* Percentage of farming households is based on constituency estimates as per the national demographic survey of 2016.

In addition, one farmer group from Ohangwena region, and four agricultural extension staff were sampled for FGDs and as key informants respectively, to triangulate the findings. The farmers group was a formal cooperative formed by the Ministry of Agriculture, Water and Forestry through the Japan's International Cooperation Agency (JICA) project to promote crop production using conservation agriculture. The farming households were mobilized and sampled with assistance from the Agricultural extension staff based at the regional and constituency offices. The interviews for the sampled households were conducted at their homesteads after explaining the objectives of the study and getting their consent. Demographic characteristics and ethnicity of the sampled households were compiled.

Three key research questions were asked during the interviews with research participants. The first question was for the respondents to describe the types of spider plant found in their areas. The description included the colour, hairiness, local names and other agro-morphological features to determine the diversity of the species. The respondents were then asked to describe the agro-ecological niche where spider plant was found growing in abundance, including associated cropping systems and further to explain how they managed the spider plant when it grew either in the wild or in their fields. This was an open-ended question, and thus based on the explanation given, a score was assigned to indicate levels of domestication. We used a model of seven scores [16] as follows: Level 0: Wildlife species; Level 1: species just spared in the fields during field works; Level 2: Species spared in the fields but benefit from some care for its growth; Level 3: species transplanted from nature to the cultivated fields or home gardens; Level 4: Species well cultivated and reproduced; Level 5: Species cultivated with some selection activities; Level 6: Pests and diseases are known as well as their means of control. Finally, the respondents were asked to determine domestication trends of spider plant over the years and the underlying factors.

2.2. Data Analysis

Data from household interviews were subjected to descriptive and chi-square tests using IBM SPSS Statistics for Windows, Version 20.0. Armonk, NY: IBM Corp. Depending on the type of data, either means or frequencies were generated and presented in tables and graphs. A Shapiro-Wilk's test ($p > 0.05$) [17] and measures of skewness and kurtosis z-values were used to test normality of data on abundance of spider plant and associated cropping systems. In addition, tests for equality of variances were done using non-parametric Levene test ($p > 0.05$) [18]. Qualitative data from FGDs and key informant interviews were summarized and analyzed based on themes [19,20].

3. Results

3.1. Demographic Characteristics of the Interviewed Households

In this study, married respondents comprised 44%, while the single (never married or were divorced) constituted 49% and the remaining six percent were widows. The percentage of farming households within each constituency that were sampled, from the seven sociolinguistic groups, ranged from 7% for Chokwe to 22% for Kwanyama (Table 2). The average number of years in formal school ranged from one year for Chokwe sociolinguistic group to 8.7 years for Kwambi (Table 2) with an overall mean of 6.6 years.

Table 2. Proportion of interviewed farming households, their education levels and mean ages across the sociolinguistic groups and regions of northern Namibia.

Category		Proportion of Respondents (%)	Education Level (Years)	Mean Age (Years)
Sociolinguistic group	Chokwe	7	1.0	46.9
	Kwambi	20	8.7	50.4
	Kwangali	13	5.6	34.0
	Kwanyama	22	6.2	55.7
	Mbadja	9	8.4	44.8
	Mbalantu	11	6.2	55.7
	Ndonga	18	7.4	53.2
Regions	Kavango	20	4.0	38.5
	Ohangwena	21	6.0	56.4
	Omusati	20	7.2	50.8
	Oshana	20	8.7	51.4
	Oshikoto	19	7.7	52.5
	Overall	100	6.6	50.0

The mean age of the farming households was 50.0 years, with a standard error of 1.7. Kwangali had the youngest group of respondents (34.0 years) while Kwanyama and Mbalantu had the oldest, each with an average of 55.7 years (Table 2). In terms of regions, Kavango had relatively younger participants (38.5 years) while Ohangwena had on average older participants (56.4 years) in the survey (Table 2). The primary occupation of the respondents was farming (90%) followed by business (7%), scholars (2%) and finally formal employment (1%).

3.2. Diversity of Spider Plant

In literature, the spider plant is categorized into four morphotypes based on pigmentation, and these are green, purple, pink and violet [21]. These four morphotypes can either be globulous or have hairs with varying density. In this study, plants with green stems and green petioles (Figure 2a,b) were the only morphotypes of the spider plant reported to exist in the five regions. No respondent mentioned having seen or used purples spider plant morphotypes (Figure 2c,d).

(a)

(b)

Figure 2. *Cont.*

(c)   (d)

Figure 2. Some of the spider plant morphotypes, (**a**) green stem without trichomes, (**b**) green stem with trichomes, (**c**) purple stem without trichomes, and (**d**) purple stem with trichomes.

In terms of hairiness, 48% reported having seen and used globulous spider plants while 52% reported knowing spider plants with dense trichomes or hairs (Table 3). Most respondents from Omusati and Kavango regions reported having seen spider plant with dense trichomes while while the majority of the respondents from Oshana and Ohangwena reported that they saw and used globulous spider plant.

Table 3. Percentage of respondent reporting the hairiness of the green spider plant across the study regions.

Region	Globulous	Dense Hair
Omusati (n = 20)	15	85
Oshana (n = 20)	80	20
Ohangwena (n = 21)	71	29
Oshikoto (n = 19)	68	32
Kavango (n = 18)	6	94
Overall (n = 98)	48	52

The spider plant was locally called Ombidi (n = 42) or Omboga (n = 38) in Omusati, Oshana, Ohangwena and Oshikoto regions while in West Kavango (n = 20) it was known as Mpungu. Abundance of spider plants in the wild, in crop fields and homestead was reported to be associated with soils of high organic manure ($p < 0.001$) particularly cow dung (75%) (Figure 3). The study also found that the abundance of spider plant was not confined to a specific cropping system ($p < 0.001$) (60% of the respondents) (Figure 3). Normality tests for both abundance and associated cropping systems using skewness and kurtosis produced the z-values in between −1.96 and 1.96 suggesting normal distribution.

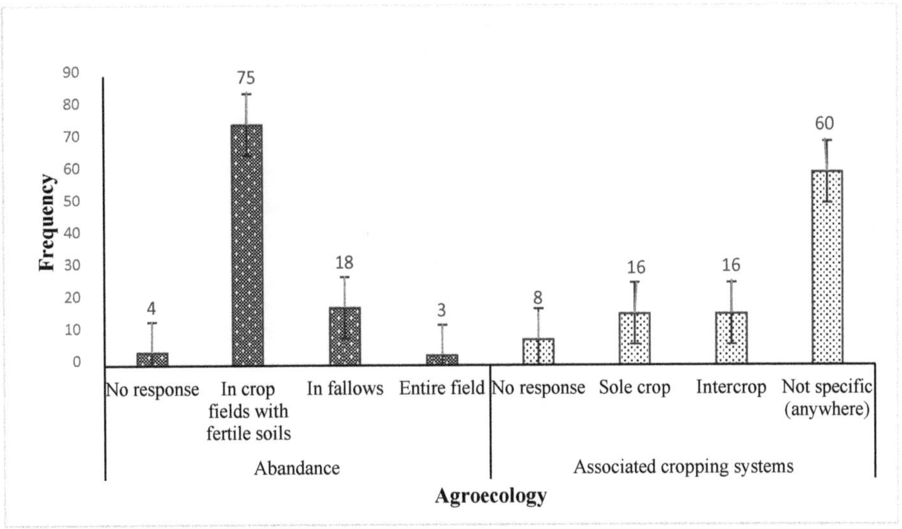

Figure 3. Frequency of respondents indicating abundance of spider plant and associated cropping systems of northern Namibia.

The respondents reported that the green form of spider plant was only found during the rainy season. Farmers harvest the green leaves when they are abundant in the rainy season and dry them either as pellets or bleached vegetable to preserve for the dry season.

3.3. Domestication Levels and Trends

The results showed that there was some domestication of spider plants taking place in the study areas. About 46% of the farmers did not remove the spider plant when working in the fields but allowed it to grow for harvesting, and the other 54% took some care to enable the spider plant to grow properly. The two scenarios corresponded to levels one and two of domestication, and the scores translate to a domestication level of 1.54 ± 0.501. This implied that the species was spared in the fields during field work and more often benefitted from some care for its growth. There were no reports of active domestication by farmers such as transplanting spider plants from the wild into fields and gardens, deliberate cultivation, including the selection of seeds from particular plants for reproduction, and use of means to control pests and diseases.

There was variation in the level of attention paid by farmers from different sociolinguistic groups. All farmers amongst the Chokwe and Kwangali groups provided some care to promote growth of the plant while the Mbalantu group just spared the crop during field work as illustrated in Figure 4 below.

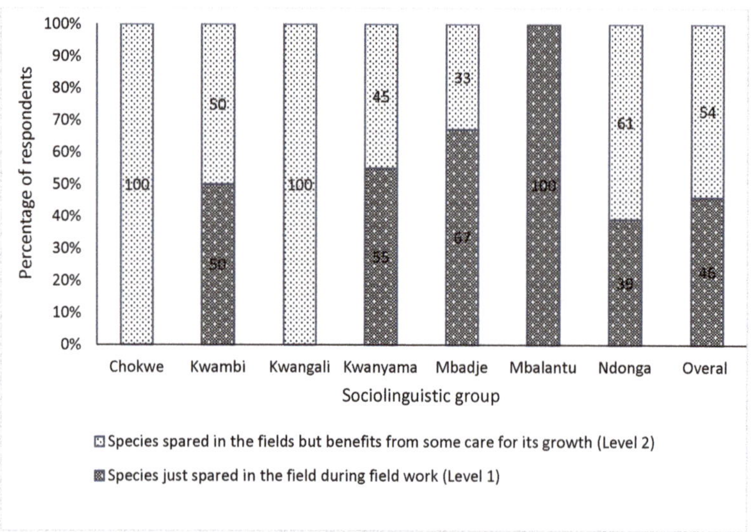

Figure 4. Levels of domestication of spider plant across different sociolinguistic groups of northern Namibia.

Trends of domestication significantly differed across the sociolinguistic groups (χ^2 (12, N = 98) = 46.9. $p < 0.001$) (Figure 5) and regions (χ^2 (12, N=100) = 47.8, $p < 0.001$) (Figure 6). In general, a decline trend in domestication was reported across the regions except in Kavango west where the respondents reported that the trend was going upwards ($p < 0.001$). Chokwe and Kwangali sociolinguistic groups reported increasing trends while the rest of the sociolinguistic groups, which were from Omusati, Oshana, Ohangwena and Oshikoto, reported a general downward trend. There was the correlation in domestication trends between the location of sociolinguistic group and the regions.

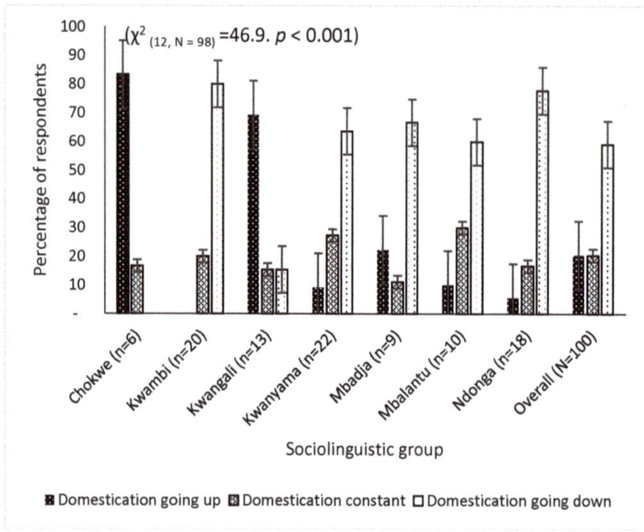

Figure 5. Trends of domestication of the spider plant across the different sociolinguistic groups of Northern Namibia.

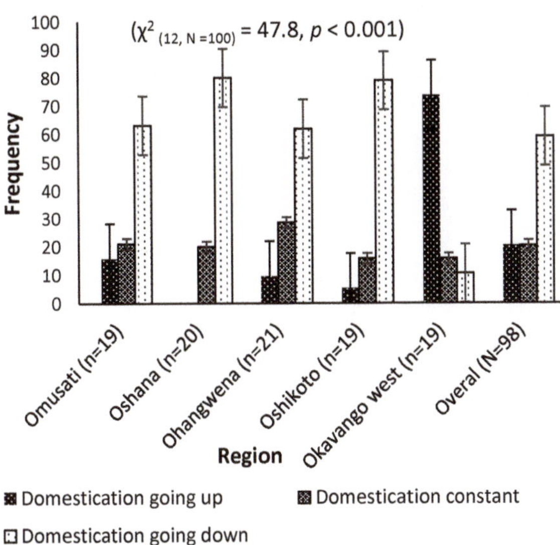

Figure 6. Trends of domestication of the spider plant across the regions of Northern Namibia.

The main reason for the downward trend was reported to be due to declining soil fertility, as the species was believed to be associated with fertile soils (Table 4). Poor rainfall, associated with drought conditions was the other factor leading to the declining trends in the domestication of spider plant.

Table 4. Reasons given by the respondents for the observed trends in spider plant domestication.

Trend of Domestication	Reasons for the Trend	Frequency
	Non Responses	**38**
Downward trends	Drought/poor rainfall	13
	Infertile soil	19
	Water lodging/high rainfall	1
Upward trends	Fertile soil	18
	Good rainfall	4
	Seeds are broadcast or left to dry in the field	4
	Adapt to many environments	3

4. Discussion

4.1. Diversity of Spider Plant in Northern Namibia

The study showed a lack of genetic diversity and ecological threats to the use of what is regarded as a useful crop. However, in view of the link between cow manure and spider plant growth alongside widespread cattle keeping across the regions, and adaptation to different cropping systems, there is an opportunity for agricultural extension investment and assistance to farmers to grow spider plant.

Genetic diversity has been reported to be the key to achieving the gains in production and productivity [9] through the enhancement of plant domestication and genetic improvement. Assessment of spider plant diversity in this study was based on stem and petiole pigmentation, and trichomes density. The study showed that only green stem, green petiole morphotypes, with and without trichomes existed in the study region.

The findings on pigmentation might imply limited opportunities for identifying superior accessions with the desired traits for use in genetic improvement. In addition, the results may also reflect the levels of knowledge of the species amongst the farmers, which might be limited by the fact that the vegetable was considered as a wild crop which did not undergo exploitation through domestication. Above all, pigmentation is just one of the several morphological characters which are used to distinguish genetically diverse accession. The findings of this study contrasted other studies which reported existence of four morphotypes in countries such as Kenya and Benin [21,22] and a mixture of green and purple colours in farmers' cultivars in Kenya [23]. In Namibia, it was reported that spider plant grew naturally particularly in Oshikoto, Oshana, Omusati, Ohangwena, Kavango East and West, Kunene and Zambezi regions but did not report the diversity of the morphotypes [14].

In contrast with green accessions, purple accessions are reported to be associated with higher nutrient density [23] which have health benefits, and also contain phytochemicals which confer resistance to insect pests [24]. The current diversity status in the study regions, therefore, might imply limited options in accessing the diversity of nutraceutical benefits from spider plant amongst the sociolinguistic groups. This limited diversity could also have a negative impact in designing and consequent implementation of breeding programs that seek to maximize genetic diversity with the aim of breeding ideal varieties, with desirable traits. Nonetheless, there is a possibility of the existence of the other morphotypes but this could only be confirmed by following up this study with field collection, identification and genetic characterization of the species across the five regions. Farmers' knowledge might only be limited to the green petiole, green stem morphotypes.

The second morphological character reported was the existence of spider plants with and without trichomes. Trichomes play important roles in pest resistance through either physical obstruction or production of phytochemicals which are toxic to herbivores [25]. The phytochemicals produced in the trichomes, as reported in plants such as *Plectranthus ornatus* [26] are also reported to impact on nutritional value of the plant species. The existence of spider plants with trichomes, therefore, offers an opportunity for identifying accessions with pest resistance potential and consequently enhance domestication.

Thirdly, the study identified a strong association of spider plant with organic matter, particularly cow manure. This was consistent with the findings of other researchers [14] who reported abundance of spider plants where manure or household refuse accumulated. This offers an opportunity in designing best-fit agronomic practices that would maximize the abundance and production as one of the ways of stimulating domestication of the species. Farmers in the study regions keep a lot of cattle, as such it should be easy to produce enough cattle manure which could be used for the production of spider plant. Furthermore, spider plant abundance was found to be widely spread across different cropping systems thus giving a wide choice for farmers to integrate spider plant production in different cropping systems. This also suggested that the spider plant did not compete negatively with associated crops in the different cropping arrangements. This finding, however needs to be followed by controlled agronomic studies to quantify the interactions of different cropping arrangements with spider plant and establish the optimum organic manure quantity and associated nutrient levels that would maximize production.

4.2. Domestication of Spider Plant

Domestication and scaling-up the adoption of indigenous vegetable species with high nutraceutical potential is one of the promising strategies that countries, that want to generate and sustain broad-based wealth, need to embrace. Understanding of domestication syndrome is considered as the starting point for the developing new and orphan crops [27]. The finding of this study suggested that spider plant was at early stages of domestication (level 2) and this was consistent with the finding of other researchers [28] who found that spider plant at level two of domestication in Gbede village in Benin. This implied that the vegetable remains neglected and thus calling for prompt action owing to its importance. One of the ways of accelerating domestication is through the use of best-fit agronomic practices that improve growth and leaf yield [29]. Generally, domestication is considered to reduce the genetic diversity hence creating a domestication bottleneck [30]. According to the theory of evolution low genetic diversity is believed to expose the species to the increased risk of extinction. This is because of the narrow genetic base which limits the capability of the species to survive changing agro-climatic conditions. Researchers have identified some of the genes underlying domestication and diversification [31] which can be exploited to adequately understand the domestication trends and diversity of spider plant at genetic level. In this study, both diversity and domestication levels were low probably because only farmers' responses were used and these could be limited to their level of knowledge.

Because of the perceived knowledge gap, it is proposed that follow-up studies should be done during the rainy season to collect and identify accessions growing both on farmers' fields and in the wild. The proposed study would embrace participatory domestication techniques [32] covering a range of species that have the potential of meeting diverse market requirements and domestic needs including nutritional and medicinal qualities. In addition, the proposed study would integrate innovative domestication with processing and commercialization as proposed by Leakey & Asaah [33] in order to make progress in the cultivation of the underutilized species. It was anticipated that the study would add more information on the available morphotypes and open up opportunities for further breeding activities.

4.3. Implications for Research and Development

This research has generated a wealth of knowledge for modelling future research aimed at promoting the domestication of spider plant. In order to develop breeding program for orphan crops such as spider plant, cultivar development [34] is key and this is dependent of diversity of the genetic materials, the potential of the material to adapt to wide range of environmental bottleneck and willingness of the farmers to domesticate. Therefore, the current diversity and domestication continuum of spider plant in Namibia provide opportunities for devising ideal genetic enhancement approaches for supporting de novo domestication. We found low diversity and downward trends of domestication, implying that the species is at the risk of genetic erosion. Since the trends in domestication were different across different sociolinguistic groups, identification of the underlying cultural factors need to be investigated. Furthermore, trends in domestication of spider plant will considerably depend on the benefits emanating from its use and also the mode of harvesting amongst other reasons. For example, uprooting the whole plant before flowering prevents seed dispersal hence leads to reduced diversity of the species. A participatory study on the appropriate methods of harvesting and utilization amongst different sociolinguistic groups is therefore recommended. We further recommend investigation on modes of conservation and production technologies to enhance domestication of the species. Use of participatory domestication methods to genetically improve orphan crops [35] promises to be a useful approach. At policy level, the roles indigenous vegetables such as spider plant, play in complimenting the major crops need to be recognized and mainstreamed in the planning framework to promote domestication. In Namibia, horticultural research only started in 1995 but did not include indigenous vegetables. The main focus was to test the suitability of varieties from South Africa in Namibian agro ecologies [36].

The proposed approaches will serve as models for designing and implementing research and development of ideal genotypes that respond to the needs of clients. Breeding efforts for orphaned vegetables are still at their infancy levels. The limited diversity of the species implies limited options for identifying candidate accessions for genetic improvement, as such researchers may need to widen the sources of breeding materials from other areas. Nevertheless, the positive association of spider plant with fertile soils offers both an opportunity and a challenge. Namibia is generally dry with poor soils which do not support optimum production of spider plant. However, Namibian farmers rear a lot of livestock which produce manure which can be used in the production of spider plant hence enhancing its domestication. Residues from spider plant can also potentially be used as feed for livestock. Finally creating awareness on the importance of domesticating spider plant through trainings and information campaigns will potentially enhance the use of spider plant and consequently its domestication.

5. Conclusions

In northern Namibia, diversity of the species was limited to green petiole and green stem morphotypes but this needs to be confirmed with a follow-up study which should aim at collecting, curating, identifying and characterizing the species and its wild relatives. The species are still in the early stages of domestication and the abundance was reported to be decreasing due declining soil fertility, hence calling for urgent action. In-depth understanding of cultural influence towards utilization of the species, including identification of preferred traits and production constraints constitute key pre-breeding stimulants aimed at popularizing the domestication of the Spider plant. In this study, we identified possible threats to the growth of the plant, which are key to informing further research to investigate options to improve domestication of the plant in northern Namibia. This constitutes a key step towards improving knowledge of the spider plant and its utility as a food crop.

Author Contributions: B.C. conceptualized the research design and the methodological approach, conducted the investigation, curated, and analyzed data, and wrote the original draft. L.A. and E.G.A.-D. reviewed the methodology and results, supervised the research work and edited the manuscript. J.S. and K.K. supervised, reviewed and edited the manuscript B.T. proofread, edited, and provided support with the ethical clearances and logistics. All authors have read and agreed to the published version of the manuscript.

Funding: This paper was part of the Ph.D. research funded by the Intra-Africa Academic Mobility Scheme of the European Union (EU).

Acknowledgments: We acknowledge the support from the EU for funding the research. The authors would also like to thank students and staff of the University of Namibia who played critical roles in logistics and actual data collection. Staff of the Ministry of Agriculture, Forestry and Water Development from Omusati, Oshana, Ohangwena, Oshikoto and Kavango West regions, and town council staff from Outapi, Oshakati and Ongwediva town councils are also acknowledged for providing logistical support during data collection.

Conflicts of Interest: The authors declare no conflict of interest.

References

1. Dansi, A.; Vodouhe, R.; Azokpota, P.; Yedomonhan, H.; Assogba, P.; Adjatin, A.; Loko, Y.; Dossou-Aminon, I.; Akpagana, K. Diversity of the neglected and underutilized crop species of importance in Benini. *Sci. World J.* **2012**, *2012*, 932947. [CrossRef]

2. Onyango, C.; Kunyanga, O.E.; Narla, R.; Kimenju, J. Current status on production and utilization of spider plant (Cleome gynandra L.) an underutilized leafy vegetable in Kenya. *Genet. Resour. Crop Evol.* **2013**, *60*, 2183–2189. [CrossRef]

3. van den Heever, E.; Venter, S. Nutritional and medicinal properties of Cleome gynandra. In *International Conference on Indigenous Vegetables and Legumes. Prospectus for Fighting Poverty, Hunger and Malnutrition, Hyderabad, India, 12–15 December 2006*; International Society for Horticultural Science: Leuven, Belgium, 2007.

4. Silue, D. *Spider Plant: An Indigenous Species with Many Uses*; AVRDC—World Vegetable Center: Arusha, Tanzania, 2009.

5. AOCC. *Progress report for African Crops Consortium*; World Agroforestry Centre: Nairobi, Kenya, 2017.

6. Steve, C.; Ben, M.; Percy, C. Indigenous Green Leafy Vegetables, Namibia's Natural Power Foods: How You Can Grow and Use Them. Available online: www.the-eis.com/data/literature/Indigenous%20Green%20Leafy%20Vegetables.pdf (accessed on 9 October 2019).

7. DAFF. *Cleome Production Guidelines*; Department of Agriculture, Forestry and Fisheries: Pretoria, South Africa, 2010.

8. Kuoppala, S.K. Reviving Tradition at the Olufuko Festival 2016: Girls' Initiation Ritual in Contemporary Namibia. Master's Thesis, University of Namibia, Windhoek, Namibia, 2018.

9. Cobb, J.; DeClerk, G.; Greenberg, A.; Clark, R.; McCouch, S. Next-generation phenotyping: Requirements and strategies for enhancing our understanding of genotype-phenotype relationships and its relevance to crop improvement. *Theor. Appl. Genet.* **2013**, *126*, 867–887. [CrossRef]

10. Maundu, P.; Achigan-Dako, E.; Morimoto, Y. Biodiversity of African Vegetables. In *African Indigenous Vegetables in Urban Agriculture*; Earthscan: London, UK, 2009; pp. 65–104.

11. Acqauuh, G. Plant domestication. In *Principles of Plant Genetics and Breeding*, 2nd ed.; John Wiley and Sons Ltd.: Chichester, UK, 2012; pp. 186–198.

12. Meyer, R.; DuVal, A.; Jensen, H. Patterns and processes in crop domestication: An historical review and quantitative analysis of 203 global food crops. *New Phytol.* **2012**, *196*, 29–48. [CrossRef]

13. Sogbohossou, E.; Achigan-Dako, E.; van Andel, T.; Schiranz, M. Drivers of Management of Spider plant (Gynandropsis gynandra) across different socio-linguistic groups of Benin and Togo. *Econ. Bot.* **2018**, *72*, 411–435. [CrossRef]

14. Kolberg, H. Indigenous Namibian leafy vegetables: A literature survey and project proposal. *Agricola* **2001**, *12*, 54–60.

15. Barlett, J.E.; Kotrilik, J.; Higgins, C. Organisational Research: Determining Appropriate Sample Size in Survey Research. *Inf. Technol. Learn. Perform. J.* **2001**, *1*, 43–50.

16. Vodouhe, R.; Dansi, A. The Bringing into Cultivation Phase of the plant domestication to In Situ Conservation of Genetic Resources in Benin. *Sci. World J.* **2012**, *2012*, 176939. [CrossRef]

17. Shapiro, S.; Wilk, M. An analysis of variance test for normality (complete samples). *Biometrica* **1965**, *52*, 591–611. [CrossRef]

18. Nordstokke, D.; Zumbo, B. Anew nonparametric Levene test for equal variances. *Psiocologica* **2010**, *31*, 401–430.

19. Maguire, M.; Delahunt, B. Doing a Thematic Analysis: A Practical, Step-by-Step Guide for Learning and Teaching Scholars. *AISHE-J.* **2017**, *3*, 3351–33514.

20. Braun, V.; Clarke, V. Using thematic analysis in psychology. *Qual. Res. Psychol.* **2006**, *3*, 77–101. [CrossRef]

21. Wasonga, D.; Ambuko, J.; Chemining'wa, G.; Odeny, D.; Crampton, B. Morphological characterization and selection of Spider plant (Cleome gynandra) accessions from Kenya and South Africa, Asian J. *Agric. Sci.* **2015**, *7*, 36–44.

22. K'Opondo, F.; van Rheenen, H.A.; Muasya, R. Assessment of Genetic Variation of Selected Spiderplant (Cleome gynandra L) Morphotypes from Western Kenya. *Afr. J. Biotechnol.* **2009**, *8*, 4325–4332.

23. Omondi, E.; Engels, C.; Nambafu, G.; Schreiner, M.; Neugart, S.; Abukutsa-Onyango, M.; Winkelmann, T. Nutritional compound analysis and morphological characterization of Spider plant (Cleome gynandra)- an African indigenous leafy vegetable. *Food Res. Int.* **2017**, *100*, 284–295. [CrossRef]

24. Chweya, J.A.; Mnzava, N.A. Cat's whiskers (Cleome gynandra L.). In *Promoting the Conservation and Use of Underutilized and Neglected Crops*; Heller, J., Engels, J., Mammer, K., Eds.; Plant Genetic and Crop Plant Research, Gatersleben/International Plant Genetic Resources Institute: Rome, Italy, 1997.

25. Morais-Filho, J.; Romeo, G. Plant glandular trichomes mediate protective mutualism in a spider-plant system. *Ecol. Entomol.* **2010**, *35*, 485–494. [CrossRef]

26. Ascensao, L.; Mota, L.; Castro, M. Grandular trichomes on the leaves and flowers of Plectranthus ornatus: Morphology, distribution and histochemistry. *Ann. Bot.* **1999**, *84*, 437–447. [CrossRef]

27. Dawson, K.; Powell, W.; Hendre, P.; Bancic, J.; Hickey, M.; Kindt, R.; Hoad, S.; Hale, I.; Jamnadass, R. The role of genetics in mainstreaming the production of new and orphan crops to diversify food systems and support human nutrition. *New Phytol.* **2019**, *224*, 37–54. [CrossRef]

28. Vodouhe, R.; Dansi, A.; Avohou, H.; Kpeki, B.; Azihou, F. Plant domestication and its contributions to in siu conservation of genetic resources in Benin. *Int. J. Biodivers. Conserv.* **2011**, *3*, 40–56.

29. Houdegbe, C.; Sogbohossou, E.; Achigan-Dako, E. Enhancing growth and yield in Gynandropsis gynandra (L.) Briq. (Cleomaceae) using agronomic practices to accelerate crop domestication. *Sci. Hortic.* **2018**, *233*, 90–98. [CrossRef]

30. Cornelius, J.P.; Clement, C.R.; Weber, J.C.; Montes, C.S.; Van Leeuwen, J.; Ugarte-Guerra, L.; Ricse-Tembladera, A.; Arevalo-Lopez, A. The trade-off between genetic gain and conservation in a participatory improvement programme: The case of Peach palm (Bactris gasipaes KUNTH). *For. Trees Livelihoods* **2006**, *16*, 17–34. [CrossRef]

31. Meyer, R.S.; Purugganan, M.D. Evolution of crop species: Genetics of domestication and diversification. *Nat. Rev. Genet.* **2013**, *14*, 840–852. [CrossRef] [PubMed]

32. Schreckenberg, K.; Awono, A.; Degrande, A.; Mbosso, C.; Ndoye, O.; Tchoundjeu, Z. Domesticating indigenous fruit trees as a contribution to poverty reduction. *For. Trees Livelihoods* **2006**, *16*, 35–51. [CrossRef]

33. Leakey, R.R.B.; Asaah, E.K. *Underutilised Species as the Backbone of Multifunctional Agriculture—The Next Wave of Crop Domestication*; Acta horticulture; International Society for Horticultural Science: Leuven, Belgium, 2013.

34. Sogbohossou, E.O.D.; Achigan-Dako, E.; Maundu, P. A roadmap for breeding orphan leafy vegetable species: A case study of Gynandropsis gynandra (Cleomaceae). *Hortic. Res.* **2018**, *5*, 1–15. [CrossRef]

35. Jamnadass, R.H.; Dawson, I.K.; Franzel, S.; Leakey, R.R.B.; Mithofer, D.; Akinnifesi, F.K.; Tchoundjeu, Z. Improving livelihoods and nutrition in sub-Saharan Africa through the promotion of indigenous and exotic fruit production in smallholders' agroforestry systems: A review. *Int. For. Rev.* **2011**, *13*, 338–354. [CrossRef]
36. NR International. *Horticultural Production and Marketing in the Kavango Region*; NR International: Windhoek, Namibia, 2001.

 *agronomy*

Article

Neglected and Underutilized Fruit Species in Sri Lanka: Prioritisation and Understanding the Potential Distribution under Climate Change

Sujith S. Ratnayake [1,2,*], Lalit Kumar [2] and Champika S. Kariyawasam [2]

1   Climate Change Secretariat, Ministry of Mahaweli Development and Environment, Colombo 10120, Sri Lanka
2   Ecosystem Management, School of Environmental and Rural Science, University of New England, Armidale, NSW 2351, Australia; lkumar@une.edu.au (L.K.); ckariyaw@myune.edu.au (C.S.K.)
*   Correspondence: gefsecsrilanka@gmail.com; Tel.: +61-266-773-5239

Received: 27 November 2019; Accepted: 19 December 2019; Published: 25 December 2019

Abstract: Neglected and underutilized fruit species (NUFS) can make an important contribution to the economy, food security and nutrition requirement for Sri Lanka. Identifying suitable areas for cultivation of NUFS is of paramount importance to deal with impending climate change issues. Nevertheless, limited studies have been carried out to assess the impact of climate change on the potential distribution of NUFS. Therefore, we examined the potential range changes of NUFS in a tropical climate using a case study from Sri Lanka. We prioritized and modeled the potentially suitable areas for four NUFS, namely *Aegle marmelos*, *Annona muricata*, *Limonia acidissima* and *Tamarindus indica* under current and projected climates (RCP 4.5 and RCP 8.5) for 2050 and 2070 using the maximum entropy (Maxent) species distribution modeling (SDM) approach. Potentially suitable areas for NUFS are predicted to decrease in the future under both scenarios. Out of the four NUFS, *T. indica* appears to be at the highest risk due to reduction in potential areas that are suitable for its growth under both emissions scenarios. The predicted suitable area reductions of this species for 2050 and 2070 are estimated as >75% compared to the current climate. A region of potentially higher climatic suitability was found around mid-county for multiple NUFS, which is also predicted to decrease under projected climate change. Further, the study identified high-potential agro-ecological regions (AERs) located in the mid-country's wet and intermediate zones as the most suitable areas for promoting the cultivation of NUFS. The findings show the potential for incorporating predictive modeling into the management of NUFS under projected climate change. This study highlights the requirements of climate change adaptation strategies and focused research that can increase the resilience of NUFS to future changes in climate.

Keywords: climate change scenarios; climate suitability; fruit selection index; Maxent; species distribution modeling

---

1. Introduction

Sri Lanka is predominantly an agrarian-based country with world-renowned unique agricultural diversity [1]. Though the climate is generally tropical, differences can be observed across the country due to changes in rainfall and elevation [2]. Sri Lanka is divided into three distinct climatic zones primarily based on the annual rainfall: a wet zone (2500–5000 mm), a dry zone (1250–1900 mm) and an intermediate zone (1900–2500 mm) [3]. The island has been defined into 46 unique agro-ecological regions (AERs) (Figure S1) based on features such as rainfall, elevation, soil type, landform, land use and relief [4]. Sri Lanka is enriched with high biodiversity despite its small size of 65,610 km^2 [2]. The variety of fruit species that are cultivated in different AERs is an important component of this biodiversity. This is evident by the over 237 recorded fruit species representing 56 plant families [5]. There are more than 60 varieties of underutilized fruit crops in Sri Lanka that are grown particularly in marginal environments [6]. These fruits are generally known as neglected and underutilized fruit species (NUFS) with under-exploited potentials for contributing to improving livelihoods (i.e., food security, income generation and health) and ecosystem stability [7,8]. Underutilized is generally used to refer to species whose potential has not been fully realized and exploited. In the Sri Lankan context, NUFS are still significant for local food and nutrition security and traditional medicine. Some species are widely distributed globally but may occupy locally confined niches [9]. Considering the current demand for NUFS due to health benefits and socio-cultural use, future research should be focused on the characterization and genetic conservation of NUFS [9,10]. These plants possess rich genetic diversity with unprecedented potential to improve the quality and resilience of future crops, and can be used by plant breeders for future crop improvements [11,12]. NUFS have received little consideration or have been completely ignored by agricultural researchers, plant breeders, policymakers, extension workers and farmers, although some research work has been carried out for collection characterization and evaluation of these species [13,14]. Increasing global population and varying dietary requirements are likely to put pressure on food and agriculture in the future [15]. Thus, more diversified agricultural and food systems are needed to cater these growing demands [13]. Recent studies show that NUFS have very good potential to address the food, nutrition and income security of rural peoples living in drought-prone areas [16]. Thus, promoting these NUFS to make them more "consumer-friendly" and "commercial" can be considered a powerful means of achieving sustainable development goal 2 (SDG 2) (i.e., reduction of poverty and malnutrition). They have the potential to enhance the diversity of food systems and, further, to make agricultural production systems less vulnerable to climate change, as these plants are hardy and resilient to such changes [13,17–19]. They can thrive in harsh climatic conditions and nutrient-poor degraded habitats [20]. As such, these species have been increasingly recognized for the future adaptation of food production to climate change [8].

The selected four NUFS—*Limonia acidissima* L., *Aegle marmelos* (L.) Correa, *Annona muricata* L. and *Tamarindus indica* L.—are terrestrial trees that are widely distributed in tropical regions including the Indian subcontinent [21–24]. All four species are mainly propagated by seeds [14,25]. They have tremendous potential for medicinal purposes and as food resources. Generally, all parts of these plants (i.e., leaf, bark, fruit, seed and root) have ethnomedicinal importance and are widely used in traditional medicine to treat various illnesses [21,24,26,27]. For instance, *L. acidissima* contains various pharmacological properties such as hepatoprotective, anti-diabetic, anti-tumor, anti-ulcerative, wound-healing and anti-microbial activities, and is used to cure asthma, tumors, cardiac debility, hepatitis and wounds [24,28,29]. *A. marmelos* possesses antiviral, antifertility, radioprotective, anticancer, chemopreventive, antidiarrheal, ulcer healing, diuretic, antigenotoxic, antimicrobial, radioprotective, antipyretic and anti-inflammatory properties and thus it is a remedy for a range of diseases [30]. *A. muricata* is used for curing hypertension, headaches, fever, rheumatism, diabetes, insomnia, parasite infection, diarrhea, dysentery and many more conditions [26,31]. Extract of *A. muricata* leaves is used as an alternative therapy for cancer as it causes apoptosis of the liver cancer cells [32]. Likewise, *T. indica* is used for abdominal pain, wound healing, constipation, inflammation, respiratory problems, diarrhea and dysentery [25,27,33].

Temperature and rainfall in the South Asian region are projected to increase significantly by 2100 under climate change [34]. Multi-model ensemble prediction carried out by the Department of Meteorology Sri Lanka indicates that the negative rainfall anomaly, especially in the dry zone of Sri Lanka, can be expected in both medium emissions (RCP 4.5) and high emissions (RCP 8.5) scenarios for 2050 and 2070. Further, multi-model ensemble prediction indicates an increasing trend of the minimum and maximum temperatures for 2050 and 2070 [35]. As Sri Lanka is an island highly vulnerable to extreme climatic events, the potential consequences of climate change that can impact the distribution of species could be significant [36]. Potential reduction of rainfall and increase of temperature in the dry zone may increase vulnerability of low country dry zone ecosystems. Such adverse changes can reduce agricultural productivity and leave farming households struggling for their livelihoods. Further, climate changes in the future may potentially impact pollinator species, which could have an additional constraint on potential area suitability of NUFS, as wild plants are greatly dependent on pollinators for fruit production [37].

In Sri Lanka, priority setting has not been carried out systematically to improve conservation and sustainable use of NUFS [3]. More specifically, the environmental factors that influence or limit distribution of these species have not been investigated. Understanding the current and potential distribution of NUFS is essential to meeting the future challenges of global food security [10]. However, there has been no comprehensive study undertaken with the purview of examining the potential distribution of these NUFS under climate change. We modeled the distribution of four NUFS in Sri Lanka under current and future (RCP 4.5 and RCP 8.5) climates for 2050 and 2070. The objectives of this study were to (i) prioritize NUFS in Sri Lanka to be considered into national food and nutrition security programs; (ii) model potential distribution of four priority NUFS and calculate range changes under projected climate change and (iii) define areas that potentially support multiple NUFS in the current climate and in scenarios of global warming. Our study is significant as it is the first study of this kind in Sri Lanka to use species distribution modeling (SDM) concepts to identify the potential distribution of NUFS under projected climate change. This information can be used for identifying future climate-suitable areas for promoting the cultivation of NUFS.

2. Materials and Methods

2.1. Priority Neglected and Underutilized Fruit Species (NUFS) Selection

Williams and Haq [38] developed a set of criteria consisting of statements for selecting priority crops from a particular geographical region. In an expert consultation meeting held on 20 December 2017 at the Plant Genetic Resources Centre, 40 experts from relevant agencies (Government Departments and research stations, academia, non-governmental organizations and leaders of community-based organizations) discussed and reviewed the above criteria. These criteria were modified and developed from 26 statements as per the study purpose and country context (Table 1). To calculate the Fruit Selection Index (FSI), we used the following formula adapted from Jayasinghe-Mudalige and Henson [39].

$$\text{FSI} = \sum_{a=1}^{s} a_{is} \cdot \text{U}s/\text{aU} \tag{1}$$

U is a set of statements (U = U_1, U_2, ..., U_s) that explains different factors to be considered for prioritizing NUFS in this study. The group of experts (*i*) provided an integer score (a) for each statement (e.g., a_{i1}, a_{i2}, ... , a_{is} on U_1, U_2, ... , U_s, respectively). Each statement was marked as "YES" or "NO" for each NUFS by considering the availability of factors in each given statement. If the statement was marked as "YES", a score was given based on degree of importance: high (H) = 3, moderate (M) = 2 and low (L) = 1. The maximum score that could be obtained by any statement for any NUFS was 3. This score was multiplied by the number of statements given in a category to get the maximum potential score (aU). Sum of the scores provided by statements in a category was divided by the maximum potential score of that category to obtain the index value for this category. The total of seven index values was divided by the number of categories (seven) to get the FSI (Table S1). The prioritized four NUFS were considered for further analysis through SDM.

Table 1. Twenty-six statement criteria used to prioritize neglected and underutilized fruit species (NUFS) in Sri Lanka.

Category No.	Category Name	Statement No.	Statement
1	Research and policy framework	U_1	Importance of national food production and food security programs
		U_2	Importance of national and regional agriculture research system
2	Germplasm and agro-ecology	U_3	Availability of germplasm
		U_4	Current genetic conservation status
		U_5	Potential demand for germplasm
		U_6	Adaptation to local climate and soil
3	Acceptability	U_7	Local preferences/consumption
		U_8	Rural income generation
4	Uses	U_9	Nutritional value and health benefits
		U_{10}	Cultural acceptance and consumer preferences
		U_{11}	Potential diversification for products
		U_{12}	Multiple uses (wood value, medicinal value, etc.)
5	Production and practices	U_{13}	Wide adaptability
		U_{14}	Cropping systems suitability
		U_{15}	Satisfies need for crop diversification
		U_{16}	Pest/disease situation
		U_{17}	Production and technology
		U_{18}	Seasonality
		U_{19}	Availability of planting material
		U_{20}	Local Knowledge
6	Post-harvest	U_{21}	Possibility of storage
		U_{22}	Processing technology
		U_{23}	Products in relation to markets
7	Market and value chain	U_{24}	Access to market
		U_{25}	Potential value addition
		U_{26}	Potential export processing

2.2. Species Occurrence Data

For this study, 781 occurrence records of four prioritized NUFS were extracted from a database developed as a result of field exploration and stakeholder consultation surveys conducted in four major donor-funded projects implemented by the Department of Agriculture and the Ministry of Environment, Sri Lanka during 2003–2018 (see Table S2 for details). During field and stakeholder surveys of these projects, geo-referenced NUFS occurrence data had been recorded from various sources, research and development institutes of the Department of Agriculture, national herbarium, experts and community consultations. Figure 1 shows the distribution of occurrences of four NUFS that were used for the study.

Spatial sampling bias is an issue in predictive modeling that causes spatial autocorrelation, resulting in poor model performance [40]. In order to reduce the effects of spatial sampling bias, occurrences were filtered in ArcMap using one of the predictor variables, enabling each grid cell to have only one occurrence record in the geographic areas where the study species are distributed (Table S3). To address the effect of sampling bias in species and background data, we used bias files by constraining the background data to have the same bias as the occurrence data [41]. In this case, we limited background point selections to the districts where NUFS occurrence data were distributed.

This provided Maxent a background file with a similar bias to the sample data in order to fine tune the Maxent models [42].

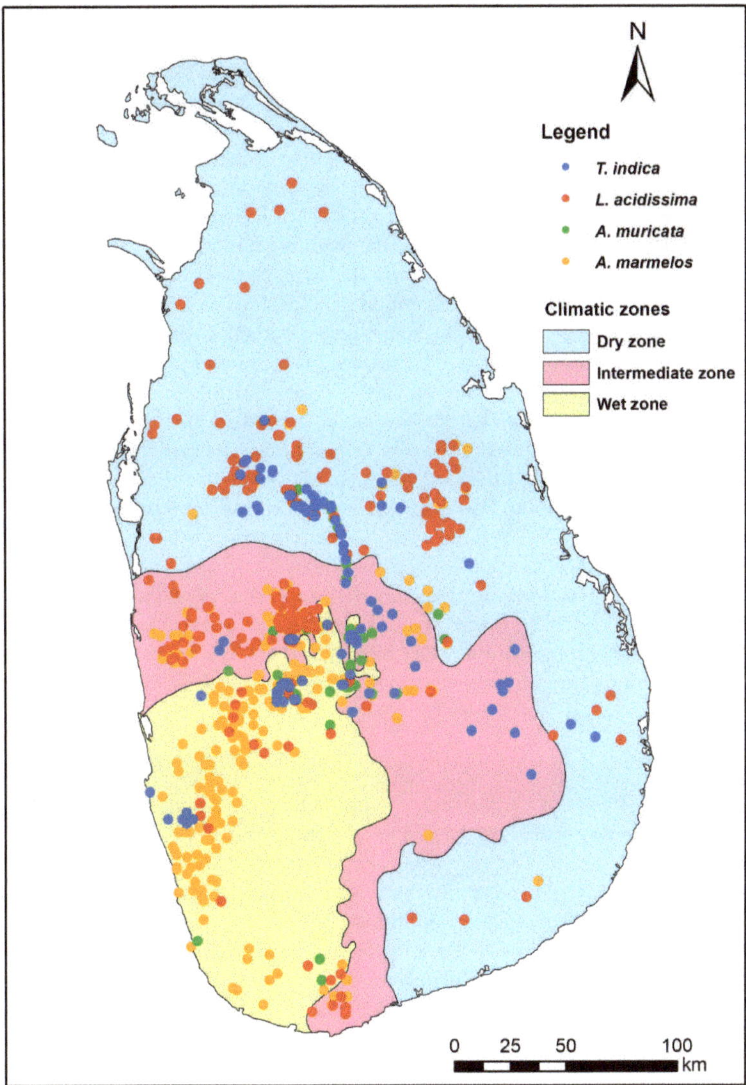

Figure 1. Distribution of occurrence records for selected priority NUFS (*Limonia acidissima*, *Aegle marmelos*, *Annona muricata* and *Tamarindus indica*) in different climatic zones of Sri Lanka.

2.3. Environmental Variables

We downloaded 19 bioclimatic variables with a spatial resolution of 30 arc-seconds (~1 km^2) from the Worldclim website for current (representative of 1960–1990) and future climates (http://www.worldclim.org/) [43]. We used the MIROC5 atmosphere-ocean general circulation model (GCM) to make future projections under medium emissions (RCP 4.5) and high emissions (RCP 8.5) scenarios for 2050 and 2070. MIROC5 has been tested by previous empirical studies for its efficiency of climate change simulations, particularly for South Asia [44–47], and, furthermore, broadly used in recent SDM studies in that region [48–51]. Multicollinearity testing was conducted to remove highly correlated variables. We used the "removeCollinearity" function of the Package "virtualspecies" (version 1.4-4) in R to remove correlated variables [52]. This method of variable selection has recently been used in SDM literature [48]. The analysis resulted in seven groups of intercorrelated variables at Pearson's correlation coefficients (r) $\geq$ 0.7 (Figure S2). While selecting variables from groups of intercorrelated variables, we were careful to pick ecologically meaningful predictors that reasonably respond to climate change and species distribution. The selected bioclimatic variables contained tolerance limits of temperature and precipitation that impose constraints on their distribution. Thus, seven nonredundant bioclimatic variables with relevance to NUFS distribution were chosen for their potential significance based on our knowledge and published literature [53–55]. The variables selected included mean diurnal range (BIO2), maximum temperature of the warmest month (BIO5), minimum temperature of the coldest month (BIO6), annual precipitation (BIO12), precipitation of driest month (BIO14), precipitation seasonality (BIO15) and precipitation of coldest quarter (BIO19). The importance of the contribution of ecologically meaningful non-climatic variables has been frequently discussed in predictive modeling literature [56,57]. We enriched the chosen bioclimatic data with selected ecophysiologically meaningful non-climatic variables: elevation (dem), aspect (direction of slope), soil and land cover. Ecophysiologically significant non-climatic variables increase the predictive power of species distribution models of plants; thus, failure to include these potentially important variables in models can result in inaccurate predictions restraining predictive capacity [58]. These variables are key determinants of plant distribution, and the importance of soil variables has been particularly identified for distribution of fruit species [59]. We downloaded global land cover data at a 300-m spatial resolution from the European Space Agency GlobCover Portal (http://due.esrin.esa.int/page_globcover.php). We received elevation, aspect and soil data used by Kadupitiya, et al. [60]. We resampled those high-resolution raster layers and changed the resolution same as with bioclim layers (~1 km^2) by providing the same parameters using ArcGIS toolbox. Therefore, we had 11 identical variables altogether of the same resolution, projection and extent for modeling the four priority NUFS in Sri Lanka.

2.4. Maxent Modeling

Maximum Entropy Species Distribution Modeling (Maxent) software version 3.4.1 was employed for the study [61]. Maxent is based on the maximum entropy principle—in other words, the most spread out or closest to uniform distributions [62]. We selected the Maxent modeling technique for several reasons: (i) Maxent is relatively more robust and frequently outperforms many other well-established presence-only modeling algorithms [63–67]; (ii) Maxent is less influenced by spatial errors in sample data and performs well with geographically biased occurrence data [68,69]; and (iii) Maxent generates a continuous model output of relative probability of presence, allowing fine distinctions in binary predictions [62–64]. Further, Maxent is widely used in SDM literature for various applications, particularly to identify potential areas of suitability [66]. We selected "do jackknife to measure variable importance", "create response curves", "make pictures of predictions" and "write background prediction". Logistic output was selected for easy interpretation of the outcome [70,71]. In the logistic output, the relative probability of presence is illustrated by a linear scale of values from 0 to 1, where 0.5 represents typical presence localities (see Phillips [72] for details). We selected cross-validation with 10 replicates, where occurrences are divided into 10 equal-size parts and all occurrences are used for model validation to efficiently generate a more accurate result [67,72]. Auto features were used as recommended (>80 training samples) except in the *A. muricata* model, where we used linear, quadratic and hinge features due to limited training samples [73]. We increased the number of iterations to 1000, enabling the algorithm to get close to convergence for smooth prediction [62] while the other parameters were kept at default. Default values are set in Maxent based on performance across a wide range of taxa to receive an optimum model output and hence recommended to use when several models are run simultaneously [71]. Models were run individually for four prioritized NUFS with the selected subsets of variables under current and future climate changes.

2.5. Model Performance

In SDM literature, model robustness is frequently verified using discrimination metrics such as the threshold independent area under the receiver operating characteristic curve (AUC) criterion and the threshold dependent true skill statistic (TSS) [62,74]. Therefore, the accuracy of prediction of selected NUFS was primarily assessed by the AUC and TSS. However, both these measures have criticisms due to their dependency on prevalence [52,74]. Therefore, relative satisfactory performance in both measures was considered while selecting models. The AUC ranged from 0 to 1 while the TSS ranged from −1 to +1. Upper levels of both measures signify the perfect prediction of presences and absences [41,74]. In general, an AUC value of <0.7 indicates poor performance; a value of 0.7–0.9 indicates moderate performance; and a value > 0.9 indicates high performance [75,76]. Similarly, TSS values that are <0.4 are considered poor, 0.4–0.8 moderate and >0.8 very good [77]. Additionally, we tested the relative performance of models using two alternative metrics: sensitivity and specificity.

2.6. Potential Area of Prediction

The Maxent generated model output, which is an average output of the replicated runs [72], was imported into ArcMap (ArcGIS version 10.4.1) for suitability analysis. Maximizing the sum of the sensitivity and specificity logistic threshold was used to discriminate presences and absences, as this approach is suggested for models that use presence-only and background data (i.e., Maxent) [78]. The suitable area distribution was visualized for each NUFS under current climates and under two emissions scenarios for 2050 and 2070. The "reclassify" tool of ArcGIS was used to visualize binary presence–absence maps showing areas above the threshold as suitable and areas below the threshold as not suitable for potential spread of species. In each case, the potentially suitable area of spread was calculated in km^2 using the ArcMap tool field calculator.

A combined raster was generated by merging four classified layers using the spatial analyst toolbox of ArcMap. We received five climatic suitability classes as very low (0 NUFS), low (1 NUFS), moderate (2 NUFS), high (3 NUFS) and very high (4 NUFS). The very high class represented the highest number of NUFS that overlapped in a given place, whereas the very low category represented the areas not suitable for any of the given NUFS. The suitability maps were developed for future climatic scenarios as well. The classes used for classifying the combined raster were uniform among all maps. Thus, five combined maps of climatic suitability were developed under current and future climatic scenarios.

3. Results

The value of the FSI ranged from 0–1, indicating that the NUFS with the highest value was the best fruit to be selected. Table 2 shows 30 NUFS grown in Sri Lanka based on the scores generated by the FSI. Accordingly, *L. acidissima* L. (Rutaceae) ranked the highest followed by *A. marmelos* L. (Rutaceae), *A. muricata* L. (Annonaceae), *Phyllanthus emblica* L. (Phyllanthaceae) and *T. indica* L. (Fabaceae). Evaluation of the performance of NUFS models was found to be acceptable for *L. acidissima*, *A. marmelos*, *A. muricata*, and *T. indica* to analyze the impact of climate change on potential distribution (Table S2). The performance of the *P. emblica* model was found to be not satisfactory and thus eliminated.

Table 2. Priority setting of 30 NUFS of Sri Lanka based on the results of the fruit selection index (FSI).

Rank	Scientific Name	Family	Common Name	Local Name	FSI
1	*Limonia acidissima*	Rutaceae	Wood apple	Divul	0.696
2	*Aegle marmelos*	Rutaceae	Bael, bhel	Beli	0.674
3	*Annona muricata*	Annonaceae	Soursop	Katuannoda	0.648
4	*Phyllanthus emblica*	Phyllanthaceae	Indian gooseberry	Nelli	0.629
5	*Tamarindus indica*	Fabaceae	Tamarind	Siyambala	0.595
6	*Citrus reticulata*	Rutaceae	Mandarine	Dodam	0.564
7	*Psidium guajava*	Myrtaceae	Guava	Pera	0.535
8	*Syzygium aqueum*	Myrtaceae	Water apple	Jambu	0.438
9	*Garcinia quaesita*	Clusiaceae	Brindle berry	Goraka	0.403
10	*Dialium ovoideum*	Fabaceae	Velvet tamarind	Gal siyambala	0.401
11	*Mangifera indica.*	Anacardiaceae	Mango	Mee amba	0.399
12	*Citrus grandis*	Rutaceae	Pumello	Jambola	0.399
13	*Psidium catlleianum*	Myrtaceae	Cherry guava	Cherry pera	0.351
14	*Flacourtia inermis*	Salicaceae	Lovi/batoko plum	Lovi	0.351
15	*Pouteria campechiana*	Sapotaceae	Canistel	Lavulu	0.341
16	*Elaeocarpus serratus*	Elaeocarpaceae	Ceylon olive	Veralu	0.337
17	*Lansium domesticum*	Meliaceae	Langsat	Gaduguda	0.337
18	*Flacourtia indica*	Salicaceae	Ramontchi	Uguressa	0.326
19	*Manilkara zapota*	Sapotaceae	Sapodilla	Sapodilla	0.309
20	*Ziziphus mauritiana*	Rhamnaceae	Chinese date	Masan	0.309
21	*Averrhoa carambola*	Oxalidaceae	Carambola	Kamaranga	0.253
22	*Psidium spp.*	Myrtaceae	Guava	Jam pera	0.233
23	*Syzygium cumini*	Myrtaceae	Black plum	Dan	0.222
24	*Cynometra cauliflora*	Fabaceae	Nam nam	Nam nam	0.201
25	*Carissa spinarum*	Apocynaceae	Conkerberry	Karamba	0.201
26	*Manilkara haxandra*	Sapotaceae	Ceylon iron wood	Palu	0.191
27	*Grewia tiliifolia*	Tiliaceae	Dhaman	Damuna	0.139
28	*Euphoria longana*	Malvaceae	Longan	Mora	0.128
29	*Schleichera oleosa*	Sapindaceae	Ceylon oak	Kon	0.128
30	*Drypetes sepiaria*	Putranjivaceae	–	Weera	0.128

The analysis of variable contributions table of the Maxent model provides estimates of relative contributions of the environmental variables. Figure 2 shows the relative percentage contributions of the three highest contributing environmental variables to the Maxent models for four NUFS. The figure presented is a result of current climate modeling. However, under future climates, the same variables a show similar performance. Therefore, these are the major environmental variables that influence climatic suitability for the fruit species under present and future climates. Among the variables, soil contributed the most to the *A. muricata* and *L. acidissima* models, whereas precipitation of driest month (BIO14) and minimum temperature of the coldest month (BIO6) contributed the highest to *A. marmelos* and *T. indica* models, respectively. Overall, soil appeared to be an influencing environmental variable to the distribution pattern of all four species. The relative contribution of aspect variables to the model prediction was negligible. Jackknife testing also confirmed soil as the leading variable in predicting the potential distribution of *A. muricata* and *L. acidissima*. Precipitation seasonality (BIO15) and mean diurnal temperature range (BIO2) were the most contributing variables in *A. marmelos* and *T. indica* models, respectively. The aspect contribution was negligible in all four models (Figure S3).

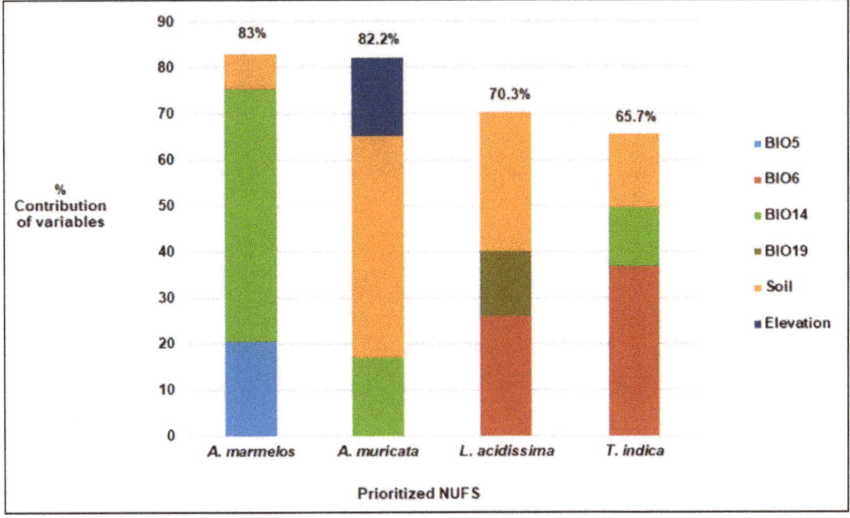

Figure 2. Relative importance (%) of the highest contributing environmental variables in the current climate (maximum temperature of the warmest month (BIO5), minimum temperature of the coldest month (BIO6), precipitation of the driest month (BIO14) and precipitation of the coldest quarter (BIO19)) to the Maxent models of each of the selected priority NUFS in Sri Lanka: *Limonia acidissima*, *Aegle marmelos*, *Annona muricata* and *Tamarindus indica*.

Figure 3 shows the projected potential distribution maps for the four NUFS species under current climate and climate change scenarios. We visualized the potential areas of suitability for each NUFS above the selected threshold of maximum training sensitivity plus the specificity logistic threshold (Figure S4). Accordingly, the wet and intermediate climatic zones of Sri Lanka were predicted as highly suitable for *A. marmelos* distribution. The potentially suitable area of *A. marmelos* under the current climate was 10,621 km^2 (Figure 4, Table S4) and the majority of this potentially suitable areas was located around Kandy, Kegalle, Colombo, Gampaha, Kalutara and Kurunegala districts (see Figure S5 for a district map). The potential area of the suitability of this NUFS is predicted to decrease in the future, particularly around the Kurunegala district. Under the current climate, the potentially suitable area for *A. muricata* was estimated to be 7579 km^2, involving mainly the central parts of Sri Lanka representing the Kandy, Kurunegala, Matale and Kegalle districts as well as parts of the Matara and Ratnapura districts. Under the medium-emissions scenario, this suitable area is predicted to contract by 2050 and increase again by 2070. Under the high-emissions scenario, the potentially suitable area of this NUFS is predicted to decrease (63%) continuously until 2070. The potentially suitable area around the Kandy district is predicted to be relatively stable under climate change. In the current climate, the *L. acidissima* model predicted a potentially suitable area of 13,135 km^2 mainly around the Anuradhapura, Kurunegala, Polonnaruwa and Vavunia districts. This potential area of suitability is predicted to increase by 2050 under the medium-emissions scenario; however, it is predicted to decrease again by 2070. Under the high-emissions scenario, the potentially suitable area of *L. acidissima* is predicted to contract prominently (72%) by 2070, limiting this NUFS only to the Kurunegala district and a few adjacent areas. *T. indica* had the lowest potentially suitable area of 5173 km^2 under the current climate, including predominantly Anuradhapura, Matale, and Badulla districts and some scattered areas adjacent to these districts. According to the projected maps of suitability, climate change is predicted to influence the potential distribution of this species in the future, resulting in dramatic area contraction. Under the medium-emissions scenario, the predicted suitable area reductions of *T. indica* for 2050 and 2070 were estimated as 78% and 94% compared to the current climate, whereas under the high-emissions scenario these figures were 82% and 89%, respectively. By 2070, the potential suitability of this NUFS is predicted to be mostly restricted to around the Kandy district and surrounding areas under both emissions scenarios. Overall, all four NUFS are predicted to decrease in the potential area of suitability under medium- and high-emissions scenarios in the future.

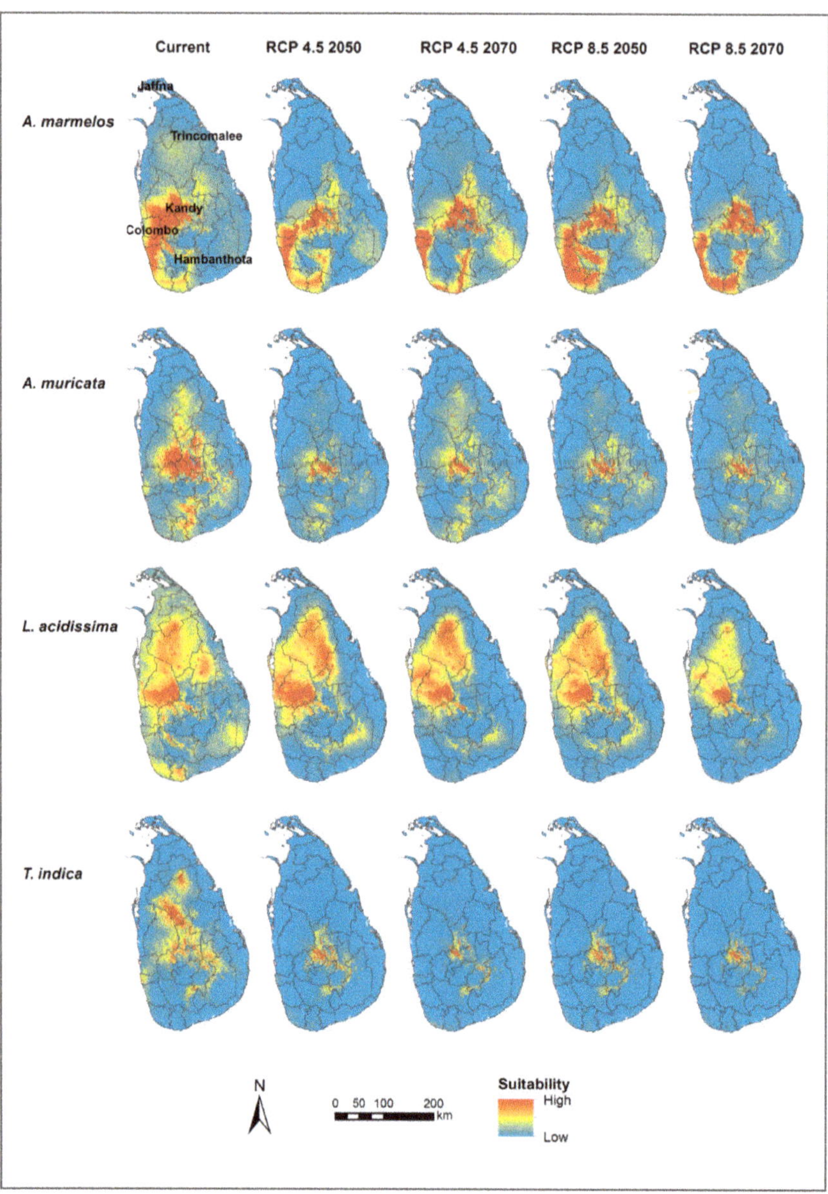

Figure 3. Projected potential distribution maps for the selected priority NUFS *Limonia acidissima*, *Aegle marmelos*, *Annona muricata* and *Tamarindus indica* in Sri Lanka under medium-emissions (RCP 4.5) and high-emissions (RCP 8.5) scenarios for 2050 and 2070. The map colors indicate 0 (blue) to 1 (red) probability of occurrence.

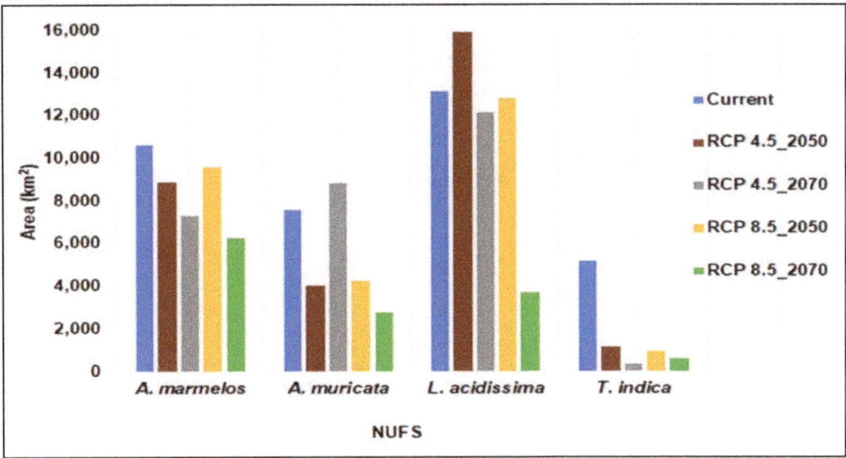

Figure 4. Projected suitable areas (km^2) of the selected priority NUFS *Limonia acidissima*, *Aegle marmelos*, *Annona muricata* and *Tamarindus indica* in Sri Lanka under the current climate, medium-emissions (RCP 4.5) and high-emissions (RCP 8.5) scenarios for 2050 and 2070.

The combined map of climatic suitability of four NUFS (Figure 5) under the current climate and future scenarios shows a region potentially suitable for the establishment of multiple NUFS. This area lies around the mid-county, particularly the area adjoining the Kurunegala, Kandy, Matale and Kegalle districts. This area is projected to reduce significantly and confine to around the Kandy district in the future under climate change.

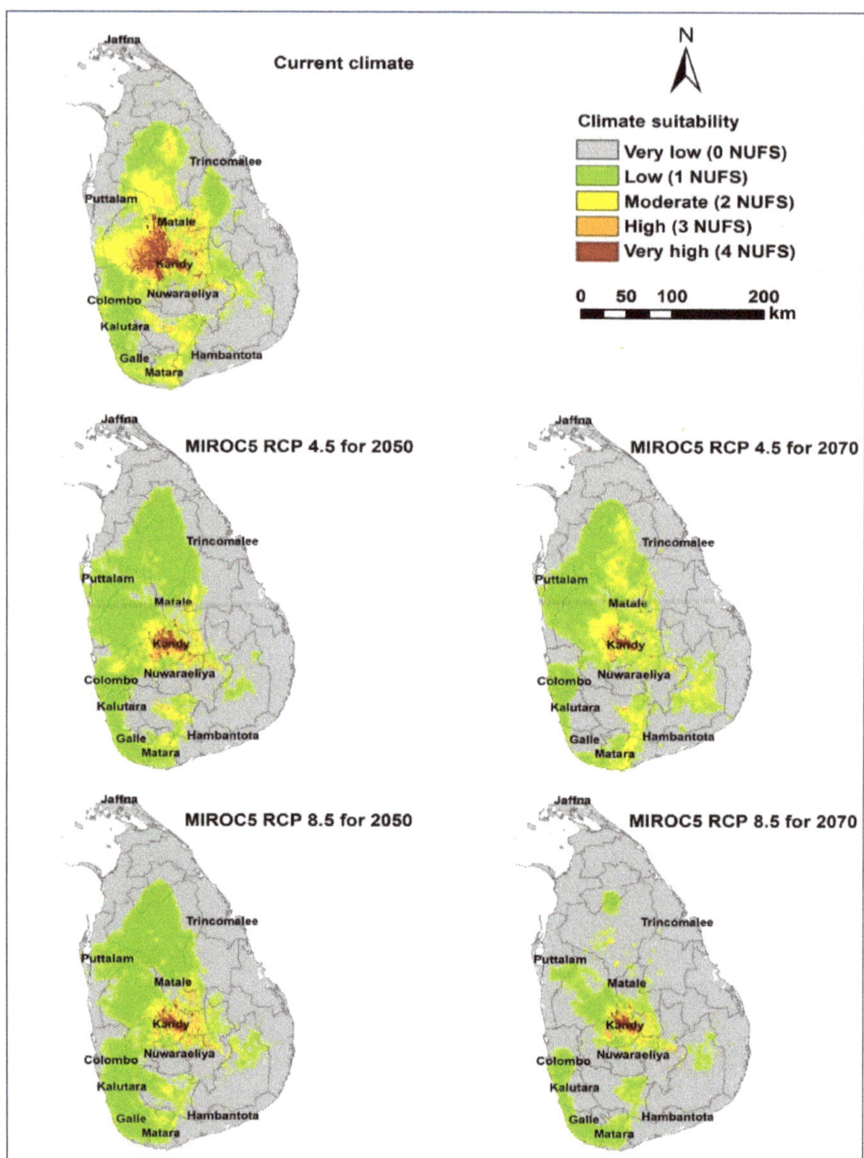

Figure 5. Projected climatic suitability for the establishment of multiple NUFS in Sri Lanka under the current climate, medium-emissions (RCP 4.5) and high-emissions (RCP 8.5) climatic scenarios for 2050 and 2070.

4. Discussion

Using a case study from Sri Lanka, we showed that the climatic suitability of NUFS will decrease in the future in tropical countries as a result of climate change. This may have implications on global food security, human nutrition and livelihood. The study provides important information about the habitat requirements of these NUFS for future prospecting. Both analysis of variable contribution and Jackknife testing of Maxent revealed that soil is an important environmental variable for determining the potential distribution of all four evaluated NUFS. We used soil as one of the variables to determine the likely areas that these NUFS can spread, but soil is not a factor that is impacted by climate. Among the other influencing environmental variables, precipitation of the driest month (BIO14), mean diurnal range (mean of monthly (max temp–min temp) (BIO2), minimum temperature of the coldest month (BIO6), precipitation seasonality (coefficient of variation) (BIO5), elevation (dem) and maximum temperature of the warmest month (BIO5) were significant.

A. muricata grows up to 1000 m elevation in areas with warm humid climates, with an average temperature of 18 °C and a rainfall greater than 1500 mm [23,79]. The potentially suitable areas of *A. muricata* lie mainly in the central parts of Sri Lanka (in and around the Kandy district) as the prevailing climatic conditions in this area are ideal for its growth. Furthermore, the soil types prevailing around the Kandy district (i.e., reddish brown latosolic soils, red-yellow podsolic soils and immature brown loams) can be suitable for *A. muricata* cultivation. *L. acidissima* is reported to grow from coastal areas to an elevation of about 450 m in a range of soil conditions where the mean annual temperature is 30 °C and mean annual rainfall is in the range 1250–1900 mm [14]. The model predicted a potentially suitable area for *L. acidissima* in the low country dry zone, which receives similar climatic conditions. *L. acidissima* is adapted to tolerate drought conditions, and thus is appropriate for the drier areas similar to conditions prevailing in the dry zone of Sri Lanka. This plant can grow well in reddish-brown earths and red-yellow podsolic soils prevailing in the dry zone. In Sri Lanka, *A. marmelos* is widely distributed in dry, intermediate and wet zones [14]. It grows well in well-drained soils up to 1200 m in elevation where the mean annual temperature range is from −6 to 48 °C and mean annual rainfall is 570–2000 mm [80]. The model suggests that the potential area lies predominantly around the Kandy, Kegalle, Kurunegala, Gampaha and Kalutara districts in the current climate. Under future scenarios, the species is predicted to move southward and be confined to the wet zone areas, suggesting that the species will no longer tolerate the projected increased temperature. *T. indica* is a light-demanding tree that grows well in dry climates and needs an evenly distributed mean annual rainfall of 500–1500 mm as well as a maximum annual temperature ranging from 33 to 37 °C with a minimum of 9.5 to 20 °C [22]. The plant grows in a wide range of soils [14]. In the current climate, *T. indica* is predicted to be distributed around north-central and central parts of Sri Lanka. However, the potential distribution is predicted to contract, move inward and be restricted around central areas of the country in the future. The projected high-temperature increases in the dry and intermediate zones may potentially limit the distribution of this plant in those areas. Such information is useful for developing measures for conservation and sustainable utilization of NUFS in order to enhance their support to global food crises.

Our results imply that potentially suitable areas of all four NUFS are predicted to decrease by more than 40% by 2070 compared to the current climate under a high-emissions scenario (Figure 4). Out of the four NUFS, *T. indica* is predicted to be at the highest risk of suitable area reduction in terms of percentage area loss compared to the current climate, suggesting the requirement of immediate management concerns. Moreover, potentially suitable areas of all four NUFS are predicted to be confined to the relatively cooler high elevation areas (i.e., central highlands) of the country. Likewise, the projected maps of multiple species establishments show an area of potential NUFS concentration around the mid-country that is projected to decrease under climate change. NUFS are ecologically responsive to climate and thus their potential distribution is limited by their capacity to tolerate climatic changes [14]. Our findings show that NUFS grown in the low country's dry and intermediate zones are predicted to be at a relatively higher risk. The trend analysis conducted in Sri Lanka using long-term climate data has revealed a decreasing trend of the diurnal temperature range (i.e., the minimum temperature of the country is increasing faster than the maximum temperature) and the annual rainfall of the country, prominently in the dry zone [35]. Therefore, such projected temperature and rainfall changes mediates extreme climate conditions and can result in serious impacts on the potential distribution of NUFS, shifting and limiting their suitability to higher altitudes.

Spatial distribution of NUFS has a direct relationship with the agro-ecology of the country [10]. The number of potentially suitable AERs for these four NUFS under current climates is predicted to reduce in the future, and is even more prominent under the high-emissions scenario by 2070 (Table S5). This implies that the capacity of these NUFS to spread in a heterogeneous agro-ecological environment will be reduced in future. This study found that AERs located in the wet zone mid-country and intermediate zone mid-country are the most suitable areas for growing NUFS with the adverse impacts of future climate change. Therefore, other important factors that may affect the cultivation of these species can be explored with major emphasis on these potential areas. However, climate change can result in changes in species ecology [81]; therefore, the boundaries of existing AERs have to be redefined according to future climate changes. Accordingly, suitable areas in re-defined AERs will have to be computed and those changes should be considered in future applications.

NUFS have become popular and high in demand in Sri Lanka due to the increased popularity of their medicinal properties. Generally, these plants are not commonly cultivated in Sri Lanka and are mainly harvested from home gardens and the wild. The production of these NUFS in Sri Lanka has been in decline due to the fragmentation of habitats for development projects and the impacts of climate change. Per-capita consumption of fruits in Sri Lanka is still well below the recommended level of the daily average of 40 g [6]. Micronutrient deficiencies have also become a growing health problem [82]. Under these circumstances, NUFS can be a very good substitute for combating "hidden hunger" and the most cost-effective means of alleviating vitamins, minerals and other micronutrient deficiencies. NUFS are generally seasonal fruits and the production and quality of NUFS is directly affected by changing climatic conditions. Therefore, identification of climatically suitable areas is important to promote the cultivation of NUFS. This is important to assure a smooth and continuous supply of NUFS to the market as demand increases.

Modeling studies have shown that the potential area of suitability for NUFS will decrease in the future, especially in developing countries [83]. Our results corroborate these findings and reaffirm that climate change will result in substantial losses to the future survival of NUFS, food security and well-being. Meanwhile, agro-climatologists propose introducing NUFS as an alternative to staple crops considering their resilience and natural adaptation to adverse climatic conditions [84]. Modeling studies are intended to provide critical spatial and temporal data as well as information that helps scientific decision-making and produces policy and regulatory frameworks and smart-solutions to promote commercial cultivation and the sustainable utilization of NUFS across the country. However, limited studies have been undertaken globally to identify the potential ranges of NUFS under climate change [8]. Though there are uncertainties, SDM is widely acknowledged as the best applicable technique for predicting the potential distribution of species [85]. Hitherto, no comprehensive study has been carried out, although there is very good potential to utilize SDM for the strategic management of NUFS in Sri Lanka.

Results of this analysis encourage developing climate change adaptation strategies to reduce the vulnerability of NUFS to climate change and sustain their cultivation in growing areas. We suggest developing species-specific strategies to improve the resilience of NUFS to climate change as species response to climate change can differ [85]. Local scientists and policymakers will have to be vigilant about the future climate change impacts and evaluate the genetic resources in genebanks and in the wild for the introduction of improved climate-resistant NUFS varieties. Policymakers and decision-makers should identify the crucial role that NUFS can play as an important source of global food and nutrition security, and take appropriate actions to promote and cultivate NUFS in climatically suitable areas.

5. Conclusions

Using Sri Lanka as a case study, we have shown the impacts of projected climate change on NUFS distribution in a tropical climate. The findings of the present study suggest that climate change particularly increases the vulnerability of NUFS and shrinks their potentially suitable areas significantly. Evidently, this can result in serious implications for future food security, human nutrition and well-being. Our results also support the general understanding that tropical species would shift to cooler areas in high altitudes under climate change. *T. indica* is predicted to be at the greatest risk and thus immediate conservation and management actions are needed. Our findings also indicate that dry and intermediate zones will suffer the highest suitable area losses from projected climate changes. Therefore, we highlight the requirements of climate change mitigation and adaption strategies with a clear focus on the targeted areas, which should be well-communicated among researchers, policymakers and decision-makers. Further, researchers should be encouraged to develop climate-resilient NUFS varieties in order to adapt to adverse climate challenges in vulnerable areas.

Supplementary Materials: The following are available online at http://www.mdpi.com/2073-4395/10/1/34/s1, Figure S1: Agro-ecological regions of Sri Lanka; Figure S2. Results of multicolinearity analysis of 19 bioclimatic variables; Figure S3: Results of Jackknife of regularized training gain for selected priority NUFS, *Aegle marmelos* (a), *Annona muricata* (b), *Limonia acidissima* (c) and *Tamarindus indica* (d) in Sri Lanka; Figure S4: Potentially suitable area of prediction above the selected threshold (maximum training sensitivity plus specificity) of selected priority NUFS, *Limonia acidissima*, *Aegle marmelos*, *Annona muricata* and *Tamarindus indica* in Sri Lanka under medium-emissions (RCP 4.5) and high-emissions (RCP 8.5) scenarios for 2050 and 2070; Figure S5: District map of Sri Lanka; Table S1: Prioritization of NUFS; Table S2: Details of NUFS occurrences used in the study; Table S3: Evaluation of model performances of NUFS, *Aegle marmelos*, *Annona muricata*, *Limonia acidissima*, *Tamarindus indica* and *Phyllanthus emblica* in Sri Lanka; Table S4: Projected suitable area of prediction and percentage area change of the selected priority NUFS under medium-emissions (RCP 4.5) and high-emissions (RCP 8.5) scenarios for 2050 and 2070.; Table S5: Agro-ecological regions (AERs) potentially suitable for selected priority NUFS in Sri Lanka under medium-emissions (RCP 4.5) and high-emissions (RCP 8.5) scenarios for 2050 and 2070.

Author Contributions: Conceptualization, S.S.R.; L.K. and C.S.K.; methodology, S.S.R.; L.K. and C.S.K.; software, S.S.R. and C.S.K.; formal analysis, S.S.R. and C.S.K.; writing—original draft preparation, S.S.R.; writing—review and editing, S.S.R.; L.K. and C.S.K.; supervision, L.K. All authors have read and agreed to the published version of the manuscript.

Funding: This research received no external funding.

Acknowledgments: We would like to thank H.K. Kadupitiya for providing the climate data used in this study. We are also grateful to the Global Environment Facility (GEF)-funded projects "In-situ conservation of crop wild relatives through enhanced Information management and field application" (2004–2010), "Mainstreaming biodiversity conservation and sustainable use for improved human nutrition and well-being" (2011–2018) and "Mainstreaming agrobiodiversity conservation and use in Sri Lankan agro-ecosystems for livelihoods and adaptation to climate change" (2011–2018), implemented by the Department of Agriculture and Ministry of Environment and SL-USA Cooperative germplasm development project on underutilized fruit species for exploration, collection, conservation and characterization of underutilized fruits 2003–2008, implemented by the Department of Agriculture, for providing geo-referenced occurrence data of studied species.

Conflicts of Interest: The authors declare no conflict of interest.

References

1. Galhena, D.H.; Freed, R.; Maredia, K.M. Home gardens: A promising approach to enhance household food security and wellbeing. *Agric. Food Secur.* **2013**, *2*, 8. [CrossRef]

2. MoFE. *Biodiversity Conservation in Sri Lanka: A Framework for Action*; Ministry of Forestry and Environment: Battaramulla, Sri Lanka, 1999.

3. MoMD&E. *National Biodiversity Strategic Action Plan 2016–2022. Colombo, Sri Lanka*; Ministry of Mahaweli Development and Environment: Colombo, Sri Lanka, 2016.

4. Punyawardena, B.V.R. *Rainfall of Sri Lanka and Agroecological Zones*; Department of Agriculture: Peradeniaya, Sri Lanka, 2008.

5. Pushpakumara, G.; Silva, P. *Agrobiodiverisity in Sri Lanka*; Biodiversity Secretariat, Ministry of Environment and Natural Resources: Battaramulla, Sri Lanka, 2009.

6. Dahanayake, N. Some neglected and underutilized fruit-crops in Sri Lanka. *Int. J. Sci. Res. Publ.* **2015**, *5*, 1–7.

7. Haq, N. Fruits for the Future in Asia. In *Proceedings of a Regional Consultation Meeting on Utilization of Tropical Fruit Trees in Asia*; Crops for the Future: Bangkok, Thailand, 2002.

8. Padulosi, S.; Heywood, V.; Hunter, D.; Jarvis, A. Underutilized species and climate change: Current status and outlook. In *Crop Adaptation to Climate Change*; Yadav, S.S., Redden, R.J., Hatfield, J.L., Eds.; Blackwell Publishing Ltd.: Oxford, UK, 2011; pp. 507–521.

9. Padulosi, S.; Hodgkin, T.; Williams, J.; Haq, N. 30 Underutilized crops: Trends, challenges and opportunities in the 21st century. In *Managing Plant Genetic Diversity*; Engels, J.M.M., Ramanatha Rao, V., Brown, A.H.D., Jackson, M.T., Eds.; CAB International Publishing: Wallingford, UK, 2002; pp. 323–338.

10. Delêtre, M.; Gaisberger, H.; Arnaud, E. *Agrobiodiversity in Perspectives—A Review of Questions, Tools, Concepts and Methodologies*; Bioversity International: Rome, Italy, 2012; p. 77.

11. Mabhaudhi, T.; Chimonyo, V.G.P.; Hlahla, S.; Massawe, F.; Mayes, S.; Nhamo, L.; Modi, A.T. Prospects of orphan crops in climate change. *Planta* **2019**, *250*, 695–708. [CrossRef] [PubMed]

12. Gruber, K. Agrobiodiversity: The living library. *Nature* **2017**, *544*, S8–S10. [CrossRef] [PubMed]

13. Padulosi, S.; Thompson, J.; Rudebjer, P. *Fighting Poverty, Hunger and Malnutrition with Neglected and Underutilized Species: Needs, Challenges and the Way Forward*; Bioversity International: Rome, Italy, 2013.

14. Pushpakumara, D.K.N.G.; Gunasena, H.P.M.; Singh, V.P. *Underutilized Fruit Trees in Sri Lanka Volume 1*; World Agroforestry Centre, South Asia Office: New Delhi, India; National Multipurpose Tree Species Research Network of Sri Lanka; Sri Lanka Council for Agriculture Research Policy; Asian Centre for Underutilized Crops in Sri Lanka: New Delhi, India, 2007.

15. Castañeda-Álvarez, N.P.; Khoury, C.K.; Achicanoy, H.A.; Bernau, V.; Dempewolf, H.; Eastwood, R.J.; Guarino, L.; Harker, R.H.; Jarvis, A.; Maxted, N. Global conservation priorities for crop wild relatives. *Nat. Plants* **2016**, *2*, 16022. [CrossRef]

16. Chivenge, P.; Mabhaudhi, T.; Modi, A.; Mafongoya, P. The potential role of neglected and underutilised crop species as future crops under water scarce conditions in Sub-Saharan Africa. *In. J. Environ. Res. Public Health* **2015**, *12*, 5685–5711. [CrossRef]

17. Bandula, A.; Jayaweera, C.; De Silva, A.; Oreiley, P.; Karunarathne, A.; Malkanthi, S. Role of underutilized crop value chains in rural food and income security in Sri Lanka. *Procedia Food Sci.* **2016**, *6*, 267–270. [CrossRef]

18. Li, X.; Siddique, K.H.M. *Future Smart Food-Rediscovering Hidden Treasures of Neglected and Underutilized Species for Zero Hunger in Asia*; Food & Agriculture Organisation of the United Nations Bangkok: Bangkok, Thailand, 2018; pp. 51–59.

19. Malkanthi, S. Importance of Underutilized Crops in Thanamalwila Divisional Secretariat Division in Monaragala District in Sri Lanka. *J. Agric. Sci. Sri Lanka* **2017**, *12*. [CrossRef]

20. Singh, A.; Dubey, P.K.; Chaurasia, R.; Dubey, R.K.; Pandey, K.K.; Singh, G.S.; Abhilash, P.C. Domesticating the Undomesticated for Global Food and Nutritional Security: Four Steps. *Agronomy* **2019**, *9*, 491. [CrossRef]

21. Brijesh, S.; Daswani, P.; Tetali, P.; Antia, N.; Birdi, T. Studies on the antidiarrhoeal activity of Aegle marmelos unripe fruit: Validating its traditional usage. *BMC Complement. Altern. Med.* **2009**, *9*, 47. [CrossRef]

22. El-Siddig, K.; Gunasena, H.P.M.; Prasad, B.A.; Pushpakumara, D.K.N.G.; Ramana, K.V.R.; Vijayanand, P.; Williams, J.T. *Tamarind: Tamarindus Indica. Vol. 1*; Southampton Centre for Underutilised Crops: Southampton, UK, 2006.

23. Pareek, O.P.; Sharma, S. *Underutilized Fruits and Nuts Vol.2*; Aavishkar Publishers Distributors: Jaipur, India, 2009.

24. Ilaiyaraja, N.; Likhith, K.; Babu, G.S.; Khanum, F. Optimisation of extraction of bioactive compounds from Feronia limonia (wood apple) fruit using response surface methodology (RSM). *Food Chem.* **2015**, *173*, 348–354. [CrossRef] [PubMed]

25. Jama, B.; Mohamed, A.; Mulatya, J.; Njui, A. Comparing the "Big Five": A framework for the sustainable management of indigenous fruit trees in the drylands of East and Central Africa. *Ecol. Indic.* **2008**, *8*, 170–179. [CrossRef]

26. Moghadamtousi, S.; Fadaeinasab, M.; Nikzad, S.; Mohan, G.; Ali, H.; Kadir, H. *Annona muricata* (Annonaceae): A review of its traditional uses, isolated acetogenins and biological activities. *Int. J. Mol. Sci.* **2015**, *16*, 15625–15658. [CrossRef] [PubMed]

27. Kuru, P. Tamarindus indica and its health related effects. *Asian Pac. J. Trop. Biomed.* **2014**, *4*, 676–681. [CrossRef]

28. Pradhan, D.; Tripathy, G.; Patanaik, S. Anticancer activity of *Limonia acidissima* Linn (Rutaceae) fruit extracts on human breast cancer cell lines. *Trop. J. Pharm. Res.* **2012**, *11*, 413–419. [CrossRef]

29. Ilango, K.; Chitra, V. Wound healing and anti-oxidant activities of the fruit pulp of *Limonia acidissima* Linn (Rutaceae) in rats. *Trop. J. Pharm. Res.* **2010**, *9*, 223–230. [CrossRef]

30. Rahman, S.; Parvin, R. Therapeutic potential of *Aegle marmelos* (L.)—An overview. *Asian Pac. J. Trop. Dis.* **2014**, *4*, 71–77. [CrossRef]

31. Adewole, S.O.; Caxton-Martins, E.A. Morphological changes and hypoglycemic effects of *Annona muricata* linn.(annonaceae) leaf aqueous extract on pancreatic β-cells of streptozotocin-treated diabetic rats. *Afr. J. Biomed. Res.* **2006**, *9*. [CrossRef]

32. Liu, N.; Yang, H.L.; Wang, P.; Lu, Y.C.; Yang, Y.J.; Wang, L.; Lee, S.C. Functional proteomic analysis revels that the ethanol extract of *Annona muricata* L. induces liver cancer cell apoptosis through endoplasmic reticulum stress pathway. *J. Ethnopharmacol.* **2016**, *189*, 210–217. [CrossRef]

33. Havinga, R.M.; Hartl, A.; Putscher, J.; Prehsler, S.; Buchmann, C.; Vogl, C.R. *Tamarindus indica* L.(Fabaceae): Patterns of use in traditional African medicine. *J. Ethnopharmacol.* **2010**, *127*, 573–588. [CrossRef]

34. MoE. *Sri Lanka's Second National Communication on Climate Change*; Ministry of Environment: Colombo, Sri Lanka, 2011.

35. Jayawardena, S.; Dharshika, T.; Herath, R. Observed climate trends, future climate change projections and possible impacts for Sri Lanka. *'Neela Haritha' Clim. Chang. Mag. Sri Lanka* **2017**, *2*, 144–151.

36. Eckstein, D.; Hutfils, M.; Winges, M. *Global Climate Risk Index 2019—Who Suffers Most From Extreme Weather Events? Weather-Related Loss Events in 2017 and 1998 to 2017*; Germanwatch: Bonn, Germany, 2018.

37. Potts, S.G.; Biesmeijer, J.C.; Kremen, C.; Neumann, P.; Schweiger, O.; Kunin, W.E. Global pollinator declines: Trends, impacts and drivers. *Trends Ecol. Evol.* **2010**, *25*, 345–353. [CrossRef] [PubMed]

38. Williams, J.T.; Haq, N. *Global Research on Underutilised Crops; an Assessment of Current Activities and Proposals for Enhanced Cooperation*; ICUC: Southampton, UK, 2002.

39. Jayasinghe-Mudalige, U.K.; Henson, S. Economic incentives for firms to implement enhanced food safety controls: Case of the Canadian red meat and poultry processing sector. *Rev. Agric. Econ.* **2006**, *28*, 494–514. [CrossRef]

40. Boria, R.A.; Olson, L.E.; Goodman, S.M.; Anderson, R.P. Spatial filtering to reduce sampling bias can improve the performance of ecological niche models. *Ecol. Model.* **2014**, *275*, 73–77. [CrossRef]

41. Phillips, S.J.; Dudík, M.; Elith, J.; Graham, C.H.; Lehmann, A.; Leathwick, J.; Ferrier, S. Sample selection bias and presence-only distribution models: Implications for background and pseudo-absence data. *Ecol. Appl.* **2009**, *19*, 181–197. [CrossRef]

42. Young, N.; Carter, L.; Evangelista, P.A. MaxEnt Model v3. 3.3 e Tutorial (ArcGIS v10). 2011. Available online: http://ibis.colostate.edu/webcontent/ws/coloradoview/tutorialsdownloads/a_maxent_model_v7.pdf (accessed on 10 March 2019).

43. Hijmans, R.J.; Cameron, S.E.; Parra, J.L.; Jones, P.G.; Jarvis, A. Very high resolution interpolated climate surfaces for global land areas. *Int. J. Clim.* **2005**, *25*, 1965–1978. [CrossRef]

44. Watanabe, M.; Suzuki, T.; O'ishi, R.; Komuro, Y.; Watanabe, S.; Emori, S.; Takemura, T.; Chikira, M.; Ogura, T.; Sekiguchi, M. Improved climate simulation by MIROC5: Mean states, variability, and climate sensitivity. *J. Clim.* **2010**, *23*, 6312–6335. [CrossRef]

45. Sharmila, S.; Joseph, S.; Sahai, A.; Abhilash, S.; Chattopadhyay, R. Future projection of Indian summer monsoon variability under climate change scenario: An assessment from CMIP5 climate models. *Glob. Planet. Chang.* **2015**, *124*, 62–78. [CrossRef]

46. Mishra, V.; Kumar, D.; Ganguly, A.R.; Sanjay, J.; Mujumdar, M.; Krishnan, R.; Shah, R.D. Reliability of regional and global climate models to simulate precipitation extremes over India. *J. Geophys. Res. Atmos.* **2014**, *119*, 9301–9323. [CrossRef]

47. Sperber, K.R.; Annamalai, H.; Kang, I.-S.; Kitoh, A.; Moise, A.; Turner, A.; Wang, B.; Zhou, T. The Asian summer monsoon: An intercomparison of CMIP5 vs. CMIP3 simulations of the late 20th century. *Clim. Dyn.* **2013**, *41*, 2711–2744. [CrossRef]

48. Kariyawasam, C.S.; Kumar, L.; Ratnayake, R.S.S. Invasive Plant Species Establishment and Range Dynamics in Sri Lanka under Climate Change. *Entropy* **2019**, *21*, 571. [CrossRef]

49. Lamsal, P.; Kumar, L.; Aryal, A.; Atreya, K. Invasive alien plant species dynamics in the Himalayan region under climate change. *Ambio* **2018**, *47*, 697–710. [CrossRef] [PubMed]

50. Su, J.; Aryal, A.; Nan, Z.; Ji, W. Climate change-induced range expansion of a subterranean rodent: Implications for rangeland management in Qinghai-Tibetan Plateau. *PLoS ONE* **2015**, *10*, e0138969. [CrossRef] [PubMed]

51. Aryal, A.; Shrestha, U.B.; Ji, W.; Ale, S.B.; Shrestha, S.; Ingty, T.; Maraseni, T.; Cockfield, G.; Raubenheimer, D. Predicting the distributions of predator (snow leopard) and prey (blue sheep) under climate change in the Himalaya. *Ecol. Evol.* **2016**, *6*, 4065–4075. [CrossRef] [PubMed]

52. Leroy, B.; Delsol, R.; Hugueny, B.; Meynard, C.N.; Barhoumi, C.; Barbet-Massin, M.; Bellard, C. Without quality presence–absence data, discrimination metrics such as TSS can be misleading measures of model performance. *J. Biogeogr.* **2018**, *45*, 1994–2002. [CrossRef]

53. Boubli, J.; De Lima, M. Modeling the geographical distribution and fundamental niches of *Cacajao* spp. and Chiropotes israelita in Northwestern Amazonia via a maximum entropy algorithm. *Int. J. Primatol.* **2009**, *30*, 217–228. [CrossRef]

54. Ibanez, I.; Silander, J., Jr.; Allen, J.M.; Treanor, S.A.; Wilson, A. Identifying hotspots for plant invasions and forecasting focal points of further spread. *J. Appl. Ecol.* **2009**, *46*, 1219–1228. [CrossRef]

55. O'donnell, J.; Gallagher, R.V.; Wilson, P.D.; Downey, P.O.; Hughes, L.; Leishman, M.R. Invasion hotspots for non-native plants in Australia under current and future climates. *Glob. Chang. Biol.* **2012**, *18*, 617–629. [CrossRef]

56. Beaumont, L.J.; Hughes, L.; Poulsen, M. Predicting species distributions: Use of climatic parameters in BIOCLIM and its impact on predictions of species' current and future distributions. *Ecol. Model.* **2005**, *186*, 251–270. [CrossRef]

57. Elith, J. Predicting Distributions of Invasive Species 2015. Available online: https://arxiv.org/ftp/arxiv/papers/1312/1312.0851.pdf (accessed on 12 February 2019).

58. Mod, H.K.; Scherrer, D.; Luoto, M.; Guisan, A. What we use is not what we know: Environmental predictors in plant distribution models. *J. Veg. Sci.* **2016**, *27*, 1308–1322. [CrossRef]

59. Heumann, B.W.; Walsh, S.J.; Verdery, A.M.; McDaniel, P.M.; Rindfuss, R.R. Land suitability modeling using a geographic socio-environmental niche-based approach: A case study from northeastern Thailand. *Ann. Assoc. Am. Geogr.* **2013**, *103*, 764–784. [CrossRef] [PubMed]

60. Kadupitiya, H.K.; De Silva, S.H.S.A.; Dilshani, D.G.S.; Weerasinghe, P. Distribution of soil organic carbon in Sri Lanka: A GIS based modeling approach. *Trop. Agric.* **2018**, *166*, 91–106.

61. Phillips, S.J.; Anderson, R.P.; Dudík, M.; Schapire, R.E.; Blair, M.E. Opening the black box: An open-source release of Maxent. *Ecography* **2017**, *40*, 887–893. [CrossRef]

62. Phillips, S.; Anderson, R.P.; Schapire, R.E. Maximum entropy modelling of species geographic distributions. *Ecol. Model.* **2006**, *190*, 231–259. [CrossRef]

63. Wisz, M.S.; Hijmans, R.J.; Li, J.; Peterson, A.T.; Graham, C.; Guisan, A.; Group, N.P.S.D.W. Effects of sample size on the performance of species distribution models. *Divers. Distrib.* **2008**, *14*, 763–773. [CrossRef]

64. Phillips, S.J.; Dudík, M.; Schapire, R.E. A maximum entropy approach to species distribution modeling. In Proceedings of the Twenty-First International Conference on Machine Learning (ACM), Banff, AB, Canada, 4–8 July 2004; p. 83.

65. Hernandez, P.A.; Graham, C.H.; Master, L.L.; Albert, D.L. The effect of sample size and species characteristics on performance of different species distribution modeling methods. *Ecography* **2006**, *29*, 773–785. [CrossRef]

66. Elith, J.; Phillips, S.J.; Hastie, T.; Dudík, M.; Chee, Y.E.; Yates, C.J. A statistical explanation of MaxEnt for ecologists. *Divers. Distrib.* **2011**, *17*, 43–57. [CrossRef]

67. Merow, C.; Smith, M.J.; Silander, J.A. A practical guide to MaxEnt for modeling species' distributions: What it does, and why inputs and settings matter. *Ecography* **2013**, *36*, 1058–1069. [CrossRef]

68. Baldwin, R. Use of maximum entropy modeling in wildlife research. *Entropy* **2009**, *11*, 854–866. [CrossRef]

69. Graham, C.H.; Elith, J.; Hijmans, R.J.; Guisan, A.; Townsend Peterson, A.; Loiselle, B.A.; Group, N.P.S.D.W. The influence of spatial errors in species occurrence data used in distribution models. *J. Appl. Ecol.* **2008**, *45*, 239–247. [CrossRef]

70. Phillips, S.J. Transferability, sample selection bias and background data in presence-only modelling: A response to Peterson et al.(2007). *Ecography* **2008**, *31*, 272–278. [CrossRef]

71. Phillips, S.J.; Dudík, M. Modeling of species distributions with Maxent: New extensions and a comprehensive evaluation. *Ecography* **2008**, *31*, 161–175. [CrossRef]

72. Phillips, S.J. A Brief Tutorial on Maxent 2017. Available online: http://biodiversityinformatics.amnh.org/open_source/maxent/ (accessed on 20 February 2019).

73. Cheng, A.; Mayes, S.; Dalle, G.; Demissew, S.; Massawe, F. Diversifying crops for food and nutrition security—A case of teff. *Biol. Rev.* **2017**, *92*, 188–198. [CrossRef] [PubMed]

74. Allouche, O.; Tsoar, A.; Kadmon, R. Assessing the accuracy of species distribution models: Prevalence, kappa and the true skill statistic (TSS). *J. Appl. Ecol.* **2006**, *43*, 1223–1232. [CrossRef]

75. Swets, J.A. Measuring the accuracy of diagnostic systems. *Science* **1988**, *240*, 1285–1293. [CrossRef]

76. Peterson, A.T.; Soberón, J.; Pearson, R.G.; Anderson, R.P.; Martínez-Meyer, E.; Nakamura, M.; Araújo, M.B. *Ecological Niches and Geographic Distributions (MPB-49)*; Princeton University Press: Princeton, NJ, USA, 2011.

77. Pramanik, M.; Paudel, U.; Mondal, B.; Chakraborti, S.; Deb, P. Predicting climate change impacts on the distribution of the threatened Garcinia indica in the Western Ghats, India. *Clim. Risk Manag.* **2018**, *19*, 94–105. [CrossRef]

78. Liu, C.; Newell, G.; White, M. On the selection of thresholds for predicting species occurrence with presence-only data. *Ecol. Evol.* **2016**, *6*, 337–348. [CrossRef]

79. Pinto, A.C.Q.; Cordeiro, M.C.R.; Andrade, S.R.M.; Ferreira, F.R.; Filgueiras, H.A.C.; Alves, R.E.; Kinpara, D.I. *Annona Species*; International Centre for Underutilised Crops: Southampton, UK, 2005; p. 261.

80. Orwa, C.; Mutua, A.; Kindt, R.; Jamnadass, R.; Anthony, S. *Agroforestree Database: A Tree Reference and Selection Guide Version 4.0*; World Agroforestry Centre: Nairobi, Kenya, 2009.

81. Walther, G.-R.; Post, E.; Convey, P.; Menzel, A.; Parmesan, C.; Beebee, T.J.; Fromentin, J.-M.; Hoegh-Guldberg, O.; Bairlein, F. Ecological responses to recent climate change. *Nature* **2002**, *416*, 389–395. [CrossRef]

82. Weerahewa, J.; Rajapakse, C.; Pushpakumara, G. An analysis of consumer demand for fruits in Sri Lanka. 1981–2010. *Appetite* **2013**, *60*, 252–258. [CrossRef]

83. Abel, C.; Dumisani, K.; Precious, M.; Upenyu, N.D.R. Assessing the impact of climate change on the suitability of rainfed flu-cured tobacco (*Nicotiana tobacum*) production in Zimbabwe. In Proceedings of the 1st Climate Science Symposium of Zimbabwe, Cresta Lodge, Harare, Zimbabwe, 19–21 June 2013.

84. Rao, N.; Shahid, M.; Al Shankiti, A.; Elouafi, I. Neglected and underutilized species for food and income security in marginal environments. *Middle East Hortic. Summit* **2013**, *1051*, 91–103. [CrossRef]

85. Kariyawasam, C.S.; Kumar, L.; Ratnayake, S.S. Invasive Plants Distribution Modeling: A Tool for Tropical Biodiversity Conservation with Special Reference to Sri Lanka. *Trop. Conserv. Sci.* **2019**, *12*, 1–12. [CrossRef]

Article

Moderation of Inulin and Polyphenolics Contents in Three Cultivars of *Helianthus tuberosus* L. by Potassium Fertilization

Anna Michalska-Ciechanowska [1], Aneta Wojdyło [1], Bożena Bogucka [2,*] and Bogdan Dubis [2]

[1] Department of Fruit, Vegetable and Plant Nutraceutical Technology, Faculty of Biotechnology and Food Science, Wrocław University of Environmental and Life Sciences, Chełmońskiego 37, 51-630 Wrocław, Poland; anna.michalska@upwr.edu.pl (A.M.-C.), aneta.wojdylo@upwr.edu.pl (A.W.)

[2] Department of Agrotechnology, Agricultural Production Management and Agribusiness, University of Warmia and Mazury in Olsztyn Oczapowskiego 8, 10-719 Olsztyn, Poland; bogdan.dubis@uwm.edu.pl

* Correspondence: bozena.bogucka@uwm.edu.pl; Tel.: +48-089-523-34-65; Fax: +48 (089)-523-32-43

Received: 30 October 2019; Accepted: 10 December 2019; Published: 13 December 2019

Abstract: Jerusalem artichoke, a widely consumed edible, is an excellent source of inulin and selected phytochemicals. However, the improvement of its chemical composition by potassium fertilization has not yet been studied. Thus, the aim of the study was to evaluate the effect of different potassium (K) fertilization levels (K_2O 150 kg ha^{-1}, 250 kg ha^{-1}, 350 kg ha^{-1}) on the content of inulin; profile and changes in polyphenolic compounds; and the antioxidant capacity, including on-line ABTS antioxidant profiles of freeze-dried tubers originated from Violette de Rennes, Topstar, and Waldspindel cultivars. Inulin content was highest in the early maturing *cv.* Topstar. The application of 350 kg ha^{-1} of K fertilizer rates during the growth of *cv.* Topstar increased the inulin content of tubers by 13.2% relative to the lowest K fertilizer rate of 150 kg ha^{-1}. In *cv.* Violette de Rennes, inulin accumulation increased in response to the fertilizer rate of 250 kg ha^{-1}. A further increase in K fertilizer rates had no effect on inulin content. The inulin content of *cv.* Waldspindel was not modified by any of the tested K fertilizer rates. Thus, the accumulation of the inulin was cultivar-dependent. In the cultivars analyzed, 11 polyphenolic compounds were identified and polyphenolic compound content was affected by the applied rate of potassium fertilizer, which was dependent on the cultivar. Chlorogenic acid was the predominant phenolic acid in all cultivars, and it accounted for around 66.4% of the identified polyphenolic compounds in *cv.* Violette de Rennes and for around 77% of polyphenolic compounds in *cv.* Waldspindel and Topstar.

Keywords: traditional crop varieties; Jerusalem artichoke; inulin; fertilization; polyphenols; antioxidant capacity

---

1. Introduction

Jerusalem artichoke (*Helianthus tuberosus* L.) is a plant with a long history of cultivation that has been making a revival as an edible vegetable in recent years. Jerusalem artichoke is a minor crop, which is why Food and Agriculture Organization (FAO) and EUROSTAT statistics for its cultivated area are not available. Jerusalem artichoke tubers contain inulin, which increases the concentration of cell fluids and confers resistance to very low temperatures (−30 °C). Inulin is a fructan with hypoglycemic properties, and it is used as a dietary supplement in diabetes management on account of its low energy value. According to the literature, Jerusalem artichoke exerts therapeutic effects by stabilizing sugar blood levels; reducing cholesterol levels; regulating blood pressure; protecting the liver and kidneys; enhancing the absorption of calcium, magnesium, and iron; and preventing osteoporosis [1–3]. The discussed plant promotes the elimination of toxic metabolites, boosts immunity, relieves stress, and improves concentration [4].

Roberfroid et al. [5] analyzed chicory inulin and found that all fructans are well fermented by gut bifidobacteria, which contributes to their anticarcinogenic properties. Inulin induces a 10-fold increase in the Lactobacillus population, which is why it suppresses appetite, regulates the passage rate of digesta, and stimulates the immune system [3,6]. Inulin consumed in a daily dose of approximately 20 g does not produce adverse side effects (such as bloating). Inulin consumption is estimated at 3–11 g in Europe and 1–4 g in the United States [7]. Jerusalem artichoke contains Si, Zn, Mn, Se, K, and Mg, and when consumed regularly, it stimulates pancreatic cells to produce insulin. The discussed plant is also characterized by unique antiparasitic activity in the treatment of giardiasis.

Jerusalem artichoke tubers are a rich source of phytochemicals. Tubers contain 221 mg of phenolic compounds in 100 g of fresh matter, and the content of phenolic acids with antioxidant properties is estimated at 16.6% on a dry matter basis. According to Petkova et al. [8], Jerusalem artichoke flour delivers health benefits due to its high total polyphenolic content. Tchoné et al. [9] identified 22 phenolic compounds in Jerusalem artichoke tubers, with a predominance of chlorogenic acid. According to Cieślik and Filipiak-Florkiewicz [10] and Florkiewicz et al. [11], the nutritional composition of Jerusalem artichoke renders it highly suitable for the production of functional foods. Similar to potatoes, Jerusalem artichokes are consumed cooked, roasted, or fried. The plant can be processed into flour, chips, salads, and additives for the production of desserts, ice cream, and fruit preserves [10–13]. Jerusalem artichoke has a high yield potential of approximately 90 t ha^{-1} tubers [10,14]. Research has demonstrated that the yield and biological value of Jerusalem artichoke tubers are influenced by cultivar, production technology, and harvest date [15,16]. Most studies of Jerusalem artichoke cultivation have focused on the effect of nitrogen fertilization [17–20]. Sawicka [17], who investigated three Jerusalem artichoke cultivars and four nitrogen fertilization levels, observed differences in the chemical composition of tubers depending on cultivar. Fertilizer rates higher than 100 kg N ha^{-1} decreased the biological value of tubers. Another study by Sawicka et al. [19] revealed that rational mineral fertilization, in particular with nitrogen, contributes to the high nutritional value of Jerusalem artichoke. The highest content of macronutrients in the aboveground parts of plants was noted in plots fertilized with 50 kg N ha^{-1}. Gao et al. [20] also reported that the tuber and biomass yield of *H. tuberosus* was highest at a fertilizer rate of 50 kg N ha^{-1}. Praznik et al. [16] and Matias et al. [18] demonstrated that the agronomic performance of Jerusalem artichoke was affected by harvest date rather than by different levels of NPK fertilization. Research shows that the yield and biological value of Jerusalem artichoke tubers are determined by cultivar, cultivation technology, and harvest date [15,18]. However, the influence of potassium (K) fertilization on the inulin content and the composition of polyphenolic compounds in different Jerusalem artichoke cultivars has not been investigated to date. In view of the above, the aim of this study was to determine the effect of different K fertilizer rates (150 kg ha^{-1}, 250 kg ha^{-1}, and 350 kg ha^{-1}; K$_2$O) on the content of inulin and polyphenolic compounds, and the antioxidant capacity of Jerusalem artichoke cultivars Topstar, Violette de Rennes, and Waldspindel in order to improve the nutrition value of such widely consumed edibles.

2. Materials and Methods

2.1. Materials

A field experiment was conducted at the Agricultural Experiment Station in Tomaszkowo (53° 42′ N, 20° 26′ E, Poland). The experiment had a two-level factorial design with randomized blocks. The first experimental factor was cultivar. The following German cultivars were investigated: Topstar, an early maturing, edible cultivar with yellow-brown tubers and high yields; Violette de Rennes, a medium-late maturing, edible cultivar with red tubers; and Waldspindel, a medium-late maturing cultivar with red tubers, used in the production of herbal supplements and in the distilling industry (all cultivars were obtained from Topinambur Manufaktur, an organic farm in Germany). The second experimental factor was the rate of mineral K fertilizer applied to soil: (I) K$_2$O 150 kg ha^{-1}, (II) K$_2$O 250 kg ha^{-1}, or (III) K$_2$O 350 kg ha^{-1} (50% potassium sulfate). Nitrogen and phosphorus fertilizers were applied once before planting (80 kg N ha^{-1}; urea (46%), 70 kg P$_2$O$_5$ ha^{-1}; triple superphosphate (46%), CaO 90 kg ha^{-1}). Organic fertilizers were not applied. Jerusalem artichoke was planted in the last ten days of April 2016 at a depth of 6–8 cm, 30 cm apart, with row spacing of 75 cm. Tubers were harvested in the last ten days of October 2016. Fertilizer rates were determined based on the results of an experiment conducted in Germany during 1994–2001 [21].

The experiment was established on Haplic Luvisol loamy sand [22]. Composite soil samples were obtained from each plot at a depth of 20 cm to determine the chemical properties of soil. The soil pH was 4.12, and soil nutrient levels were determined at 81.5 mg kg^{-1} P (Egner–Riehm method), 107 mg kg^{-1} K (Egner–Riehm method), and 31 mg kg^{-1} Mg (Atomic Absorption Spectrophotometry-ASS) [23,24]. Pesticides were not used during the experiment.

2.1.1. Sample Preparation

Jerusalem artichoke (*Helianthus tuberosus* L.) tubers (12 tubers from each fertilization treatment) were washed, peeled, and cut into 0.5 cm cubes. The cubes were freeze-dried in the Alpha 1-2LD Plus freeze-drier (Martin Christ GmbH, Osterode am Harz, Germany) (50 h; −72 °C) [25,26]. After freeze-drying, the samples were pulverized in a laboratory grinder, vacuum-packed, and stored (−20 °C) until analyses. The dry matter of the samples was on average 24.8%.

2.1.2. Identification of Inulin

Inulin content was determined by the methods proposed by Megazyme [26] and Topolska et al. [27]. The results were expressed in grams per 100 g of freeze-dried samples (*n* = 2).

2.1.3. Identification and Quantification of Polyphenolic Compounds

Polyphenolic compounds were identified in extracts prepared according to the method proposed by Wojdyło et al. [28]. Identification was performed in the Acquity Ultraperformance Liquid Chromatograpy system [29] equipped with a photodiode sensor (PDA, UPLC) (Waters Corp., Milford, MA, USA) and G2 QToF Micromass spectrometer (Waters, Manchester, UK) with electrospray ionization (ESI). Polyphenols were separated on an UPLC BEH column (1.7 μL, 2.1 × 100 mm; Waters Corp., Milford, MA, USA) at a temperature of 30 °C.

Polyphenols were quantified in the Acquity Ultraperformance LC system [29]. The retention times (t_R) of polyphenolic compounds in tubers were compared against commercial standards (Table 1). Standard curves for chlorogenic, neochlorogenic, cryptochlorogenic, and ferulic acids were developed over a concentration range of 0.05 to 5 mg mL^{-1} (r^2 = 0.9998). Polyphenol concentrations were determined in a series of replicates and expressed in mg kg^{-1} on a dry matter (DM) basis.

Table 1. Polyphenols identified in Jerusalem artichoke tubers by LC/MS QToF (t_R: retention time).

Compound	t_R	λ_{max} (nm)	MS (*m/z*)	MS/MS (*m/z*)
Unidentified	2.50	246/263	359.05	297.23/281.42
Neochlorogenic acid	2.79	329	353.07	191.02/135.04
Chlorogenic acid	3.75	329	353.07	191.02/135.04
3,4-Dicaffeoylquinic acid	4.05	325	515.03	353.06/191.02/173.05/135.12
3,5-Dicaffeoylquinic acid	4.62	326	515.03	353.06/191.02
Cryptochlorogenic acid	4.72	324	353.07	191.02
p-Coumaroylquinic acid	5.95	311	337.03	191.02/173.05
Caffeic acid	6.19	323	179.04	136.06
Feruoylquinic acid	6.95	325	3670.5	191.0.2/172.06
Caffeoyl-glucoside acid	7.13	327	341.04	179.04/161.0.7
1,5-Dicaffeoylquinic acid	7.25	325	515.03	354.03/191.02/179.04
4,5-Dicaffeoylquinic acid	7.46	327	515.03	191.02/179.04/173.03/135.07

2.1.4. Antioxidant Capacity

The antioxidant capacity of Jerusalem artichokes was determined in water–methanol extracts of freeze-dried tubers (1:4; v/v) according to the method described by Wojdyło et al. [28]. Antioxidant capacity was measured in the ABTS radical scavenging activity assay (TEAC ABTS) [30] and the Ferric Reducing Antioxidant Potential (FRAP) assay [31]. The results were expressed in Trolox equivalents per 100 g DM ($n = 3$).

2.1.5. Antioxidant Profiling by On-Line HPLC-PDA with Post-Column Derivatization with ABTS

The antioxidant activity of polyphenolic compounds in Jerusalem artichokes was determined by on-line HPLC-PDA coupled with post-column derivatization with ABTS according to the method proposed by Kusznierewicz et al. [32] and described by Tkacz et al. [33].

2.2. Statistical Analysis

The results were processed statistically by one-way analysis of variance (ANOVA) (Statistical 10.0, StatSoft, Tulsa, OK, USA). Significant differences ($p \leq 0.05$) between samples were determined by Tukey's test.

3. Results and Discussion

3.1. Inulin

Jerusalem artichoke tubers are most abundant in inulin between mid-October and December. In successive months, polyfructose content decreases and the concentration of simple sugar increases. The accumulation of oligofructans in tubers is influenced by weather conditions [34].

The inulin content of the evaluated cultivars ranged from 45.94 to 60.85 g 100 g^{-1} of freeze-dried samples (Table 2). Similar results were noted by Cieślik et al. [35] and Florkiewicz et al. [11] in Polish cultivars Albik and Rubik (41.4–50.4 g·100 g^{-1}). The coefficient of variation did not exceed 20% in the tested cultivars, which indicates that inulin content is a stable trait. Inulin concentration was most stable in *cv.* Waldspindel.

Table 2. The effect of cultivar and potassium fertilizer rate on the inulin content of Jerusalem artichoke tubers (g·100g^{-1} freeze-dried sample).

Cultivar	Potassium Fertilizer Rate			Mean	Coefficient of Variation
	150 kg ha^{-1}	250 kg ha^{-1}	350 kg ha^{-1}		
Violette de Rennes	45.94 [d]	50.51 [bc]	51.68 [bc]	49.38 ± 3.03 [c]	5.59
Waldspindel	53.18 [b]	51.79 [bc]	52.00 [bc]	52.49 ± 0.75 [b]	2.74
Topstar	52.83 [b]	49.26 [cd]	60.85 [a]	54.31 ± 5.94 [a]	9.76
Average	50.65 [b]	50.52 [b]	55.01 [a]		

Note: a, b, c, d = values in columns marked with the same letters do not differ significantly at $p \leq 0.05$ (Tukey's test, analysis of variance (ANOVA)).

An analysis of K fertilization levels revealed that inulin content was highest in response to the highest fertilizer rate (350 kg K$_2$O ha^{-1}). The analyzed cultivars responded differently to higher rates of K fertilizer. According to Sawicka [17,36], Jerusalem artichoke tubers are characterized by cultivar-dependent variations in nutrient levels. Matias et al. [18] found no differences in inulin content in response to higher rates of NPK fertilizers (Level 1: 54, 108, 162; Level 2: 108, 216, 324) and concluded that unlike harvest date, fertilization has a minor effect on tuber yields.

Inulin accumulation was significantly higher in the early-maturing *cv.* Topstar relative to medium-late maturing cultivars. Higher rates of K fertilizer exerted the greatest influence on the inulin content of Jerusalem artichoke *cv.* Topstar. The inulin content of freeze-dried tubers increased by 8.02 g ·100 g^{-1} when the fertilizer rate was increased by 200 kg K$_2$O ha^{-1}.

Violette de Rennes was characterized by the lowest inulin content and the smallest variations in insulin levels. In this cultivar, the inulin content of freeze-dried tubers increased by 4.57 g·100g^{-1} when the fertilizer rate was increased from 150 kg K$_2$O ha^{-1} to 250 kg K$_2$O ha^{-1}. The inulin content of the medium-late *cv.* Waldspindel, which is used in the production of herbal supplements and in the distilling industry, did not change in response to higher rates of K fertilizer.

3.2. Polyphenols

The polyphenols content of Jerusalem artichoke tubers is influenced by cultivar [37], harvest date, and storage conditions [38]. According to Terzić et al. [39] and Kapusta et al. [37], differences in the concentrations of polyphenolic compounds are also genetically conditioned. The effect of various rates of K fertilizer on the content of polyphenolic compounds in Jerusalem artichoke tubers has not been investigated to date. In the present study, the total content of the polyphenolic compounds identified in the analyzed cultivars of Jerusalem artichoke ranged from 1477 to 1801 mg kg^{-1} DM. Similar results were reported by Kapusta et al. [37] (Table 3).

In treatments fertilized with 150 kg K ha^{-1}, polyphenol levels were lowest in *cv.* Violette de Rennes and highest in *cv.* Topstar. In *cv.* Violette de Rennes, total polyphenolic content increased by 5% and 13% when the rate of K fertilizer was increased from 150 kg ha^{-1} to 250 and 350 kg ha^{-1}, respectively. Similar observations were made by Kavalcova et al. [40] in onions, where polyphenol levels measured spectrophotometrically increased with a rise in K fertilizer rate. In this regard, it was concluded that the content of polyphenolic compounds could be a varietal trait in Jerusalem artichoke. In *cv.* Waldspindel and Topstar, polyphenol levels decreased by 6.4% and 1.5% on average, respectively, when K fertilizer rate was increased by 100 and 200 kg ha^{-1}. These findings suggest that the potassium-stimulated synthesis of polyphenolic compounds is affected by the unique chemical composition of different cultivars of Jerusalem artichoke.

Chlorogenic acid was the predominant polyphenolic compound in Jerusalem artichoke tubers and leaves [41]. The average content of chlorogenic acid was estimated at 66.4% in *cv.* Violette de Rennes and at 77% in *cv.* Waldspindel and Topstar, regardless of K fertilizer rate. The content of chlorogenic acid was lowest in *cv.* Violette de Rennes and highest in *cv.* Topstar, regardless of K fertilizer rate. In *cv.* Violette de Rennes, the content of chlorogenic acid increased by 5% and 15% in response to K fertilizer rates of 250 kg ha^{-1} and 350 kg ha^{-1}, respectively. In *cv.* Waldspindel and Topstar, the content of chlorogenic acid decreased by 6% and 10% on average when the K fertilizer rate was increased to 250 and 350 ha^{-1}, respectively. The above findings indicate that cultivar and chemical composition significantly affect the potassium-induced synthesis of chlorogenic acid.

Table 3. The effect of potassium fertilizer rates on the polyphenol content of three Jerusalem artichoke cultivars (mg kg^{-1} dm).

Cultivar	Rate (kg K ha^{-1})	Chlorogenic acid	1,5-dicaffeoylquinic acid	3,4-dicaffeoylquinic acid	3,5-dicaffeoylquinic acid	Neochlorogenic acid	Crypto-chlorogenic	Caffeoyl-glucoside acid	p-coumaroyl-quinic acid	Feruoylquinic acid	Caffeic acid	4,5-dicaffeoylquinic acid	Total polyphenols
Violette de Rennes	150	1029.2 ± 16.22 [ab]	156.33 ± 0.62 [c]	96.54 ± 1.71 [abc]	73.82 ± 0.25 [cd]	57.47 ± 1.02 [ab]	23.15 ± 0.03 [d]	15.7 ± 0.36 [de]	11.79 ± 0.34 [b]	5.35 ± 0.57 [bc]	4.33 ± 1.1 [a]	3.64 ± 0.15 [bc]	1477
	250	996.89 ± 33.58 [a]	230.51 ± 13.31 [d]	89.39 ± 2.66 [a]	91.69 ± 3.06 [e]	55.07 ± 3.41 [ab]	23.62 ± 1.86 [d]	25.6 ± 1.7 [f]	15.49 ± 0.86 [c]	7.5 ± 3.24 [c]	4.7 ± 1.61 [a]	3.93 ± 0.18 [c]	1545
	350	1132.74 ± 8.43 [bc]	271.97 ± 2.35 [e]	100.79 ± 0.2 [bcd]	82.01 ± 0.14 [de]	57.51 ± 0.58 [ab]	17.74 ± 0.43 [abc]	45.73 ± 0.22 [g]	16.24 ± 0.8 [c]	6.35 ± 0.56 [c]	5.84 ± 0.01 [ab]	6.05 ± 0.63 [d]	1743
Waldspindel	150	1262.59 ± 57.44 [d]	82.95 ± 7.4 [a]	104.26 ± 3.19 [cde]	64.23 ± 6.83 [bc]	54.22 ± 1.56 [ab]	16.19 ± 0.25 [ab]	12.35 ± 1.07 [cd]	1.78 ± 0.16 [a]	0.84 ± 0.37 [ab]	9.94 ± 0.12 [cd]	2.1 ± 0.64 [ab]	1612
	250	1195.68 ± 9.61 [cd]	72.73 ± 0.06 [a]	112.45 ± 1.24 [ef]	58.79 ± 1.39 [b]	58.88 ± 0.49 [ab]	20.19 ± 1.82 [bcd]	12.46 ± 0.11 [c]	2.35 ± 0.28 [a]	0.77 ± 0.54 [ab]	10.44 ± 0.39 [cd]	2.34 ± 0.52 [abc]	1547
	350	1153.11 ± 13.13 [bcd]	82.29 ± 0.97 [a]	108.91 ± 2.12 [de]	61.87 ± 0.83 [bc]	60.17 ± 1.39 [bc]	22.68 ± 2.79 [cd]	17.57 ± 0.62 [e]	2.54 ± 0.12 [a]	0.28 ± 0.09 [a]	11.7 ± 0.43 [d]	2.94 ± 0.52 [bc]	1524
Topstar	150	1394.84 ± 14.5 [e]	120.22 ± 2.86 [b]	117.62 ± 1.72 [f]	62.34 ± 7.66 [bc]	66.13 ± 1.65 [c]	16.1 ± 1.14 [ab]	5.35 ± 0.29 [a]	2.37 ± 0.25 [a]	9.71 ± 1.23 [c]	6.25 ± 0.39 [ab]	0.81 ± 0.04 [a]	1802
	250	1266.77 ± 65.72 [de]	116.08 ± 4.83 [b]	111.32 ± 0.36 [ef]	41.03 ± 0.61 [a]	53.52 ± 0.56 [a]	30.92 ± 0.75 [e]	5.51 ± 0.17 [a]	1.45 ± 0.07 [a]	7.99 ± 0.01 [c]	8.24 ± 0.36 [bc]	0.88 ± 0.02 [a]	1644
	350	1246.34 ± 0.63 [cd]	134.52 ± 14.42 [bc]	95.48 ± 3.16 [ab]	61.83 ± 1.59 [bc]	55.11 ± 2.11 [ab]	14.15 ± 0.23 [a]	8.62 ± 0.01 [b]	3.03 ± 0.21 [a]	5.41 ± 0.59 [bc]	12.28 ± 0.23 [d]	0.78 ± 0.64 [a]	1620

Note: average ± standard deviation; a, b, c, d, e, f; g: values in columns marked with the same letters do not differ significantly at $p \leq 0.05$ (Tukey's test, ANOVA).

Four isomers of dicaffeoylquinic acid (1,5-, 3,4-, 3,5-, and 4,5-dicaffeoylquinic acids) were identified in Jerusalem artichoke tubers in this study, which is consistent with the results reported by Kapusta et al. [37]. Here, 1,5-dicaffeoylquinic acid, 3,4-dicaffeoylquinic acid, and 3,5-dicaffeoylquinic acid accounted for approximately 25% of the identified polyphenolic compounds in *cv.* Violette de Rennes and for 16.5% of the identified polyphenolic compounds in *cv.* Waldspindel and Topstar, regardless of K fertilizer rate. In *cv.* Violette de Rennes, the content of 1,5-dicaffeoylquinic acid and 3,4-dicaffeoylquinic acid increased with a rise in K fertilization levels and was highest in response to the rate of 350 kg K ha^{-1}. In *cv.* Waldspindel, the concentrations of the above compounds did not change considerably in response to increasing rates of K fertilizer. In *cv.* Topstar, the content of above components decreased when the K fertilizer rate was increased from 150 kg ha^{-1} to 250 and 350 kg ha^{-1}. The analyzed cultivars also contained neochlorogenic acid, which accounted for 3.5% of all polyphenolic compounds on average. The variations in the concentration of neochlorogenic acid across fertilizer treatments were cultivar-dependent, and they differed from those observed in the content of dicaffeoylquinic acids. Cryptochlorogenic acid was also identified in the analyzed Jerusalem artichoke cultivars. On average, cryptochlorogenic acid accounted for 1.2% of the identified polyphenolic compounds regardless of cultivar and K fertilizer rate. The content of cryptochlorogenic acid was highest in *cv.* Topstar and it more than doubled in response to a K fertilizer rate of 250 kg ha^{-1}. The content of caffeoyl-glucoside acid and caffeic acid increased in all cultivars in response to a K fertilizer rate of 350 kg ha^{-1}. The concentration of *p*-coumaroyl-quinic acid was highest in *cv.* Violette de Rennes, and it increased with a rise in K fertilizer rate. A reverse trend was observed in the remaining cultivars, which indicates that polyphenol synthesis is a varietal trait in Jerusalem artichoke.

3.3. Antioxidant Capacity

Antioxidant capacity ranged from 0.87 to 3.28 µmol Trolox kg^{-1} DM in the ABTS radical scavenging activity assay (Table 4).

Table 4. The effect of potassium fertilizer rates on the antioxidant capacity of three Jerusalem artichoke cultivars (µmol Trolox kg^{-1}).

Cultivar	Rate (kg K ha^{-1})	TEAC ABTS	FRAP
Violette de Rennes	150	1.55 ± 0.08 [ab]	2.07 ± 0.09 [de]
	250	2.65 ± 0.21 [cd]	1.81 ± 0.1 [cd]
	350	3.28 ± 0.39 [d]	1.95 ± 0.08 [cde]
Waldspindel	150	0.88 ± 0.12 [a]	1.28 ± 0.12 [b]
	250	0.87 ± 0.12 [ab]	0.82 ± 0.02 [a]
	350	1.16 ± 0.09 [ab]	0.57 ± 0.02 [a]
Topstar	150	1.37 ± 0.08 [ab]	2.13 ± 0.06 [e]
	250	1.58 ± 0.14 [b]	0.59 ± 0.01 [a]
	350	2.45 ± 0.07 [c]	1.71 ± 0.09 [c]

Note: average ± standard deviation; a, b, c, d, e: values in columns marked with the same letters do not differ significantly at $p \leq 0.05$ (Tukey's test, ANOVA).

Antioxidant capacity measured by TEAC ABTS assay was the highest in *cv.* Violette de Rennes and the lowest in *cv.* Waldspindel. Considerable differences in the antioxidant capacity of different Jerusalem artichoke cultivars were also reported by Catana et al. [42]. The antioxidant capacity of *cvs.* Violette de Rennes, Waldspindel, and Topstar increased by approximately 53%, 24%, and 44%, respectively, when K fertilizer rate was increased from 150 kg ha^{-1} to 350 kg ha^{-1}. The content of polyphenolic compounds was correlated with antioxidant capacity measured in the ABTS assay only in *cv.* Violette de Rennes ($r = 0.998$).

The chromatographic profile of phenolic components in Jerusalem artichoke before and after derivatization of the negative control (ABTS reagent) is presented in Figure 1. Chlorogenic acid exerted the strongest effect on the antioxidant potential of Jerusalem artichokes, followed by 3,5-caffeoylquinic acid, 3,4-caffeoylquinic acid, 1,5-caffeoylquinic acid, and neochlorogenic acid.

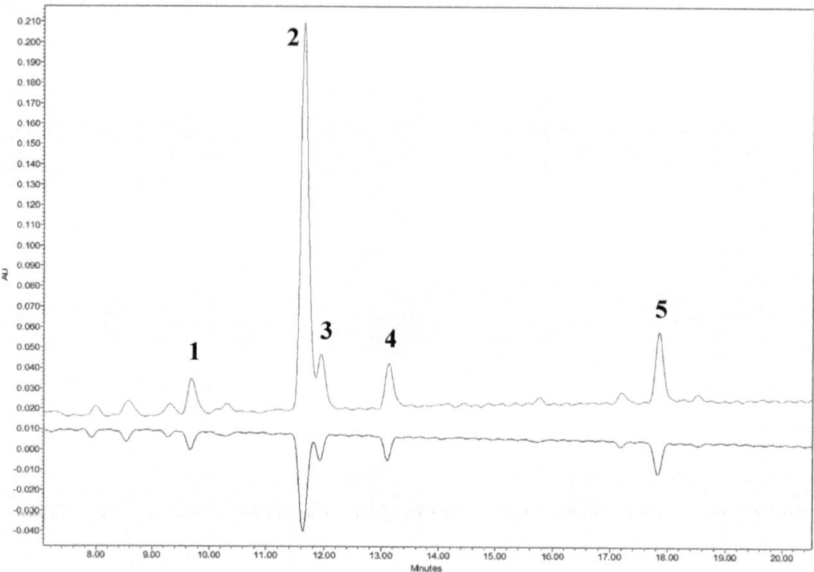

Figure 1. Standard UV chromatograms (blue line) and on-line ABTS antioxidant profiles (black line) of Jerusalem artichoke. Peaks: **1**: neochlorogenic acids; **2**: chlorogenic acid; **3**: 1,5-dicaffeoylquinic acid; **4**: 3,4-dicaffeoylquinic acid; **5**: 3,5-dicaffeoylquinic acid.

In summary, no significant correlations were found between the concentrations of individual polyphenolic compounds in Jerusalem artichoke tubers and the antioxidant capacity of the extracts determined spectrophotometrically in the ABTS radical scavenging assay.

On the other hand, the FRAP values indicated that the highest antioxidant capacity was noted for *cv.* Topstar in the treatment supplied with K fertilizer at 150 kg ha^{-1}. The increase of the K content during fertilization led to the decrease in antioxidant capacity measured by FRAP. Thus, the ability of compounds able to reduce the Fe ion in Jerusalem artichoke can be moderated by the K levels.

4. Conclusions

Based on the results obtained, it was concluded that among *cv.* Violette de Rennes, Waldspindel, and Topstar, the inulin accumulation was significantly higher in the early-maturing *cv.* Topstar. Higher rates of K fertilizer exerted the greatest influence on the inulin content of Jerusalem artichoke in this cultivar. This led to the increase of inulin content of 4.4 g 100 g^{-1} when the K fertilizer rate was increased from 150 kg K$_2$O ha^{-1} to 350 kg K$_2$O ha^{-1}.

Eleven polyphenolic compounds were identified in 3 cultivars of Jerusalem artichoke. The content of polyphenolic compounds ranged from 1.5 to 1.8 g kg^{-1} DM of tuber samples, and it was influenced by the rate of K fertilizer. Chlorogenic acid was the predominant phenolic acid in all cultivars, and it accounted for around 66.4% of the identified polyphenolic compounds in *cv.* Violette de Rennes and for around 77% of polyphenolic compounds in *cv.* Waldspindel and Topstar. Four isomers of dicaffeoylquinic acid were also identified in the evaluated tubers, and 1,5-dicaffeoylquinic acid was the predominant isomer. The content of the remaining compounds varied across cultivars and K fertilization treatments. Chlorogenic acid, 3,5-, 3,4-, 1,5-caffeoylquinic acids, and neochlorogenic acid had the strongest influence on antioxidant potential measured by the ABTS on-line profiling method.

Taking the above into consideration, fertilization with selected microelements of edibles, including Jerusalem artichoke, could be a new strategy for the improvement of the nutritional value of such plants. Nevertheless, the polyphenolic compounds are stress metabolites, and their content in the plants can be modified by the type of the fertilizer as well as the quantity applied; thus, numerous aspects should be considered in order to provide a thorough recommendation for a single polyphenolic component [43].

Author Contributions: Conceptualization, B.B.; methodology, B.B., A.M.-C., and A.W.; formal analysis, B.B., A.M.-C., A.W.; writing—original draft preparation, B.B., A.M.-C., A.W., and B.D.

Funding: This research was co-funded by Ministry of Science and Higher Education in the frame of the program entitled 'Regional Initiative of Excellence' for the years 2019-2022, Project No. 010/RID/2018/19, amount of funding 12.000.000 PLN.

Acknowledgments: The publication is the result of the research group activity: 'Plants4food'.

Conflicts of Interest: The authors declare no conflicts of interest.

References

1. Georgescu, L.A.; Stoica, I. Studies concerning the dynamic of enzyme hydrolyze on the Jerusalem artichoke (*Helianthus tuberosus* L.) inulin. *Anal. Univ. Dunarea Galati VI Food Technol.* **2005**, *1*, 77–81.

2. Zaky, E.A. Physiological response to diets fortified with Jerusalem artichoke tubers (*Helianthus tuberosus* L.) powder by diabetic rats. *Am. Eurasian J. Agric. Environ. Sci.* **2009**, *5*, 682–688.

3. Cieślik, E.; Gębusia, A. Topinambur (*Helianthus tuberosus* L.)—Bulwa o właściwościach prozdrowotnych. *Postęp. Nauk Rol.* **2010**, *3*, 91–103. (In Polish)

4. Young-Jun, L.; Myoung-gi, L.; Seok-Yeong, Y.; Won-Byong, Y.; Ok-Hwan, L. Changes in physicochemical characteristics and antioxidant activities of Jerusalem artichoke tea infusions resulting from different production processes. *Food Sci. Biotechnol.* **2014**, *23*, 1885–1892.

5. Roberfroid, M.B.; Van Loom, J.; Gibson, G.R. The bifidogenic nature of chicory inulin and its hydrolysis products. *J. Nutr.* **1998**, *128*, 11–19. [CrossRef] [PubMed]

6. El-Hofi, A.A. Technological and biological uses of Jerusalem artichoke powder and resistant starch. *Ann. Agric. Sci. Moshtohor* **2005**, *43*, 279–291.

7. Roberfroid, M.B. Inulin-type fructans: Functional food ingredients. *J. Nutr.* **2007**, *137*, 2493–2502. [CrossRef]

8. Petkova, N.; Ivanov, I.; Denev, P.; Pavlov, A. Bioactive substance and free radical scavenging activities of flour from Jerusalem Artichoke (*Helianthus tuberosus* L.) tubers—A comparative study. *Turk. J. Agric. Nat. Sci.* **2014**, *2*, 1773–1778.

9. Tchoné, M.; Bärwald, G.; Annemüller, G.; Fleischer, L.G. Separation and identification of phenolic compounds in Jerusalem artichoke (*Helianthus tuberosus* L.). *Sci. Aliment.* **2006**, *26*, 394–408. [CrossRef]

10. Cieślik, E.; Filipiak-Florkiewicz, A. Prospective usage of Jerusalem artichoke (*Helianthus tuberosus* L.) for producing functional food. *Żywność-Nauka Technologia Jakość* **2007**, *1*, 73–81. (In Polish)

11. Florkiewicz, A.; Cieślik, E.; Filipiak-Florkiewicz, A. Wpływ odmiany i terminu zbioru na skład chemiczny bulw topinamburu (*Helianthus tuberosus* L.). *Żywność-Nauka Technologia Jakość* **2007**, *3*, 71–81. (In Polish)

12. Cieślik, E.; Kopeć, A.; Praznik, W. Functional properties of fructan. In Proceedings of the Ninth Seminar on Inulin, Budapest, Hungary, 18–19 April 2002; Volume 27.

13. Mystkowska, I.; Zarzecka, K. Wartość odżywcza i prozdrowotna słonecznika bulwiastego (*Helianthus tuberosus* L.). *Post. Fitoter.* **2013**, *2*, 123–126. (In Polish)

14. Baldini, M.; Danuso, F.; Monti, A.; Amaducci, M.T.; Stevanato, P.; Mastro, G. Chichory and Jerusalem artichoke productivity in different areas of Italy, in relation to water availability and time of harvest. *Ital. J. Agron.* **2006**, *2*, 291–307. [CrossRef]

15. Góral, S. Zmienność morfologiczna i plonowanie wybranych klonów słonecznika bulwiastego –topinambur (*Helianthus tuberosus* L.). *Hodowla Roślin i Nasiennictwo* **1998**, *2*, 6–11. (In Polish)

16. Praznik, W.; Cieślik, E.; Filipiak, A. The influence of harvest time on the content of nutritional components in tubers of Jerusalem artichoke (*Helianthus tuberosus* L.). In Proceedings of the Seventh Seminar on Inulin, Leuven, Belgium, 22–23 January 1998; pp. 154–157.

17. Sawicka, B. Wpływ nawożenia azotem na wartość biologiczną bulw *Helianthus tuberosus* L. *Roczniki Akademii Rolniczej w Poznaniu* **2000**, *313*, 447–451. (In Polish)

18. Matias, J.; Gonzales, J.; Cabanillas, J.; Royano, L. Influence of NPK fertilisation and harvest date on agronomic performance of Jerusalem artichoke crop in the Guadiana Basin (Southwestern Spain). *Ind. Crop. Prod.* **2013**, *48*, 191–197. [CrossRef]

19. Sawicka, B.; Kalembasa, D.; Skiba, D. Variability in macroelement content in the aboveground part of *Helianthus tuberosus* L. at different nitrogen fertilization levels. *Plant Soil Environ.* **2015**, *61*, 158–163. [CrossRef]

20. Gao, K.; Zhu, T.X.; Wang, Q.B. Nitrogen fertilization, irrigation, and harvest times affect biomass and energy value of *Helianthus tuberosus* L. *J. Plant Nutr.* **2016**, *39*, 1906–1914.

21. Stolzenburg, K. Anbau und Verwertung von Topinambur (*Helianthus tuberosus* L.). *Informationen fur Pflanzenproduktion Sonderheft* **2002**, *1*, 10.

22. IUSS Working Group WRB. World Reference Base for Soil Resources. In *World Soil Resources Reports*, 2nd ed.; FAO: Roma, Italy, 2006.

23. Houba, V.J.G.; Van der Lee, J.J.; Novozamsky, I. Soil analysis procedure: other procedure. *J. Soil Sci. Plant Nut.* **1995**.

24. Lachman, J.; Hamouz, K.; Orsák, M.; Pivec, V.; Hejtmánková, K.; Pazderů, K.; Čepl, J. Impact of selected factors—Cultivar, storage, cooking and baking on the content of anthocyanins in coloured-flesh potatoes. *Food Chem.* **2012**, *133*, 1107–1116. [CrossRef]

25. Nemś, A.; Pęksa, A.; Kucharska, A.Z.; Sokół-Łętowska, A.; Kita, A.; Drożdż, W.; Hamouz, K. Anthocyanin and antioxidant activity of snacks with coloured potato. *Food Chem.* **2015**, *172*, 175–182. [CrossRef] [PubMed]

26. Sustainable Chemistry Solutions, AOAC (Method 999.03), AACC (Method 32-32.01), CODEX (Typ III Method). In *Enzyme Sources Guide*; Megazyme International Ireland: Bray Business Park, Bray, Co.: Wicklow, Ireland, 2014; p. 11.

27. Topolska, K.; Filipiak-Florkiewicz, A.; Florkiewicz, A.; Cieślik, E. Fructan stability in strawberry sorbets in dependence on their source and the period of storage. *Eur. Food Res. Technol.* **2017**, *243*, 701–709. [CrossRef]

28. Wojdyło, A.; Oszmiański, J.; Bielicki, P. Polyphenolic composition, antioxidant activity, and polyphenol oxidase (PPO) activity of quince (*Cydonia oblonga Miller*) varieties. *J. Agric. Food Chem.* **2013**, *61*, 2762–2772. [CrossRef] [PubMed]

29. Wojdyło, A.; Nowicka, P.; Laskowski, P.; Oszmiański, J. Evaluation of sour cherry (*Prunus cerasus* L.) fruits for their polyphenol content, antioxidant properties, and nutritional components. *J. Agric. Food Chem.* **2014**, *62*, 12332–12345. [CrossRef] [PubMed]

30. Re, R.; Pellegrini, N.; Proteggente, A.; Pannala, A.; Yang, M.; Rice-Evans, C. Antioxidant activity applying an improved ABTS radical cation decolorization assay. *Free Radic. Biol. Med.* **1999**, *26*, 1231–1237. [CrossRef]

31. Benzie, I.F.; Strain, J.J. Ferric reducing/antioxidant power assay: Direct measure of total antioxidant activity of biological fluids and modified version for simultaneous measurement of total antioxidant power and ascorbic acid concentration. *Method Enzymol.* **1999**, *299*, 15–27.

32. Kusznierewicz, B.; Piasek, A.; Bartoszek, A.; Namieśnik, J. The optimisation of analyical parameters for routine profiling of antioxidants in complex mixtures by HPLC coupled-postcolumn derivatisation. *Phytochem. Anal.* **2011**, *22*, 392–402. [CrossRef]

33. Tkacz, K.; Wojdyło, A.; Nowicka, P.; Turkiewicz, I.; Golis, P. Characterization *in vitro* potency of biological active fractions of seeds, skins and flesh from selected *Vitis vinifera* L. cultivars and interspecific hybrids. *J. Funct. Foods* **2019**, *56*, 353–363. [CrossRef]

34. Chekroun, M.B.; Amzile, J.; El Yachioui, M.; El Haloui, N.E.; Prevost, J. Qualitative and quantitative development of carbohydrate reserves during the biological cycle of Jerusalem artichoke (*Helianthus tuberosus* L.) tubers. *N. Z. J. Crop. Horticult. Sci.* **1994**, *22*, 31–37. [CrossRef]

35. Cieślik, E.; Filipiak, A.; Filipiak-Florkiewicz, A. Wpływ terminu zbioru na zawartość węglowodanów w bulwach topinamburu (*Helianthus tuberosus* L.). *Żywienie Człowieka i Metabolizm* **2003**, *30*, 1076–1080. (In Polish)

36. Sawicka, B. Słonecznik bulwiasty (*Helianthus tuberosus* L.)- Biologia, Hodowla, Znaczenie Użytkowe. *Wyd. UP Lublin* **2016**, 223. (In Polish)

37. Kapusta, I.; Krok, E.; Jamro, D.; Cebulak, T.; Kaszuba, J.; Salach, R. Identyfication and quantification of phenolic compounds from Jerusalem artichoke (*Helianthus tuberosus* L.) tubers. *J. Food Agric. Environ.* **2013**, *11*, 601–606.

38. Danilcenko, H.; Jariene, E.; Slepetiewne, A.; Sawicka, B.; Zaldariene, S. The distribution of bioactive compounds in the tubers of organically grown Jerusalem artichoke (*Helianthus tuberosus* L.) during the growing period. *Acta Sci. Pol. Hortorum Cultus* **2017**, *16*, 97–107. [CrossRef]

39. Terzić, S.; Atlagić, J.; Maksimo, I.; Zeremski, T.; Petrović, S.; Dedić, B. Influence of photoperiod on vegetation phases and tuber development in topinambur (*Helianthus tuberosus* L.). *Arch. Biol. Sci.* **2012**, *64*, 175–182. [CrossRef]

40. Kavalcova, P.; Bystricka, J.; Toth, T.; Volnova, B.; Kopernicka, M.; Harangozo, L. Potassium and its effect on the content of polyphenols in onion (*Allium cepa* L.). *J. Microbiol. Biotechnol. Food Sci.* **2015**, *4*, 74–77. [CrossRef]

41. Chen, F.; Long, X.; Liu, Z.; Shao, H.; Liu, L. Analysis of phenolic acids of Jerusalem artichoke (*Helianthus tuberosus* L.) responding to salt-stress by liquid chromatography/tandem mass spectrometry. *Sci. World J.* **2014**, *2014*, 1–8.

42. Catana, L.; Catana, M.; Iorga, E.; Lazar, A.G.; Lazar, M.A.; Teodorescu, R.; Asanica, A.; Belc, N.; Iancu, A. Valorification of Jerusalem artichoke tubers (*Helianthus tuberosus* L.) for achieving of functional ingredient with high nutritional value. *Sciendo* **2018**, *1*, 276–283. [CrossRef]

43. Heimler, D.; Romani, A.; Ieri, F. Plant polyphenol content, soil fertilization and agricultural management: A review. *Eur. Food Res. Technol.* **2017**, *43*, 1107–1115. [CrossRef]

Article

Growing Conditions Affect the Phytochemical Composition of Edible Wall Rocket (*Diplotaxis erucoides*)

Carla Guijarro-Real [1], Ana M. Adalid-Martínez [1], Katherine Aguirre [2], Jaime Prohens [1], Adrián Rodríguez-Burruezo [1] and Ana Fita [1,*]

[1] Instituto de Conservación y Mejora de la Agrodiversidad Valenciana (COMAV), Universitat Politècnica de València, 46022 Valencia, Spain; carguire@etsia.upv.es (C.G.-R.); anadmar@alumni.upv.es (A.M.A.-M.); jprohens@btc.upv.es (J.P.); adrodbur@upvnet.upv.es (A.R.-B.)
[2] Facultade de Ciências Exatas e da Engenharia, Universidade da Madeira, 9000-082 Funchal, Portugal; katymi1102@hotmail.com
* Correspondence: anfifer@btc.upv.es; Tel.: +34-963-87-9418

Received: 20 November 2019; Accepted: 6 December 2019; Published: 7 December 2019

Abstract: Wall rocket (*Diplotaxis erucoides*) is a wild vegetable with the potential to become a crop of high antioxidant quality. The main bioactive compounds include ascorbic acid (AA), sinigrin, and a high content of total phenolic compounds (TP). It also accumulates nitrates. Since these compounds are affected by environmental conditions, adequate crop management may enhance its quality. Eleven accessions of wall rocket were evaluated under field and greenhouse conditions during two cycles (winter and spring) and compared to *Eruca sativa* and *Diplotaxis tenuifolia* crops. The three species did not differ greatly. As an exception, sinigrin was only identified in wall rocket. For the within-species analysis, the results revealed a high effect of the growing system, but this was low among accessions. The highest contents of AA and TP were obtained under field conditions. In addition, the levels of nitrates were lower in this system. A negative correlation between nitrates and antioxidants was determined. As a counterpart, cultivation in the field–winter environment significantly decreased the percentage of humidity (87%). These results are of relevance for the adaptation of wall rocket to different growing conditions and suggest that the field system enhances its quality. The low genotypic differences suggest that intra-species selections in breeding programs may consider other aspects with greater variation.

Keywords: ascorbic acid; *Diplotaxis erucoides*; field; greenhouse; new crops; nitrates; sinigrin

1. Introduction

Modern societies have become increasingly aware of the importance of diet as part of a healthy lifestyle. Thus, many consumers look for additional health benefits to be obtained from specific foods, which are known as functional foods [1,2]. On the other hand, some consumers are demanding products with new and differentiated aromas and tastes to enrich daily dishes and increase the culinary experience [3]. These demands offer an opportunity for the enhancement of wild edible plants (WEPs). In fact, several WEPs have high bioactive properties and may be considered as potential functional foods [4,5], while they have also differentiated organoleptic characteristics that are highly appreciated [6]. Apart from the direct harvest from the wild, a promising strategy for such revalorization could be domestication and adaptation into cultivation systems; this is an alternative that offers several advantages such as better yields, uniformity and accessibility [3].

Mediterranean cultures have a rich ethnobotanic knowledge and tradition in the consumption of WEPs, as has been compiled in many works (e.g., [7–9]). These reports show a great diversity of WEPs that have potential as new crops, including the edible *Diplotaxis erucoides* (L.) DC. (wall rocket). Wall rocket is an annual plant from the *Brassicaceae* family, broadly distributed along the Mediterranean areas of Europe and Africa to the Middle East [7]. Considered as a weed for many crops, the species is also appreciated as a wild vegetable for its tender leaves, with a characteristic pungent flavor, and also for its flowers as decorative elements. Wall rocket is eaten fresh or cooked, added to salads, soups, pasta dishes or even fried in omelettes [8,9]. One commercial variety of wall rocket is currently available (var. Wasabi, Shamrock Seed Company, Inc.), but as far as we know, its cultivation is negligible.

Wall rocket is taxonomically related to the popular rocket crops *Eruca sativa* Mill. (salad rocket) and *D. tenuifolia* (L.) DC. (wild rocket). These crops accumulate large amounts of phytochemicals including vitamin C, phenolic compounds and glucosinolates [10] and could therefore be considered for in vivo models and clinical assays addressed to test their potential bioactive properties. In fact, both vitamin C and phenolic compounds are potent antioxidants against plant oxidative stress, and such compounds could be also involved in reducing the risk of different illnesses such as cardiovascular diseases, hepatotoxicity and general inflammation risks [11–14]; in addition, vitamin C is an essential microelement with antiscorbutic activity [15]. Glucosinolates (GSLs) are secondary metabolites from *Brassicaceae* and other families within the *Brassicales* order [16]. The enzymatic hydrolysis of GSLs releases volatile compounds that are responsible of the bitter and pungent flavor of *Brassicaceae* species [17]. In addition, different compounds in this class have been analyzed in terms of their potential health benefits using in vivo models, as reviewed by Dinkova-Kostova and Kostov [18]. Together with the bioactive compounds, rocket crops also accumulate high amounts of nitrates [19,20], considered to be antinutrients with potential health risks [21]. Thus, maximum levels are established for the commercial production of rocket crops (*E. sativa* and *Diplotaxis* sp.) and other vegetables in Europe [22].

Rocket crops are cultivated under field and greenhouse conditions [23] and can be grown in the Mediterranean regions for most of the year. As a result, these crops are subjected to growth under variable agronomic and environmental factors that include temperature, length and incidence of sunlight, irrigation, soil type, or time of harvest, among others. These environmental changes can affect the accumulation of bioactive compounds [24]. For instance, an increase of light intensity and photoperiod can decrease the content of nitrates [25]. Stresses such as heat shock, chilling or high light conditions activate the accumulation of protective phytochemicals such as ascorbic acid or phenolic compounds [26]. Abiotic stresses such as growing under non-optimal temperature conditions can increase the content of glucosinolates as well [27].

Although these are general behaviors, information related to the effect of cultivation on wall rocket is scarce. Ceccanti et al. [4] suggested that, as part of the breeding programs of wild edible plants into new crops, it is important to study the proper cultivation practices to allow large-scale, high-yield production and, at the same time, ensure a good-quality product including nutritional quality. In this sense, testing different growing environments may lead to identifying the most adequate conditions for the development of wall rocket as a crop. In addition, the use of a local germplasm may help to establish the crop in the Mediterranean regions, as these materials would have been naturally selected for its adaptation to these conditions [28]. Thus, the current study aimed to analyze the effect of growing systems (greenhouse and field) and cycles (winter and spring) on selected compounds of relevance including ascorbic acid, sinigrin and nitrates, as well as the content of total phenolics, for pre-selected accessions of wall rocket derived from local germplasm. In addition, accessions of *D. tenuifolia* and *E. sativa* were used as reference materials with the aim of contextualizing the values obtained for wall rocket, as the acceptance of a new, differentiated crop will also depend on its recognisable differences from other crops. Overall, the study allows us to gain a general insight of the behaviour of wall rocket as a crop. Moreover, the current study can be useful for establishing a basis for the future exploitation of this emerging crop, which may have a high added value due to its content of bioactive compounds.

2. Materials and Methods

2.1. Plant Material and Cultivation

Ten pre-selected accessions of wall rocket and four commercial cultivars of rocket species were evaluated in the experiment. The pre-selected accessions corresponded to the second generation seedlings from wild populations collected in the Valencian Community (Spain) (Table S1). Seeds are conserved at the Universitat Politècnica de València (UPV, Valencia, Spain), where a domestication program is being developed. The commercial cultivars (from Shamrock Seed Co., Salinas, CA, USA) included the species *D. tenuifolia* (var. SSC2402 and var. Wild Rocket), *E. Sativa* (var. S. Rocket SSC2965), and the only commercial variety of *D. erucoides* that, to our knowledge, is currently available (var. Wasabi).

The experiments were performed at the UPV following the same experimental design as described in Guijarro-Real et al. [29]. Thus, two independent growing cycles were evaluated: the late autumn–winter season (hereafter called the winter season) and late winter–early spring season (hereafter called the spring season). In each cycle, assays were simultaneously carried out in two cultivation systems: a heated glasshouse (39°29′0″ N, 0°20′26″ W) and an experimental field under an anti-pest mesh (39°28′56″ N, 0°20′11″ W).

First of all, seeds were treated with a pre-germinative treatment in order to break the possibly secondary dormancy and increase the germination uniformity [30]. Thus, seeds were treated with commercial sodium hypochlorite 2.5% (v/v) for 5 min plus gibberellic acid 100 ppm (Duchefa Biochemie, Haarlem, The Netherlands) for 24 h. Treated seeds were sown in commercial Neuhaus Humin-substrat N3 substrate (Klasmann-Deilmann Gmbh, Geeste, Germany) and placed in a growing chamber with long day conditions (16/8 h, 25 °C) for two days. For materials used in the greenhouse system, sowing was directly performed in 40×25 cm^2 trays, in which plants remained for the entire experiment; plants used for the field system were instead sown in seedling trays.

Two days after being sown, trays were moved to a greenhouse. Trays used for the greenhouse system remained in these conditions during the entire experiment. In contrast, plants used in the field system were allowed to grow in the greenhouse until the appearance of the second true leaf and then were transplanted to the field until the end of the experiment. In both the greenhouse and field systems, the same experimental design was followed: a complete randomized block design with five blocks, with each block including one replicate of 30 plants per accession. This totals 8400 plants used for the experiments performed.

2.2. Preparation of Samples

All plants in each replicate were harvested together as a pool, except for plants with visible growing damages (e.g., a very small size compared to the average of the block) that were discarded. Samples were processed on the same day as harvesting. One fresh sub-sample was used for the analysis of ascorbic acid, and the rest were frozen at −80 °C and then lyophilized. The difference between the weight before and after lyophilization was used to calculate the percentage of moisture. The lyophilized material was powdered with a commercial grinder and stored in darkness until being analyzed for total phenolics, sinigrin and nitrates. All results were expressed as contents per each 100 g of fresh weight (FW) using the percentage of moisture for conversion, as this result provides a more appropriate value considering that the product is eaten raw.

2.3. Traits Measured

The content of ascorbic acid (AA) was measured according to Cano and Bermejo [31] with slight modifications. Briefly, 1.0 g of fresh material was homogenized with 5 mL of cold meta-phosphoric acid 3.0% (v/v) for 1 min using a mortar. The aqueous phase was filtered through a 0.22 μm PVDF filter and analyzed on a HPLC 1220 Infinity LC System (Agilent Technologies; Santa Clara, CA, USA) using a BRISA C$_{18}$ column (150 mm × 4.6 mm i.d., 3 μm particle size; Teknokroma; Barcelona, Spain). The mobile phase consisted of methanol: 1% acetic acid (5:95) for 15 min at a flow rate of 1 mL min^{-1}. The injection volume was 5 µL, and quantification was performed at 254 nm using an external standard calibration of *L*-ascorbic acid (Sigma-Aldrich, Saint Louis, MO, USA).

The content of sinigrin (SIN) was determined as described by Grosser and van Dam [32] with slight modifications. Firstly, 0.1 g of powdered samples was heated for 2 min at 75 °C using a Termoblock TD150 P2 (Falc Instruments, Treviglio, Italy) for myrosinase inactivation [33]. Extraction was then performed using 1 mL of methanol 70% (v/v) for 15 min at 75 °C. After centrifugation, the supernatant was collected. The extraction step was repeated with 1 mL of methanol 70% (v/v) for another 15 min at 75 °C. Both supernatants were mixed and injected into an SPE column containing a DEAE Sephadex anion exchanger (A-25, Sigma-Aldrich, Saint Louis, MO, USA) activated with 20 mM sodium acetate buffer (pH 5.5) and incubated with 20 µL of diluted sulfatase (Sigma-Aldrich, Saint Louis, MO, USA) overnight. Desulphonated sinigrin was eluted with 500 µL plus 500 µL of milliQ water and analyzed using the same HPLC apparatus as for AA analysis and a Luna® Omega C_{18} column (150 mm × 4.6 mm i.d., 3 µm particle size; Phenomenex, Torrance, CA, USA). The mobile phases consisted of acetonitrile (A) and water (B), with the following gradient: from 98% A to 65% A in 35 min, then equilibrated for 5 min to the initial conditions. The injection volume was 10 µL and the flow rate was 0.75 mL min^{-1}. Quantification was performed at 229 nm using desulphoned sinigrin hydrate (PhytoPlan, Heidelberg, Germany) as an external standard.

The content of total phenolics (TP) was determined according to the Folin–Ciocalteu procedure [34] as in Guijarro-Real et al. [35]. For that, 0.125 g of lyophilised material was extracted with 5 mL of acetone 70% (v/v) containing acetic acid 0.5% (v/v) for 24 h under continuous stirring. Aliquots of 65 µL were incubated with 500 µL of diluted Folin-Ciocalteu (1:10; Scharlab S.L., Sentmenat, Spain) for 5 min; then, 500 µL of sodium carbonate 60 g L^{-1} was added and incubated for other 90 min. Quantification was performed at 765 nm in a iMark™ Microplate Reader spectrophotometer (Bio-Rad, Hercules, CA, USA). Chlorogenic acid (Sigma-Aldrich) was used as an external standard and the results were expressed as mg of chlorogenic acid equivalents (mg CAE 100 g^{-1} FW).

Finally, the content of nitrates was determined using a nitrate-selective ion (Crison Instruments S.A., Alella, Barcelona, Spain), with an extraction protocol adapted from Egea-Gilabert et al. [36]. Nitrates from 0.1 g were extracted with 50 mL of distilled water for 15 min under continuous stirring and stabilized with 1 mL of 2 M diammonium sulfate $((NH_4)_2\ SO_4)$ buffer at the moment of measurement using the nitrate-selective ion.

2.4. Data Analysis

Data were subjected to a fixed effects model analysis of variance [37] using the Statgraphics Centurion XVII v.17.2 (Statpoint Technologies, Inc., Warrenton, VA, USA). Two different analyses were performed: (1) a comparison among materials from different species, and (2) a comparison among accessions of wall rocket. For the analysis of species, the average values for accession considering the five replicates per environment were used as data. Data were then submitted to a multivariate analysis of variance (ANOVA) and the effects of species (S, corresponding to three levels: wall rocket, wild rocket, salad rocket), environment (E, four levels: greenhouse in winter, field in winter, greenhouse in spring, field in spring) and the S × E interaction were tested. The linear model applied was

$$X_{ijk} = \mu + S_i + E_j + (S \times E)_{ij} + e_{ij(k)}$$

where X_{ijk} is the value for accession k of species i and environment j, μ is the general mean, S_i is the effect of the species i, E_j is the effect of the environment j, $(S \times E)_{ij}$ is the effect of the interaction between species i and environment j, and $e_{ij(k)}$ is the residual error of the accession k. Mean values and standard error were obtained for the three species and significant differences determined using the Student–Newman–Keuls multiple range test ($p = 0.05$).

The analysis of wall rocket aimed to study the presence of differences among accessions and/or among systems, considering each growing cycle independently [29]. Thus, individual data were submitted to a multivariate analysis of variance (ANOVA) and the effects of accession (A, eleven accessions), growing system (GS, two levels: greenhouse, field) and the A × GS interaction were tested. The linear model applied was

$$X_{ijkl} = \mu + A_i + B_{j(ik)} + GS_j + (A \times GS)_{ik} + e_{ijk(l)}$$

where X_{ijkl} is the value for replicate l of accession i in block j and growing system k, μ is the general mean, A_i is the effect of the genotype i, $B_{j(ik)}$ is the effect of block j for accession i and system k, GS_j is the effect of the growing system j, $(A \times GS)_{ik}$ is the effect of the interaction between accession i and system k, and $e_{ijk(l)}$ is the residual error of the replicate l. Mean values and standard errors were obtained, and significant differences among environments were determined according to the LSD test ($p = 0.05$). Accessions were ranked for their average values of AA, TP, SIN and NO_3^- within each environment, where high levels of AA, TP and SIN and low levels of NO_3^- were a positive trait. These ranks were then used to obtain a global ranking table for the eleven accessions of wall rocket. Finally, the Spearman rank coefficients of correlation (ρ) were calculated for phenotypic ($n = 44$) and environmental ($n = 213$) correlations.

3. Results

3.1. Differences among Materials of Different Species

Wall rocket was compared to the reference materials including two accessions of *D. tenuifolia* and one accession of *E. sativa* in terms of the percentage of moisture and contents of AA, TP and NO_3^-. Differences in the contents of SIN were not analyzed because this compound was only present in wall rocket. A significant effect of the environment was determined in the four traits evaluated (Table 1). This factor was the main contributor to the total sum of squares in all cases, with values ranging between 52.8% (NO_3^-) and 72.6% (TP). On the contrary, the species factor was only significant for the percentage in moisture and the content in AA. Their contribution was, in any case, lower than 10.5%. Finally, a significant S × E interaction was determined for all traits except for the content of AA, and the effect of this interaction accounted for up to 17.7% of the total sum of squares (Table 1).

Table 1. Sum of squares (in percentage, %) and degrees of freedom (*d.f*) for the effects of species (S) with three levels: wall rocket ($n = 44$), wild rocket ($n = 8$) and salad rocket ($n = 4$); environment (E) with four levels: greenhouse–winter, field–winter, greenhouse–spring, and field–spring; S × E interaction and residuals for the percentage in moisture and the content in ascorbic acid (AA), total phenolics (TP) and nitrates (NO_3^-).

Trait	Sum of Squares (%)			
	S	E	S × E	Residual
d.f	2	3	6	44
Moisture	4.21 *	56.22 *	17.69 ***	21.88
AA	10.22 ***	59.10 **	5.57 ns	25.12
TP	0.83 ns	72.61 **	7.74 *	18.82
NO_3^-	3.86 ns	52.78 *	13.98 **	29.38

ns, *, ** and *** indicate no significant or significant at $p < 0.05$, 0.01 and 0.001, respectively.

The average values for each trait are represented in Figure 1. The percentage of moisture was close to 90.0% for the three species, with salad rocket displaying the highest values on average. Both wild rocket and wall rocket significantly decreased the moisture of leaves under the field–winter environment, while the percentage in salad rocket was stable for the environments tested. Regarding the content of AA, wall rocket accumulated the highest value on average (70.02 mg AA 100 g^{-1}), at approximately 30% greater than wild rocket. The effect of the environment was similar in the three species, with the greenhouse environments providing the lowest values. As an exception, the accumulation of AA in salad rocket was not affected by the growing system (field or greenhouse) during the spring cycle (Figure 1).

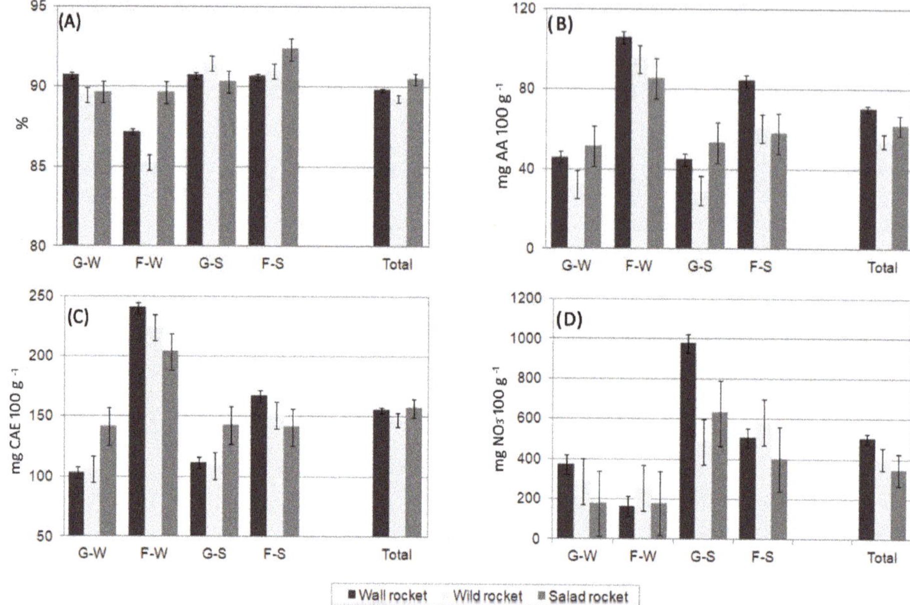

Figure 1. Mean values ± SE for the traits determined in each environment tested (greenhouse–winter, G-W; field–winter, F-W; greenhouse–spring, G-S; field–spring, F-S) for wall rocket ($n = 11$), wild rocket ($n = 2$) and salad rocket ($n = 1$), and global average values: (**A**) Percentage of moisture (%); (**B**) Content of ascorbic acid (AA, expressed as mg in 100 g^{-1} FW); (**C**) Content of total phenolics (TP, expressed as mg of chlorogenic acid equivalents, CAE, in 100 g^{-1} FW); (**D**) Content of nitrates (NO$_3^-$, expressed as mg in 100 g^{-1} FW).

Although no significant differences were established among species for the contents of TP and NO$_3^-$, both traits were affected by the S × E interaction (Table 1). The content of TP was not significantly affected by the environment for salad rocket, while the field environments increased the estimated TP in the other species (Figure 1). Finally, none of the three species showed total average values above 600 mg NO$_3^-$ 100 g^{-1}. However, the spring environments significantly increased the accumulation of these ions, with the maximum value obtained for wall rocket growing in the greenhouse–spring environment.

3.2. Variation among Wall Rocket Accessions

3.2.1. Effects of Accession, Growing System and Interaction

The effects of accession (A), growing system (GS) and A × GS interaction were independently analyzed for each growing cycle (Table 2). The winter cycle was highly affected by the growing system for all traits except for the content of NO$_3^-$. The contribution of this factor to the total sum of squares ranged between 16.5% (NO$_3^-$) and 81.1% (TP); moreover, this factor was the greatest contributor to the percentage of moisture, AA and TP (>50%). On the contrary, the contribution of the growing system to the total sum of squares was lower during the spring cycle, with percentages significantly decreasing for all traits (Table 2). Moreover, during this cycle, its effect was only significant for the contents of AA and TP. As in the winter cycle, it remained the main contributor to the total sum of squares for both AA and TP, accounting for 52.8% and 57.3%, respectively.

Table 2. Sum of squares (in percentage, %) and degrees of freedom (*d.f*) for the effects of accession (A, *n* = 11), growing system (GS, field or greenhouse), A × GS interaction, block and residuals for the percentage in moisture, ascorbic acid (AA), total phenolics (TP), sinigrin (SIN) and nitrates (NO_3^-) evaluated in the eleven accessions of wall rocket during the winter and spring cycles.

Cycle	Trait	A	GS	A × GS	Block	Residual
d.f		*10*	*1*	*10*	*8*	*80*
Winter	Moisture	3.63 ns	60.62 ***	2.92 ns	6.36	26.47
	AA	3.81 *	63.37 ***	3.26 ns	14.92	14.64
	SIN	10.00 *	35.89 **	4.19 ns	11.44	38.48
	TP	1.21 ns	81.11 ***	1.33 ns	5.42	10.93
	NO_3^-	2.84 ns	16.55 ns	3.04 ns	34.22	43.35
Spring	Moisture	1.72 ns	0.07 ns	6.14 ns	47.30	44.76
	AA	2.75 ns	52.85 ***	7.00 *	13.63	23.77
	SIN	11.63 ns	5.10 ns	14.96 ns	9.06	59.25
	TP	2.97 ns	57.35 **	3.54 *	22.76	13.38
	NO_3^-	9.53 ns	5.04 ns	9.49 ns	32.99	42.96

ns, *, ** and *** indicate no significance or significance at $p < 0.05$, 0.01 and 0.001, respectively.

On the other hand, the effect of accession was not significant for most traits in any of the cycles (Table 2). In fact, this factor only affected the contents of AA and SIN during the winter cycle, accounting for 3.8% and 10.0% of the total sum of squares, respectively. In a similar way, the A × GS interaction effects were mostly non-significant (Table 2). As an exception, an interaction effect was determined during the spring cycle for the contents in AA and TP.

3.2.2. Effects of Accession, Growing System and Interaction

The average values and dispersion for the different traits evaluated in each environment are summarized in Figure 2. Results were compared between systems for each cycle, while the indirect effect of the growing period was also evaluated by comparing within systems.

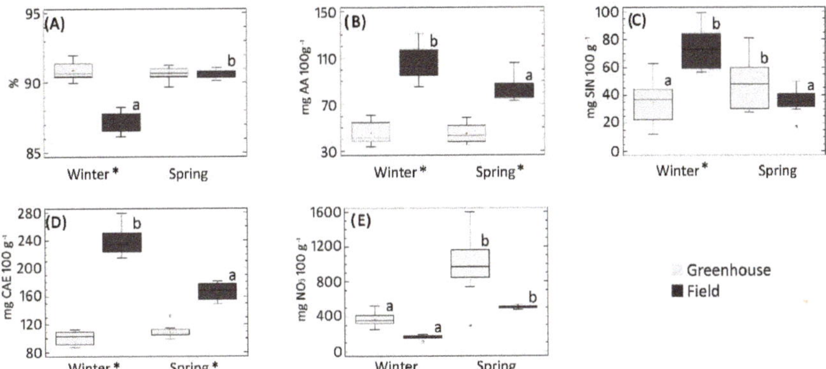

Figure 2. Box and whisker plot reflecting the average values (marked as +) and distribution for the traits evaluated in the accessions of wall rocket (*n* = 11) growing under greenhouse or field systems, during the winter and spring cycles. (**A**) Percentage of moisture (%). (**B**) Content of ascorbic acid (AA, expressed as mg in 100 g^{-1} FW). (**C**) Content of sinigrin (SIN, expressed as mg in 100 g^{-1} FW). (**D**) Content of total phenolics (TP, expressed as mg of chlorogenic acid equivalents, CAE, in 100 g^{-1} FW). (**E**) Content of nitrates (NO_3^-, expressed as mg in 100 g^{-1} FW). * indicates significant differences between systems within cycles according to the LSD test (*p* = 0.05). Different letters indicate significant differences between cycles within systems according to the LSD test (*p* = 0.05).

The percentage of moisture was only affected by the field–winter cycle. Plants growing under these conditions reduced the accumulation of water in leaf tissues in approximately 4% in comparison with the other environments tested, where the percentage in moisture was around 90.7%. On the contrary, the contents of AA, TP and SIN significantly increased when plants grew in the field-winter environment. Values under these conditions were more than two-fold greater with respect to the greenhouse system (Figure 2). A similar performance was found during the spring cycle for the contents of AA and TP but not for the content of SIN. Thus, the accumulation of AA and TP also increased for plants growing in the field system in the second cycle, but differences among systems were lower in this case.

An indirect effect of the growing period was also found (Figure 2). Plants growing in the greenhouse displayed the least differences between cycles. In this system, the cycle only affected the levels of SIN and NO_3^-, with plants growing in spring displaying the highest contents. Thus, the accumulation of NO_3^- displayed a 2.6-fold increase with respect to the winter cycle (Figure 2). In fact, this environment provided the maximum levels of NO_3^- considering the four environments (974 mg 100 g^{-1} FW). Regarding the field system, the results indicated that all traits were influenced by the growing cycle (Figure 2). Plants growing during the winter cycle had higher contents of AA, TP and SIN. The greatest increase was found for the levels of SIN, corresponding to a two-fold increase (35.4 mg 100 g^{-1} vs. 73.0 mg 100 g^{-1} FW). On the contrary, the winter cycle resulted in a reduction of the levels in NO_3^-. In fact, plants growing in this environment displayed the lowest accumulation (164 mg NO_3^- 100 g^{-1} FW) (Figure 2).

The accessions were ranked considering their bioactive properties as well as the levels of NO_3^- along the four environments tested (Table 3). The high content of bioactive compounds was considered as a positive trait, while the accumulation of NO_3^- was considered as negative. The commercial cv. Wasabi ranked second together with accession DER055-1. The first in rank was DER001-1, although it had a low rank position for the levels of NO_3^-. DER006-1was also very close to the commercial cultivar. On the contrary, accessions DER064-1 and DER085-1 had the lowest scores.

Table 3. Average values and coefficient of variation (CV, %) for the contents of ascorbic acid (AA), sinigrin (SIN), total phenolics (TP) and nitrates (NO_3^-) evaluated in the 11 accessions of wall rocket across the four environments tested, and overall rank. Contents are expressed as mg 100 g^{-1} FW, and the levels of total phenolics are expressed as equivalents of chlorogenic acid (CAE). $n = 4$.

Accession	n	AA		SIN		TP		NO_3^-		Rank
		Mean	CV	Mean	CV	Mean	CV	Mean	CV	
DER001-1	4	70.21	35.6	67.96	31.7	163.88	39.9	573.32	73.4	1
DER055-1	4	77.45	40.1	48.17	18.9	153.33	35.3	470.74	63.7	2.5
cv. Wasabi	4	73.61	54.5	47.93	51.1	168.10	48.7	319.33	49.8	2.5
DER006-1	4	68.91	40.9	55.29	47.2	158.84	39.3	494.37	76.0	4
DER031-1	4	67.70	32.9	59.76	38.8	149.84	39.3	527.23	87.2	5.5
DER089-1	4	64.21	54.3	44.01	48.9	143.00	39.5	542.74	61.3	5.5
DER081-1	4	64.50	41.2	48.52	38.2	152.74	40.8	656.95	97.8	7
DER045-1	4	62.30	44.9	43.10	52.4	141.96	39.3	493.40	55.1	8
DER067-1	4	69.55	55.2	33.99	55.9	159.66	51.1	438.57	54.2	9
DER064-1	4	79.71	47.5	37.40	56.6	155.16	46.8	522.47	75.2	10
DER085-1	4	71.82	39.6	39.62	29.8	148.97	38.2	502.18	66.6	11
Total	44	70.00	40.1	47.79	42.7	154.13	37.3	503.75	67.5	

3.2.3. Correlation between Nutritional Traits

All Spearman rank phenotypic correlations among traits were highly significant (Table 4). A positive correlation was found between the percentage of moisture and the levels of NO_3^- in the tissue ($\rho = 0.498$). On the contrary, both traits were negatively correlated with the content of bioactive compounds. Among them, the highest correlation coefficients were obtained among moisture and bioactive compounds, while the correlations with the content in NO_3^- were $\rho < -0.56$. On the contrary, positive correlations were found among bioactive compounds, with the content of AA–TP having the highest coefficient ($\rho = 0.921$) (Table 4).

Table 4. Phenotypic (above the symmetry axis, $n = 44$) and environmental (below the symmetry axis, $n = 213$) Spearman rank correlations (ρ) between the percentage in moisture, ascorbic acid (AA), total phenolics (TP), sinigrin (SIN), and nitrates (NO_3^-) determined in the accessions of wall rocket.

	Moisture	AA	TP	SIN	NO_3^-
Moisture		−0.7032 ***	−0.7832 ***	−0.7820 ***	0.4981 ***
AA	−0.5712 ***		0.9209 ***	0.4554 **	−0.5549 ***
TP	−0.8546 ***	0.6488 ***		0.6211 ***	−0.4898 **
SIN	−0.5878 ***	0.3670 ***	0.5545 ***		−0.3489 *
NO_3^-	0.2374 **	−0.2149 **	−0.3252 ***	−0.2489 ***	

ns, *, ** and *** indicate no significant or significant at $p < 0.05$, 0.01 and 0.001, respectively.

Similar results were obtained for the analysis of environmental correlations, with lower coefficients being generally found in this case (Table 4). The greatest decrease was found for the moisture –NO_3^- correlation ($\rho = 0.237$). A high reduction in the ρ coefficient comparing environmental vs. phenotypic correlations was also determined for the AA–NO_3^- content (−0.215 vs. −0.555, respectively). Finally, a moderate environmental correlation was found for the AA–TP ($\rho = 0.649$) (Table 4).

4. Discussion

Wall rocket is a common weed in the Mediterranean regions. However, it is also appreciated as a wild edible vegetable [38] and therefore has the potential to become a new crop. The present work aimed at studying the effect of different cultivation environments on the nutritional quality of pre-selected accessions of wall rocket. Due to the close phylogenetic relationships and similarities in terms of growth and commercial use, wall rocket may be potentially cultivated in similar environments as the already established rocket crops. Thus, it might be produced in field or greenhouse systems, although soil-less systems may also be available [23,39]. The greenhouse and field environments differ in several factors such as temperature, light intensity, air humidity or the effect of rains, among others [40]. These factors can also differ between growing cycles. Environmental factors have been proven to affect the leaf morphology of wall rocket [29]. In addition, environmental factors have been proven to influence the accumulation of secondary metabolites and nitrates in different crops [41–43]. Thus, the quality of wall rocket in terms of bioactive compounds and nitrates may be affected as well.

As phytochemicals, the contents of AA, TP, SIN and NO_3^- were evaluated. Only the reduced form of vitamin C was evaluated since we previously concluded that the AA form represented around 90% of the total vitamin C in wall rocket materials [44]. Differences between wall rocket and the reference materials were moderate and only significant for the content of AA and the percentage of moisture. Thus, the traits analyzed in the present work were not useful enough to clearly separate among them. These results are in contrast to our previous work evaluating the leaf morphology in the three species, where materials were clearly differentiated by the shape and size of leaves [29]. In consequence, the exploitation of distinctiveness in wall rocket with a commercial purpose could focus on other traits such as visual traits. As exception, wall rocket had as a distinctive trait the accumulation of SIN as main glucosinolate, since this compound was neither determined in *E. sativa* nor *D. tenuifolia* materials. These findings are in agreement with previous works comparing the glucosinolate profile of the three species [23,45]. Nevertheless, different profiles in other wall rocket materials, characterized for the absence of SIN, have been identified as well [45,46]. Discrepancies may correspond to differences related to the origin of materials, inter-subspecies differences—i.e., the analysis of *D. erucoides* subsp. *erucoides* or subsp. *longisiliqua* materials—as suggested by D'Antuono et al. [45], or they may even correspond to inter-specific crosses.

In a second analysis, the 11 accessions of wall rocket were compared among systems under two different growing periods. It has been previously observed that the growing period in a short-cycle species such as wall rocket can affect its morphology [29,47]. Moreover, the accumulation of different compounds such as nitrates can be affected by the growing period as well, as Bonasia et al. [48] described for wild rocket. Our results showed a low contribution of the accession effect to the total sum of squares, together with a general absence of significance. This result was an indicator of low nutritional variation among the accessions analyzed. The lack of variation may be due to the close geographic origin of materials, as original populations were collected in a relatively small territory. On the other hand, the low variation found may be due in part to a high intra-population variability considering that no homogenization efforts have been addressed, as suggested by the residual effect. However, a commercial cv. was included, which is assumed to be obtained from different populations, presumably with a different origin, and to show a high degree of uniformity. Thus, these low differences may also correspond to the low variation of wall rocket as a species in terms of bioactive compound contents and NO_3^- accumulation capacity. Nevertheless, some accessions could be considered in new programs as promising materials, including, for example, accessions DER055-1 and DER006-1. The latter was, in fact, also selected by its morphology as a promising material [29].

A comparison of different environments demonstrated a high effect on the final phytochemical composition of the product. Plants growing in the field during the winter cycle experienced the most extreme environment, both considering the two systems (greenhouse vs. field during the winter cycle) and the different growing periods (winter vs. spring in the field system). High adverse conditions took place during this growing cycle with remarkable low temperatures, so plants were subjected to high abiotic stresses. Abiotic stress increases the levels of reactive oxygen species, causing oxidative stress in plants [49]. As part of the defense response to this possible oxidative damage in plant tissues, the content of secondary metabolites such as AA and phenolic compounds can increase as well. Oh et al. [26] found that plants of lettuce exposed to cold stress increased the accumulation of protective metabolites by the activation of genes involved in their biosynthesis. In the same way, it has been observed among *Brassicaceae* that plants accumulate a greater content of glucosinolates when they grow under non-optimal temperatures [27], as our results suggest. In particular, it has been observed that a decrease in temperature can increase the accumulation of glucosinolates [50,51]. By contraposition, the filed-winter environment resulted in the accumulation of the lowest content in NO_3^-, which is also of interest for a commercial purpose. Light intensity has been positively correlated to nitrate reductase activity and a consequent lower accumulation of NO_3^- [48]. This season-dependence explains the different maximum limits established for lettuce and rocket crops in the European Union [22]. However, our experiment was conducted in two consecutive cycles with similar light exposure, and therefore light differences may be not be high enough to affect the reductase activity. Thus, other physiological processes may be related to this different accumulation.

The field–spring environment also provided a product accumulating high levels of pytochemicals of interest (AA, SIN, TP) and low levels of NO_3^-. The content of AA was significantly higher than previously described by Salvatore et al. [52], who found that mature, wild plants of wall rocket on average accumulated 13.9 mg AA 100 g^{-1}. Values were also comparable or even greater than levels of vitamin C (VC) previously described for wild rocket by Spadafora et al. [10] (22 mg 100 g^{-1} FW) or Durazzo et al. [24] (21–81 mg 100 g^{-1} FW). The accumulation of SIN, however, did not reach the levels previously described for wall rocket by Di Gioia et al. [23], with an average level of 11.6 mg g^{-1} DW. In addition, the content of NO_3^-, although greater in this cycle, was below the maximum limit of 7000 mg kg^{-1} established for the commercialization of rocket crops harvested before April [22]. Comparable or slightly lower levels have been previously found in cultivated rocket crops [43], usually ranging between 3500–4500 mg kg^{-1} but reaching 7349 mg kg^{-1}. Levels described for non-cultivated wall rocket would be between 2000–2500 mg kg^{-1} [53,54], suggesting that the species tends to increase this accumulation under cultivated conditions. Finally, the increased percentage of moisture was reflected in a greater visual appearance and less coriaceous aspect—traits that are essential for consumer acceptance.

In contrast, the greenhouse system may not be adequate for the commercial production of wall rocket according to our results. Heated greenhouses are used to provide a more appropriate and stable temperature for plant growth compared to field conditions, but also affect other factors such as wind, air humidity, solar radiation, the effect of rains and storms or crop management [40]. Thus, our results suggest that growing wall rocket under greenhouse conditions would enhance the homogenization of most of the traits evaluated but provide a product of lower quality, especially in terms of AA and TP contents. Moreover, plants in this system accumulated high levels of NO_3^-, which in the spring cycle exceeded the limits established [22] and made the product obtained not commercially acceptable.

Finally, phenotypic coefficients of correlations were greater than the environmental ones. These results indicated that the different factors evaluated had a similar effect on materials on average; however, the residuals among those traits had a lower correlation. The high correlation between AA and TP may be explained by their antioxidant function in the plants and their accumulation in response to environmental stresses [55]. However, both AA and TP had lower correlations with the content of SIN. As with AA and phenolic phytochemicals, the accumulation of glucosinolates in plant tissues is part of the plant response mechanism against abiotic stress conditions and can therefore be affected by environmental factors such as light intensity, season or fertilization [16,27,41], but it is also highly related to the biotic stress by pests and pathogens [16,27]. The combined effect of both biotic and abiotic stress could be responsible for this lower correlation. On the other hand, a negative correlation between the antioxidant phytochemicals and the percentage of moisture was found. This may correspond to a concentration effect in the tissues [41], although it could be also related to the plant behavior and defense system against cold stress. Król et al. [56] found that leaves of grapevine developed under cold stress reduce their percentage of moisture, although the total phenolics were also decreased; on the contrary, Oh et al. [26] found that the exposure of lettuce to chilling conditions increased the total phenolics against oxidative damage. Moreover, the negative correlation with these compounds and the content of NO_3^- has been previously observed [48], in agreement with our results. Finally, a positive correlation between the percentage of moisture and the levels of NO_3^- was found, as extensively observed in many species [57]. This positive correlation between both traits is related to the osmotic effect of NO_3^- ions, meaning that its accumulation increases the capacity of tissues to retain water [20,57].

5. Conclusions

This work aimed to evaluate the most adequate conditions for the establishment of wall rocket as a new crop. Our results indicated that growing this vegetable under field conditions would enhance the accumulation of AA and TP in the final product. Moreover, the accumulation of NO_3^- was reduced in this environment compared to the greenhouse system. Among all environments, the field–winter system resulted in the lowest content in NO_3^-, which is a trait of high interest for a commercial purpose, but also the lowest percentage of moisture, with this reducing the visual quality and presumably consumer acceptance. Thus, our results suggest that stressful conditions such as low temperatures in winter may not be adequate for commercial production in an unprotected field. In this sense, the use of crop thermal blankets may reduce such stress.

The low variability of the phytochemicals among accessions of wall rocket may reflect the low genotypic differences among the selected materials or at a species level. Moreover, the levels found in wall rocket did not clearly differ from the reference crops. As an exception, wall rocket had, as a distinctive trait, the presence of sinigrin as its main glucosinolate unlike the other species, as previously described [23,45]. These results increase the information available for the species and are of relevance for breeding programs and future commercial strategies, suggesting that the promotion of the distinctiveness of this new crop among the other rocket crops should focus on other aspects such as visual quality or flavor instead of bioactive traits.

Supplementary Materials: The following are available online at http://www.mdpi.com/2073-4395/9/12/858/s1, Table S1: Geographical location of the original ten wild populations of wall rocket.

Author Contributions: Conceptualization, A.F., A.R.-B. and J.P.; data curation, A.M.A.-M., C.G.-R. and K.A.; formal analysis, A.F., C.G.-R. and J.P.; investigation, A.M.A.-M., C.G.-R. and K.A.; methodology, A.F., C.G.-R. and J.P.; project administration, A.F., C.G.-R., and J.P.; resources, A.F., A.R.-B. and J.P.; supervision, A.F., A.R.-B. and J.P.; validation A.M.A.-M., and C.G.-R.; visualization, A.F., A.R.-B., C.G.-R., and J.P.; writing—original draft preparation, A.F., C.G.-R. and J.P.; writing—review and editing, A.F., A.R.-B. and J.P.

Funding: This research received no external funding.

Acknowledgments: C.G. is grateful to the Ministerio de Educación, Cultura y Deporte of Spain (MECD) for the financial support by means of a predoctoral FPU grant (FPU14-06798). The authors also thank Ms. E. Moreno and Ms. M.D. Lerma for their help in the field tasks.

Conflicts of Interest: The authors declare no conflict of interest.

References

1.   Nutrition Society. Scientific concepts of functional foods in Europe. Consensus document. *Br. J. Nutr.* **1999**, *81*, S1–S27. [CrossRef]

2.   Pinela, J.; Carocho, M.; Dias, M.I.; Caleja, C.; Barros, L.; Ferreira, I.C.F.R. Wild plant-based functional foods, drugs, and nutraceuticals. In *Wild Plants, Mushrooms and Nuts: Functional Food prOperties and Applications*; Ferreira, I.C.F.R., Morales, P., Barros, L., Eds.; John Wiley & Sons, Ltd.: Chichester, UK, 2017; pp. 315–352.

3.   Pinela, J.; Carvalho, A.M.; Ferreira, I.C.F.R. Wild edible plants: Nutritional and toxicological characteristics, retrieval strategies and importance for today's society. *Food Chem. Toxicol.* **2017**, *110*, 165–188. [CrossRef] [PubMed]

4.   Ceccanti, C.; Landi, M.; Benvenuti, S.; Pardossi, A.; Guidi, L. Mediterranean wild edible plants: Weeds or "New functional crops"? *Molecules* **2018**, *23*, 2299. [CrossRef] [PubMed]

5.   Guarrera, P.M.; Savo, V. Perceived health properties of wild and cultivated food plants in local and popular traditions of Italy: A review. *J. Ethnopharmacol.* **2013**, *146*, 659–680. [CrossRef] [PubMed]

6.   Serrasolses, G.; Calvet-Mir, L.; Carrió, E.; D'Ambrosio, U.; Garnatje, T.; Parada, M.; Vallès, J.; Reyes-García, V. A matter of taste: Local explanations for the consumption of wild food plants in the Catalan Pyrenees and the Balearic Islands. *Econ. Bot.* **2016**, *70*, 176–189. [CrossRef]

7.   Pignone, D.; Martínez-Laborde, J.B. Diplotaxis. In *Wild Crop Relatives: Genomic and Breeding Resources. Oilseeds*; Kole, C., Ed.; Springer: Berlin, Germany, 2011; pp. 137–147.

8.   Guarrera, P.M.; Savo, V. Wild food plants used in traditional vegetable mixtures in Italy. *J. Ethnopharmacol.* **2016**, *185*, 202–234. [CrossRef]

9.   Licata, M.; Tuttolomondo, T.; Leto, C.; Virga, G.; Bonsangue, G.; Cammalleri, I.; Gennaro, M.C.; Bella, S.L. A survey of wild plant species for food use in Sicily (Italy)-results of a 3-year study in four Regional Parks. *J. Ethnobiol. Ethnomed.* **2016**, *12*, 12. [CrossRef]

10. Spadafora, N.D.; Amaro, A.L.; Pereira, M.J.; Müller, C.T.; Pintado, M.; Rogers, H.J. Multi-trait analysis of post-harvest storage in rocket salad (*Diplotaxis tenuifolia*) links sensorial, volatile and nutritional data. *Food Chem.* **2016**, *211*, 114–123. [CrossRef]

11. Adikwu, E.; Deo, O. Hepatoprotective effect of vitamin C (ascorbic acid). *Pharmacol. Pharm.* **2013**, *4*, 84–92. [CrossRef]

12. Ashor, A.W.; Siervo, M.; Mathers, J.C. Vitamin C, antioxidant status, and cardiovascular aging. In *Molecular Basis of Nutrition and Aging: A Volume in the Molecular Nutrition Series*; Malavolta, M., Mocchegiani, E., Eds.; Academic Press: Cambridge, MA, USA, 2016; pp. 609–619. [CrossRef]

13. Procházková, D.; Boušová, I.; Wilhelmová, N. Antioxidant and prooxidant properties of flavonoids. *Fitoterapia* **2011**, *82*, 513–523. [CrossRef]

14. Panche, A.N.; Diwan, A.D.; Chandra, S.R. Flavonoids: An overview. *J. Nutr. Sci.* **2016**, *5*, e47. [CrossRef] [PubMed]

15. Mandl, J.; Szarka, A.; Bánhegyi, G. Vitamin C: Update on physiology and pharmacology. *Br. J. Pharmacol.* **2009**, *157*, 1097–1110. [CrossRef] [PubMed]

16. Gols, R.; van Dam, N.M.; Reichelt, M.; Gershenzon, J.; Raaijmakers, C.E.; Bullock, J.M.; Harvey, J.A. Seasonal and herbivore-induced dynamics of foliar glucosinolates in wild cabbage (*Brassica oleracea*). *Chemoecology* **2018**, *28*, 77–89. [CrossRef]

17. Bell, L.; Oruna-Concha, M.J.; Wagstaff, C. Identification and quantification of glucosinolate and flavonol compounds in rocket salad (*Eruca sativa*, *Eruca vesicaria* and *Diplotaxis tenuifolia*) by LC–MS: Highlighting the potential for improving nutritional value of rocket crops. *Food Chem.* **2015**, *172*, 852–861. [CrossRef]

18. Dinkova-Kostova, A.T.; Kostov, R.V. Glucosinolates and isothiocyanates in health and disease. *Trends Mol. Med.* **2012**, *18*, 337–347. [CrossRef]

19. Santamaria, P. Nitrate in vegetables: Toxicity, content, intake and EC regulation. *J. Sci. Food Agric.* **2006**, *86*, 10–17. [CrossRef]

20. Schiattone, M.I.; Viggiani, R.; Di Venere, D.; Sergio, L.; Cantore, V.; Todorovic, M.; Perniola, M.; Candido, V. Impact of irrigation regime and nitrogen rate on yield, quality and water use efficiency of wild rocket under greenhouse conditions. *Sci. Hortic.* **2018**, *229*, 182–192. [CrossRef]

21. Habermeyer, M.; Roth, A.; Guth, S.; Diel, P.; Engel, K.; Epe, B.; Fürst, P.; Heinz, V.; Humpf, H.U.; Joost, H.G.; et al. Nitrate and nitrite in the diet: How to assess their benefit and risk for human health. *Mol. Nutr. Food Res.* **2015**, *59*, 106–128. [CrossRef]

22. European Comission. Commission Regulation (EU) No. 1258/2011 of 2 December 2011 amending Regulation (EC) No. 1881/2006 as regards maximum levels for nitrates in foodstuffs. *Off. J. Eur. Union* **2011**, *L320*, 15–17.

23. Di Gioia, F.; Avato, P.; Serio, F.; Argentieri, M.P. Glucosinolate profile of *Eruca sativa*, *Diplotaxis tenuifolia* and *Diplotaxis erucoides* grown in soil and soilless systems. *J. Food Compos. Anal.* **2018**, *69*, 197–204. [CrossRef]

24. Durazzo, A.; Azzini, E.; Lazzè, M.; Raguzzini, A.; Pizzala, R.; Maiani, G. Italian wild rocket [*Diplotaxis tenuifolia* (L.) DC.]: Influence of agricultural practices on antioxidant molecules and on cytotoxicity and antiproliferative effects. *Agriculture* **2013**, *3*, 285–298. [CrossRef]

25. Cavaiuolo, M.; Ferrante, A. Nitrates and glucosinolates as strong determinants of the nutritional quality in rocket leafy salads. *Nutrients* **2014**, *6*, 1519–1538. [CrossRef]

26. Oh, M.M.; Carey, E.E.; Rajashekar, C.B. Environmental stresses induce health-promoting phytochemicals in lettuce. *Plant Physiol. Biochem.* **2009**, *47*, 578–583. [CrossRef]

27. Björkman, M.; Klingen, I.; Birch, A.N.E.; Bones, A.M.; Bruce, T.J.A.; Johansen, T.J.; Meadow, R.; Molmann, J.; Seljasen, R.; Smart, L.E.; et al. Phytochemicals of *Brassicaceae* in plant protection and human health-Influences of climate, environment and agronomic practice. *Phytochemistry* **2011**, *72*, 538–556. [CrossRef]

28. Sogbohossou, E.O.D.; Achigan-Dako, E.G.; Maundu, P.; Solberg, S.; Deguenon, E.M.S.; Mumm, R.H.; Hale, I.; Van Deynze, A.; Schranz, M.E. A roadmap for breeding orphan leafy vegetable species: A case study of *Gynandropsis gynandra* (*Cleomaceae*). *Hortic. Res.* **2018**, *5*, 2. [CrossRef]

29. Guijarro-Real, C.; Prohens, J.; Rodriguez-Burruezo, A.; Fita, A. Potential of wall rocket (*Diplotaxis erucoides*) as a new crop: Influence of the growing conditions on the visual quality of the final product. *Sci. Hortic.* **2019**, *258*, 108778. [CrossRef]

30. Guijarro-Real, C.; Adalid-Martínez, A.M.; Gregori-Montaner, A.; Prohens, J.; Rodríguez-Burruezo, A.; Fita, A. Factors affecting germination of *Diplotaxis erucoides* and their effect on selected quality properties of the germinated products. *Sci. Hortic.* **2019**. [CrossRef]

31. Cano, A.; Bermejo, A. Influence of rootstock and cultivar on bioactive compounds in citrus peels. *J. Sci. Food Agric.* **2011**, *91*, 1702–1711. [CrossRef]

32. Grosser, K.; van Dam, N.M. A straightforward method for glucosinolate extraction and analysis with High-Pressure Liquid Chromatography (HPLC). *J. Vis. Exp.* **2017**, *121*, e55425. [CrossRef]

33. Pasini, F.; Verardo, V.; Cerretani, L.; Caboni, M.F.; D'Antuono, L.F. Rocket salad (*Diplotaxis* and *Eruca* spp.) sensory analysis and relation with glucosinolate and phenolic content. *J. Sci. Food Agric.* **2011**, *91*, 2858–2864. [CrossRef]

34. Singleton, V.; Rossi, J. Colorimetry of total phenolics with phosphomolybdic phosphotungstic acid reagents. *Am. J. Enol. Viticult.* **1965**, *16*, 144–158. [CrossRef]

35. Guijarro-Real, C.; Prohens, J.; Rodriguez-Burruezo, A.; Adalid-Martínez, A.M.; López-Gresa, M.P.; Fita, A. Wild edible fool's watercress, a potential crop with high nutraceutical properties. *PeerJ* **2019**, *7*, e6296. [CrossRef] [PubMed]

36. Egea-Gilabert, C.; Ruiz-Hernández, M.V.; Parra, M.Á.; Fernández, J.A. Characterization of purslane (*Portulaca oleracea* L.) accessions: Suitability as ready-to-eat product. *Sci. Hortic.* **2014**, *172*, 73–81. [CrossRef]

37. Gomez, K.A.; Gomez, A.A. *Statistical Procedures for Agricultural Research*, 2nd ed.; John Wiley & Sons, Inc.: Manila, Phillipines, 1984; p. 704. ISBN 978-0-471-87092-0.

38. Parada, M.; Carrió, E.; Vallès, J. Ethnobotany of food plants in the Alt Empordà region (Catalonia, Iberian Peninsula). *J. Appl. Bot. Food Qual.* **2011**, *84*, 11–25.

39. Egea-Gilabert, C.; Fernández, J.A.; Migliaro, D.; Martínez-Sánchez, J.J.; Vicente, M.J. Genetic variability in wild vs. cultivated *Eruca vesicaria* populations as assessed by morphological, agronomical and molecular analyses. *Sci. Hortic.* **2009**, *121*, 260–266. [CrossRef]

40. Figàs, M.R.; Prohens, J.; Raigón, M.D.; Pereira-Dias, L.; Casanova, C.; García-Martínez, M.D.; Rosa, E.; Soler, E.; Plazas, M.; Soler, S. Insights into the adaptation to greenhouse cultivation of the traditional Mediterranean long shelf-life tomato carrying the alc mutation: A multi-trait comparison of landraces, selections, and hybrids in open field and greenhouse. *Front. Plant Sci.* **2018**, *9*, 1774. [CrossRef]

41. Bell, L.; Wagstaff, C. Enhancement of glucosinolate and isothiocyanate profiles in *Brassicaceae* crops: Addressing challenges in breeding for cultivation, storage, and consumer-related Traits. *J. Agric. Food Chem.* **2017**, *65*, 9379–9403. [CrossRef]

42. Colonna, E.; Rouphael, Y.; Barbieri, G.; De Pascale, S. Nutritional quality of ten leafy vegetables harvested at two light intensities. *Food Chem.* **2016**, *199*, 702–710. [CrossRef]

43. Weightman, R.; Huckle, A.; Ginsburg, D.; Dyer, C. Factors influencing tissue nitrate concentration in field- grown wild rocket (*Diplotaxis tenuifolia*) in southern England. *Food Addit. Contam.* **2012**, *29*, 1425–1435. [CrossRef]

44. Guijarro-Real, C.; Rodriguez-Burruezo, A.; Prohens, J.; Adalid-Martínez, A.M.; Fita, A. Influence of the growing conditions in the content of vitamin C in *Diplotaxis erucoides*. *Bull. Univ. Agric. Sci. Vet. Med. Cluj Napoca Hortic.* **2017**, *74*, 144–145. [CrossRef]

45. D'Antuono, L.F.; Elementi, S.; Neri, R. Glucosinolates in *Diplotaxis* and *Eruca* leaves: Diversity, taxonomic relations and applied aspects. *Phytochemistry* **2008**, *69*, 187–199. [CrossRef] [PubMed]

46. Bennett, R.N.; Rosa, E.A.S.; Mellon, F.A.; Kroon, P.A. Ontogenic profiling of glucosinolates, flavonoids, and other secondary metabolites in *Eruca sativa* (*Salad rocket*), *Diplotaxis erucoides* (*Wall rocket*), *Diplotaxis tenuifolia* (*Wild rocket*), and *Bunias orientalis* (*Turkish rocket*). *J. Agric. Food Chem.* **2006**, *54*, 4005–4015. [CrossRef] [PubMed]

47. Sans, F.X.; Masalles, R.M. Life-history variation in the annual arable weed *Diplotaxis erucoides* (Cruciferae). *Can. J. Bot.* **1994**, *72*, 10–19. [CrossRef]

48. Bonasia, A.; Lazzizera, C.; Elia, A.; Conversa, G. Nutritional, biophysical and physiological characteristics of wild rocket genotypes as affected by soilless cultivation system, salinity level of nutrient solution and growing period. *Front. Plant Sci.* **2017**, *8*, 300. [CrossRef]

49. Fita, A.; Rodríguez-Burruezo, A.; Boscaiu, M.; Prohens, J.; Vicente, O. Breeding and domesticating crops adapted to drought and salinity: A new paradigm for increasing food production. *Front. Plant Sci.* **2015**, *6*, 978. [CrossRef]

50. Kissen, R.; Eberl, F.; Winge, P.; Uleberg, E.; Martinussen, I.; Bones, A.M. Effect of growth temperature on glucosinolate profiles in *Arabidopsis thaliana* accessions. *Phytochemistry* **2016**, *130*, 106–118. [CrossRef]

51. Steindal, A.L.H.; Rdven, R.; Hansen, E.; Mlmann, J. Effects of photoperiod, growth temperature and cold acclimatisation on glucosinolates, sugars and fatty acids in kale. *Food Chem.* **2015**, *174*, 44–51. [CrossRef]

52. Salvatore, S.; Pellegrini, N.; Brenna, O.V.; Del Rio, D.; Frasca, G.; Brighenti, F.; Tumino, R. Antioxidant characterization of some sicilian edible wild greens. *J. Agric. Food Chem.* **2005**, *53*, 9465–9471. [CrossRef]

53. Bianco, V.V.; Santamaria, P. Nutritional value and nitrate content in edible wild species used in Southern Italy. *Acta Hortic.* **1998**, *476*, 71–90. [CrossRef]

54. Disciglio, G.; Tarantino, A.; Frabboni, L.; Gagliardi, A.; Giuliani, M.M.; Tarantino, E.; Gatta, G. Qualitative characterisation of cultivated and wild edible plants: Mineral elements, phenols content and antioxidant capacity. *Ital. J. Agron.* **2017**, *12*, 1036. [CrossRef]

55. Orsini, F.; Maggio, A.; Rouphael, Y.; De Pascale, S. "Physiological quality" of organically grown vegetables. *Sci. Hortic.* **2016**, *208*, 131–139. [CrossRef]

56. Król, A.; Amarowicz, R.; Weidner, S. The effects of cold stress on the phenolic compounds and antioxidant capacity of grapevine (*Vitis vinifera* L.) leaves. *J. Plant Physiol.* **2015**, *189*, 97–104. [CrossRef]

57. Cárdenas-Navarro, R.; Adamowicz, S.; Robin, P. Nitrate accumulation in plants: A role for water. *J. Exp. Bot.* **1999**, *50*, 613–624. [CrossRef]

Article

Nutraceutical Uses of Traditional Leafy Vegetables and Transmission of Local Knowledge from Parents to Children in Southern Benin

Amandine D. M. Akakpo and Enoch G. Achigan-Dako *

Laboratory of Genetics, Horticulture & Seed Science, Faculty of Agronomic Sciences,
University of Abomey-Calavi, Abomey-Calavi 01BP526, Benin; amandineakakpo@gmail.com
* Correspondence: enoch.achigandako@uac.bj

Received: 28 October 2019; Accepted: 16 November 2019; Published: 26 November 2019

Abstract: This study assessed differences on the uses and transmission of traditional knowledge (TK) about three traditional leafy vegetables (*Crassocephalum crepidioides* (Juss. ex Jacq.) S. Moor, *Launaea taraxacifolia* (Willd.) Amin ex C. Jeffrey, and *Vernonia amygdalina* Del.) of the Asteraceae family over two generations in three villages: Adjohoun, Dangbo, and Pobè (southern Benin). Individual semi-structured ethnobotanical interviews of 360 respondents were conducted in the villages with young girls, boys, and their two parents. The relative frequency of citation, use value, and Jaccard similarity index were used for data analyses. *Vernonia amygdalina* was the most commonly known and used vegetable in all villages, while *L. taraxacifolia* was confined to Pobè. Factors such as village of survey, generation, and gender affected the use value of the species, but the patterns of recognition and cultivation were species-specific. Leaves were the most used plant part. Traditional knowledge was largely acquired from parents (90% of citation), and both mothers and fathers transmitted a similar amount of knowledge to their progenies. The knowledge on *V. amygdalina* was transmitted to a larger scale than knowledge of *C. crepidioides* and *L. taraxacifolia*. Irrespective of the species, transmission of TK was higher in Pobè. Gender and generation knowledge dynamic hypothesis is species-specific. TK transmission was species-specific too and may be linked to the local importance and use of those resources. These findings will inform strategies and programs for the sustainable use and conservation of leafy vegetables in local communities and national research and development institutions.

Keywords: *Crassocephalum crepidioides*; ethnobotany; generations; knowledge dynamics; *Launaea taraxacifolia*; use value; *Vernonia amygdalina*

1. Introduction

Nutraceutical plants loosely allude to species that are used as food or parts of food, and provide medicinal or health benefits—including the prevention and treatment of diseases—based on the perceived health properties of those plants [1]. For instance, several vegetables [2–4] and aromatic plants [5] were recognized as nutraceutical crops and are actively sought out for direct consumption or to be added to meals for their therapeutic or nutrient-rich properties. Those species exhibit specific bioactive compounds that can be used to target a range of diseases. In addition, they contain valuable nutritive elements, such as vitamins, minerals, essential oils, and antioxidants, that play a vital role in healthcare while serving as foods [6]. Recent ethnobotanical studies revealed that vegetable plants are by-and-large consumed by local people in Africa [3,7,8]. In Benin, 245 species belonging to 62 families are consumed as traditional vegetables. Among them, Asteraceae have the widest diversity with 29 species [9,10]. In addition, 41 species, including 13 leafy vegetables, have been listed as neglected and underutilized among which *Crassocephalum crepidioides* e (Voucher ID: Ganvié: Zon 369 at Benin National Herbarium), *Launaea taraxacifolia* (Voucher ID: Grand-Popo: Lisowski 0-1141 at Benin

National Herbarium), and *Vernonia amygdalina* (Voucher ID: Massi: Ayichédéhou 434 at Benin National Herbarium). These vegetables are currently among the top ten most consumed leafy vegetables in local communities [11], and exhibit high nutrient content (Table 1), but are neglected by research and development organizations. Moreover, they have difficult sexual reproduction and seeds are scarcely conserved in any national or international germplasms. In southern Benin, *C. crepidioides* is considered to be on the decline and among the most endangered wild edible plants [12] similarly to *L. taraxacifolia* [13]. These species are also exploited in traditional healthcare systems, but have been less explicitly documented with respect to their nutraceutical uses.

The knowledge of use value of the three vegetables is relevant not only for economic promotion, but more importantly they are element of the local people's culture which unfortunately is eroded through globalization and westernization [3,14,15] while questions arose about how social dynamics and human traits affect plant selection [3,16,17].

The transmission of traditional knowledge across generations has received increasing interest given the ongoing adverse effects in rural areas, whereby young people are increasingly disinterested in traditional or rural beliefs and life styles [14,15,18]. While this issue has been central in conservation biology and the management of plant genetic resources and agrobiodiversity [14,15,18], there are only fewer case studies in West African regions, despite their instrumental importance. Cultural knowledge transmission can occur between individuals of different generations. However, it most often occurs within genealogy (vertical transmission), as is the case from parent to child [19]. Indeed, parents, share their context with their children, and hence play a crucial role in transmitting knowledge in early life [20]. It may also occur between individuals of the same generation, even with different genealogy. However, vertical transmission i.e., across generations, in particular from parents to children, has been found to be highly conservative [14].

This study was therefore conducted to address the following questions: (i) what traditional knowledge is held on the uses of *C. crepidioides*, *L. taraxacifolia*, and *V. amygdalina*, three leafy vegetables of the Asteraceae family? (ii) How does this knowledge change across villages, age groups, and species? (iii) How does the knowledge change across generations, in particular, from parents to their children?

There are several evidences that the use of species varies according to informants' age and gender [21,22]. At the same time, other evidences indicated a neutral relationship [23], suggesting a context-specific relationship [24]. As a result, many recent studies have advocated for species-specific analyses for optimal science-based decision making and policies [24]. In addition, not all species have the same importance for local people, some are more important than others and are therefore the subject of special attention [25]. Among the three species, *V. amygdalina* is the most common in both rural and urban areas. The other two species are relatively less common [7]. Therefore, we predict to detect a significant gender and age relationship across localities for these two species and a neutral relationship for *V. amygdalina*. In addition, the transmission of local knowledge may depend on the importance of the species. The more a species is important, the more people hold and share knowledge on its uses. As a result, we hypothesized that the transmission of local knowledge from parents to children is species-specific. In particular, we predict to detect better knowledge transmission for *V. amygdalina*, while a relatively lower knowledge transmission for the other two species may prevail.

Table 1. Nutrient content of *Vernonia amygdalina*, *Launaea taraxacifolia*, and *Crassocephalum crepidioides*, three leafy vegetables consumed in southern Benin (values are collated from Yang and Keding [4]; Denton [26]; Stadlmayr et al. [27]; Adebisi [28]; Fomum [29]).

Nutrients	*V. amygdalina*	*L. taraxacifolia*	*C. crepidioides*
Energy (Kcal)	52–62	44	64
Water (g)	82.6	84.3	79.9
Protein (g)	5.2	3.2	3.2
Fat (g)	0.4	0.8	0.7
Carbohydrate (g)	8.5–10	8.3	14.0
Fiber (g)	1.5	2	1.9
Ash (g)	1.7	-	-
Ca (mg)	145	326	260
Fe (mg)	5.0	-	-
P (mg)	67.0	58	52
Vit C (mg)	51	-	-

2. Materials and Methods

The study was carried out in three villages of southern Benin with different sociolinguistic groups where the three species were reported [30]. We considered sociolinguistic group as "a group in which a member inherits a common language of communication and shares social attributes such as customs, history, and food habits" [9]. Villages included Adjohoun, Dangbo, and Pobè (Figure 1). The district of Adjohoun is located in the Department of Ouémé, 32 km north of Porto-Novo. Its total area is nearly 308 km^2. Its population is about 74,956 inhabitants [31]. It is composed of 48% men and about 52% women. Adjohoun is mainly rural and more than 80% of the active population is employed in agriculture. The district is mainly populated by the Wémè sociolinguistic group and to a lesser extent by Goun and Yoruba. The district of Dangbo is also located in the Department of Ouémé and covers about 149 km^2 with a population size of about 95,908 [31], of which 49% are men and 51% are women. The main sociolinguistic group is Aïzo. The third district is Pobè, located in the Department of Plateau. It covers about 400 km^2. The population of Pobè is around 123,740 inhabitants and composed of 48% men and 52% women. The main sociolinguistic groups is Holly with a minority of Nagot people [32]. All three sociolinguistic groups belong to the 'Kwa' linguistic categories. However, Aïzo and Wémè belong to the 'gbe' subcategory dominated by the Fon/Adja group, while Holly belongs to the 'ede' subcategory dominated by the Yoruba linguistic group [32]. There are differences among sociolinguistic groups and these are related to customs and beliefs, traditional religions, and practices. Communities sharing the same cultural background are likely to have convergent patterns of plant knowledge and use [33], with the exceptions that people of the same sociocultural background living in contrasting phytogeographical areas may also have different plant knowledge [9].

The three municipalities are under a tropical humid climate. Two dry seasons (mid-November to mid-March and mid-July to mid-September) that alternate with two rainy seasons (mid-March to mid-July and mid-September to mid-November). The native vegetation is composed of fallows and small semi-deciduous forest patches often of less than 5 ha, most of which are sacred groves. There are also some swamp forests especially along the Ouémé valley.

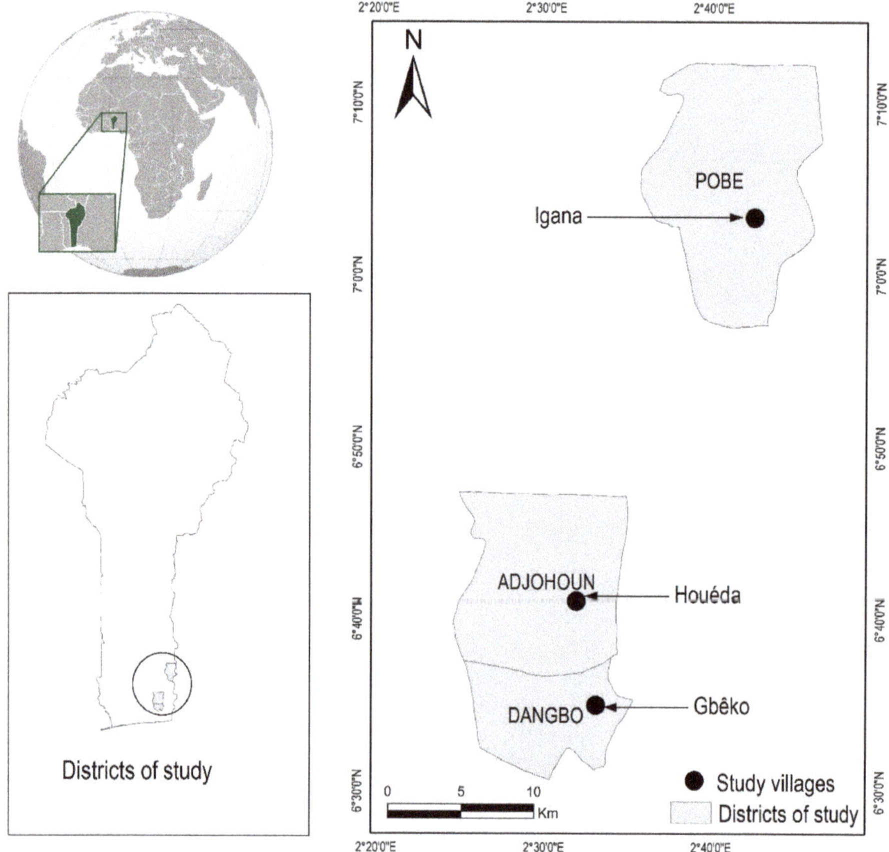

Figure 1. Map showing the municipalities and the villages where the study was carried out.

The objective of the study was first explained to village authorities and also to each of the informants to have their consent for participation prior to interviews. Only individuals that consented to participate in the study were considered. Forty young people aged 15–18 years (20 girls and 20 boys) and their two parents (fathers and mothers) were randomly selected in each village for a total of 120 informants per village, totalling 360 informants for the whole study. The random selection was implemented using the table of random numbers applied to a list of target young people previously established for each village. Young respondents with a single parent were not considered in the sample. The study was conducted in the village of Gbéko (in Dangbo) dominated by the Aïzo sociolinguistic group, the village of Houèda (in Adjohoun) dominated by the Wémè sociolinguistic group, and the village of Igana (in Pobè) dominated by the Holly sociolinguistic group (Table 2). These communities were selected based on previous reports on their distribution [30,32]. In addition, a pilot survey was carried out in each community to confirm the production of at least one of the three traditional leafy vegetables targeted and the good representation of the different sociolinguistic groups. Data were collected through individual interviews with the assistance of experienced local translators that were previously trained. Primary data included: (i) sociodemographic data, (ii) recognition and cultivation of the three leafy vegetable species, (iii) knowledge on the uses of the three species, and (iv) sources of knowledge held by the informants.

Table 2. Sociodemographic characteristics of informants (village, age, gender, and sociolinguistic group), number of informants surveyed, and their instruction levels. [†] mean (standard error).

Villages	Age Categories	Gender	Number	Age (years) [†]	Sociolinguistic Groups (%)				Instruction Levels			
					Wémè	Aïzo	Goun	Holly	Illiterate	Primary	Secondary	University
Adjohoun	Parent	Men	40	48.05 (1.94)	87.5	0.0	12.5	0.0	47.5	40.0	7.5	5.0
		Women	40	39.95 (1.04)	85.0	0.0	12.5	2.5	75.0	25.0	0.0	0.0
	Child	Boys	20	16.40 (0.27)	85.0	0.0	15.0	0.0	0.0	35.0	65.0	0.0
		Girls	20	16.30 (0.34)	90.0	0.0	10.0	0.0	0.0	15.0	85.0	0.0
Dangbo	Parent	Men	40	48.08 (1.78)	0.0	100.0	0.0	0.0	40.0	40.0	17.5	2.5
		Women	40	44.95 (1.26)	7.5	92.5	0.0	0.0	75.0	22.5	2.5	0.0
	Child	Boys	20	16.45 (0.23)	0.0	100.0	0.0	0.0	0.0	0.0	100.0	0.0
		Girls	20	16.80 (0.29)	0.0	100.0	0.0	0.0	25.0	20.0	55.0	0.0
Pobè	Parent	Men	40	46.11 (1.35)	0.0	0.0	0.0	100.0	80.0	7.5	12.5	0.0
		Women	40	42.44 (1.36)	0.0	0.0	0.0	100.0	95.0	2.5	2.5	0.0
	Child	Boys	20	15.62 (0.15)	0.0	0.0	0.0	100.0	0.0	14.29	85.71	0.0
		Girls	20	16.26 (0.30)	0.0	0.0	0.0	100.0	5.26	10.53	84.21	0.0

Descriptive statistics based on the relative frequency of citation (RFC, %) was used to describe the pattern of recognition and cultivation of the target traditional leafy vegetables.

The analysis of the knowledge of the uses of the vegetables was based on four ethnobotanical indices. Total use value (UV_{Total}), food use value (UV_{Food}), medicinal use value ($UV_{Medicinal}$), plant part use value (PPUV), and the relative frequency of citation (RFC) of each specific use were reported. The use value was adapted from Etongo et al. [34], while PPUV was adapted from Gomez-Beloz [35] and RFC was similar to the fidelity level of Friedman et al. [36].

The total use value (UV_{Total}) of the species s is the average total number of uses reported by all informants for that species, which is the sum of the average number of uses reported in the food use category (UV_{Food}) and the average number of uses reported in the medicinal use category ($UV_{Medicinal}$) as described in Equations (1) and (2).

$$UV_{Total} = UV_{Food} + UV_{Medicinal} \tag{1}$$

$$UV_{Total} = \sum_{i=1}^{n} \frac{UR_{Food_i}}{n} + \sum_{i=1}^{n} \frac{UR_{Medicinal_i}}{n} \tag{2}$$

where $UR_{Food\,i}$ is the total number of food uses reported for species s by informant i, $UR_{Medicinal\,i}$ is the total number of medicinal uses reported for species s by informant I, and n is the number of informants in the sociocultural category considered for computation.

To test whether the use value of a species correlated with generation (children versus parents), gender (men and women) and village (Igana, Houéda, Gbéko), a Poisson generalized linear model (GLM) was performed with generation, gender, and village as predictors and use values as response variables. The full model (i.e., with all possible interactions) was first established. Then, the most parsimonious model (i.e., the one with the minimum significant predictors) was searched using backward elimination based on likelihood ratio tests.

The plant part use value (PPUV) measures the use value of each plant part for each species. It is the average number of uses reported for that plant part for a given species. For a plant part k of a species s, $PPUV_{k,s}$ was computed as follows:

$$PPUV_{k,s} = \sum_{i=1}^{n} \frac{UR_{k,s,i}}{n} \tag{3}$$

where $UR_{k,s,i}$ is the number of reported uses by the informant i for the plant part k of the species s.

PPUV was used to identify the most used plant part for each species. To test whether the PPUV depends on the plant part, a Poisson GLM was also used.

The relative frequency of citation (RFC) is a measure of how much informants agree on a specific use (i.e., the consensus of informants on that use or knowledge). For each specific use m of each species s, RFC (RFC_{ps}) was computed as the number of informants who cited the specific use m ($n_{m,s}$) divided by the total number of informants for the species s (N_s) [36].

$$RFC_{m,s}(\%) = \frac{n_{m,s}}{N_s} \times 100 \tag{4}$$

The transmission of local knowledge from parents to child was assessed by two means. First, frequency analysis was used to describe the sources of knowledge (from who the knowledge was obtained). Second, the Jaccard similarity index was calculated to quantify the degree of similarity in knowledge held by children (distinguishing boys and girls) with that of their parents (distinguishing fathers and mothers) as indicated in Equation (5).

$$\text{Jaccard similarity} = \frac{a}{a+b+c} \times 100 \tag{5}$$

where a is the number of uses common to a child (boy or girl, taken separately) and the parent (mother or father, taken separately), b is the number of uses reported only by the parent, and c is the number of uses reported only by the child.

Two-sample T-tests were used to compare child/parent knowledge similarities for each species in each village. An analysis of variance (ANOVA) on log-transformed similarities (to meet ANOVA assumptions of normality and homoscedasticity) was performed to test whether the degree of shared knowledge varied with the gender of the child (boy *vs* girl), gender of parent (mother *vs* father), and village. The initial full model including main and interaction effects was simplified to minimum adequate using a stepwise selection. All statistical analyses were performed in R version 3.3.2 [37].

3. Results

3.1. Sociodemographic Characteristics of Informants

Mothers were generally younger (39.95 ± 1.04 to 44.95 ± 1.26 years old) than fathers (46.11 ± 1.35 to 48.08 ± 1.78 years old). Girls and boys had similar ages, 15.62 ± 0.15 years and 16.80 ± 0.29 years, respectively. In addition, most of the children were educated, in contrast to their parents (100%, 87.5%, and 97.4% educated children against 38.8%, 42.5%, and 12.5% educated parents in Adjohoun, Dangbo, and Pobè, respectively) (Table 2). Most of the children had a secondary school level of education (≥ 55%) (Table 2). The illiteracy rate among parents was higher in Pobè than in the other two villages (Table 2). Irrespective of the village, illiteracy rate was higher (≥75%) among mothers than fathers (Table 2). Agriculture and retailing were the main activities of the parents and occupied 46% and 14% of the respondents, respectively. Children mostly went to school (greater than 60%).

3.2. Pattern of Recognition and Cultivation of Traditional Leafy Vegetables

The recognition and cultivation patterns exhibited different topology and changed mainly across villages and species (Figure 2). Overall, recognition and cultivation perception of the three leafy vegetables changed among villages, species, and informant generations. *Vernonia amygdalina* was the most recognized and cultivated, while *L. taraxacifolia* was the least recognized and cultivated.

We observed that *V. amygdalina* was by-and-large recognized by children and parents in all three villages irrespective of the gender and generation. *C. crepidioides* and *L. taraxacifolia* were recognized by children and parents only in Pobè (100% relative frequency of citation), a Holly sociolinguistic community. In Dangbo (populated by Wémè), parents showed a citation frequency higher than 80% for *L. taraxacifolia* and *C. crepidioides*; about 30% of children informants only cited the two species in that village. In Adjohoun (Aïzo community), the citation frequencies were even lower for those two species for both parents and children (<50% and <20%, respectively) compared with Dangbo and Pobè.

Cultivation citations showed different patterns. *Vernonia amygdalina* was cultivated in all three villages by both children and parents with no gender difference, though in Pobè the cultivation citation frequencies were lower than that in the other two villages. *Crassocephalum crepidioides* was cultivated by very few parents in the Aïzo sociolinguistic group in Adjohoun; in Dangbo, it was cultivated by fewer young informants (less than 20%) and a relatively moderate number of parents (about 50%). In the Holly community in Pobè, *C. crepidioides* was cultivated by a higher number of children and parents, irrespective of the gender. Citation frequencies of the cultivation of *L. taraxacifolia* were overall low in Pobè and nil in the other two villages.

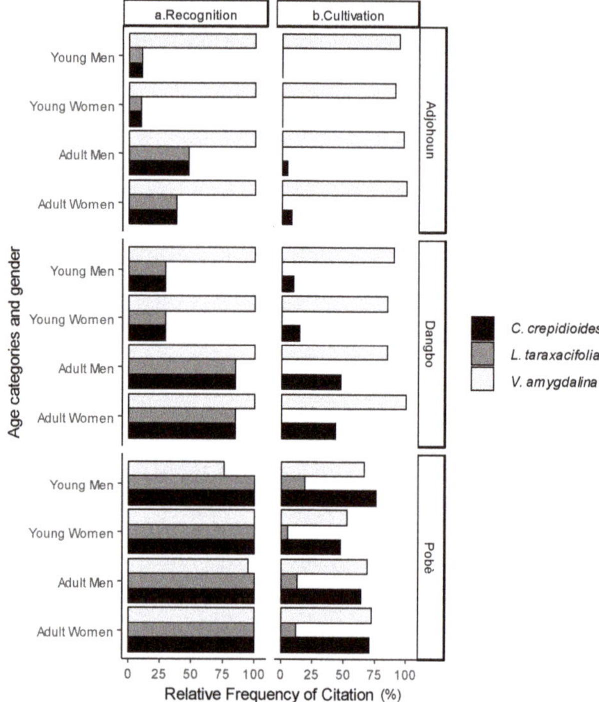

Figure 2. Pattern of recognition and cultivation of *C. crepidioides*, *L. taraxacifolia*, and *V. amygdalina* among informants in Pobè (Holly), Dangbo (Wémè), and Adjohoun (Aïzo).

3.3. Use Value and Diversity of Utilizations in C. crepidioides, V. amygdalina, and L. taraxacifolia

3.3.1. Use Value of Species Across Generation, Gender, and Village

- *Crassocephalum Crepidioides*

The total use value of *C. crepidioides* was 0.23 ± 0.05 in Adjohoun, 0.51 ± 0.05 in Dangbo, and 0.91 ± 0.04 in Pobè. The total use value of this species was 0.63 ± 0.04 for adults, 0.38 ± 0.05 for young informants, 0.53 ± 0.04 for women, and 0.57 ± 0.05 for men. There was no significant relationship between gender and total use value of *C. crepidioides*. However, young informants had a significantly lower use value compared to parents ($p < 0.003$, Table 3, Figure 3). In addition, the total use value of *C. crepidioides* differed among villages; Pobè had the highest use value ($p < 0.001$), while Dangbo had a non-significantly different use value although higher than the value of Adjohoun. A similar trend was noticed for the food use value (Table 3, Figure 3). The medicinal use value of *C. crepidioides* was higher for men than women (Table 3, Figure 3). Overall, the food use value of *C. crepidioides* outweighed the medicinal use value (up to 96% of the species total use value).

- *Launaea Taraxacifolia*

The total use value of *L. taraxacifolia* differed only among villages, with Pobè having the highest value (0.48 ± 0.05 against 0.06 ± 0.03 for Adjohoun and 0.01 ± 0.01 for Dangbo) (Figure 3, Table 3). A similar trend was noticed for the food use value. The medicinal use value, however, was similar between a child and parent, and men and women, though women had a relatively higher use value (0.22 ± 0.04 for women against 0.15 ± 0.03 for men). The medicinal use value was similar between gender, generation, and villages ($p > 0.05$). Overall, the food use value of *L. taraxacifolia* also outweighed the medicinal use value (>89% of the species total use value) although the latter exhibited more cited specific uses (Table 4).

Table 3. Relationship between total (UV$_{Total}$), food (UV$_{Food}$), and medicinal (UV$_{Medicinal}$) use values and informant gender, generation, and location (village): Summary of Poisson Generalized Linear Models.

Terms in the Final Model	UV$_{Total}$			UV$_{Food}$			UV$_{Medicinal}$		
	est	se	*p*	est	se	*p*	est	se	*p*
C. crepidioides									
Gender: women	-	-	-	-	-	-	−1.55	0.33	<0.001
Generation: young	−0.25	0.08	0.003	−0.24	0.09	0.007	-	-	-
Village: Dangbo	0.08	0.11	0.483	0.15	0.12	<0.001	-	-	-
Village: Pobè	0.74	0.14	<0.001	0.89	0.15	<0.001	-	-	-
L. taraxaciolia									
Gender: men	-	-	-	-	-	-	-	-	-
Generation: young	-	-	-	-	-	-	-	-	-
Village: Dangbo	−3.17	1.63	0.052	−1.71	0.69	0.014	-	-	-
Village: Pobè	2.30	0.87	0.008	−0.61	0.51	0.229	-	-	-
V. amygdalina									
Gender: women	-	-	-	0.04	0.02	0.026	-	-	-
Generation: young	−0.08	0.03	0.014	-	-	-	−0.16	0.07	0.019
Village: Dangbo	−0.09	0.04	0.034	−0.03	0.02	0.277	−0.24	0.09	0.009
Village: Pobè	−0.27	0.04	<0.001	−0.07	0.02	0.004	−0.56	0.08	<0.001

Reference levels: men for gender, adult for generation, and Adjohoun for village; est = estimate, se = standard error, *p* = probability of the test, - means that the factor was not selected in the parsimonious model.

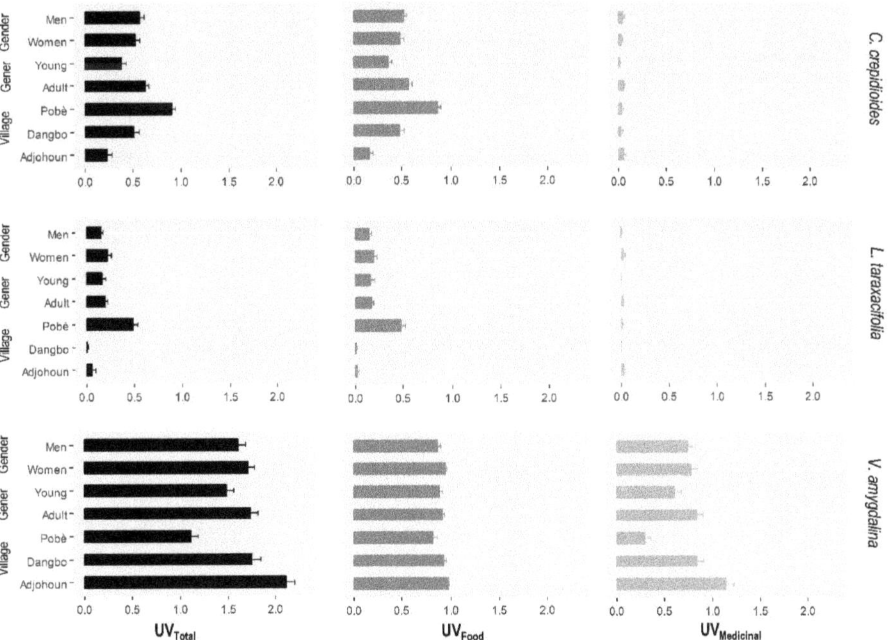

Figure 3. Variation of food, medicinal, and total use values of the three leafy vegetables in relation to location, generation (abbreviated as gener), and gender. Bars represent mean value and the error bars are the corresponding standard errors.

Table 4. Use categories, description of utilization, and relative frequency of citation (RFC, %) of three traditional leafy vegetables as reported by informants in Adjohoun, Dangbo, and Pobè.

Species	Local Names [†]	Plant Part	Use Categories	Description of Food and Medicinal Uses Reported	RFC (%)		
					Adjohoun	Dangbo	Pobè
C. crepidioides	Gbolo (Wémè) Gbolo/Akogbo (Aïzo) Gbolo (Holly)	Leaves	Food	Consumed as a sauce	**15.83**	**49.17**	**88.33**
				Toast of the leaves with seasoning in leaves of *Musa* spp. L.	0.83	0.00	0.00
			Medicinal	Consumed as a sauce to kill intestinal worms	0.83	0.83	0.00
				Consumed as a sauce to treat aches	0.00	0.00	0.83
				Consumption as a sauce or an infusion in an antibiotic	1.67	1.67	0.00
				Eating the leaves daily treats rheumatism	0.83	0.00	0.00
				Trituration and consumption of the juice for tonsillitis	0.83	0.00	0.00
				Trituration with salt and lemon is used to treat ulcer	0.00	0.83	0.00
				Use of an infusion in infants' bath to treat fever	0.00	0.00	2.50
L. taraxacifolia	Awonto (Wémè) Awonto (Aïzo) Efognanri/Gnanri (Holly)	Leaves	Food	Consumed as a sauce	**1.67**	**0.83**	**47.50**
			Medicinal	Trituration added with palm oil to rub the body to treat fever	0.83	0.00	0.00
				Trituration extract on wound to induce healing and reduce scarring	0.83	0.00	0.00
				Trituration extract as a drink for fever	0.00	0.00	0.83
		Stem	Medicinal	Seasoning the stem with salt and leaving it closer to the fire stimulates the appetite of patients	0.83	0.00	0.00
V. amygdalina	Alomangbo (Wémè) Aloman (Aïzo) Eyouro (Holly)	Leaves	Food	Consumed as a sauce	**97.50**	**92.50**	**81.67**
			Medicinal	Trituration extract used to treat tooth pains	0.83	0.00	0.00
				Trituration extract used to treat burns and snake bites and to induce healing	2.50	0.83	0.83
				Trituration added with palm oil to rub the body to treat fever	**25.83**	**18.33**	**0.00**
				Trituration extract added to lemon juice and used as a drink to treat fever	1.67	2.50	0.83
				Trituration with salt to brush teeth for anorexia	3.33	7.50	0.00
				Trituration extract as a drink to treat stomach pains	**15.00**	**4.17**	**2.50**
				Trituration extract as a drink to treat ulcer	0.83	0.00	0.00
				Trituration extract as a drink to treat fever	5.83	5.83	2.50
				Slightly softened leaves with salt to swallow for treating tonsillitis	1.67	3.33	0.00
				Brush the teeth with the leaves to treat bad breath	13.33	0.00	0.83

Table 4. *Cont.*

Species	Local Names [†]	Plant Part	Use Categories	Description of Food and Medicinal Uses Reported	RFC (%)		
					Adjohoun	Dangbo	Pobè
				Trituration added with palm oil is used to rub the body to treat measles	5.83	0.00	0.00
				Trituration with leaves of *Ocimum gratissimum* and salt to treat stomach pains	0.83	1.67	0.00
				Extract with salt and lemon used as a drink to treat headaches	0.83	0.00	0.00
				Extract with salt is used to treat fever	2.50	0.00	0.00
				Extract is used to treat measles	10.83	1.67	0.00
				Extract is used as a drink to treat chickenpox	2.50	0.83	0.00
				Ethanolic extract is given to children to treat chickenpox	0.83	0.83	0.00
				Consumption as a sauce for anorexia	1.67	0.00	3.33
				Extract is used as a drink to kill worms	3.33	5.83	0.00
				Extract with petroleum and palm oil plus seven or nine pinches of sand and rub it on children's body to heal fever	0.83	0.00	0.00
V. amygdalina	Alomangbo (Wémè) Aloman (Aïzo) Eyouro (Holly)	Leaves	Medicinal	Extract with lemon juice used to treat stomach aches	2.50	0.00	0.00
				The leaves are wrapped in banana's leaves with a little salt and then heated in the ash to collect the juice in a spoon that is given to children to treat horrible stomach aches	0.83	0.00	0.00
				Extract added with palm oil is used as a drink against vomiting	0.83	1.67	0.00
				Extract is used as a drink to treat hypertension	0.83	0.83	0.00
				Extract is used as an antibiotic	0.00	0.83	0.00
				Macerate and drink it to treat malaria	0.00	1.67	0.00
				Triturate + palm oil + "red capsule" for scarring	0.00	0.83	0.00
				Eat leaves uncooked to treat mouth wounds	0.00	0.83	0.00
				Triturate and drink the juice to treat cough	0.83	0.83	3.33
				Triturate the leaves and take the bath with it to treat fever	0.00	1.67	0.00
				Triturate and drink the juice to treat scabby penis	0.00	0.83	0.00
				Triturate and squeeze the juice in the ears to treat ear aches	0.00	0.83	0.00
				Macerate the leaves with immature papaya and drink it to treat malaria	0.00	0.83	0.00
				Triturate and drink the juice for treating hemorrhages	0.00	0.83	0.00
				Triturate the leaves in the shower water to treat scabies and measles	0.00	0.00	1.67

Table 4. *Cont.*

Species	Local Names [†]	Plant Part	Use Categories	Description of Food and Medicinal Uses Reported	RFC (%)		
					Adjohoun	Dangbo	Pobè
V. amygdalina	Alomangbo (Wémè) Aloman (Aïzo) Eyouro (Holly)	Leaves	Medicinal	Triturate and drink the juice to treat cold	0.00	0.00	2.50
				Triturate and drink the juice to treat allergies	0.00	0.00	1.67
				Triturate and drink the juice against vomiting	0.00	1.67	2.50
		Stem	Medicinal	Toothbrush for tonsillitis	1.67	0.83	0.00
				Remove the first layer and put salt on the stem then brush with it to treat cold	0.83	0.00	0.00
				Against anorexia and bad breath	2.50	0.00	0.00
				Dry powder is used to treat stomach aches	0.83	0.00	0.83
				Sore throat	0.00	0.83	0.00
				Stomach aches	0.00	2.50	0.00
				Nausea	0.00	1.67	0.00
				Snake bites	0.00	0.83	0.00
				Chew stem + *Aframomum melegueta* K. Schum. and swallow the juice only to treat diarrhoea	0.00	1.67	0.00
				Infusion of stem + *Citrus limon* (L.) Burm. f to treat malaria	0.00	0.00	0.83
		Root	Medicinal	Cut the root and put it in a bottle with an alcoholic drink to treat cough	0.83	0.00	0.00
				Cut the root and put it in a bottle with an alcoholic drink to treat malaria	0.83	0.00	0.00

[†] names in parentheses are local languages for the local names.

- *Vernonia amygdalina*

Contrary to the two previous vegetables, the use value of *V. amygdalina* was the lowest in Pobè (1.11 ± 0.08 use report per informant), while it was higher in Adjohoun (2.11 ± 0.09 use report per informant). Children also had a significantly lower total use value compared to adults (Table 3, Figure 3). For the food use, the same trend was observed among villages. However, the use value for women (0.77 ± 0.06) was significantly higher ($p = 0.026$) than that of men (0.73 ± 0.07). For the medicinal use value, the trend was the same as that observed for the species total use value. The food use value outweighed the medicinal use value (52% to 79% of the species total use value) (Table 3).

In summary, V. *amygdalina* was more used in Adjohoun, but relatively less used in Pobè. Conversely, the use values of C. *Crepidioides* and L. *taraxacifolia* were higher in Pobè, but relatively lower in Adjohoun and Dangbo (Figure 3).

3.3.2. Plant Part Use Value (PPUV), Diversity, and Category of Uses

Leaves and stem were the two plant parts used for either food or traditional medicine. The plant part use value varied greatly among vegetables. Irrespective of the vegetable, leaves had a higher plant part use value. Specifically, *V. amygdalina* had the highest leaf PPUV (1.59 ± 0.05), while *L. taraxacifolia* had the lowest leaf PPUV (0.18 ± 0.02). The stem of *L. taraxacifolia* and *C. crepidioides* were almost not used (PPUV < 0.006). The stem PPUV of *V. amygdalina* (0.26 ± 0.02) was six times lower compared to its leaf PPUV (1.59 ± 0.05).

Table 4 shows the diversity of the specific uses of the species per plant part, the traditional preparation methods, and their relative frequency of citation per study community.

Nine specific uses were mentioned for *C. crepidioides*, five for *L. taraxacifolia*, and 52 for *V. amygdalina*. The food uses mainly consist of the consumption of the leaves as a sauce. The leaves of *C. crepidioides* were consumed alone, or as a sauce or roasted with seasoning, in a *Musa* spp L. leaf. The leaves of *L. taraxacifolia* were also consumed alone as a sauce. As for the leaves of *V. amygdalina*, they were consumed either alone or together with other leafy vegetables, such as *Amaranthus* spp. L., *Celosia argentea* L. var. *argentea*, *Solanum macrocarpon* L., and *Ocimum gratissimum* L. The leaves were consumed either fresh or dried or powdered in a *Solanum lycopersicum* L., *Arachis hypogaea* L., or *Elaeis guineensis* Jacq. sauce.

Crassocephalum crepidioides was used against seven ailments; *L. taraxacifolia* was used against four ailments, while *V. amygdalina* was used to treat 48 ailments (Table 4), i.e., approximately seven and twelve times higher than *C. crepidioides* and *L. taraxacifolia*, respectively.

Gastric disorders (e.g., constipation, stomach pain, tooth pain, angina, buccal and intestinal pain, nausea, ulcer, vomiting, diarrhea, and bad breath), respiratory disorders (e.g., cough and cold), fevers, malaria, measles, chickenpox, hypertension, scabies, wound healing, scalds, snake bites, and lack of appetite were treated by the leaves of *V. amygdalina* (Table 4). Tiredness, stiffness, rheumatism, fevers, angina, intestinal disorders, and ulcer were treated by the eaves *C. crepidioides* (Table 4). The leaves of *L. taraxacifolia* were used to treat fevers and wound healing (Table 4).

The stems of *L. taraxacifolia* were used to treat the lack of appetite during illness, while the stems of *V. amygdalina* were used as toothbrushes, but also against ailments such as gastric disorders (stomach pain, angina, nausea, diarrhea, and bad breath), respiratory disorders (cold), lack of appetite, snake bites, and malaria.

These plant parts were often used alone or together with lemon, salt, palm oil, or a plant part of other plant species such as *Ocimum gratissimum* L. and *Carica papaya* L.

3.4. Transmission of Traditional Knowledge on Target Vegetables

The knowledge that the informants hold on the uses of selected vegetables was acquired from their parents (usually more than 90% of informants, Table 5). The other sources (e.g., social network and formal instruction) were negligible (often less than 15%) (Table 5).

Table 5. Frequency of citation of the sources of knowledge acquisition on the use of three traditional leafy vegetables as reported by informants in Adjohoun, Dangbo, and Pobè.

Villages	Generation	Gender	Sources of Knowledge				
			Parents	Grandparents	Friends	Training	Others
Adjohoun	Adults	Men	92.5	10.0	2.5	0.0	2.5
		Women	65.0	5.0	12.5	5.0	12.5
	Child	Boys	100.0	0.0	5.0	0.0	0.0
		Girls	95.0	15.0	0.0	0.0	0.0
Dangbo	Adults	Men	95.0	97.5	0.0	0.0	0.0
		Women	100.0	0.0	0.0	0.0	0.0
	Child	Boys	100.0	0.0	0.0	0.0	0.0
		Girls	100.0	5.0	0.0	0.0	0.0
Pobè	Adults	Men	100.0	17.5	12.5	0.0	0.0
		Women	97.5	15.0	22.5	0.0	0.0
	Child	Boys	100.0	14.3	14.3	0.0	0.0
		Girls	94.7	21.1	31.6	0.0	0.0

Results from the analysis of the knowledge similarity between boys or girls and the two parents (mother and father) based on Jaccard similarity coefficients showed that for *C. crepidioides* and *L. taraxacifolia*, no significant difference ($p > 0.05$, Figure 4) was detected between the children and the parents' knowledge in any village. For those two leafy vegetables, the knowledge transmitted by mothers was similar to the one transmitted by fathers. Likewise, child knowledge similarity with that of parents did not differ significantly between girls and boys ($p = 0.685$ for *C. crepidioides*, $p = 0.426$ for *L. taraxacifolia*) indicating a similar level of knowledge transmission to boys and girls. Differences were only observed among villages. For *C. crepidioides*, child knowledge similarity with that of parents was at least five times higher in Pobè (Jaccard similarity = 0.74 ± 0.05) than in the other villages ($p < 0.01$). There was practically no knowledge of *L. taraxacifolia* in Adjohoun and Dangbo and no exchange between children and parents. However, a significant difference was observed between boys and girls whose knowledge of *L. taraxacifolia* was more similar to that of the parents (0.64 ± 0.09); boys and parents' knowledge similarity was comparatively lower (Jaccard similarity = 0.31 ± 0.09, Figure 4).

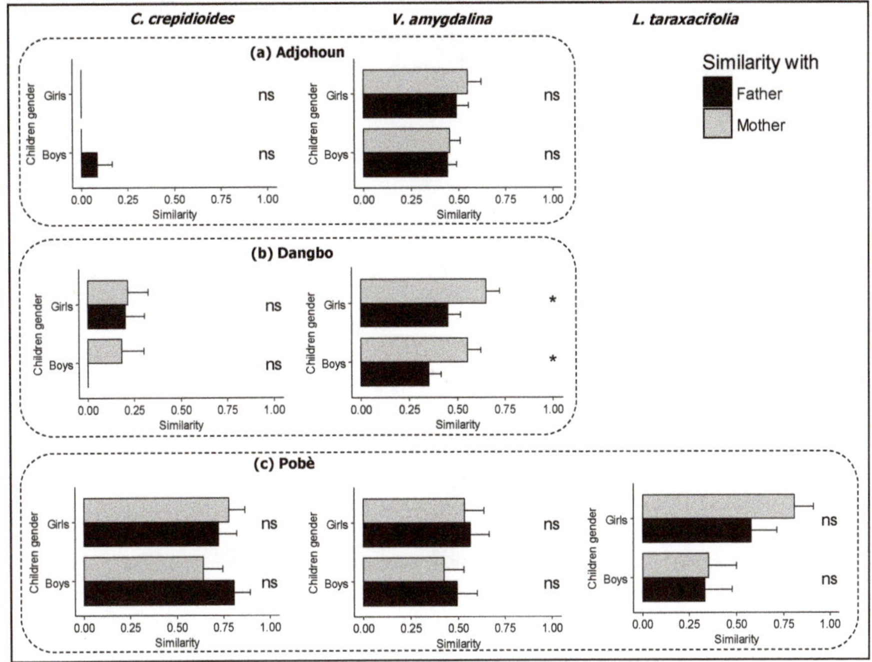

Figure 4. Boys and girls related knowledge similarities with parents in Adjohoun, Pobè, and Dangbo (South Benin). The bars represent mean value and the error bars are the corresponding standard errors. Ns and * are statistical significance of the two-sample t-test comparing children's knowledge similarity with fathers and mothers' knowledge.

For *V. amygdalina*, we observed a significant difference ($p < 0.05$) between the Jaccard similarity index of children's knowledge and that of fathers' and mothers' in Dangbo (Figure 4). In that village, boys and girls hold a different amount of knowledge with that of their parents (mothers and fathers, separately).

In Pobè, knowledge transmission on *V. amygdalina* and *C. crepidioides* was overall the same for both girls and boys, while it was higher for girls than boys for *L. taraxacifolia* (Figure 4). Irrespective of the vegetable species, about 50% of parents' knowledge was transmitted to children (Figure 4).

In Adjohoun and Dangbo, differences were more perceptible among species. Our results indicated that knowledge on *V. amygdalina* was better transmitted (0.45–0.65). No knowledge on *L. taraxacifolia* was transmitted in our respondents from Adjohoun and Dangbo (Figure 4).

4. Discussion

4.1. Pattern of Recognition and Cultivation of Leafy Vegetables Among Informants: Variation Across Generations, Genders, and Villages

Our results suggested a differential pattern in the recognition and cultivation of the three leafy vegetables across villages, generation, and gender, as the studied vegetables were more recognized in Pobè than in Dangbo and Adjohoun. This differential pattern could be due to the natural distribution of the species and food habits (the species may occur without being used). All three species were reported in the flora of Benin [30] to occur in the study areas (phytodisctricts of Pobè and Ouèmé). *Vernonia amygdalina*, the most popular vegetable, has historically been used in all parts of the country [9]. It is also cultivated in home gardens and in urban and peri-urban agriculture. The availability hypothesis, according to which, plants are used because they are more accessible or locally abundant [16,38], may explain this pattern of recognition. *Vernonia amygdalina* as a widely distributed and cultivated vegetable [30], was more familiar to local farmers compared to the other two species. *Crassocephalum crepidioides* was cultivated by only adults at Adjohoun and Dangbo. *Launaea taraxacifolia* was cultivated only in Pobè by all categories of informants. A higher proportion of older informants cultivated *L. taraxacifolia* and *C. crepidioides* in Dangbo and Adjohoun, while neither age-specific nor gender-specific patterns were observed in Pobè. That finding suggested that older informants had more knowledge on the propagation of *L. taraxacifolia* and *C. crepidioides* in Dangbo and Adjohoun, while in Pobè younger and older informants, irrespective of their gender, held similar knowledge on all species.

The observed recognition pattern may also be due to the species status (i.e., wild or cultivated), the ability for propagation, or market opportunities. The informants in Adjohoun ascertained that *C. crepidioides* is a cultivated or wild species, while informants at Dangbo and Pobè thought that the species is a cultivated vegetable but may also be wild. *Launaea taraxacifolia* was said to be a wild or cultivated species, but *V. amygdalina* was reported as a cultivated and domesticated species. These results are congruent with several previous studies. *Vernonia amygdalina* is reported as a cultivated plant species, available in all seasons and the 4th of the top five traditional leafy vegetables of importance in Benin after *Solanum macrocarpon*, *Corchorus olitorius* L., and *Amaranthus cruentus* L. [7]. Both *C. crepidioides* and *L. taraxacifolia* are considered wild and available only in the rainy season in contrast to *V. amygdalina*, but undergoing a domestication process which is reported to be at the phase of bringing these species into cultivation [7,39,40].

Therefore, knowledge on species propagation is the primary difficulty for farmers. All informants reported that *V. amygdalina* can be propagated by simple cuttings. The other two species were propagated differently, mainly through seeds. However, according to farmers, the seeds of *C. crepidioides* and *L. taraxacifolia* are not available for production because seeds drop off at maturity and also cannot be stored for long time as they die quickly. Because informants in the village of Pobè were able to cultivate all species, in particular *L. taraxacifolia* and *C. crepidioides*, further research on the reproduction biology of those vegetables will be beneficial. Additional reason explaining the observed patterns in the cultivation of three vegetables may be linked to market opportunities which are closely linked to consumption habits because the choice of species for cultivation is dependent on the availability of sales market, the market value [12], and profitability of the species [3].

4.2. Use Value and Use Diversity of The Three Leafy Vegetables in Relation to Village, Gender, and Generation

4.2.1. Use Value of Study Species: Relationship with Informants' Generation, Gender, and Village

The fact that the total use value of *V. amygdalina* was higher than that of *C. crepidioides* and *L. taraxacifolia* is congruent with the availability hypothesis which predicts that plants are more known and used because they are more accessible (including easily cultivable) or locally abundant [16,38]. Regarding species-specific patterns, older informants held more knowledge on the uses of *V. amygdalina* and *C. crepidioides* than younger ones, informants at Pobè held more knowledge on *L. taraxacifolia* and *C. crepidioides*, while informants at Dangbo held more knowledge on *V. amygdalina*. The differences between villages can be explained by the availability hypothesis, but also by the local importance of the species. The fact that older informants knew more than younger ones is congruent with the theory that knowledge accumulation on the uses of plants is a life-long process [41]. Therefore, the longer you live, the more you know. This finding contradicts with our prediction of neutral relationships of gender, age, and village with the use value of *V. amygdalina*. This finding is certainly due to the fact that older informants held more knowledge on the medicinal uses than younger informants.

By disaggregating the overall use value into food use value and medicinal use value, we found a similar trend as the total use value, except that women knew more than men on the food uses of *V. amygdalina* which is congruent with the labor division in households [42] where women are responsible for food in the study villages. With respect to this assumption, and because men are mostly responsible for health issues in households, one should expect that men hold more knowledge on medicinal uses [43], which was only true for *C. crepidioides*. This may be due to the fact that either the species are not practically used (case of *L. taraxacifolia* and *C. crepidioides*) or that the medicinal uses are so common that everyone knows the uses which may be true for *V. amygdalina*.

Finally, more than 90% of the total use value for *L. taraxacifolia* and *C. crepidioides* was due to food uses, while about 60% of the total use value of *V. amygdalina* was due to food use. Therefore, the nutraceutical value was more tangible for *V. amygdalina* than the others which were mostly food leafy vegetables. Leaves were the most used and this confirmed that they are all leafy vegetables. However, the stem of *V. amygdalina* was also used for medicinal purposes and as a toothbrush.

4.2.2. Diversity of Specific Uses Per Category of Uses in Relationship to Villages

The number of uses cited by informants was about 10 and 6 times higher for *V. amygdalina* (fifty-two) than for *C. crepidioides* (nine) and *L. taraxacifolia* (five), respectively. The relative frequency of citation of the use of leaves for food were higher for all three species indicating a consensus on their use as food. The use of leaves as food was mentioned in several previous studies, including Adjatin, et al. [44], Achigan-Dako et al. [10], and Dairo and Adanlawo [40] for *C. crepidioides*; Arawande et al. [45] and Adebisi [28] for *L. taraxacifolia*; and Achigan-Dako et al. [10] for *V. amygdalina*. The leaves may be consumed alone or seasoned with other food ingredients to increase the taste and flavor. Such ingredients in this study included *Musa* spp leaf, salt, *Citrus limon* for *C. crepidioides*; palm oil, and salt for *L. taraxacifolia*; and palm oil, salt, and *O. gratissimum* leaves for *V. amygdalina*. *Vernonia amygdalina* was used against a number of health disorders seven and twelve times higher than that of *C. crepidioides* and *L. taraxacifolia*, respectively. Apart from the uses of *V. amygdalina* against fever in children, stomach pains, anorexia, fever, bad breath, and measles, where the relative frequency of citation was greater than 5% (5%–26%, Table 4), for all other cited health disorders, the RFC was low (1%–2%) to very low (<1%) indicating a relative lack of consensus on this knowledge.

4.3. Sources and Transmission of Traditional Knowledge on Leafy Vegetables from Parent to Children: Children Sex-Related Knowledge Similarities

Traditional knowledge on the traditional leafy vegetables were mostly acquired from parents. Similar results were previously reported in Argentina [15] and Canada [18] and referred to as vertical knowledge transmission [46]. The transmission of traditional plant knowledge begins at an early age, as a family custom, in which women play a predominant role [15].

As predicted, the study revealed that the local knowledge transmission depends on the species, but also on the village. Knowledge transmission was not influenced neither by parent gender nor by the sex of the young for *V. amygdalina*, certainly because of its availability and commonness. This suggests that the more a species is common and used, the more knowledge people hold on its uses, which are effectively transmitted. Knowledge transmission was, however, higher for girls in the case of *L. taraxacifolia*, certainly because *L. taraxacifolia* was mostly used for food and because of labour division most of the knowledge was transmitted to girls by both the parents, but mostly by mothers. In addition, knowledge transmission differed among villages, with higher transmission at Pobè for *C. crepidioides* likely, because it was most used and known there. This can also be a particularity of the "Holly" known for its intense relations with the flora in Benin [47].

5. Conclusions

Understanding the dynamics of traditional knowledge (TK) and uses of locally adapted but underutilized species is critical for planning actions relevant for agricultural diversification and conservation of resources. The three studied species are important traditional vegetables, but *V. amygdalina* was the most commonly known and used in all villages, while *L. taraxacifolia* was confined to the village Pobè. Village, generation, and gender had significant effects on the use value, but the patterns or recognition were globally species-specific. Leaves were the most used plant part confirming the three species as leafy vegetables. The TK was acquired from parents, and both mothers and fathers transmitted a similar amount of knowledge to their progenies. The knowledge on *V. amygdalina* was better transmitted and was followed by *C. crepidioides* and *L. taraxacifolia*. We conclude that gender and generation knowledge dynamic hypothesis is species-specific and that TK transmission is also species-specific and may be linked to the local importance and use of resources. Due to their nutraceutical value and short cultivation cycle (tree months for first harvest), we propose to increase the production and utilization of these plants and encourage seed production and distribution. Promoting these species in home, school, or allotment gardening could be a good starting point. National agricultural services in partnerships with research institutions can also work together to develop and popularize the best agronomic practices for these species. For this utilization to be sustainable, research on genetic improvement of the species will be instrumental. Such studies are currently lacking or insufficient.

Author Contributions: Conceptualization, A.D.M.A. and E.G.A.-D.; methodology, AD.M.A. and E.G.A.-D.; software, A.D.M.A; validation, A.D.M.A. and E.G.A.-D.; formal analysis, A.D.M.A. and E.G.A.-D.; investigation, A.D.M.A. and E.G.A.-D.; resources, E.G.A.-D.; data curation, A.D.M.A; writing—A.D.M.A.; writing—review and editing, A.D.M.A. and E.G.A.-D.; visualization, A.D.M.A; supervision, E.G.A.-D.; project administration, E.G.A.-D.; funding acquisition, E.G.A.-D.

Funding: This work was funded by the Laboratory of Genetics, Horticulture, and Seed Science (GBioS).

Acknowledgments: Authors are grateful to the local communities of Igana (Pobè), Houèda (Adjohoun), and Gbéko (Dangbo) for having shared their knowledge. We are also grateful to Djamal Ayifimi and Armel Gouvoeke for their help in the field. Contributions from Fernand Sohindji are acknowledged.

Conflicts of Interest: The authors declare no conflicts of interest.

References

1. Ambrose-Oji, B. Urban Food Systems and African Indigenous Vegetables: Defining the Spaces and Places for African Indigenous Vegetables in Urban and Peri-Urban Agriculture Urban food systems and trends in vegetable production in urban. In *African Indigenous Vegetables in Urban Agriculture*; Routledge: London, UK, 2009; pp. 33–66.

2. Sogbohossou, E.D.; Achigan-Dako, E.G.; van Andel, T.; Schranz, M.E. Drivers of management of spider plant (*Gynandropsis gynandra*) across different socio-linguistic groups in Benin and Togo. *Econ. Bot.* **2018**, *72*, 411–435. [CrossRef]

3. Sogbohossou, O.E.; Achigan-Dako, E.G.; Komlan, F.A.; Ahanchede, A. Diversity and differential utilization of *Amaranthus* spp. along the urban-rural continuum of southern Benin. *Econ. Bot.* **2015**, *69*, 9–25. [CrossRef]

4. Yang, R.-Y.; Keding, G.B. Nutritional contributions of important African indigenous vegetables. In *African Indigenous Vegetables in Urban Agriculture*; Routledge: London, UK, 2009; pp. 137–176.

5. Guidi, L.; Landi, M. Aromatic plants: Use and nutraceutical properties. *Novel Plant Bioresour. Appl. Food Med. Cosmet.* **2014**, 303–345. [CrossRef]

6. Vihotogbé, R.; Sossa-Vihotogbé, C.; Achigan-Dako, G. Safety of Botanical Ingredients in Personal Healthcare: Focus on Africa. *Novel Plant Bioresour. Appl. Food Med. Cosmet.* **2014**, *28*, 395–408.

7. Dansi, A.; Adjatin, A.; Adoukonou-Sagbadja, H.; Faladé, V.; Yedomonhan, H.; Odou, D.; Dossou, B. Traditional leafy vegetables and their use in the Benin Republic. *Genet. Resour. Crop Evol.* **2008**, *55*, 1239–1256. [CrossRef]

8. Segnon, A.C.; Achigan-Dako, E.G. Comparative analysis of diversity and utilization of edible plants in arid and semi-arid areas in Benin. *J. Ethnobiol. Ethnomed.* **2014**, *10*, 80. [CrossRef]

9. Achigan-Dako, E.G.; N'Danikou, S.; Assogba-Komlan, F.; Ambrose-Oji, B.; Ahanchede, A.; Pasquini, M.W. Diversity, geographical, and consumption patterns of traditional vegetables in sociolinguistic communities in Benin: Implications for domestication and utilization. *Econ. Bot.* **2011**, *65*, 129. [CrossRef]

10. Achigan-Dako, E.G.; Pasquini, M.W.; Assogba Komlan, F.; N'danikou, S.; Yédomonhan, H.; Dansi, A.; Ambrose-Oji, B. *Traditional vegetables in Benin*; Institut National des Recherches Agricoles du Bénin, Imprimeries du CENAP: Cotonou, Benin, 2010.

11. Dansi, A.; Vodouhè, R.; Azokpota, P.; Yedomonhan, H.; Assogba, P.; Adjatin, A.; Loko, Y.; Dossou-Aminon, I.; Akpagana, K. Diversity of the neglected and underutilized crop species of importance in Benin. *Sci. World J.* **2012**, *2012*, 932947. [CrossRef]

12. N'Danikou, S.; Achigan-Dako, E.G.; Wong, J.L. Eliciting local values of wild edible plants in Southern Benin to identify priority species for conservation. *Econ. Bot.* **2011**, *65*, 381–395. [CrossRef]

13. Sanoussi, F.; Ahissou, H.; Dansi, M.; Hounkonnou, B.; Agre, P.; Dansi, A. Ethnobotanical investigation of three traditional leafy vegetables [*Alternanthera sessilis* (L.) DC. *Bidens pilosa* L. *Launaea taraxacifolia* Willd.] widely consumed in southern and central Benin. *J. Biodivers. Environ. Sci.* **2015**, *6*, 187–198.

14. Eyssartier, C.; Ladio, A.H.; Lozada, M. Cultural transmission of traditional knowledge in two populations of north-western Patagonia. *J. Ethnobiol. Ethnomed.* **2008**, *4*, 25. [CrossRef]

15. Lozada, M.; Ladio, A.; Weigandt, M. Cultural transmission of ethnobotanical knowledge in a rural community of northwestern Patagonia, Argentina. *Econ. Bot.* **2006**, *60*, 374–385. [CrossRef]

16. Voeks, R.A. Disturbance pharmacopoeias: Medicine and myth from the humid tropics. *Ann. Assoc. Am. Geogr.* **2004**, *94*, 868–888.

17. Voeks, R.A. Are women reservoirs of traditional plant knowledge? Gender, ethnobotany and globalization in northeast Brazil. *Singap. J. Trop. Geogr.* **2007**, *28*, 7–20. [CrossRef]

18. Ruddle, K. The transmission of traditional ecological knowledge. In *Traditional Ecological Knowledge: Concepts and Cases*; Canadian Museum of Nature and IDRC: Ottawa, ON, Canada, 1993; pp. 17–31.

19. Boesch, C.; Tomasello, M. Chimpanzee and Human Cultures. *Curr. Anthropol.* **1998**, *39*, 591–604.

20. Aunger, R. The life history of culture learning in a face-to-face society. *Ethos* **2000**, *28*, 445–481. [CrossRef]

21. Guimbo, I.D.; Mueller, J.G.; Larwanou, M. Ethnobotanical knowledge of men, women and children in rural Niger: A mixed-methods approach. *Ethnobot. Res. Appl.* **2011**, *9*, 235–242. [CrossRef]

22. Souto, T.; Ticktin, T. Understanding interrelationships among predictors (age, gender, and origin) of local ecological knowledge. *Econ. Bot.* **2012**, *66*, 149–164. [CrossRef]

23. Lykke, A.; Kristensen, M.; Ganaba, S. Valuation of local use and dynamics of 56 woody species in the Sahel. *Biodivers. Conserv.* **2004**, *13*, 1961–1990. [CrossRef]

24. Gaoue, O.G.; Coe, M.A.; Bond, M.; Hart, G.; Seyler, B.C.; McMillen, H. Theories and major hypotheses in ethnobotany. *Econ. Bot.* **2017**, *71*, 269–287. [CrossRef]

25. Avohou, H.T.; Vodouhe, R.S.; Dansi, A.; Kpeki, B.; Bellon, M. Ethnobotanical factors influencing the use and management of wild edible plants in agricultural environments in Benin. *Ethnobot. Res. Appl.* **2012**, *10*, 571–592.

26. Denton, O.A. Crassocephalum crepidioides (Benth.) S. Moore. In *Plant Ressources of Tropical Africa 2: Vegetables*; Grubben, G.J.H., Denton, O.A., Messiaen, C.-M., Schippers, R.R., Lemmens, R.H.M.J., Oyen, L.P.A., Eds.; PROTA Foundation: Wageningen, The Netherlands; Backhuys Publishers: Leiden, The Netherlands; CTA: Wageningen, The Netherlands, 2014.

27. Stadlmayr, B.; Charrondiere, U.R.; Addy, P.; Samb, B.; Enujiugha, V.N.; Bayili, R.G.; Fagbohoun, E.G.; Smith, I.F.; Thiam, I.; Burlingame, B. *Composition of selected foods from West Africa*; Food and Agriculture Organization: Rome, Italy, 2010; pp. 13–14.

28. Adebisi, A. Launaea Taraxacifolia (Willd) Amin ex C Jeffrey {Internet} Record from Protabase. In *PROTA (Plant Resources of Tropical Africa/Ressources Vegetales de l'Afrique Tropicale)*; Grubben, G.J.H., Denton, O.A., Eds.; PROTA Foundation: Wageningen, The Netherlands; Backhuys Publishers: Leiden, The Netherlands; CTA: Wageningen, The Netherlands, 2004.

29. Fomum, F.U. Vernonia amygdalina Delile. In *Plant Ressources of Tropical Africa 2: Vegetables*; Grubben, G.J.H., Messiaen, C.-M., Schippers, R.R., Lemmens, R.H.M.J., Oyen, L.P.A., Eds.; PROTA Foundation: Wageningen, The Netherlands; Backhuys Publishers: Leiden, The Netherlands; CTA: Wageningen, The Netherlands, 2004; pp. 543–546.

30. Akoègninou, A.; Van der Burg, W.; Van der Maesen, L.J.G. *Flore analytique du Bénin*; Backhuys Publishers: Wageningen, The Netherlands, 2006.

31. Économique, I.N.D.L.S.E.D.L.A.; International, I. *Enquête Démographique et de Santé du Bénin 2011–2012: Rapport de synthèse*; INSAE et ICF International Calverton: Calverton, MD, USA, 2013.

32. CENALA. *Atlas et études Sociolinguistiques du Bénin*; Agence Intergouvernementale de la Francophonie: Cotonou, Benin, 2003.

33. Sogbohossou, E.D.; Achigan-Dako, E.G.; Maundu, P.; Solberg, S.; Deguenon, E.M.; Mumm, R.H.; Hale, I.; Van Deynze, A.; Schranz, M.E. A roadmap for breeding orphan leafy vegetable species: A case study of *Gynandropsis gynandra* (Cleomaceae). *Hortic. Res.* **2018**, *5*, 1–15. [CrossRef]

34. Etongo, D.; Djenontin, I.N.S.; Kanninen, M.; Glover, E.K. Assessing use-values and relative importance of trees for livelihood values and their potentials for environmental protection in Southern Burkina Faso. *Environ. Dev. Sustain.* **2017**, *19*, 1141–1166. [CrossRef]

35. Gomez-Beloz, A. Plant use knowledge of the Winikina Warao: The case for questionnaires in ethnobotany. *Econ. Bot.* **2002**, *56*, 231–241. [CrossRef]

36. Friedman, J.; Yaniv, Z.; Dafni, A.; Palewitch, D. A preliminary classification of the healing potential of medicinal plants, based on a rational analysis of an ethnopharmacological field survey among Bedouins in the Negev Desert, Israel. *J. Ethnopharmacol.* **1986**, *16*, 275–287. [CrossRef]

37. R Core Team, P. *R: A Language and Environment for Statistical Computing [Computer Software Manual]*; R Core Team, R Foundation for Statistical Computing: Vienna, Austria, 2016.

38. De Albuquerque, U.P. Re-examining hypotheses concerning the use and knowledge of medicinal plants: A study in the Caatinga vegetation of NE Brazil. *J. Ethnobiol. Ethnomed.* **2006**, *2*, 30. [CrossRef]

39. Adjatin, A.; Dansi, A.; Eze, C.; Assogba, P.; Dossou-Aminon, I.; Akpagana, K.; Akoègninou, A.; Sanni, A. Ethnobotanical investigation and diversity of Gbolo (*Crassocephalum rubens* (Juss. ex Jacq.) S. Moore and *Crassocephalum crepidioides* (Benth.) S. Moore), a traditional leafy vegetable under domestication in Benin. *Genet. Resour. Crop Evol.* **2012**, *59*, 1867–1881. [CrossRef]

40. Dairo, F.; Adanlawo, I. Nutritional quality of Crassocephalum crepidioides and Senecio biafrae. *Pak. J. Nutr.* **2007**, *6*, 35–39.

41. Hanazaki, N.; Herbst, D.F.; Marques, M.S.; Vandebrock, I. Evidence of the shifting baseline syndrome in ethnobotanical research. *J. Ethnobiol. Ethnomed.* **2013**, *9*, 75. [CrossRef]

42. Begossi, A.; Hanazaki, N.; Tamashiro, J.Y. Medicinal plants in the Atlantic Forest (Brazil): Knowledge, use, and conservation. *Hum. Ecol.* **2002**, *30*, 281–299. [CrossRef]

43. Dovie, D.B.; Witkowski, E.; Shackleton, C.M. Knowledge of plant resource use based on location, gender and generation. *Appl. Geogr.* **2008**, *28*, 311–322. [CrossRef]

44. Adjatin, A.; Dansi, A.; Badoussi, E.; Sanoussi, A.; Dansi, M.; Azokpota, P.; Ahissou, H.; Akouegninou, A.; Akpagana, K.; Sanni, A. Proximate, mineral and vitamin C composition of vegetable Gbolo [*Crassocephalum rubens* (Juss. ex Jacq.) S. Moore and *C. crepidioides* (Benth.) S. Moore] in Benin. *Int. J. Biol. Chem. Sci.* **2013**, *7*, 319–331. [CrossRef]

45. Arawande, J.O.; Amoo, I.A.; Lajide, L. Chemical and phytochemical composition of wild lettuce (*Launaea taraxacifolia*). *J. Appl. Phytotechnol. Environ. Sanit.* **2013**, *2*, 25–30.

46. Reyes-García, V.; Broesch, J.; Calvet-Mir, L.; Fuentes-Peláez, N.; McDade, T.W.; Parsa, S.; Tanner, S.; Huanca, T.; Leonard, W.R.; Martínez-Rodríguez, M.R. Cultural transmission of ethnobotanical knowledge and skills: An empirical analysis from an Amerindian society. *Evol. Hum. Behav.* **2009**, *30*, 274–285. [CrossRef]

47. Codjia, J.T.C.; Vihotogbe, R.; Lougbegnon, T. Phytodiversité des légumes-feuilles locales consommées par les peuples Holli et Nagot de la région de Pobè au sud-est du Bénin. *Int. J. Biol. Chem. Sci.* **2009**, *3*. [CrossRef]

Article

Recovery of Wheat Heritage for Traditional Food: Genetic Variation for High Molecular Weight Glutenin Subunits in Neglected/Underutilized Wheat

Juan B. Alvarez * and Carlos Guzmán

Departamento de Genética, Escuela Técnica Superior de Ingeniería Agronómica y de Montes,
Edificio Gregor Mendel, Campus de Rabanales, Universidad de Córdoba, CeiA3, ES-14071 Córdoba, Spain;
carlos.guzman@uco.es
* Correspondence: jb.alvarez@uco.es

Received: 2 October 2019; Accepted: 13 November 2019; Published: 14 November 2019

Abstract: Club wheat (*Triticum aestivum* L. ssp. *compactum* (Host) Mackey), macha wheat (*T. aestivum* L. ssp. *macha* (Dekapr. & A.M. Menabde) Mackey) and Indian dwarf wheat (*T. aestivum* L. ssp. *sphaerococcum* (Percival) Mackey) are three neglected or underutilized subspecies of hexaploid wheat. These materials were and are used to elaborate modern and traditional products, and they could be useful in the revival of traditional foods. Gluten proteins are the main grain components defining end-use quality. The high molecular weight glutenin subunit compositions of 55 accessions of club wheat, 29 accessions of macha wheat, and 26 accessions of Indian dwarf wheat were analyzed using SDS-PAGE. Three alleles for the *Glu-A1* locus, 15 for *Glu-B1* (four not previously described), and four for *Glu-D1* were detected. Their polymorphisms could be a source of genes for quality improvement in common wheat, which would permit both their recovery as new crops and development of modern cultivars with similar quality characteristics but better agronomic traits.

Keywords: electrophoresis; genetic resources; neglected hexaploid wheat; seed-storage proteins

---

1. Introduction

Wheat is an important crop that has been associated with human food for many centuries [1]. It is the basis for a diverse range of products, mainly bread, noodles, pasta, and beer, which are present in most diets worldwide. In some cases, the same wheat type is used for all four different products depending on the geographical or cultural area [2]. Up until the Industrial Revolution, all baking processes were carried out by hand, which permitted the use of wheat varieties with rheological properties greatly different to current wheat varieties. Nevertheless, the use of machinery in baking processes forced people to look for varieties with very specific qualities [3], neglecting the traditional wheats mainly because of their lower yields and, in many cases, their unsuitability for mechanized production.

In recent times, in many places throughout the world, the search for more balanced and healthier diets has strengthened the return to traditional products [4]. However, paradoxically, one of the main problems is the need to use the flour of modern cultivars to develop these old products, and this is not successful in all cases because the modern cultivars have characteristics adapted to new uses. In this context, recovery of the materials that were traditionally used to develop these products has proven to be key in this revival.

In addition, some studies have suggested that the wheat breeding programs centered on high-yield cultivars could have eroded the genetic variability from the quality traits among and within cultivars [5]. This has given great importance to the search for species that could be useful in contributing genes for wheat quality improvement [6]. Wheat relatives are considered to be interesting sources of new alleles for these traits that could increase the crop's genetic basis [7]. Among these relatives, the wild relatives as well as the old varieties and landraces of the current or ancient wheats of all ploidic levels are included. Utilization of these latter materials as gene sources is advantageous, compared to the wild relatives, because they are easy to cross with modern wheat and there is little linkage drag of unwanted traits, which results from their high degree of domestication [8].

Wheat quality is associated with three main grain components: endosperm storage proteins related to gluten visco-elastic properties [9], starch synthases related to starch [10], and puroindolines related to grain hardness [11]. The endosperm storage proteins are divided in two main groups, gliadins and glutenins, according to their molecular characteristics [12]. Glutenins are also divided into high molecular weight (HMWGs) and low molecular weight (LMWGs) subunits [13,14]. HMWGs are coded at the *Glu-1* loci located on the long arm of group-1 homologous chromosomes, whereas the *Glu-3* loci that code for the LMWGs and the *Gli-1* loci that controls synthesis of ω-, γ-, and some β-gliadins are located on the short arm. On the short arm of group-6 homologous chromosomes, the *Gli-2* loci that code mainly for components present in the α region and some β-gliadins are located [15]. Among the endosperm storage proteins, the best studied are the HMWG subunits coded at the *Glu-A1*, *Glu-B1*, and *Glu-D1* loci on the long arm of group-1 homologous chromosomes in common wheat [12]. Each locus consists of two linked genes that code for two types of HMWG subunits, with different mobilities in SDS-PAGE, named *x*- and *y*-types [16].

Within the hexaploid species, over the last decade, our research group has conducted several studies on the genetic diversity for endosperm storage proteins, waxy proteins, and puroindolines in Spanish and Mexican landraces of common wheat (*Triticum aestivum* L. ssp. *aestivum*) [17,18] and Spanish spelt wheat (*T. aestivum* L. ssp. *spelta* (L.) Thell.) [19–22]. Recently, other neglected or underutilized wheat subspecies have been screened for genes related to quality traits, including club wheat (*T. aestivum* L. ssp. *compactum* (Host) Mackey) (important in the Pacific Northwest region in the USA but not in the rest of the world) and Indian dwarf wheat (*T. aestivum* L. ssp. *sphaerococcum* (Percival) Mackey), both included within the naked wheat group as common wheat, and macha wheat (*T. aestivum* L. ssp. *macha* (Dekapr. & A.M. Menabde) Mackey) from the same hulled wheat group as spelt wheat. Our data obtained with these species showed notable variability for waxy proteins (granule-bound starch synthase I, E.C. 2.4.11.11), detecting new allelic variants for starch synthase not previously described in common wheat [23]. However, variability studies on the endosperm storage proteins in these species have been scarce, showing low variation [24,25].

The main goal of this survey was to evaluate the polymorphisms of the seed storage proteins present in a collection of hexaploid wheats, club wheat, macha wheat, and Indian dwarf wheat, collected in their natural distribution areas. The variation of these wheats for endosperm storage proteins could be a good source of quality genes for common wheat breeding, increasing the wheat genetic background together with the development of new cultivars.

2. Materials and Methods

2.1. Plant Material

Fifty-five accessions of club wheat, 29 accessions of macha wheat, and 26 accessions of Indian dwarf wheat obtained from the National Small Grain Collections (Aberdeen, ID, USA) were analyzed in this study (Tables S1, S2 and S3). At least five grains for each accession were analyzed to detect the possible intra-accession variability.

The HMWGs alleles were designated according to Payne and Lawrence [26]. Several cultivars of durum (cv. *Lobeiro*: 1, 14 + 15) and common wheat (cv. *Chinese Spring*: null, 7 + 8, 2 + 12, cv. *Cheyenne*: 2*, 7 + 9, 5 + 10, and cv. *Frondoso*: 2*, 13 + 19, 2 + 12) were used as standards to compare and classify the detected subunits in the analyzed species.

2.2. Protein Extraction and SDS-PAGE Electrophoretic Analysis

Proteins were extracted from crushed endosperm. Before glutenin solubilization, the gliadins were extracted with a 1.5 M dimethylformamide aqueous solution following a double-wash with 50% (v/v) propan-1-ol at 60 °C for 30 min with agitation every 10 min. Glutenin was solubilized with 250 µL of buffer containing 50% (v/v) propan-1-ol, 80 mM Tris-HCl (pH 8.5), and 2% (w/v) dithiothreitol at 60 °C for 30 min. After centrifugation, 200 µL of the supernatant was transferred to a new tube, mixed with 3 µL of 4-vinylpyridine, and incubated for 30 min at 60 °C. The samples were precipitated with 1 ml of cold-acetone. The dried pellet was solubilized in buffer containing 625 mM Tris-HCl (pH 6.8), 2% (w/v) SDS, 10% (v/v) glycerol, 0.02% (w/v) bromophenol blue, and 2% (w/v) dithiothreitol in a 1:5 ratio (mg/ µL) to wholemeal.

Reduced and alkylated glutenin subunits were fractionated by electrophoresis in vertical SDS-PAGE slabs in a discontinuous Tris-HCl–SDS buffer system (pH: 6.8/8.8) at a polyacrylamide concentration of 8% (w/v, C: 1.28%). The Tris-HCl/glycine buffer system of Laemmli [27] was used. Electrophoresis was performed at a constant current of 30 mA/gel at 18 °C for 45 min after the tracking dye migrated off the gel. Gels were stained overnight with 12% (w/v) trichloroacetic acid solution containing 5% (v/v) ethanol and 0.05% (w/v) Coomassie Brilliant Blue R-250. De-staining was carried out with tap water.

2.3. Statistical Analysis

Allelic frequencies for the *Glu-A1*, *Glu-B1*, and *Glu-D1* loci were calculated for each subspecies. The classification of Marshall and Brown [28] was used for evaluating the distribution of alleles by their presence as frequent (≥5%), rare (≤5%), and very rare (≤1%). To assess potential genetic erosion, the number of alleles per locus (*A*), the effective number of alleles per locus (*Ne*), and Nei's diversity index (*He*) were measured [29,30].

3. Results

3.1. Variation for HMWGs

The HMWG compositions of all accessions of each subspecies (club, macha, and Indian dwarf wheat) were analyzed. A representative sample of the variability detected for the HMWGs in each subspecies is shown in Figure 1.

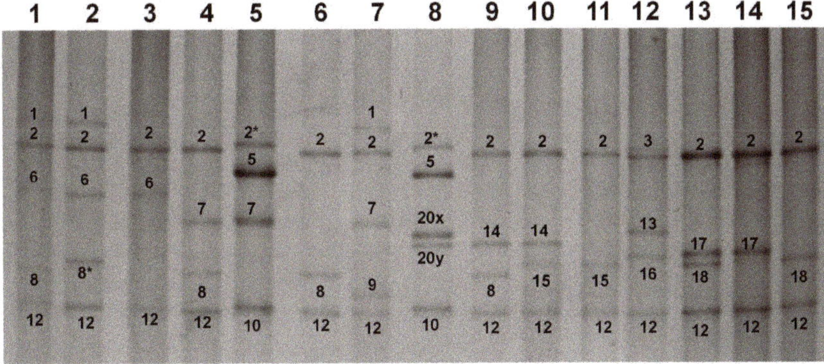

Figure 1. SDS-PAGE of representative variation for high molecular weight glutenin subunits (HMWGs) found in the collection of club (lanes 1, 3, 5–8, and 10–13), macha (lanes 2, 4, and 9), and Indian dwarf wheat (lanes 14 and 15).

Twenty-one allelic variants (3 at the *Glu-A1* locus, 15 at *Glu-B1*, and 3 at *Glu-D1*) were detected in the evaluated accessions (Table 1). Four out of the 15 for *Glu-B1* locus were novel. The distribution of these alleles in each subspecies was unequal.

In club wheat, three alleles were found for the *Glu-A1* locus, with the *Glu-A1a* allele being the least frequent (Table 1). The *Glu-B1* locus was more variable with 11 alleles, although only 3 of them showed frequencies above the average value that should have occurred if the distribution was random (mean value = 9.1%). The rest of the alleles were classified as rare according to the Marshall and Brown classification [28]. One of these rare alleles (null+15, Figure 1 lane 15) was also novel, detected only in accession PI 157920, and we propose to tentatively name this *Glu-B1ck* following the order of the Wheat Gene Catalogue [31]. For the *Glu-D1* locus, one allele (*Glu-D1a*, subunits 2 + 12) was clearly hegemonic, being present in 78.2% of the accessions evaluated.

Table 1. Allelic frequency for *Glu-A1*, *Glu-B1*, and *Glu-D1* loci of the evaluated accessions.

Allele	Subunit	No. of Samples (Frequency (%))		
		Club Wheat (*n* = 55)	**Macha Wheat** (*n* = 29)	**Indian Dwarf Wheat** (*n* = 26)
Glu-A1				
a	1	8 (14.5)	5 (17.2)	-
b	2*	27 (49.1)	6 (20.7)	5 (19.2)
c	null	20 (36.4)	18 (62.1)	21 (80.8)
Glu-B1				
a	7 + null	3 (5.5)	-	-
b	7 + 8	21 (38.2)	17 (58.6)	1 (3.8)
c	7 + 9	4 (7.3)	5 (17.2)	4 (15.4)
d	6 + 8	3 (5.5)	-	-
e	20x + 20y	12 (21.8)	2 (6.9)	4 (15.4)
f	13 + 16	2 (3.6)	-	-
h	14 + 15	1 (1.8)	-	-
i	17 + 18	2 (3.6)	1 (3.4)	11 (42.3)
aj	null + 8	1 (1.8)	-	-
am	null + 18	-	-	3 (11.5)
an	6 + null	5 (9.1)	-	-
	null + 15	1 (1.8)	-	-
	14 + 8	-	3 (10.3)	-
	6 + 8*	-	1 (3.4)	-
	17 + null	-	-	3 (11.5)
Glu-D1				
a	2 + 12	43 (78.2)	19 (65.5)	26 (100.0)
b	3 + 12	5 (9.1)	4 (13.8)	-
d	5 + 10	7 (12.7)	6 (20.7)	-

Similar to club wheat, macha wheat presented three alleles for the *Glu-A1* locus, although in this case, one (*Glu-A1c*) was three times more frequent that the other two (Table 1). For *Glu-B1*, more than half of the materials presented the *Glu-B1b* allele, whereas two were rare and found in only one accession. Therefore, two novel alleles were detected in this subspecies (subunits 14 + 8 and 6 + 8*, Figure 1 lanes 9 and 2, respectively). The first allele (subunits 14 + 8) was present in three accessions (PI 272554, PI 278660, and PI 290507), whereas the second (6 + 8*) was only found in accession PI 428177. We propose to tentatively name these alleles *Glu-B1cl* and *Glu-B1cm*, respectively, following the current order in the Wheat Gene Catalogue [31]. The variation for the *Glu-D1* locus was largest for the three subspecies, showing three alleles, one hegemonic (subunits 2 + 12) and the other two with similar frequencies.

The variation for the *Glu-A1* locus was low in Indian dwarf wheat, with only two alleles found and one representing more of 80% of the material (Table 1). The *Glu-B1* locus showed some variation (six alleles); one of them (subunit 17+null, Fig.1 lane 14) was novel, detected in three accessions (CItr 4531, PI 272581, and PI 282452) and we propose to name it *Glu-B1cn*. In contrast, the materials were homogenous at the *Glu-D1* locus. In this subspecies, only the *Glu-B1b* allele (subunits 7+8) can be considered rare.

When the three loci were evaluated together, the number of combinations was highly variable among the three subspecies analyzed (Table 2), with eight in Indian dwarf wheat and 26 in club wheat.

Table 2. Frequencies of the HMW glutenin subunit compositions found among accessions analyzed.

Club Wheat				Macha Wheat			
Glu-A1	*Glu-B1*	*Glu-D1*	N	*Glu-A1*	*Glu-B1*	*Glu-D1*	N
1	20x + 20y	2 + 12	2	1	6 + 8*	2 + 12	1
1	13 + 19	3 + 12	1	1	7 + 8	2 + 12	3
1	6 + 8	2 + 12	2	1	7 + 9	2 + 12	1
1	7 + 8	5 + 10	1	2*	14 + 8	null + 12	1
1	7 + 9	2 + 12	1	2*	7 + 8	null + 12	4
1	7 + 9	5 + 10	1	2*	7 + 9	null + 12	1
2*	6 + null	2 + 12	1	null	20x + 20y	3 + 12	1
2*	7 + null	2 + 12	2	null	20x + 20y	5 + 10	1
2*	7 + null	5 + 10	1	null	17 + 18	2 + 12	1
2*	20x + 20y	2 + 12	7	null	14 + 8	2 + 12	1
2*	20x + 20y	5 + 10	2	null	14 + 8	3 + 12	1
2*	13 + 19	2 + 12	1	null	7 + 8	2 + 12	5
2*	14 + 15	2 + 12	1	null	7 + 8	5 + 10	5
2*	7 + 8	2 + 12	9	null	7 + 9	2 + 12	1
2*	7 + 8	3 + 12	2	null	7 + 9	3 + 12	2
2*	7 + 9	2 + 12	1				
null	6 + null	2 + 12	4	**Indian Dwarf Wheat**			
null	null + 8	2 + 12	1	*Glu-A1*	*Glu-B1*	*Glu-D1*	N
null	null + 15	2 + 12	1	2*	17 + null	2 + 12	3
null	20x + 20y	3 + 12	1	2*	null + 18	2 + 12	1
null	17 + 18	2 + 12	2	2*	17 + 18	2 + 12	1
null	6 + 8	2 + 12	1	null	null + 18	2 + 12	2
null	7+8	2 + 12	6	null	20x + 20y	2 + 12	4
null	7 + 8	3 + 12	1	null	17 + 18	2 + 12	10
null	7 + 8	5 + 10	2	null	7 + 8	2 + 12	1
null	7 + 9	2 + 12	1	null	7 + 9	2 + 12	4

In each subspecies, the most frequent combination also differed, and in some cases the most frequent one in one subspecies was the least in another. In club wheat, although there was a great number of combinations, any of them can be considered hegemonic, and the most frequent combination was 2*, 7 + 8, and 2 + 12, which appeared in 9 of the 55 accessions. This combination was only found in three accessions of macha wheat, but it was absent in Indian dwarf wheat. A similar situation occurred with the most frequent combination in Indian dwarf wheat (null, 17 + 18, and 2 + 12), which was only detected in two accessions of club wheat and one of macha wheat.

The *Glu-1* quality score [32] for this last combination was associated with low gluten quality (score = 6), while the first combination (2*, 7 + 8, and 2 + 12) had a higher value and was associated with medium gluten quality (score = 8). Only one club wheat accession showed the highest score (10 for 1, 7 + 8, and 5 + 10) according to the scale of Payne et al. [32] developed for use in modern breeding programs targeting industrial bread-making processes.

3.2. Genetic Diversity

Some genetic parameters measured in each subspecies are shown in Table 3. These parameters detected important genetic erosion both in club wheat and macha wheat. The *Ne* values were especially significant for the *Glu-B1* locus in both subspecies, with values lower than 43% of the allelic variation detected (*A*). However, the genetic diversity (*He*) of this locus was high, possibly related to the fact that no hegemonic allele was detected in these subspecies. Nevertheless, the low frequency of most of the alleles suggested that these could easily be missed due to genetic drift effects.

Table 3. Genetic parameters for the *Glu-1* loci in the evaluated subspecies.

Subspecies	Locus	*A*	*Ne*	*He*
Club wheat	*Glu-A1*	3	2.53	0.605
	Glu-B1	11	4.62	0.783
	Glu-D1	3	1.57	0.364
	Mean	5.67	2.91	0.584
Macha wheat	*Glu-A1*	3	2.18	0.542
	Glu-B1	6	2.56	0.609
	Glu-D1	3	2.04	0.509
	Mean	4.00	2.26	0.553
Indian dwarf wheat	*Glu-A1*	2	1.45	0.310
	Glu-B1	6	3.93	0.746
	Glu-D1	1	1.00	0.000
	Mean	3.00	2.13	0.352

A: number of alleles; *Ne*: effective number of alleles; and *He*: genetic diversity.

These differences were slightly lower in Indian dwarf wheat, for which only the *Glu-B1* locus showed certain variability. In contrast, the *Glu-D1* locus was similar for all accessions evaluated (*Glu-D1a* allele = 2 + 12).

4. Discussion

Since the mid-twentieth century, the development of high-yield wheat cultivars has led to the replacement of local varieties, old varieties, or ancient wheat by modern cultivars [5]. Many of these materials remain stored in gene banks, being used in many cases as resources to generate new cultivars. However, new food movements have led to some of these neglected or underutilized crops beginning to be used as sources in traditional products. Unfortunately, for years, because of the lack of appropriate flours, these 'traditional' products have been made with modern flours that must be conditioned for these uses; consequently, whether these old materials would be better for making these traditional products remains unknown. For this reason, their analysis is necessary to determine the quality traits associated with optimal adaptation for this traditional food. This would permit both their recovery as new crops and the development of modern cultivars with similar quality characteristics but better agronomic traits. In any case, it is important to indicate that these ancient or neglected wheats cannot substitute for modern wheat; both types are clearly complementary within a more varied diet. Among these neglected wheats, the three subspecies evaluated in this study are options to independently explore, and they can be used as sources of agronomic traits. In this respect, Indian dwarf wheat has been evaluated as a potential source of stripe rust resistance [33] and salt tolerance [34]. However, the aspects related to flour quality have been seldom studied, mainly because most studies aimed to obtain new wheat cultivars with strong gluten that could be used for flour enrichment. In this context, these old wheats are not the best candidates. The interest in these wheat types has its origin in the recovery of traditional products performed with old techniques [35]. For this reason, comparisons with modern standards should be made cautiously.

Numerous studies carried out with wide collections of wheat have shown the high variation for HMWGs in this crop [9]. The high level of polymorphisms in these genes is a consequence of their physiological role. During the germination process, the seed storage proteins are a source of amino acid residues in the synthesis of new proteins needed for plant development [36]. Apparently, these proteins have no catalytic function, which has meant that changes in their amino acid composition or size has had no effect on plant viability and permits that the mutations can be easily fixed, generating wide polymorphisms. Numerous variants of these proteins have been detected in wheat, although many appear at very low frequencies, which implies a high risk of loss due to genetic drift processes.

However, the relationship of these proteins with food products has resulted in some alleles being fixed and some discarded, as the flour of these genotypes has shown better adaptation to a specific product or use. For this reason, variations are lower in modern than in ancient wheat. Although our knowledge of these proteins and their roles in flour quality is relatively recent [9], it is obvious that some of these alleles were empirically fixed by the farmers and bakers over time. They selected flours adapted to traditional uses, and, consequently, the HMWG combinations that better suited these quality properties were fixed, while the rest were gradually discarded. In this respect, the Indian dwarf wheat evaluated here is a wheat type endemic to the north of India and Pakistan associated with the elaboration of flat breads such as chapati [2], which requires flour with a medium gluten strength (score = 4–6) and high extensibility. Club wheat was also traditionally used to make cookies, for which weak flour is required; now, this wheat has some commercial importance in the Pacific Northwest of the USA because of this use [37].

Many other examples about the preferred use of ancient wheat and old wheat landraces for the elaboration of traditional wheat products are found in the literature. In a recent study in Turkey, Morgounov et al. [38] showed that, in different regions across the country, farmers have access to modern cultivars but still kept growing their landraces. Their main reason to do this is because they were happy with the grain quality and its suitability for homemade products (mainly typical loaves and thin types with bread wheat, and bulgur with durum wheat). In the same study, only 30% of the farmers rated the yield of the landraces as good, which clearly indicates that, despite their higher yields, modern wheat varieties do not satisfy them because their end-use quality is inadequate. This is in agreement with Bardsley et al. [39], who explained that the landrace Kirik is retained in Northeast Anatolia villages primarily because the baking qualities in the flour are appropriate for the local bread, lavash. Further studies are necessary to determine the grain quality components and properties that led to this preference, something that has already started in that country [40]. In other cases, such as the bread named "Pane Nero di Castelvetrano" from Sicily (Italy), the association between its end-use quality and the use of a specific landrace to elaborate it has been established [41]. Castelvetrano black bread is characterized by the intense brownness of its crumb. This bread is elaborated with durum wheat, and at least 20% of the flour blend should be from the autochthonous durum cultivar Timilia [42]. In fact, this landrace has high concentration of phenolic compounds that, when coupled with its extremely high polyphenol oxidase (PPO) activity, leads to the intense brownness of crumb [43]. On the other hand, modern durum cultivars are characterized for their low PPO activity, which makes them unsuitable for the preparation of this bread type. Other Italian traditional breads such as "Pane di Altamura" or "Pagnotta del Dittaino" (which have the European mark of Protected Designation of Origin, PDO) need to also be manufactured, by law, with durum semolina or re-milled semolina of specific cultivars (Appulo, Arcangelo, Duilio, and Simeto) to confer these breads their distinctive sensory properties and longer shelf-life [42,44]. These cultivars were developed in different times of the twentieth century, but they all have in their pedigree Cappelli, an old Italian cultivar derived from the Tunisian landrace Jennah Khetifa [45].

Although the variation for seed storage proteins of these subspecies was previously evaluated by Rayfuse and Jones [24] and Xu et al. [25], notable differences between our data and these previous findings were found. The detection of alleles that they did not find, together with changes in the frequencies of numerous common alleles, were especially significant. One possible cause of these differences is the use of different polyacrylamide concentrations in separation gels. Some of these subunits (e.g., subunit 2* vs. subunit 2) are difficult to detect in SDS-PAGE gel with 10% polyacrylamide concentration (C: 2.67%). For this reason, in the current study, the subunits were separated in SDS-PAGE gels using different concentrations (T: 8%, C: 1.28%) that our previous studies had confirmed as more adequate for separation of these proteins [17–19]. In total, 3 alleles were found for *Glu-A1*, 15 for *Glu-B1* (4 of them novel), and 3 for *Glu-D1*.

Probably because of their endemic condition and local use, the observed variation was especially low in macha and Indian dwarf wheat. This high homogeneity is very notable in all materials that have been cyclically used and neglected, as the narrowing of the genetic base, and subsequent reduced selection pressure, results in the loss of spontaneous variants and fixing of the most common ones by genetic drift. Compared with other hexaploid wheat subspecies, such as spelt, the variation was similar for the *Glu-A1* and *Glu-B1* loci, whereas variation for the *Glu-D1* locus detected here was slightly lower than that found in a wide collection of this hulled wheat of Spanish origin, where up to nine alleles were detected for this locus [19]. This could be a consequence of the wider geographical area where it is grown and its more diverse uses [46].

5. Conclusions

The neglected/underutilized wheat subspecies evaluated here showed wide polymorphisms for HMWGs, including novel alleles not previously described. Although the effects of these new allelic variants on technological properties should be further evaluated, this information may be of interest to wheat breeders for choosing parents to obtain recombinant lines with different gluten properties. Nevertheless, in the context of healthier and sustainable food, and as sources of genes for quality improvement in common wheat, these subspecies could be used to develop new/old crops with good agronomic traits and optimal flour characteristics for new and traditional products.

Supplementary Materials: The following are available online at http://www.mdpi.com/2073-4395/9/11/755/s1, Table S1: Allelic composition for the *Glu-1* loci in club wheat, Table S2: Allelic composition for the *Glu-1* loci in macha wheat, Table S3: Allelic composition for the *Glu-1* loci in Indian dwarf wheat.

Author Contributions: J.B.A. conceived and designed the study. J.B.A. and C.G. performed the experiments, analyzed the data, and wrote the paper. Both authors read and approved the final manuscript.

Funding: This research was supported by grant RTI2018-093367-B-I00 from the Spanish State Research Agency (Ministry of Science, Innovation and Universities), co-financed by the European Regional Development Fund (FEDER) from the European Union. Carlos Guzman gratefully acknowledges the European Social Fund and the Spanish State Research Agency (Ministry of Science, Innovation and Universities) for financial funding through the Ramon y Cajal Program (RYC-2017-21891).

Acknowledgments: We thank the National Small Grain Collection (Aberdeen, USA) for supplying the analyzed material.

Conflicts of Interest: The authors declare no conflicts of interest.

References

1. Harlan, J.R. The early history of wheat: Earliest traces to the sack of Rome. In *Wheat Science-Today and Tomorrow*; Evans, L.T., Peacock, W.J., Eds.; Cambridge University Press: Cambridge, UK, 1981; pp. 1–29.

2. Faridi, H.; Faubion, J.M. *Wheat End-Uses Around the World*; American Association of Cereal Chemists: St. Paul, MN, USA, 1995.

3. Matz, S.L. *Bakery Technology and Engineering*; Avi Publishing Co.: Westport, CT, USA, 1960.

4. Trichopoulou, A.; Soukara, S.; Vasilopoulou, E. Traditional foods: A science and society perspective. *Trends Food Sci. Technol.* **2007**, *18*, 420–427. [CrossRef]

5. Esquinas-Alcazar, J. Protecting crop genetic diversity for food security: Political, ethical and technical challenges. *Nat. Rev. Genet.* **2005**, *6*, 946–953. [CrossRef] [PubMed]

6. Jauhar, P.P. Alien Gene Transfer and Genetic Enrichment of Bread Wheat. In *Biodiversity and Wheat Improvement*, 1st ed.; Damania, A.B., Ed.; ICARDA-A Wiley Sayce Publication: Aleppo, Syria, 1993; pp. 103–119.

7. Alvarez, J.B.; Guzmán, C. Interspecific and intergeneric hybridization as a source of variation for wheat grain quality improvement. *Theor. Appl. Genet.* **2018**, *131*, 225–251. [PubMed]

8. Zamir, D. Improving plant breeding with exotic genetic libraries. *Nat. Rev. Genet.* **2001**, *2*, 983–989. [CrossRef]

9. Wrigley, C.; Békés, F.; Bushuk, W. (Eds.) *Gliadin and Glutenin: The Unique Balance of Wheat Quality*; AACC International Press: St. Paul, MN, USA, 2006.

10. Guzmán, C.; Alvarez, J.B. Wheat waxy proteins: Polymorphism, molecular characterization and effects on starch properties. *Theor. Appl. Genet.* **2016**, *129*, 1–16. [CrossRef]

11. Morris, C.F. Puroindolines: The molecular genetic basis of wheat grain hardness. *Plant Mol. Biol.* **2002**, *48*, 633–647. [CrossRef]

12. Payne, P.I. Genetics of wheat storage proteins and the effects of allelic variation on bread-making quality. *Annu. Rev. Plant Physiol.* **1987**, *38*, 141–153. [CrossRef]

13. Singh, N.K.; Shepherd, K.W. Linkage mapping of genes controlling endosperm storage proteins in wheat. 1. Genes on the short arms of group 1 chromosomes. *Theor. Appl. Genet.* **1988**, *75*, 628–641. [CrossRef]

14. Pogna, N.E.; Autran, J.C.; Mellini, F.; Lafiandra, D.; Feillet, P. Chromosome 1B-encoded gliadins and glutenin subunits in durum wheat: Genetics and relationship to gluten strength. *J. Cereal Sci.* **1990**, *11*, 15–34. [CrossRef]

15. Metakovsky, E.V.; Novoselskaya, A.Y.; Kopus, M.M.; Sobko, T.A.; Sozinov, A.A. Blocks of gliadin components in winter wheat detected by one-dimensional polyacrylamide gel electrophoresis. *Theor. Appl. Genet.* **1984**, *67*, 559–568. [CrossRef]

16. Harberd, N.P.; Bartels, D.; Thompson, R.D. DNA restriction-fragment variation in the gene family encoding high molecular weight (HMW) glutenin subunits of wheat. *Biochem. Genet.* **1986**, *24*, 579–596. [CrossRef] [PubMed]

17. Caballero, L.; Peña, R.J.; Martín, L.M.; Alvarez, J.B. Characterization of Mexican Creole wheat landraces in relation to morphological characteristics and HMW glutenin subunit composition. *Genet. Resour. Crop Evol.* **2010**, *57*, 657–665. [CrossRef]

18. Ayala, M.; Guzmán, C.; Peña, R.J.; Alvarez, J.B. Diversity of phenotypic (plant and grain morphological) and genotypic (glutenin alleles in *Glu-1* and *Glu-3* loci) traits of wheat landraces (*Triticum aestivum*) from Andalusia (Southern Spain). *Genet. Resour. Crop Evol.* **2016**, *63*, 465–475. [CrossRef]

19. Caballero, L.; Martin, L.M.; Alvarez, J.B. Allelic variation of the HMW glutenin subunits in Spanish accessions of spelt wheat (*Triticum aestivum* ssp *spelta* L. em. Thell.). *Theor. Appl. Genet.* **2001**, *103*, 124–128. [CrossRef]

20. Caballero, L.; Martín, L.M.; Alvarez, J.B. Variation and genetic diversity for gliadins in Spanish spelt wheat accessions. *Genet. Resour. Crop Evol.* **2004**, *51*, 679–686. [CrossRef]

21. Caballero, L.; Martín, L.M.; Alvarez, J.B. Agrobiodiversity of hulled wheats in Asturias (North of Spain). *Genet. Resour. Crop Evol.* **2007**, *54*, 267–277. [CrossRef]

22. Caballero, L.; Martín, L.M.; Alvarez, J.B. Genetic diversity in Spanish populations of *Triticum spelta* L. (escanda): Example of an endangered genetic resource. *Genet. Resour. Crop Evol.* **2008**, *55*, 675–682. [CrossRef]

23. Ayala, M.; Alvarez, J.B.; Yamamori, M.; Guzmán, C. Molecular characterization of waxy alleles in three subspecies of hexaploid wheat and identification of two novel *Wx-B1* alleles. *Theor. Appl. Genet.* **2015**, *128*, 2427–2435. [CrossRef]

24. Rayfuse, L.M.; Jones, S.S. Variation at *Glu-1* loci in club wheats. *Plant Breed.* **1993**, *111*, 89–98. [CrossRef]

25. Xu, L.-L.; Li, W.; Wei, Y.-M.; Zheng, Y.-L. Genetic diversity of HMW glutenin subunits in diploid, tetraploid and hexaploid *Triticum* species. *Genet. Resour. Crop Evol.* **2009**, *56*, 377–391. [CrossRef]

26. Payne, P.I.; Lawrence, G.J. Catalogue of alleles for the complex gene loci, *Glu-A1*, *Glu-B1* and *Glu-D1* which code for high-molecular-weight subunits of glutenin in hexaploid wheat. *Cereal Res. Commun.* **1983**, *11*, 29–35.

27. Laemmli, U.K. Cleavage of structural proteins during the assembly of the head of bacteriophage T4. *Nature* **1970**, *227*, 680–685. [CrossRef]

28. Marshall, D.; Brown, A. Optimum Sampling Strategies in Genetic Conservation. In *Crop Genetic Resources for Today and Tomorrow*; Frankel, O.H., Hawkes, J.G., Eds.; Cambridge University Press: Cambridge, UK, 1975; pp. 53–70.

29. Nei, M. Genetic distance between populations. *Am. Nat.* **1972**, *106*, 283–292. [CrossRef]

30. Nei, M. Analysis of gene diversity in subdivided populations. *Proc. Natl. Acad. Sci. USA* **1973**, *70*, 3321–3323. [CrossRef]

31. McIntosh, R.A.; Yamazaki, Y.; Dubcovsky, J.; Rogers, W.J.; Morris, G.; Appels, R.; Xia, X.C. Catalogue of Gene Symbols for Wheat. 2013. Available online: http://www.shigen.nig.ac.jp/wheat/komugi/genes/macgene/2013/GeneSymbol.pdf (accessed on 23 August 2019).

32. Payne, P.I.; Nightingale, M.A.; Krattiger, A.F.; Holt, L.M. The relationships between HMW glutenin subunit composition and the bread-making quality of British-grown wheat varieties. *J. Sci. Food Agric.* **1987**, *40*, 51–65. [CrossRef]

33. Badridze, G.; Weidner, A.; Asch, F.; Börner, A. Variation in salt tolerance within a Georgian wheat germplasm collection. *Genet. Resour. Crop Evol.* **2009**, *56*, 1125–1130. [CrossRef]

34. Chen, S.-S.; Chen, G.-Y.; Chen, H.; Wei, Y.-M.; Li, W.; Liu, Y.-X.; Liu, D.-C.; Lan, X.-J.; Zheng, Y.-L. Mapping stripe rust resistance gene *YrSph* derived from *Triticum sphaerococcum* Perc. with SSR, SRAP, and TRAP markers. *Euphytica* **2012**, *185*, 19–26. [CrossRef]

35. Miedaner, T.; Longin, F. *Neglected Cereals: From Ancient Grains to Superfood*; Agrimedia: Clenze, Germany, 2016.

36. Shutov, A.D.; Vaintraub, I.A. Degradation of storage proteins in germinating seeds. *Phytochemistry* **1987**, *26*, 1557–1566. [CrossRef]

37. Peterson, C.J.; Allan, R.E.; Peterson, C.L.J. US Pacific Northwest Region. In *The World Wheat Book: A History of Wheat Breeding*; Bonjean, A.P., Angus, W.J., Eds.; Lavoisier Publishing Inc.: Paris, France, 2001; pp. 407–429.

38. Morgounov, A.; Keser, M.; Kan, M.; Küçükçongar, M.; Özdemir, F.; Gummadov, N.; Muminjanov, H.; Zuev, E.; Qualset, C.O. Wheat landraces currently grown in Turkey: Distribution, diversity, and use. *Crop Sci.* **2016**, *56*, 3112–3124. [CrossRef]

39. Bardsley, D.; Thomas, I. Valuing local wheat landraces for agrobiodiversity conservation in Northeast Turkey. *Agric. Ecosyst. Environ.* **2005**, *106*, 407–412. [CrossRef]

40. Akcura, M.; Kokten, K.; Akcacik, A.G.; Aydogan, S. Pattern analysis of Turkish bread wheat landraces and cultivars for grain and flour quality. *Turk. J. Field Crops* **2016**, *21*, 120–130. [CrossRef]

41. Giancaspro, A.; Colasuonno, P.; Zito, D.; Blanco, A.; Pasqualone, A.; Gadaleta, A. Varietal traceability of bread 'Pane Nero di Castelvetrano' by denaturing high pressure liquid chromatography analysis of single nucleotide polymorphisms. *Food Control* **2016**, *59*, 809–817. [CrossRef]

42. Pasqualone, A. Italian Durum Wheat Breads. In *Bread Consumption and Health*; Pedrosa Silva Clerici, M.T., Ed.; Nova Science Publishers: New York, NY, USA, 2012, pp. 57–79.

43. Taranto, F.; Delvecchio, L.N.; Mangini, G.; Del Faro, L.; Blanco, A.; Pasqualone, A. Molecular and physico-chemical evaluation of enzymatic browning of whole meal and dough in a collection of tetraploid wheats. *J. Cereal Sci.* **2012**, *55*, 405–414. [CrossRef]

44. Pasqualone, A.; Alba, V.; Mangini, G.; Blanco, A.; Montemurro, C. Durum wheat cultivar traceability in PDO Altamura bread by analysis of DNA microsatellites. *Eur. Food Res. Technol.* **2010**, *230*, 723–729. [CrossRef]

45. Maccaferri, M.; Sanguineti, M.C.; Donini, P.; Porcedu, E.; Tuberosa, R. A Retrospective Analysis of Genetic Diversity in Durum Wheat Elite Germplasm Based on Microsatellite Analysis: A Case Study. In *Durum Wheat Breeding: Current Approaches and Future Strategies*; Royo, C., Nachit, M.N., Di Fonzo, N., Araus, J.L., Pfeiffer, W.H., Slafer, G.A., Eds.; Haworth Press: New York, NY, USA, 2005; pp. 99–142.

46. Peña-Chocarro, L. Promoting the Conservation and Use of Underutilized and Neglected Crops. In *Situ Conservation of Hulled Wheat Species: The Case of Spain, Hulled Wheats*; Padulosi, S., Hammer, K., Heller, J., Eds.; International Plant Genetic Resources Institute: Rome, Italy, 1996; pp. 129–146.

Article

Long-Term Effects of Biochar-Based Organic Amendments on Soil Microbial Parameters

Martin Brtnicky [1,2], Tereza Dokulilova [1], Jiri Holatko [1], Vaclav Pecina [1], Antonin Kintl [3], Oldrich Latal [4], Tomas Vyhnanek [5], Jitka Prichystalova [2] and Rahul Datta [1,*]

[1]   Department of geology and pedology, Faculty of Forestry and Wood technology, Mendel University in Brno, Brno 61300, Czech Republic; martin.brtnicky@seznam.cz (M.B.); tereza.dokulilova@mendelu.cz (T.D.); jiri.holatko@centrum.cz (J.H.); vaclav.pecina@mendelu.cz (V.P.)

[2]   Department of Agrochemistry, Soil Science, Microbiology and Plant Nutrition, Faculty of AgriSciences, Mendel University in Brno, Brno 61300, Czech Republic; jitka.prichystalova@mendelu.cz

[3]   Agriculture Research, Ltd., Troubsko 66441, Czech Republic; kintl@vupt.cz

[4]   Agrovyzkum Rapotin, Ltd., Rapotin 78813, Czech Republic; oldrich.latal@vuchs.cz

[5]   Department of Plant Biology, Faculty of AgriSciences, Mendel University in Brno, Brno 61300, Czech Republic; tomas.vyhnanek@mendelu.cz

*   Correspondence: rahulmedcure@gmail.com; Tel.: +420-773990283

Received: 8 October 2019; Accepted: 9 November 2019; Published: 12 November 2019

Abstract: Biochar application to the soil has been recommended as a carbon (C) management approach to sequester C and improve soil quality. Three-year experiments were conducted to investigate the interactive effects of three types of amendments on microbial biomass carbon, soil dehydrogenase activity and soil microbial community abundance in luvisols of arable land in the Czech Republic. Four different treatments were studied, which were, only NPK as a control, NPK + cattle manure, NPK + biochar and NPK + combination of manure with biochar. The results demonstrate that all amendments were effective in increasing the fungal and bacterial biomass, as is evident from the increased values of bacterial and fungal phospholipid fatty acid analysis. The ammonia-oxidizing bacteria population increases with the application of biochar, and it reaches its maximum value when biochar is applied in combination with manure. The overall results suggest that co-application of biochar with manure changes soil properties in favor of increased microbial biomass. It was confirmed that the application of biochar might increase or decrease soil activity, but its addition, along with manure, always promotes microbial abundance and their activity. The obtained results can be used in the planning and execution of the biochar-based soil amendments.

Keywords: biomass; biochar; soil; BPLFA; FPLFA; DHA; ammonia-oxidizing bacteria

---

1. Introduction

Fertilizers used in agricultural management influence soil quality and health [1,2]. The recent rise in concerns about environmental problems caused by the excessive use of chemical fertilizers necessitates detailed studies on alternate strategies to address such hazards. Long-term use of organic amendments to the soil helps in improving several soil parameters like organic carbon, aggregate stability and crop yield, in contrast to the application of chemical fertilizers [3–5]. Organic amendments also increase soil carbon sequestration and play a decisive role in mitigating the adverse effect of climate change [3,6,7]. Independent of the type and nature of the applied organic amendment, different changes in soil properties and fertility have been observed in a broad time horizon under different pedoclimatic conditions [8]. The positive effect on total soil carbon, soil nitrogen, soil microbial biomass carbon (MBC) and dehydrogenase activity (DHA) has been observed in soils treated with manure [9,10].

Organic amendment to the soil a has long term impact on soil restoration. Miller and Miller (2000) showed that long term application of manure has greater impact on soil properties as compared to short term application [11]. Long-term application of manure causes enhanced soil physical, chemical and biological properties [12]. In another study Shindo et al. (2006) [13] reported that continuous long term application of manure to a field drastically increases fulvic, humic acids and total humus content in the soil. On the contrary, the absence of organic fertilizer input to soil results in non-complex light weight humus [14]. A literature survey on the long-term application of manure pointed out that the use of manure with a mineral fertilizer (NPK) enhances soil properties and crop yield as compared to organic amendment alone [15,16].

Biochar is a carbon-rich material produced by pyrolysis reaction under limited or no oxygen environments, often used for soil amendment and carbon sequestration [17]. Biochar amendment improves soil physicochemical and biochemical properties. It increases the soil pH and cation exchange capacity (CEC) [18], improves soil structure [19], alters soil microbial populations [20] and enhances nutrient retention [21–24]. Although biochar is recalcitrant in nature, its ability to interact with soil properties makes it a good investment in soil [25]. Long term application of biochar brings change in the physio chemical properties of the soil. That leads to an alternation in the soil bacterial community.

Many studies have reported a synergistic effect of biochar and organic fertilizer in (a) improving plant growth by nitrate-capture in co-composted biochar [23,26], (b) promoting carbon stabilization through the formation of organo-mineral complexes [27] and (c) affecting soil nutrient cycles [28]. Organic fertilizer may form a coating inside and outside the biochar particle and increase hydrophilicity, thus raising the nutrient retention [29]. However, the antagonistic or neutral effect of manure and biochar interaction has been reported in studies [30–32].

The positive effect of soil treatment with manure on total soil carbon, nitrogen, microbial biomass carbon and dehydrogenase activity in the surface soil (0–5 cm) has generally been observed [10]. However, the combined effect of manure and the other soil amendment (e.g., biochar) has not been widely studied. In a few studies in which manure-derived biochar was applied to soils, a mostly positive effect was seen [33,34], but also a converse [34–36] effect on plant growth and soil microbial diversity was observed.

This study aimed to determine and compare the long-term effect of the biochar-based organic amendments on selected soil properties (MBC, DHA and soil microbial community abundance) on the agricultural land of the temperate climate zone of Central Europe. Only a few studies have been done so far on the agricultural area of temperate soils [37–39]. Prior studies suggested that manure addition to biochar as a soil amendment might have a synergistic or antagonistic effect. We hypothesized that the supplement of manure with biochar will represent a valid strategy to further enhance the soil microbial biomass and related soil quality indicators.

2. Materials and Methods

2.1. Study Site and Rationale behind the Field Scale Experiment

Field-scale experiments were designed to evaluate the potential of biochar and its combination with manure in improving the studied soil properties in on-site conditions. Experiments were carried out at arable land (luvisols) during the cropping season from the year 2014 to 2017. Experimental plots were located in Rapotín locality, the Czech Republic, at an altitude of about 345 m a.s.l., and it comes under the temperate continental climate zone, with a mean annual temperature of 7 °C. The mean annual precipitation is about 705 mm in the area. The rainfall pattern is 400–450 mm in the vegetation season and 250–300 mm during the winter period. The experiment consisted of the application of four soil treatments that were only NPK (mineral fertilizer) as a control, NPK + cattle manure (50 t/ha), NPK + biochar (15 t/ha) and NPK + combination of manure (50 t/ha) with biochar (15 t/ha) (MB). Biochar was added at the start of the experiment, while manure was added every year. Dosage of cattle manure 50 t/ha is added as recommended by (Singh et al., 2011) [40].

Dosage of biochar 15 t/ha was chosen close to the maximum amount of biochar allowed to be amended to the arable soil on the field (Pereira et al., 2011) [41]. Variant 4 is a combination with the same dosage of both amendments i.e., cattle manure 50 t/ha and biochar 15 t/ha. The experimental area was divided as follows: three small-scale-plots (10 × 10 m) per each of four variants of soil amendment (12 small-scale-plots overall).

2.2. Soil Sampling and Preparation

At the end of third crop season, the samples for the final analyses of the application of amendments were collected in October 2017. Three spatially-independent mixed soil subsamples from each experimental variant were collected in the following way: A portion of topsoil from a depth 0–15 cm was taken by a soil drill at five spots of an experimental field and mixed in a plastic sampling bag. The samples (app. 500 g) were immediately cooled down and transported to the laboratory at 0–4 °C and homogenized by sieving the soil through a 2 mm mesh under sterile conditions [42]. Samples for the enzyme activity assays were stored at 4 °C until analyzed (within one week). Samples for qPCR and phospholipid fatty acids (PLFA) analysis were freeze-dried.

2.3. Dehydrogenase Activities

Triphenyl tetrazolium chloride-dehydrogenase activity (TTC-DHA) was used to determine microbial activity in the soil. The methodology was modified according to Tabatabi (1994) [43], based on (Casida et al., 1964) [44]: 3-gram soil sample was mixed with MgO and sealed with the standard solution (triphenyl tetrazolium chloride + distilled water). The samples were incubated in the thermostat at 37 °C for 24 h. Afterwards, triphenylformazan (TPF) was extracted from the samples using methyl alcohol, resulting in the color change of the solution. The spectrophotometer (DR 3900, Hach Lang, Duesseldorf Germany) was used to measure the color intensity at a wavelength of 485 nm. DHA was calculated according to the calibration curve and expressed in $\mu g\ TPF{\cdot}g^{-1}{\cdot}h^{-1}$.

2.4. Quantification of Microbial Biomass

The samples for PLFA analysis were extracted from the mixture of chloroform-methanol-phosphate buffer (1:2:0.8) [45]. Phospholipids were separated using solid-phase extraction cartridges (LiChrolut Si 40, Merck, Bellefonte, PA, USA). The samples were then subjected to mild alkaline methanolysis and extracted to hexane as a final solvent [46]. The free methyl esters of phospholipid fatty acids were analyzed using gas chromatography-mass spectrometry (Agilent 7890A with FID detector, Agilent Technologies, USA). The gas chromatography instrument was equipped with a split/splitless injector, and a CP Sil 88 column was used for separation (100 m, 0.25 mm i.d., 0.2 μm film thickness). The temperature program started at 80 °C and was held for 1 min in splitless mode. Then the splitter was opened, and the oven was heated to 160 °C at a rate of 20 °C·min^{-1}. The second temperature ramp was up to 225 °C at a rate of 5 °C·min^{-1}; this temperature was maintained for 12 min.

Methylated fatty acids were identified according to their mass spectra and using a mixture of chemical standards obtained from Sigma Aldrich (Merck, USA)/Matreya LLC (USA). Fungal biomass was quantified based on the 18:2ω6,9 content (FPLFA), and bacterial biomass was quantified as a sum of i14:0, i15:0, a15:0, 16:1ω7t, 16:1ω9, 16:1ω7, 10Me-16:0, i17:0, a17:0, cy17:0, 17:0, 10Me-17:0, 10Me-18:0 and cy19:0 (BPLFA). The fatty acids found in both bacteria and fungi, 15:0, 16:0 and 18:1ω7, were excluded from the analysis. The relative content of individual PLFA molecules was also calculated. The total content of all PLFA molecules (PLFAT) was used as an indicator of total microbial biomass.

2.5. Microbial Biomass Carbon

Soil Microbial Biomass Carbon (MBC) was characterized and determined by the fumigation extraction method [47], based on the lysis of microbial cells upon contact with chloroform (24 h). Sample sets were duplicated, and only one set was subjected to fumigation, followed by the extraction of K_2SO_4 and comparison of fumigated and non-fumigated samples.

2.6. DNA Extraction and Real-Time qPCR

DNA was extracted from 0.5 g of lyophilized soil with the help of a DNeasy PowerSoil Kit (Qiagen, Valencia, CA, United States). Real-time PCR was performed to quantify partial bacterial (16S rDNA) and fungal (18S rDNA) rDNA gene in soil DNA extracts. Each sample was spiked with the DNA of plasmid vector derived from pUC18 serving as an internal standard for valuation of yield efficiency and contamination with PCR inhibitors. Isolated DNA was quantified using Picodrop. SYBR-green assays were performed in a CFX96 Real-Time PCR Detection System (Bio-Rad Laboratories). The primers used were 1108F (5′ ATGGYTGTCGTCAGCTCGTG 3′) and 1132R (5′GGGTTGCGCTCGTTGC 3′) for bacteria and FF390 (5′ AICCATTCAATCGGTAIT 3′) and FR1 (5′ AICCATTCAATCGGTAIT 3′) for fungi [48]. Combination of primer was used for the quantification of ammonia-oxidizing bacteria (AOB) 16S rDNA, CTO189FA/B (5′GGAGRAAAGCAGGGGATCG 3′), CTO189FC (5′GGAGGAAAGTAGGGGATCG 3′), and RT1R (5′CGTCCTCTCAGACCARCTACTG 3′) primers at a 2:1:2 ratio [49]. DNA of pUC18-derivate (internal standard) was quantified by qPCR using SQP (5′GTTTTCCCAGTCACGAC 3′) and SQPR2 (5′CTCGTATGTTGTGTGGAA 3′) primers.

2.7. Statistics Analysis

Comparison of individual data sets was made by one-way analysis of variance (ANOVA) and comparison methods. Duncan's multiple range test was used to compare treatments means, and a ($p < 0.05$) was considered statistically significant. Two way (ANOVA) was used to find the interaction between manure and biochar on measured soil properties (MBC, DHA, microbial community abundance). All the data were analyzed by Statistica ver. 13.4.0.14 software package.

3. Results and Discussion

3.1. Dehydrogenase

Dehydrogenase enzyme catalyzes organic matter decomposition in soil by transferring H^+ from the organic substrate with the help of coenzyme such as $NAD^+/NADP^+$ [50]. It is located inside living soil bacteria (e.g., genus *Pseudomonas*) and acts as an extracellular enzyme. It also implies that the enzyme cannot be deposited extra-cellularly in the soil in its active form [51]. Therefore, a high DHA activity suggests a higher number of the bacterial community present in the soil [52].

There was a significant decrease in DHA activity for the soil samples treated with biochar as compared to the control (Figure 1A). Data concerning the effect of different types (sources and pyrolytic temperature) of biochar on the dehydrogenase activity are still limited and contradictory. Different research groups reported different effects of biochar amendments on BPLFA values i.e., positive [53,54], neutral [55], and negative [56] effects. At least one study reported that DHA activity (and C mineralization) was lower in the biochar amended soil, however-glucosidase activity and the extracted PLFA concentration was not affected by biochar treatment [57]. Mechanisms for these different responses remain unclear [58]. Biochar effect on dehydrogenase activity in soil depends on the extent of the interaction between the substrate, enzyme and biochar (e.g., sorption and desorption of substrates on the biochar surface) [59]. The substrate and enzyme are attracted toward functional groups present on the biochar surface. Sorption of the enzyme to biochar blocks the active site present on the enzyme; resulting in the reduction in dehydrogenase activity [60]. The application of biochar produced at high temperature (approx. 400 °C or more) decreases soil enzyme activity and affects the soil nutrition dynamics, as it nonselectively sorbs the enzyme as well as a substrate due to its high absorptivity [61]. This high absorptivity can be attributed to high surface area and porosity of biochar created due to high temperature.

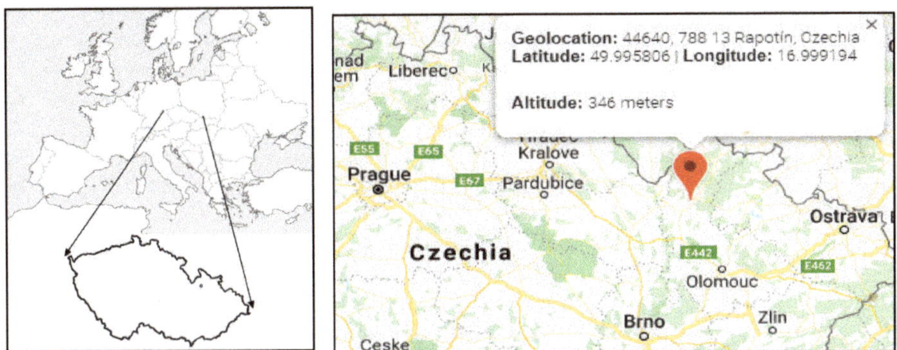

Figure 1. Geographical picture of study area.

MB amendment to biochar shows a significant increase in DHA activity as compared to biochar treatment (Figure 2A). The coapplication of manure caused a decrease in absorptive surface characteristics of biochar (sorptive sites occupancy) and hence neutralizing the negative effect of biochar. This agreed with the concept that net negative surface charge on biochar sorbs positive charge nutrients (NH_4^+, K^+, Ca^+, Fe^{++}, Cu^{++}) from manure [62]. Additionally, The sustainable and slow release of immobilized nutrient from biochar for an extended period is another reason for the increase in DHA activity [63–67]. The sorption of nutrient onto the biochar is reported to be a reversible phenomenon which favors the retention, and they are delay released into the soil [68–70].

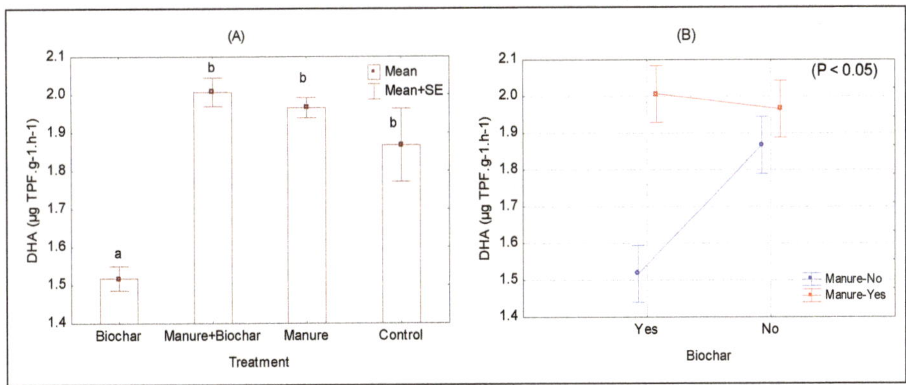

Figure 2. (**A**) DHA activity in soil amended with manure, biochar and MB. (**B**) Interaction graph of biochar and manure for DHA activity.

Two-factor ANOVA (with replication) was conducted to test the interaction effect of manure and biochar on DHA activity. The result shows that manure and biochar have a significant main effect on DHA activity (Figure 2B), whereas MB treatment shows a significant ordinal interaction on DHA activity (Figure 2B).

3.2. Soil Phospholipid Fatty Acid Analysis

PLFA, a rapid and sensitive method, was used to detect changes in the microbial community in soil. PLFAT, FPLFA and BPLFA can be viewed as indicators of total microbial biomass, fungal biomass, and bacterial biomass, respectively. Soil PLFAs analysis is a widely accepted method based on the rapid degradation of PLFAs after cell death [71].

Soil microbial community biomass represented by FPLFA was significantly higher in the sample treated with biochar, as compared to the control (Figure 3A). The highest value of FPLFA was recorded for soil treated with MB (Figure 3A). Generally, biochar is considered recalcitrant in nature [72], and microorganisms use it rarely, while mediating it in the rhizosphere as a source of nutrient for plants [73]. However, some studies suggest the existence of a labile-carbon fraction in it [31,41,74]. This labile fraction is available for the microorganism as a carbon source and supports microbial growth. This labile fraction may contain lipids, (up to 4.5%) with strong domination of glycolipids and phospholipids [75]. Additionally, biochar has been suggested to stimulate the dormant soil microorganism growth, thereby increasing the microbial biomass [76–79]. Saprotrophic fungi were shown to efficiently colonize biochar in association with decomposing fibrous organic matter [80], and the higher nutrient (namely P) content of the biochar increases its fungal colonization [81]. The unexpectedly higher FPLFA value in the sample treated with biochar (compared to manure treatment) can be explained due to the combined effect of initial biochar-derived phospholipids and the P-enhanced fungal colonization of biochar particles. This is also documented in agreement with high 18S rDNA content in the biochar-treated soil sample (Figure 5A). The synergy in co-application of manure and biochar on PLFA can be explained by their direct interaction and stimulation of microbial growth [82,83].

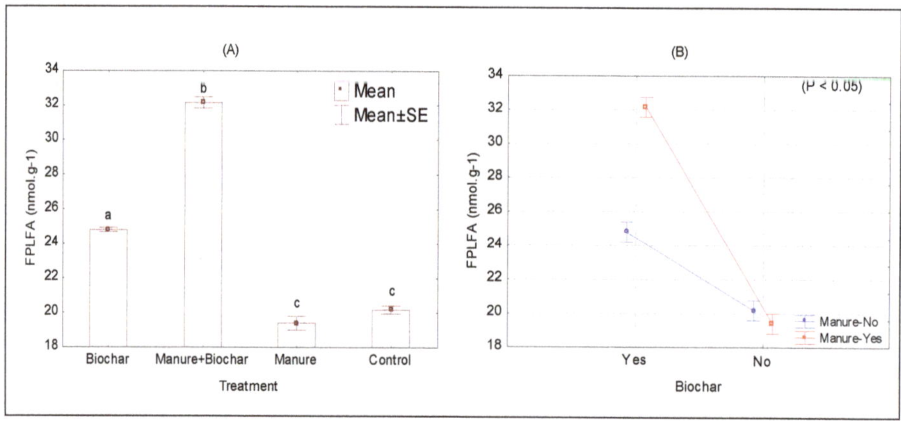

Figure 3. (**A**) Amount of FPLFA (nmol·g^{-1}) in soil amended with manure, biochar and MB. (**B**) Interaction graph of biochar and manure for FPLFA.

Two-factor ANOVA (with replication) was done to test the interaction effect of manure and biochar on the FPLFA value. The result shows that manure and biochar have a significant main effect on this FPLFA value (Figure 3B), whereas biochar and manure show significant ordinal interaction (Figure 3B).

BPLFA shows a similar response to biochar amendment (Figure 4) as FPLFA. There was a significant increase in BPLFA value following the addition of biochar. The noticeable increase in BPLFA concentration was observed when the mixture of manure with biochar is applied to the soil. BPLFA value for MB treatment was even higher than manure itself (Figure 4A).

Two-factor ANOVA (with replication) was conducted to test the interaction effect of manure and biochar on BPLFA values. The result shows that manure and biochar have a significant main effect on BPLFA values (Figure 4B), and a significant ordinal interaction was found between biochar and manure (Figure 4B).

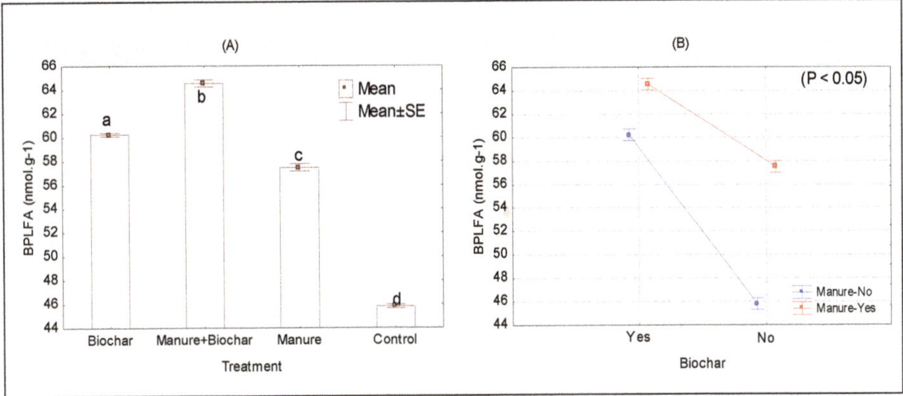

Figure 4. (**A**) Amount of BPLFA (nmol·g^{-1}) in soil amended with manure biochar and MB. (**B**) Interaction graph of biochar and manure for BPLFA.

3.3. Microbial Biomass Carbon

MBC is the main characteristic of soil organic carbon activity, so it is primarily used for the evaluation of soil quality [84]. It plays an essential role in biogeochemical cycles and is a major driver of ecosystem functioning [85,86]. Our experiment result shows a decline in the MBC value for the soil sample treated with biochar, and was lowest among all treatment (Figure 5A). Decrease in MBC upon biochar treatment has also been reported in the past [87,88]. We speculate that decline in the MBC value was either due to immobilization of carbon [89–91]. Or it could be due to the fact that biochar reduced the concentrations of nutrients through sorption and sequestration. Nutrients are protected from the microbes by their adsorption on the biochar surface. A decrease in the nutrient arability as a result reduces the microbial abundance. According to Dempster et al. (2012) [87] decrease in decomposition of SOM could be the main reason for the reduction in soil microbial biomass, upon biochar treatment. Our results appear to support this view. Several other studies show a positive effect of biochar on soil microbes [92,93]. Highest values for MBC were recorded when manure was applied in combination with biochar (Figure 5A). MBC values for biochar-only treatments are contradictory by PFLA values.

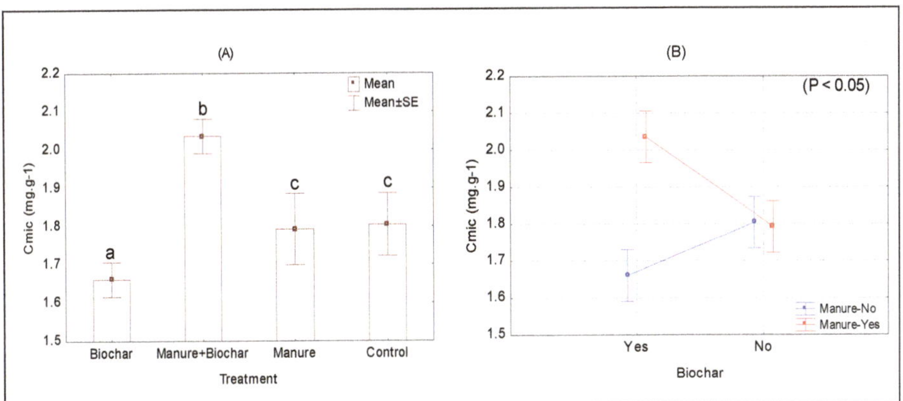

Figure 5. (**A**) MBC in soil amended with manure biochar and MB. (**B**) Interaction graph of biochar and manure for MBC.

These results indicate that chloroform fumigation used to determine microbial biomass and the decline in active cell numbers, does not always accompany with any decrease in microbial biomass. Hence, PLFAs, as a constant composition of living cell membranes, may also be unchanged when fumigation-sensitive microbial biomass decreases.

Two-factor ANOVA (with replication) was conducted to test the interaction effect of manure and biochar on MBC value. The result shows that biochar does not have any significant main effect on MBC (Figure 5B), whereas significant ordinal interaction was found between biochar and manure (Figure 5B).

3.4. DNA Extraction and Real-time qPCR

The effect of biochar, manure, and MB on the microbial community of treated soils is represented by the quantities of bacterial and fungal biomass, estimated as total 18S rDNA (fungal DNA) and 16S rDNA (bacterial DNA) content in the soil.

3.4.1. 18S rDNA

One-way analysis of variance shows a significant increase in log 18S rDNA copies relative to control on an application of biochar. The possible reason for an increase in 18S rDNA is that biochar can act as a habitat for many fungi [94–97]. Even though the extent of the hyphal colonization of biochar in soil is reported to be weak, extensive hyphal colonization of the surface of the biochar occurs, however it contrasts with a limited hyphal colonization of pores within the biochar [98]. Available nutrients like P, K and Ca on the biochar surface were possible reasons for fungal colonization [80,81], while the role of the labile-carbon fraction in the biochar was considered, similarly to the explanation of the observed FPLFA values (Topic 3.2). Small condo in biochar protects the microorganism from a natural soil predator such as mites, *Collembola* and larger (>16 μm in diameter) protozoans and nematodes [95,97,99–101]. The soil samples treated with MB showed a significant increase in log 18S rDNA copies, as compared to all treatments, including control (Figure 6). The increases in log 18S rDNA value followed the order: control < manure < biochar < MB (Figure 6A).

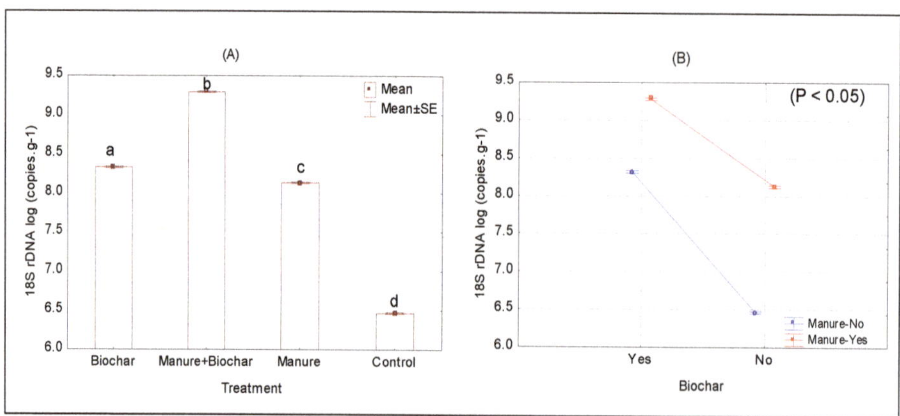

Figure 6. (**A**) 18S rDNA log (copies·g^{-1}) in soil amended with Biochar and MB. (**B**) Interaction graph of biochar and manure for 18S rDNA log (copies·g^{-1}).

The increase in log 18S rDNA copies upon MB treatment was putatively associated with a favorable environment for microbial proliferation [17]. Moreover, the functional group present on the biochar surface helps the sorption of dissolved organic carbon, decomposable organic compounds, and the chemisorption of the ammonium ion ($NH4^+$), and makes it a perfect microbial habitat [64]. Previous research showed that manure supplies macro- and micronutrients to be sorbed on the biochar surface and provide a suitable environment for fungal and other microbial growth and proliferation [27,63,102].

Two-factor ANOVA (two-way ANOVA with replication) is conducted to test the interaction effect of manure and biochar on log 18S rDNA copies. The result shows that manure and biochar have a significant main effect on log 18S rDNA copies (Figure 6B). Also, manure and biochar show significant ordinal interaction (Figure 6B).

3.4.2. 16S rDNA

Similarly, observations of an increase in log 16S rDNA copies relative to control was observed for biochar treatment and the soil treated with biochar in combination with manure, which shows the highest value of log 16S rDNA copies, even higher than the only manure treatment (Figure 7). Some authors also reported a significant increase in bacterial 16S rDNA genes abundance in samples coupled with biochar poultry-manure [94,103].

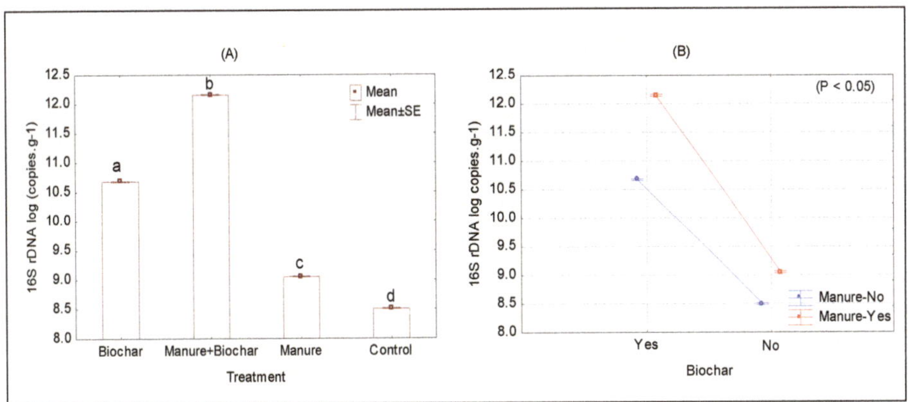

Figure 7. (**A**) 16S rDNA in soil amended with manure biochar and MB. (**B**) Interaction graph of biochar and manure for 16S, log (copies·g^{-1}).

Two-factor ANOVA (with replication) was conducted to test the interaction effect of manure with biochar on 16S rDNA gene copy. The result shows that manure and biochar have a significant main effect on the log-transformed 16S rDNA value (Figure 7B), whereas there was also a significant ordinal interaction found between biochar and manure (Figure 7B) with an increase in the 16S rDNA gene abundance, which was also observed in earlier studies [21].

The results obtained by estimation of 16S and 18S rDNA were supported by quantification of microbial biomass via phospholipidic fatty acids (PLFA). Two-factor ANOVA results of both bacterial BPLFA and fungal FPLFA show the statistical significance interactions between the amendments of biochar and manure.

3.4.3. 16S rDNA (AOB)

The effects of biochar and their combination with manure on 16S rDNA AOB copies are presented in Figure 8A. There was a significant increase in log 16S rDNA AOB copies for all the treatments compared to the control (Figure 8A). The 16S rDNA AOB value is an indicator of microbial activity in the process of nitrogen mineralization. It has been observed that the addition of organic fertilizers (e.g., compost) positively affects the microbial activity and utilization of nitrogen in contrast to the addition of the NPK fertilizer [104,105]. Among all the treatments, manure amendment with biochar shows the highest log 16S rDNA AOB copies.

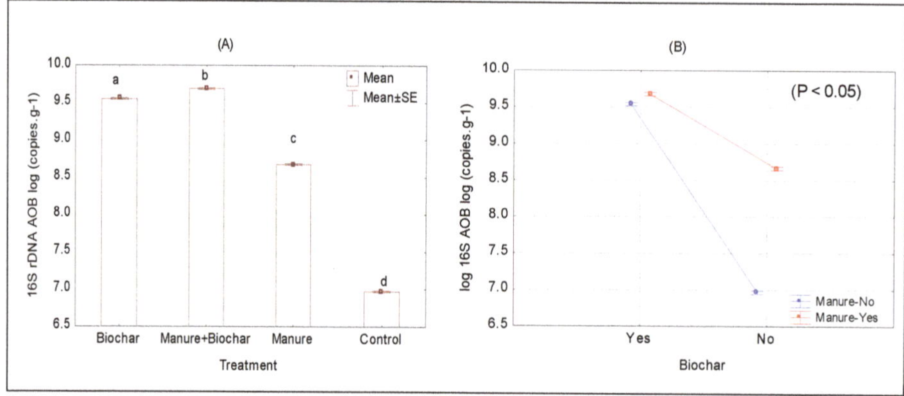

Figure 8. (**A**) 16S rDNA AOB rRNA in soil amended with biochar and combination with manure. (**B**) Interaction graph of biochar and manure for log 16S rDNA AOB copies.

Adding biochar into soils changes the soil structure [99,100] and alters soil microbial populations [79]. It is well known from the previous studies that adding biochar to soil, especially in combination with manure, can potentially alter the nitrification process in soil by affecting ammonia- and nitrite-oxidizing bacteria [106], decreasing N_2O emission [107,108] and increasing NH_4^+ storage [107].

Recently, several studies found that phenolic compounds (PHCs) and polycyclic aromatic hydrocarbons (PAHs) are retained in the biochar during the pyrolysis, and are really responsible for the inhibition of microbial activity, soil AOB and soil NO_3^- [107,109]. Previously published experimental outcomes proved that biochar addition to the soil retards the microbial nitrification mainly due to the toxicity of PHCs to AOB [110]. However, in our experiments, 16S rDNA AOB copies increase in response to the application of either biochar or its combination with manure. It was either due to lesser or no PHCs in the biochar, or increased availability of NH_4^+ sorbed on the biochar surface [62].

Two-factor ANOVA (with replication) was conducted to test the interaction effect of manure and biochar on log-transformed 16S rDNA AOB value (Figure 8B). Manure and biochar have a significant main effect on log 16S rRNA copies (Figure 8B), whereas significant ordinal interaction was found between biochar and manure (Figure 8B).

4. Conclusions

The results of this study demonstrate that the application of biochar with or without manure positively affect the fungal and bacterial biomass, as evident from the increased quantity of phospholipid fatty acid (BPLFA and FPLFA) and the DNA copy number (16S r DNA and 18S rDNA). Soil MBC and DHA activity decrease with the incorporation of biochar, but the maximum value is recorded for co-application with manure. These two properties were most affected by sorptive characteristics that might be directly dependent on the pyrolytic temperature used in biochar preparation. The results also revealed that the AOB population increased with the application of biochar and reached its maximum value when biochar is applied in combination with manure. However, further detailed studies are required to investigate the influence of biochars on nitrification and AOB community.

It can be concluded that the amendment of biochar in combination with manure changed the soil properties in three years of soil experiment in favor of increased microbial biomass. It was also confirmed that the application of biochar solely might increase or decrease soil activity, but their addition, along with manure, always promotes microbial abundance and their activity. However, quantity and sorption characteristics must be taken into account while planning the biochar-based soil amendments.

Author Contributions: Conceptualization, M.B. and J.H.; methodology, M.B., J.H.; software, R.D.; validation, M.B., J.H. and A.K.; formal analysis, T.D. and V.P.; investigation, M.B., J.H., J.P., T.V.; resources, M.B., J.H., O.L.; data curation, T.D., V.P., A.K.; writing—original draft preparation, R.D., J.H.; writing—review and editing, R.D.; visualization, J.H.; supervision, J.H., O.L., T.V.; project administration, M.B.; funding acquisition, M.B.

Funding: This research received no external funding.

Acknowledgments: This research was funded by Technology Agency of the Czech Republic project TH02030169: "Effect of biologically transformed organic matter and biochar application on the stability of productive soil properties and reduction of environmental risks" and by Technology Agency of the Czech Republic project TH03030319: "Promoting the functional diversity of soil organisms by applying classical and modified stable organic matter while preserving the soil's production properties" and the APC was funded by these projects.

Conflicts of Interest: The authors declare no conflict of interest.

References

1. Doran, J.W.; Zeiss, M.R. Soil health and sustainability: Managing the biotic component of soil quality. *Appl. Soil Ecol.* **2000**, *15*, 3–11. [CrossRef]

2. Singh, B.; Ryan, J. Managing fertilizers to enhance soil health. *Int. Fertil. Ind. Assoc. Paris Fr.* **2015**, 1–24. [CrossRef]

3. Diacono, M.; Montemurro, F. Long-term effects of organic amendments on soil fertility. In *Sustainable Agriculture*; Springer: Berlin/Heidelberg, Germany, 2011; Volume 2, pp. 761–786.

4. Zimmer, D.; Kruse, J.; Siebers, N.; Panten, K.; Oelschläger, C.; Warkentin, M.; Hu, Y.; Zuin, L.; Leinweber, P. Bone char vs. S-enriched bone char: Multi-method characterization of bone chars and their transformation in soil. *Sci. Total Environ.* **2018**, *643*, 145–156. [CrossRef] [PubMed]

5. Cheng, J.; Lee, X.; Tang, Y.; Zhang, Q. Long-term effects of biochar amendment on rhizosphere and bulk soil microbial communities in a karst region, southwest China. *Appl. Soil Ecol.* **2019**, *140*, 126–134. [CrossRef]

6. Liu, X.; Ye, Y.; Liu, Y.; Zhang, A.; Zhang, X.; Li, L.; Pan, G.; Kibue, G.W.; Zheng, J.; Zheng, J. Sustainable biochar effects for low carbon crop production: A 5-crop season field experiment on a low fertility soil from Central China. *Agric. Syst.* **2014**, *129*, 22–29. [CrossRef]

7. Liu, X.; Zhou, J.; Chi, Z.; Zheng, J.; Li, L.; Zhang, X.; Zheng, J.; Cheng, K.; Bian, R.; Pan, G. Biochar provided limited benefits for rice yield and greenhouse gas mitigation six years following an amendment in a fertile rice paddy. *Catena* **2019**, *179*, 20–28. [CrossRef]

8. Rasmussen, P.E.; Goulding, K.W.; Brown, J.R.; Grace, P.R.; Janzen, H.H.; Körschens, M. Long-term agroecosystem experiments: Assessing agricultural sustainability and global change. *Science* **1998**, *282*, 893–896. [CrossRef] [PubMed]

9. Mandal, A.; Patra, A.K.; Singh, D.; Swarup, A.; Masto, R.E. Effect of long-term application of manure and fertilizer on biological and biochemical activities in soil during crop development stages. *Bioresour. Technol.* **2007**, *98*, 3585–3592. [CrossRef] [PubMed]

10. Peacock, A.D.; Mullen, M.D.; Ringelberg, D.B.; Tyler, D.D.; Hedrick, D.B.; Gale, P.M.; White, D.C. Soil microbial community responses to dairy manure or ammonium nitrate applications. *Soil Biol. Biochem.* **2001**, *33*, 1011–1019. [CrossRef]

11. Sumner, M.E. *Handbook of Soil Science*; CRC Press: Boca Raton, FL, USA, 2000; ISBN 0-8493-3136-6.

12. Van-camp, L.; Bujarrabal, B.; Gentile, A.R.; Jones, R.J.; Montanarella, L.; Olazabal, C.; Selvaradjou, S. *Technical Working Groups Established under the Thematic Strategy for Soil Protection*; Office for Official Publications of the European Communities: Brussels, Belgium, 2004.

13. Shindo, H.; Hirahara, O.; Yoshida, M.; Yamamoto, A. Effect of continuous compost application on humus composition and nitrogen fertility of soils in a field subjected to double cropping. *Biol. Fertil. Soils* **2006**, *42*, 437–442. [CrossRef]

14. Nardi, S.; Morari, F.; Berti, A.; Tosoni, M.; Giardini, L. Soil organic matter properties after 40 years of different use of organic and mineral fertilisers. *Eur. J. Agron.* **2004**, *21*, 357–367. [CrossRef]

15. Ros, M.; Klammer, S.; Knapp, B.; Aichberger, K.; Insam, H. Long-term effects of compost amendment of soil on functional and structural diversity and microbial activity. *Soil Use Manag.* **2006**, *22*, 209–218. [CrossRef]

16. Poulain, S.; Sergeant, C.; Simonoff, M.; Le Marrec, C.; Altmann, S. Microbial Investigations in Opalinus Clay, an Argillaceous Formation under Evaluation as a Potential Host Rock for a Radioactive Waste Repository. *Geomicrobiol. J.* **2008**, *25*, 240–249. [CrossRef]

17. Lehmann, J.; da Silva, J.P., Jr.; Rondon, M.; Cravo, M.; da Silva, G.M.; Greenwood, J.; Nehls, T.; Steiner, C.; Glaser, B. Slash-and-char-a feasible alternative for soil fertility management in the central Amazon. In Proceedings of the 17th World Congress of Soil Science, Bangkok, Thailand, 14–21 August 2002; pp. 1–12.

18. Kookana, R.S.; Sarmah, A.K.; Van Zwieten, L.; Krull, E.; Singh, B. Biochar application to soil: Agronomic and environmental benefits and unintended consequences. In *Advances in Agronomy*; Elsevier: Amsterdam, The Netherlands, 2011; Volume 112, pp. 103–143. ISBN 0065-2113.

19. Jones, D.L.; Rousk, J.; Edwards-Jones, G.; DeLuca, T.H.; Murphy, D.V. Biochar-mediated changes in soil quality and plant growth in a three year field trial. *Soil Biol. Biochem.* **2012**, *45*, 113–124. [CrossRef]

20. Lehmann, J. *Biochar for Environmental Management*; Routledge: Abingdon, UK, 2015.

21. Azeem, M.; Hayat, R.; Hussain, Q.; Ahmed, M.; Imran, M.; Crowley, D.E. Effect of biochar amendment on soil microbial biomass, abundance and enzyme activity in the mash bean field. *J. Biodivers. Environ. Sci.* **2016**, *8*, 2222–3045.

22. Clough, T.; Condron, L.; Kammann, C.; Müller, C. A review of biochar and soil nitrogen dynamics. *Agronomy* **2013**, *3*, 275–293. [CrossRef]

23. Powell, S.J.; Prosser, J.I. Protection of Nitrosomonas europaea colonizing clay minerals from inhibition by nitrapyrin. *J. Gen. Microbiol.* **1991**, *137*, 1923–1929. [CrossRef] [PubMed]

24. Thies, J.E.; Rillig, M.C. Characteristics of biochar: Biological properties. In *Biochar for Environmental Management*; Routledge: Abingdon, UK, 2012; pp. 117–138.

25. Sohi, S.; Lopez-Capel, E.; Krull, E.; Bol, R. Biochar, climate change and soil: A review to guide future research. *CSIRO Land Water Sci. Rep.* **2009**, *5*, 17–31.

26. Sánchez-Monedero, M.A.; Cayuela, M.L.; Sánchez-García, M.; Vandecasteele, B.; D'Hose, T.; López, G.; Martínez-Gaitán, C.; Kuikman, P.J.; Sinicco, T.; Mondini, C. Agronomic Evaluation of Biochar, Compost and Biochar-Blended Compost across Different Cropping Systems: Perspective from the European Project FERTIPLUS. *Agronomy* **2019**, *9*, 225. [CrossRef]

27. Plaza, C.; Giannetta, B.; Fernández, J.M.; López-de-Sá, E.G.; Polo, A.; Gascó, G.; Méndez, A.; Zaccone, C. Response of different soil organic matter pools to biochar and organic fertilizers. *Agric. Ecosyst. Environ.* **2016**, *225*, 150–159. [CrossRef]

28. Ye, J.; Zhang, R.; Nielsen, S.; Joseph, S.D.; Huang, D.; Thomas, T. A Combination of Biochar–Mineral Complexes and Compost Improves Soil Bacterial Processes, Soil Quality, and Plant Properties. *Front. Microbiol.* **2016**, *7*, 372. [CrossRef] [PubMed]

29. Hagemann, N.; Joseph, S.; Schmidt, H.-P.; Kammann, C.I.; Harter, J.; Borch, T.; Young, R.B.; Varga, K.; Taherymoosavi, S.; Elliott, K.W.; et al. Organic coating on biochar explains its nutrient retention and stimulation of soil fertility. *Nat. Commun.* **2017**, *8*, 1089. [CrossRef] [PubMed]

30. Seehausen, M.; Gale, N.; Dranga, S.; Hudson, V.; Liu, N.; Michener, J.; Thurston, E.; Williams, C.; Smith, S.; Thomas, S. Is There a Positive Synergistic Effect of Biochar and Compost Soil Amendments on Plant Growth and Physiological Performance? *Agronomy* **2017**, *7*, 13. [CrossRef]

31. Zimmerman, A.R.; Gao, B.; Ahn, M.-Y. Positive and negative carbon mineralization priming effects among a variety of biochar-amended soils. *Soil Biol. Biochem.* **2011**, *43*, 1169–1179. [CrossRef]

32. Cely, P.; Tarquis, A.M.; Paz-Ferreiro, J.; Méndez, A.; Gascó, G. Factors driving the carbon mineralization priming effect in a sandy loam soil amended with different types of biochar. *Solid Earth* **2014**, *5*, 585–594. [CrossRef]

33. Kanwal, S.; Batool, A.; Ghufran, M.A.; Khalid, A. Effect of dairy manure derived biochar on microbial biomass carbon, soil carbon and Vitis vinifera under water stress conditions. *Pak. J. Bot* **2018**, *50*, 1713–1718.

34. Nguyen, B.T.; Trinh, N.N.; Le, C.M.T.; Nguyen, T.T.; Tran, T.V.; Thai, B.V.; Le, T.V. The interactive effects of biochar and cow manure on rice growth and selected properties of salt-affected soil. *Arch. Agron. Soil Sci.* **2018**, *64*, 1744–1758. [CrossRef]

35. Dodor, D.E.; Amanor, Y.J.; Attor, F.T.; Adjadeh, T.A.; Neina, D.; Miyittah, M. Co-application of biochar and cattle manure counteract positive priming of carbon mineralization in a sandy soil. *Environ. Syst. Res.* **2018**, *7*, 5. [CrossRef]

36. Liang, X.; Chen, L.; Liu, Z.; Jin, Y.; He, M.; Zhao, Z.; Liu, C.; Niyungeko, C.; Arai, Y. Composition of microbial community in pig manure biochar-amended soils and the linkage to the heavy metals accumulation in rice at harvest. *Land Degrad. Dev.* **2018**, *29*, 2189–2198. [CrossRef]

37. Bamminger, C.; Poll, C.; Sixt, C.; Högy, P.; Wüst, D.; Kandeler, E.; Marhan, S. Short-term response of soil microorganisms to biochar addition in a temperate agroecosystem under soil warming. *Agric. Ecosyst. Environ.* **2016**, *233*, 308–317. [CrossRef]

38. Hüppi, R.; Neftel, A.; Lehmann, M.F.; Krauss, M.; Six, J.; Leifeld, J. N use efficiencies and N2O emissions in two contrasting, biochar amended soils under winter wheat—Cover crop—Sorghum Rotation. *Environ. Res. Lett.* **2016**, *11*, 084013. [CrossRef]

39. Horák, J.; Kondrlová, E.; Igaz, D.; Šimanský, V.; Felber, R.; Lukac, M.; Balashov, E.V.; Buchkina, N.P.; Rizhiya, E.Y.; Jankowski, M. Biochar and biochar with N-fertilizer affect soil N2O emission in Haplic Luvisol. *Biologia (Bratisl.)* **2017**, *72*. [CrossRef]

40. Singh, J.S.; Pandey, V.C.; Singh, D.P. Coal fly ash and farmyard manure amendments in dry-land paddy agriculture field: Effect on N-dynamics and paddy productivity. *Appl. Soil Ecol.* **2011**, *47*, 133–140. [CrossRef]

41. Pereira, R.C.; Kaal, J.; Arbestain, M.C.; Lorenzo, R.P.; Aitkenhead, W.; Hedley, M.; Macias, F.; Hindmarsh, J.; Macia-Agullo, J.A. Contribution to characterisation of biochar to estimate the labile fraction of carbon. *Org. Geochem.* **2011**, *42*, 1331–1342. [CrossRef]

42. Datta, R.; Vranová, V.; Pavelka, M.; Rejšek, K.; Formánek, P. Effect of soil sieving on respiration induced by low-molecular-weight substrates. *Int. Agrophysics* **2014**, *28*, 119–124. [CrossRef]

43. Tabatabai, M.A. Soil enzymes. In *Methods Soil Anal. Part. 2—Microbiological Biochem. Prop*; Soil Science Society of America: Madison, WI, USA, 1994; pp. 775–833.

44. Casida, L.E., Jr.; Klein, D.A.; Santoro, T. Soil dehydrogenase activity. *Soil Sci.* **1964**, *98*, 371–376. [CrossRef]

45. Bligh, E.G.; Dyer, W.J. A rapid method of total lipid extraction and purification. *Can. J. Biochem. Physiol.* **1959**, *37*, 911–917. [CrossRef] [PubMed]

46. Oravecz, O.; Elhottová, D.; Krištůfek, V.; Šustr, V.; Frouz, J.; Tříska, J.; Marialigeti, K. Application of ARDRA and PLFA analysis in characterizing the bacterial communities of the food, gut and excrement of saprophagous larvae ofPenthetria holosericea (Diptera: Bibionidae): A pilot study. *Folia Microbiol. (Praha)* **2004**, *49*, 83. [CrossRef] [PubMed]

47. Vance, E.D.; Brookes, P.C.; Jenkinson, D.S. An extraction method for measuring soil microbial biomass C. *Soil Biol. Biochem.* **1987**, *19*, 703–707. [CrossRef]

48. Vainio, E.J.; Hantula, J. Direct analysis of wood-inhabiting fungi using denaturing gradient gel electrophoresis of amplified ribosomal DNA. *Mycol. Res.* **2000**, *104*, 927–936. [CrossRef]

49. Hermansson, A.; Lindgren, P.-E. Quantification of ammonia-oxidizing bacteria in arable soil by real-time PCR. *Appl. Env. Microbiol* **2001**, *67*, 972–976. [CrossRef] [PubMed]

50. Balba, M.T.; Al-Awadhi, N.; Al-Daher, R. Bioremediation of oil-contaminated soil: Microbiological methods for feasibility assessment and field evaluation. *J. Microbiol. Method.* **1998**, *32*, 155–164. [CrossRef]

51. Ukaoma, A.A.; Ukaoma, V.O.; Opara, F.N.; Osuala, F.O.U. Inhibition of Dehydrogenase Activity in Pathogenic Bacteria Isolates by Aqueous Extract of Curcuma Longa (Turmeric) Rhizome. *J. Phytopharma.* **2013**, *2*, 9–17.

52. Walls-Thumma, D. Dehydrogenase Activity in Soil Bacteria. *Dehydrogenase Act. Soil Bact. Gard. Pub* **2000**, *130633*.

53. Kumar, S.; Masto, R.E.; Ram, L.C.; Sarkar, P.; George, J.; Selvi, V.A. Biochar preparation from Parthenium hysterophorus and its potential use in soil application. *Ecol. Eng.* **2013**, *55*, 67–72. [CrossRef]

54. Park, J.H.; Choppala, G.K.; Bolan, N.S.; Chung, J.W.; Chuasavathi, T. Biochar reduces the bioavailability and phytotoxicity of heavy metals. *Plant. Soil* **2011**, *348*, 439. [CrossRef]

55. Wu, F.; Jia, Z.; Wang, S.; Chang, S.X.; Startsev, A. Contrasting effects of wheat straw and its biochar on greenhouse gas emissions and enzyme activities in a Chernozemic soil. *Biol. Fertil. Soils* **2013**, *49*, 555–565. [CrossRef]

56. Paz-Ferreiro, J.; Gascó, G.; Gutiérrez, B.; Méndez, A. Soil biochemical activities and the geometric mean of enzyme activities after application of sewage sludge and sewage sludge biochar to soil. *Biol. Fertil. Soils* **2012**, *48*, 511–517. [CrossRef]

57. Ameloot, N.; Sleutel, S.; Case, S.D.C.; Alberti, G.; McNamara, N.P.; Zavalloni, C.; Vervisch, B.; delle Vedove, G.; De Neve, S. C mineralization and microbial activity in four biochar field experiments several years after incorporation. *Soil Biol. Biochem.* **2014**, *78*, 195–203. [CrossRef]

58. Bailey, V.L.; Fansler, S.J.; Smith, J.L.; Bolton Jr, H. Reconciling apparent variability in effects of biochar amendment on soil enzyme activities by assay optimization. *Soil Biol. Biochem.* **2011**, *43*, 296–301. [CrossRef]

59. Lammirato, C.; Miltner, A.; Kaestner, M. Effects of wood char and activated carbon on the hydrolysis of cellobiose by β-glucosidase from Aspergillus niger. *Soil Biol. Biochem.* **2011**, *43*, 1936–1942. [CrossRef]

60. Czimczik, C.I.; Masiello, C.A. Controls on black carbon storage in soils. *Glob. Biogeochem. Cycles* **2007**, *21*. [CrossRef]

61. Ameloot, N.; De Neve, S.; Jegajeevagan, K.; Yildiz, G.; Buchan, D.; Funkuin, Y.N.; Prins, W.; Bouckaert, L.; Sleutel, S. Short-term CO2 and N2O emissions and microbial properties of biochar amended sandy loam soils. *Soil Biol. Biochem.* **2013**, *57*, 401–410. [CrossRef]

62. Yao, Y.; Gao, B.; Zhang, M.; Inyang, M.; Zimmerman, A.R. Effect of biochar amendment on sorption and leaching of nitrate, ammonium, and phosphate in a sandy soil. *Chemosphere* **2012**, *89*, 1467–1471. [CrossRef] [PubMed]

63. Hagemann, N.; Kammann, C.I.; Schmidt, H.-P.; Kappler, A.; Behrens, S. Nitrate capture and slow release in biochar amended compost and soil. *PLoS ONE* **2017**, *12*, e0171214. [CrossRef] [PubMed]

64. Kammann, C.I.; Schmidt, H.-P.; Messerschmidt, N.; Linsel, S.; Steffens, D.; Müller, C.; Koyro, H.-W.; Conte, P.; Joseph, S. Erratum: Plant growth improvement mediated by nitrate capture in co-composted biochar. *Sci. Rep.* **2015**, *5*, 12378. [CrossRef] [PubMed]

65. Li, W.; Huang, C.; Xie, H.; Bi, Y. Study on the release property of nitrogen in sustained-release fertilizer with carrier of bentonite. In Proceedings of the 2013 Third International Conference on Intelligent System Design and Engineering Applications, Hong Kong, China, 16–18 January 2013; IEEE: Piscataway, NJ, USA, 2013; pp. 1364–1367.

66. Lu, Q.; Feng, X.; Sun, K.; Liao, Z. Study on the use of polymer/bentonite composites for controlled release. *Plant. Nutr. Fertil. Sci.* **2005**, *2*.

67. Yuan, H.; Lu, T.; Wang, Y.; Chen, Y.; Lei, T. Sewage sludge biochar: Nutrient composition and its effect on the leaching of soil nutrients. *Geoderma* **2016**, *267*, 17–23. [CrossRef]

68. Liang, B.; Lehmann, J.; Solomon, D.; Kinyangi, J.; Grossman, J.; O'neill, B.; Skjemstad, J.O.; Thies, J.; Luizao, F.J.; Petersen, J. Black carbon increases cation exchange capacity in soils. *Soil Sci. Soc. Am. J.* **2006**, *70*, 1719–1730. [CrossRef]

69. Novak, J.M.; Busscher, W.J.; Laird, D.L.; Ahmedna, M.; Watts, D.W.; Niandou, M.A. Impact of biochar amendment on fertility of a southeastern coastal plain soil. *Soil Sci.* **2009**, *174*, 105–112. [CrossRef]

70. Van Zwieten, L.; Kimber, S.; Morris, S.; Chan, K.Y.; Downie, A.; Rust, J.; Joseph, S.; Cowie, A. Effects of biochar from slow pyrolysis of papermill waste on agronomic performance and soil fertility. *Plant. Soil* **2010**, *327*, 235–246. [CrossRef]

71. Drenovsky, R.E.; Elliott, G.N.; Graham, K.J.; Scow, K.M. Comparison of phospholipid fatty acid (PLFA) and total soil fatty acid methyl esters (TSFAME) for characterizing soil microbial communities. *Soil Biol. Biochem.* **2004**, *36*, 1793–1800. [CrossRef]

72. Skjemstad, J.O.; Reicosky, D.C.; Wilts, A.R.; McGowan, J.A. Charcoal carbon in US agricultural soils. *Soil Sci. Soc. Am. J.* **2002**, *66*, 1249–1255. [CrossRef]

73. Prendergast-Miller, M.T.; Duvall, M.; Sohi, S.P. Biochar-root interactions are mediated by biochar nutrient content and impacts on soil nutrient availability. *Eur. J. Soil Sci.* **2014**, *65*, 173–185. [CrossRef]

74. Bruun, E.W.; Hauggaard-Nielsen, H.; Ibrahim, N.; Egsgaard, H.; Ambus, P.; Jensen, P.A.; Dam-Johansen, K. Influence of fast pyrolysis temperature on biochar labile fraction and short-term carbon loss in a loamy soil. *Biomass Bioenergy* **2011**, *35*, 1182–1189. [CrossRef]

75. Kuzyakov, Y.; Bogomolova, I.; Glaser, B. Biochar stability in soil: Decomposition during eight years and transformation as assessed by compound-specific 14C analysis. *Soil Biol. Biochem.* **2014**, *70*, 229–236. [CrossRef]

76. Hamer, U.; Marschner, B.; Brodowski, S.; Amelung, W. Interactive priming of black carbon and glucose mineralisation. *Org. Geochem.* **2004**, *35*, 823–830. [CrossRef]

77. Hilscher, A.; Heister, K.; Siewert, C.; Knicker, H. Mineralisation and structural changes during the initial phase of microbial degradation of pyrogenic plant residues in soil. *Org. Geochem.* **2009**, *40*, 332–342. [CrossRef]

78. Knicker, H. How does fire affect the nature and stability of soil organic nitrogen and carbon? A review. *Biogeochemistry* **2007**, *85*, 91–118. [CrossRef]

79. Kuzyakov, Y.; Subbotina, I.; Chen, H.; Bogomolova, I.; Xu, X. Black carbon decomposition and incorporation into soil microbial biomass estimated by 14C labeling. *Soil Biol. Biochem.* **2009**, *41*, 210–219. [CrossRef]

80. Ascough, P.L.; Sturrock, C.J.; Bird, M.I. Investigation of growth responses in saprophytic fungi to charred biomass. *Isotopes Environ. Health Stud.* **2010**, *46*, 64–77. [CrossRef] [PubMed]

81. Hammer, E.C.; Balogh-Brunstad, Z.; Jakobsen, I.; Olsson, P.A.; Stipp, S.L.S.; Rillig, M.C. A mycorrhizal fungus grows on biochar and captures phosphorus from its surfaces. *Soil Biol. Biochem.* **2014**, *77*, 252–260. [CrossRef]

82. Lehmann, J.; Rillig, M.C.; Thies, J.; Masiello, C.A.; Hockaday, W.C.; Crowley, D. Biochar effects on soil biota–a review. *Soil Biol. Biochem.* **2011**, *43*, 1812–1836. [CrossRef]

83. Steinbeiss, S.; Gleixner, G.; Antonietti, M. Effect of biochar amendment on soil carbon balance and soil microbial activity. *Soil Biol. Biochem.* **2009**, *41*, 1301–1310. [CrossRef]

84. Gil-Sotres, F.; Trasar-Cepeda, C.; Leirós, M.C.; Seoane, S. Different approaches to evaluating soil quality using biochemical properties. *Soil Biol. Biochem.* **2005**, *37*, 877–887. [CrossRef]

85. Harris, J.A. Measurements of the soil microbial community for estimating the success of restoration. *Eur. J. Soil Sci.* **2003**, *54*, 801–808. [CrossRef]

86. Fontaine, S.; Barot, S.; Barré, P.; Bdioui, N.; Mary, B.; Rumpel, C. Stability of organic carbon in deep soil layers controlled by fresh carbon supply. *Nature* **2007**, *450*, 277–280. [CrossRef] [PubMed]

87. Dempster, D.; Gleeson, D.; Solaiman, Z.M.; Jones, D.; Murphy, D. Decreased soil microbial biomass and nitrogen mineralisation with Eucalyptus biochar addition to a coarse textured soil. *Plant. Soil* **2012**, *354*, 311–324. [CrossRef]

88. Li, Q.; Lei, Z.; Song, X.; Zhang, Z.; Ying, Y.; Peng, C. Biochar amendment decreases soil microbial biomass and increases bacterial diversity in Moso bamboo (Phyllostachys edulis) plantations under simulated nitrogen deposition. *Environ. Res. Lett.* **2018**, *13*, 044029. [CrossRef]

89. Zhang, H.; Voroney, R.P.; Price, G.W. Effects of Biochar Amendments on Soil Microbial Biomass and Activity. *J. Environ. Qual.* **2014**, *43*, 2104. [CrossRef] [PubMed]

90. Luo, L.; Gu, J.-D. Alteration of extracellular enzyme activity and microbial abundance by biochar addition: Implication for carbon sequestration in subtropical mangrove sediment. *J. Environ. Manag.* **2016**, *182*, 29–36. [CrossRef] [PubMed]

91. Chen, J.; Li, S.; Liang, C.; Xu, Q.; Li, Y.; Qin, H.; Fuhrmann, J.J. Response of microbial community structure and function to short-term biochar amendment in an intensively managed bamboo (*Phyllostachys praecox*) plantation soil: Effect of particle size and addition rate. *Sci. Total Environ.* **2017**, *574*, 24–33. [CrossRef] [PubMed]

92. De Tender, C.A.; Debode, J.; Vandecasteele, B.; D'Hose, T.; Cremelie, P.; Haegeman, A.; Ruttink, T.; Dawyndt, P.; Maes, M. Biological, physicochemical and plant health responses in lettuce and strawberry in soil or peat amended with biochar. *Appl. Soil Ecol.* **2016**, *107*, 1–12. [CrossRef]

93. Zhai, L.; CaiJi, Z.; Liu, J.; Wang, H.; Ren, T.; Gai, X.; Xi, B.; Liu, H. Short-term effects of maize residue biochar on phosphorus availability in two soils with different phosphorus sorption capacities. *Biol. Fertil. Soils* **2014**, *51*, 113–122. [CrossRef]

94. Ogawa, M.; Yambe, Y. Effects of charcoal on VA mycorrhiza and root nodule formations of soybean: Studies on nodule formation and nitrogen fixation in legume crops. *Bull. Green Energy Program. Group II* **1986**, *8*, 108–134.

95. Saito, M. Charcoal as a micro-habitat for VA mycorrhizal fungi, and its practical implication. *Agric. Ecosyst. Environ.* **1990**, *29*, 341–344. [CrossRef]

96. Gaur, A.; Adholeya, A. Effects of the particle size of soil-less substrates upon AM fungus inoculum production. *Mycorrhiza* **2000**, *10*, 43–48. [CrossRef]

97. Ezawa, T.; Yamamoto, K.; Yoshida, S. Enhancement of the effectiveness of indigenous arbuscular mycorrhizal fungi by inorganic soil amendments. *Soil Sci. Plant. Nutr.* **2002**, *48*, 897–900. [CrossRef]

98. Jaafar, N.M.; Clode, P.L.; Abbott, L.K. Microscopy Observations of Habitable Space in Biochar for Colonization by Fungal Hyphae From Soil. *J. Integr. Agric.* **2014**, *13*, 483–490. [CrossRef]

99. Pietikäinen, J.; Kiikkilä, O.; Fritze, H. Charcoal as a habitat for microbes and its effect on the microbial community of the underlying humus. *Oikos* **2000**, *89*, 231–242. [CrossRef]

100. Saito, M.; Marumoto, T. Inoculation with arbuscular mycorrhizal fungi: The status quo in Japan and the future prospects. In *Diversity and Integration in Mycorrhizas*; Springer: Berlin/Heidelberg, Germany, 2002; pp. 273–279.

101. Warnock, D.D.; Lehmann, J.; Kuyper, T.W.; Rillig, M.C. Mycorrhizal responses to biochar in soil–concepts and mechanisms. *Plant. Soil* **2007**, *300*, 9–20. [CrossRef]

102. Kammann, C.I.; Schmidt, H.-P.; Messerschmidt, N.; Linsel, S.; Steffens, D.; Müller, C.; Koyro, H.-W.; Conte, P.; Joseph, S. Plant growth improvement mediated by nitrate capture in co-composted biochar. *Sci. Rep.* **2015**, *5*, 11080. [CrossRef] [PubMed]

103. Lu, H.; Lashari, M.S.; Liu, X.; Ji, H.; Li, L.; Zheng, J.; Kibue, G.W.; Joseph, S.; Pan, G. Changes in soil microbial community structure and enzyme activity with amendment of biochar-manure compost and pyroligneous solution in a saline soil from Central China. *Eur. J. Soil Biol.* **2015**, *70*, 67–76. [CrossRef]

104. Elbl, J.; Záhora, J. The comparison of microbial activity in rhizosphere and non-rhizosphere soil stressed by drought. *Proc. Mendel Net* **2014**, 234–240.

105. Elbl, J.; Vaverková, M.; Adamcová, D.; Plošek, L.; Kintl, A.; LošÁk, T.; Hynšt, J.; Kotovicová, J. Influence of fertilization on microbial activities, soil hydrophobicity and mineral nitrogen leaching. *Ecol. Chem. Eng. S* **2014**, *21*, 661–675. [CrossRef]

106. Mia, S.; Van Groenigen, J.W.; Van de Voorde, T.F.J.; Oram, N.J.; Bezemer, T.M.; Mommer, L.; Jeffery, S. Biochar application rate affects biological nitrogen fixation in red clover conditional on potassium availability. *Agric. Ecosyst. Environ.* **2014**, *191*, 83–91. [CrossRef]

107. Wang, Z.; Zheng, H.; Luo, Y.; Deng, X.; Herbert, S.; Xing, B. Characterization and influence of biochars on nitrous oxide emission from agricultural soil. *Environ. Pollut.* **2013**, *174*, 289–296. [CrossRef] [PubMed]

108. Xu, H.-J.; Wang, X.-H.; Li, H.; Yao, H.-Y.; Su, J.-Q.; Zhu, Y.-G. Biochar impacts soil microbial community composition and nitrogen cycling in an acidic soil planted with rape. *Environ. Sci. Technol.* **2014**, *48*, 9391–9399. [CrossRef] [PubMed]

109. Clough, T.J.; Bertram, J.E.; Ray, J.L.; Condron, L.M.; O'Callaghan, M.; Sherlock, R.R.; Wells, N.S. Unweathered wood biochar impact on nitrous oxide emissions from a bovine-urine-amended pasture soil. *Soil Sci. Soc. Am. J.* **2010**, *74*, 852–860. [CrossRef]

110. Wang, Z.; Zong, H.; Zheng, H.; Liu, G.; Chen, L.; Xing, B. Reduced nitrification and abundance of ammonia-oxidizing bacteria in acidic soil amended with biochar. *Chemosphere* **2015**, *138*, 576–583. [CrossRef] [PubMed]

Article

Sustainable Agronomic Strategies for Enhancing the Yield and Nutritional Quality of Wild Tomato, *Solanum Lycopersicum* (l) Var *Cerasiforme* Mill

Kanagaraj Muthu-Pandian Chanthini, Vethamonickam Stanley-Raja, Annamalai Thanigaivel, Sengodan Karthi, Radhakrishnan Palanikani, Narayanan Shyam Sundar, Haridoss Sivanesh, Ramaiah Soranam and Sengottayan Senthil-Nathan *

Sri Paramakalyani Centre for Excellence in Environmental Sciences, Manonmaniam Sundaranar University, Alwarkurichi, Tirunelveli 627412, Tamil-Nadu, India; clairdelune127@gmail.com (K.M.-P.C.); stanleyraja@gmail.com (V.S.-R.); www.thani@gmail.com (A.T.); karthientomology@gmail.com (S.K.); kani.palani5@gmail.com (R.P.); krn.shyamsundar@gmail.com (N.S.S.); Sivanesh2020@gmail.com (H.S.); soranamr@gmail.com (R.S.)
* Correspondence: senthil@msuniv.ac.in; Tel.: +91-463-428-3066

Received: 30 April 2019; Accepted: 28 May 2019; Published: 13 June 2019

Abstract: Urbanization and global climate change have constrained plant development and yield. Utilization of wild gene pool, together with the application of sustainable and eco-friendly agronomic crop improvement strategies, is being focused on to tackle mounting food insecurity issues. In this aspect, the green seaweed, *Ulva flexuosa*, was assessed for plant biostimulant potential on cherry tomato, in terms of seed priming effects, nutrition and yield. SEM-EDX analysis of *U. flexuosa* presented the occurrence of cell wall elements (O, Na, Mg, S, Cl, K and Ca). The phytochemical analyses of liquid seaweed extract (EF-LSE) revealed the presence of carbohydrates, protein, phenols, flavonoids, saponins, tannins and coumarins. The EF-LSEs were found to stimulate seed germination in a dose-dependent manner, recording higher seed germination, and biomass and growth parameters. The seedlings of treated seeds altered the biochemical profile of the fruit, in terms of TSS (93%), phenol (92%), lycopene (12%) and ascorbic acid (86.8%). The EF-LSEs positively influenced fruit yield (97%). Henceforth, this investigation brings to light the plant biostimulant potential of the under-utilized seaweed source, *U. flexuosa*, to be useful as a bio fertilizer in agronomic fields for a cumulative enhancement of crop vigour as well as yields to meet the growing food demands.

Keywords: seed priming; seaweed extract; biostimulant; germination energy; seedling vigour

1. Introduction

Agriculture is facing various crises that are worsening with time. Increasing food production to meet or feed the mounting population is a foremost challenge. This can be accomplished by further use of farm lands for an overall hike in food production or technically enhancing the yields from pre-existing lands by application of fertilizers or implementation of novel approaches; for instance, precision farming systems viz., cutting-edge irrigation arrangements, and ecologically accomplishable crop revolutions [1]. Crop diseases decrease yield, resulting in a prominent crisis to food security, creating a global malnutrition spree affecting nearly 815 million people [2]. Henceforward, natural fertilizers are well thought out as probable as well as safe alternatives to chemical fertilizers [3]. Additionally, the presence of several horticultural important traits in the wild gene pool makes them suitable as potential breeding candidates for crop improvement [4]. In this aspect, plant secondary metabolites are being emphasized for their disease regulator competences and combined in more than a few defense control programs [5].

The marine ecosystem serves as a rich source of bioactive compounds, such as sulfated polysaccharides, terpenoids, phenolics, lactones, sterol and fatty acids, possessing pharmacological and plant growth-stimulating properties [6]. Seaweeds form a key portion of these bioactive natural composites, with over 9000 species, known for their biostimulator potentials. Additionally, seaweed products as biostimulants that can enhance crop production are also being focused on. Biostimulants are materials supplementary to fertilizers, which endorse plant growth at lower concentrations [7]. In addition, seaweeds are extensively applied in the fields of agriculture and horticulture to improve quality and quantity, and the results are promising [8]. Innumerable seaweeds are being applied as liquid fertilizers to upsurge crop yields, as they are rich in macro-nutrients, besides trace elements essential for the development and enrichment of plants. Commercial seaweed products are also being marketed successfully [9]. Besides being inexpensive, the seaweed extracts have surplus allelopathic chemicals that promote seed germination as well as emergence rates. Seaweed extracts are known to have a positive impact on prime stages of plant ontogenesis—starting from seed germination to seedling growth [10]. Furthermore, seaweeds are reported with higher amounts of growth hormones, attributed to their plant biostimulant activities [11].

Seed germination is a decisive procedure in plant growth, and the enrichment of germination potentialities of a seed can eventually enable a surge in crop yields, and is dependent on numerous chemical factors (soil moisture salinity, metal, mineral composition) [12]. The emergence of seed is promoted by various methods for enhanced agricultural yields, like exposing them to biostimulants or growth promoting hormones by the process of seed priming. As the very first stage of plant growth, germination is defined as an outcrop of the radicle from the tissues enfolding the seed [13]. As germination rates may vary among species, the analyses of germination rates might be directly proportional to the growth rates and consequently, their yields [14,15].

Cherry tomato, *Solanum lycopersicum* (L.) var. *cerasiforme* Mill. is a widespread, table purpose tomato variety, bearing bright red color, and small fruits. It is a probable ancestor of a cultivated tomato variety with small fruits bright red in color, resembling a cherry and tasting excellent [16]. They are also favorable candidates in breeding programs for their genetic diversity, offering the selection of parental traits along with extensive geographic ranges. [17]. With the debarring effects of crop growth promoting chemicals that alter soil ecology and have hazardous environmental and health impacts, researchers are concentrating on the allegation of naturally benign substitutes to increase yields, while offering effective crop protection. The favorable agronomic traits of cherry tomato (an intermediary genetic admixture flanked by wild currant-type tomatoes and domesticated garden tomatoes), such as higher nutrient composition, offer plans for balanced utilization to unravel indigenous complications encompassing crop adaptation to climatic variations or, to endorse functional food consumption [18].

Seed priming of native seed species can evoke ecological restoration, stimulating the expression of dormant genes responsible for the expression of favorable agronomic traits. Since the sources of natural varieties are collected from the wild and their sources are limited, there is a persistent requirement for novel methods for seed-based restoration technologies. In this aspect, seed priming could be a strategy for sustainable seedling establishment, plant growth, and restoration of native seed. Hence, this research intended to discover the bio stimulator potentials of liquid seaweed extract of green alga, *U. flexuosa*, along with screening of their phytochemical and elemental composition on cherry tomato.

2. Materials and Methods

2.1. Seed Collection and Preparation

Cherry tomato seeds, variety ATL-01-19, were purchased from Tamil Nadu Agricultural University, Coimbatore, Tamil Nadu, India. Seeds of undeviating dimensions and hue were carefully chosen for the study, surface sterilized with 0.1% mercuric chloride, washed thrice in sterile distilled water. The experiment was carried out at Sri Paramakalyani Centre for Excellence in Environmental Science, Manonmaniam Sundaranar University, Alwarkurichi, from July to August, 2018.

2.2. Seaweed Collection

The seaweed, *Ulva flexuosa* (Ulvaceae), was collected from the rocks in coastal areas at Colachel beach (Figure 1), Kanyakumari (8°14′5168″ N and 77°14′35.209″ E) during a low tide period (August 2018). It was washed in seawater several times to remove impurities, sand particles, and epiphytes, and brought to the research laboratory in taped up polythene bags. The seaweed was washed thoroughly in tap water several times, wearied and spread on blotting paper to remove excess water, and shade dried for 2 to 3 hours.

2.3. Preparation of Starter Fertilizer Solution (SFS)

Ammonium dihydrogen phosphate (Merck) was used as a starter fertilizer solution by mixing 1 mg of $NH_4H_2PO_4$ in sterile distilled water (10 mL), designated as positive control.

2.4. Preparations of Liquid Seaweed Extract (EF-LSE)

The washed seaweed was then cut into minor fragments, boiled in distilled water (100 gms/1 L) for one hour in an autoclave (121 °C, 15 psi of pressure), and filtered through a cheese cloth (double layered), yielding 890 ml of LSE. The LSE was stored at 4 °C in a refrigerator until further use. The test concentrations of EF-LSE were prepared by diluting the extract with distilled water (20%, 40%, 60%, 80% and 100%).

2.5. EF-LSE Analysis

2.5.1. Physicochemical Analysis

Physicochemical structures of the EF-LSEs—pH, electrical conductivity, and color appearance—was determined. pH was determined using a pH meter (ELICO LI 120, Hyderabad, India), whereas conductivity was done with the help of a conductivity meter (Microprocessor EC Meter 1615, Parwanoo, Himachal Pradesh, India) and expressed in ds/m. The colour of the EF-LSE was visually observed and noted.

2.5.2. Elemental Composition of EF-LSE Using X-ray–Energy Dispersive Spectroscopic (ED) Analysis

The elemental composition of *U. flexuosa* was performed using EDAX (BRUKER) to elucidate the components present in the seaweed cell wall.

2.5.3. Phytochemical and Biochemical Screening

Phytochemical [19] and biochemical screening of *U. flexuosa* was performed to analyse the quantitative amounts of phenol [20], chlorophyll [21] and protein [22] by standard procedures.

2.6. Preparation of Seeds

Tomato seeds, tomato wild relative, *Solanum lycopersicum* (L.) var. *cerasiforme* Mill, were used for all the tests. Tomato seeds, without any visible signs of infection, of uniform size, shape and colour, were carefully chosen, surface sterilized using 0.1% mercuric chloride, before and after rinsing in sterile distilled water. The seeds were used for further analysis.

To investigate the possible effects of UF-LSEs on tomato plant's vegetative growth and yield, small pot field experiments containing sterilized soil were conducted by sowing the primed seeds in a tray. The seedlings (2-3 true leaf stage) were transplanted into autoclaved pot mixture (red soil: cow dung: vermiculate at 2:1:1, w/w/w) in the surface sterilized (1% mercuric chloride) pots (15 cm diameter, 750 mL volume) at 1 seedling/pot. The pots were labeled based on the treatments. Vegetative growth parameters were analyzed by sampling from seedlings selected from randomized block designs.

2.7. Biostimulant Assays

2.7.1. Seed Bioassay

Seed Germination Test

Seed germination assay was conducted using five surface sterilised seeds per assay, replicated five times. The seeds were exposed to EF-LSE extracts (10 mL), SFS in sealed and labelled conical flasks and kept in a shaker for 12 hours. Seeds in 10 mL sterile distilled water were used as control. The seeds were then removed and spread on a filter paper to blot out the solutions at room temperature for 24 hours. The treated seeds were placed in pre-labelled sterile petri dishes (9 cm) over filter paper (Whatman No. 5) that was moistened (sterile distilled water), instantaneously taped up with parafilm (Merck) to prevent moisture loss, and incubated (25 ± 2 °C/alternative 16 h light-8 h dark). The plates were checked for radicle protrusion (>2 mm) on a daily basis (hint of germination). Ten seeds were tested for each concentration of EF-LSE.

Germination was recorded every day by counting the emerging hypocotyls. The mean germination time (MGT) was premeditated [23] by counts made on the time taken for 1%, 10%, 25%, 50%, 75% and 100% of the seeds to germinate and expressed as days.

$$MGT = \frac{\Sigma\,(n\,T)}{\Sigma\,n} \tag{1}$$

where,

n = number of newly germinated seeds at time T (25 °C)
T = hours from the beginning of the germination test
$\Sigma\,n$ = final germination
*100% will refer to the total number of seeds germinated after exposure to the highest EF-LSE concentration.

The germination percentage (GP) was calculated using the following formula:

$$GP = \frac{\text{Number of seeds germinated}}{\text{Total number of seeds}} \times 100 \tag{2}$$

Seed Germination Energy (GE) was calculated according to the following formula:

$$GE = \frac{\text{Number of germinating seeds}}{\text{No. of total seeds per test post germination for 3 days}} \times 100 \tag{3}$$

Seedling vigour index (SVI) SVI was calculated [24] by the following formula:

$$SVI = \text{Seedling length (cm)} \times \text{germination \%} \tag{4}$$

Seed Imbibition

The biomass (wet and dry weight, mg) of the seeds primed in EF-LSEs and SFS solutions for 24 hours were determined with the help of an electronic balance after oven-drying at 40 °C for two days. Seed imbibition was determined by measuring the weight of seeds (100 seeds/treatment) before and during priming with SFS and EF-LSEs at 6, 12, 24, 36 and 48 hours and plotting the water imbibition curve by determining the seed moisture content (MC) and through means of which, the seed imbibition time was calculated [25]. Seeds primed in distilled water served as control. The MC was calculated by the formula:

$$\text{Moisture content} = \frac{\text{Fresh weight} - \text{Dry weight}}{\text{Dry weight}} \times 100 \tag{5}$$

2.7.2. Growth Parameter Assay

Growth parameters such as lengths of total plant, plumule and radicle length (cm) and root-shoot lengths and ratio were measured with a Vernier calliper post 5 days and 20 days for petri plate and glasshouse tests, respectively.

2.7.3. Fruit Yield and Quality Parameter assay

The seedlings derived from primed seeds were cultured as per recommended standard procedures [26] were selected on the basis of randomised block design, performing yield and quality parameter assays, repeating each test five times. The effect of EF-LSE on the yield of cherry tomato plants was estimated by fruit weight. The quality parameters of treated cherry tomato plants were calculated in terms of total soluble solids (TSS) [27], ascorbic acid [28], lycopene [29] and phenol [20] contents.

2.8. Statistical Analysis

All the tests were repeated five times. The effect of EF-LSE on seeds was determined by analysis of variance, one-way (ANOVA), and the treatment means were compared by Tukey-family error test ($p < 0.05$) by using Minitab®17 software package (LEAD Technologies Inc., Charlotte, NC, USA).

3. Results

3.1. Seaweed Collection and Identification

The seaweeds collected from Colachel beach (Figure 1) were subjected to microscopical (Nikon Phase Contrast, Japan) and macroscopical analyses and confirmed to be Ulva flexuosa Wulfen (Ulvaceae), based on morphological characteristics and organoleptic features (Table 1). The seaweed belonging to the green alga, phylum Chlorophyta, is tubular and branched (Figure 2). The cells were observed to be arranged in transverse rows.

Table 1. Organoleptic features of *Ulva flexuosa*.

Feature	*U. flexuosa* Appearance
Habitat	Highly cosmopolitan in shallow marine or brackish habitats
Shape	Long, filamentous hollow tube thallus
Size	20 cm
Colour	Light green
Odour	Foul smelling
Taste	Salty
Base	Divided
Blade	Expanding above the stalk, which ends in a rounded tip

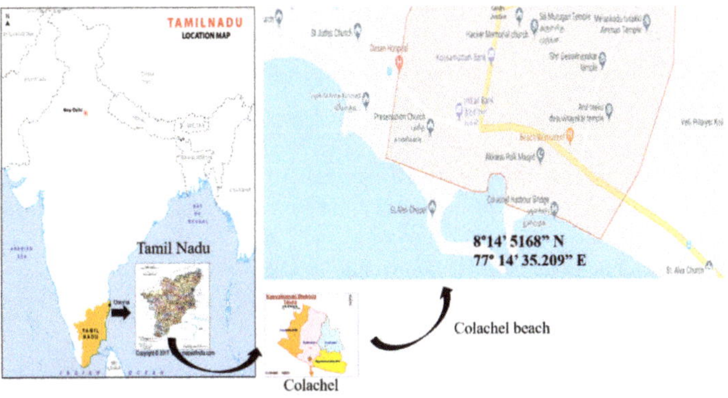

Figure 1. Sample collection site of seaweed from Colachel beach, Kanyakumari, Tamil Nadu, India.

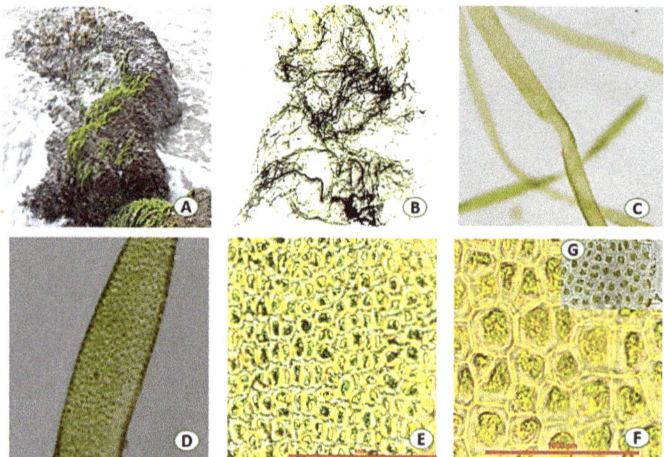

Figure 2. Macroscopic and microscopic images of *U. flexuosa*. ((**A**): Sample collection site, Colachel beach; (**B**): *U. flexuosa*; (**C,D**): Macroscopic image; Microscopic image of *U. flexuosa*-(**E**): 10×; (**F**): 40× and (**G**): 100×, showing blue coloured pigments on cell wall).

3.2. EF-LSE Analyses

3.2.1. Physicochemical Screening

The EF-LSEs appeared as pale greenish yellow in colour. The pH of the EF-LSEs of doses 20%, 40%, 60%, 80% and 100% was measured and found to be 7.160, 7.240, 7.34, 7.4 and 7.58, respectively. The pH of control was 7 and that of SFS was 4.2. The electrical conductivity (EC) of the respective EF-LSE doses was recorded as 0.96 ds m^{-1}, 1.01 ds m^{-1}, 1.08 ds m^{-1}, 1.54 ds m^{-1} and 2.8 ds m^{-1}, respectively (Table 2).

Table 2. Physico-chemical properties of LSEs, pH and EC (dS m^{-1}).

Treatments	pH	EC (dS m−1)
C	7.02 ± 0.004 [a]	-
SFS	4.2 ± 0.89 [b]	0.756 ± 0.03 [a]
EF 20%	7.160 ± 0.30 [c]	1.0180± 0.10 [b]
EF 40%	7.240± 0.26 [c]	1.0800 ± 0.15 [c]
EF 60%	7.340 ± 0.21 [cd]	1.540 ± 0.36 [c]
EF 80%	7.400 ± 0.35 [cd]	2.880 ± 0.54 [c]
EF 100%	7.580 ± 0.37 [d]	3.700 ± 0.57 [d]

Columns denoted by a different letter are significantly different at $p \leq 0.05$.

3.2.2. Elemental Composition of EF-LSE Using X-ray–EDS Analysis

The elemental composition of the seaweed elucidated via EDX analysis (Figure 3) revealed the presence of seven compounds on seaweed cell surface—oxygen, Na, Mg, S, Cl, K and Ca. Oxygen was present in higher quantities (56.25%), followed by chlorine (13.4%), sulphur (7.79), and potassium (7.52%). Magnesium (5.3%), calcium (4.89%) and sodium (4.84%) were also recorded.

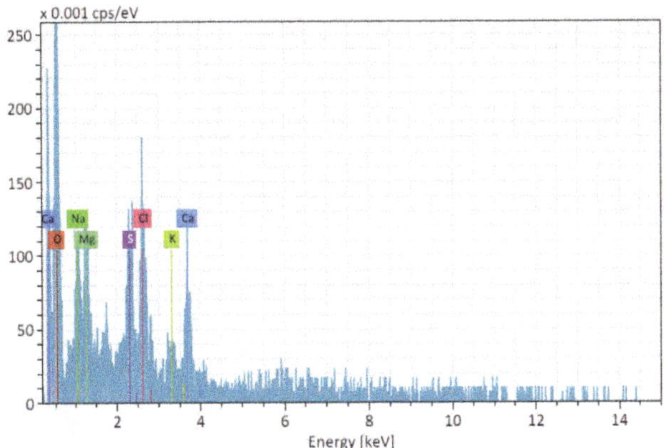

Figure 3. SEM-EDX—Energy Dispersive Spectrum of *U. flexuosa* showing the presence of various elements in their cell wall.

3.2.3. Phytochemical Screening

The phytochemical analysis of EF-LSE revealed the presence of carbohydrates, protein, phenols, flavonoids, saponins, tannins and coumarins (Table 3).

Table 3. Phytochemical composition of EF-LSE.

Phytochemicals	Inference
Alkaloids	-
Carbohydrates	+
Carboxylic acid	-
Coumarin	+
Flavonoids	+
Phenolics	+
Protein	+
Quinone	-
Saponin	+
Steroid	-
Tannin	+
Xanthoprotein	-

3.2.4. Biochemical Screening

The seaweed examined qualitatively revealed the presence of 1 mg/g of phenol, 6.1% protein, and 0.9 mg/g of total chlorophyll contents.

3.3. Biostimulant Assays

The cherry tomato seeds purchased were sown in a greenhouse, according to standard horticultural methods (Figure 4).

Figure 4. Cherry tomato plants.

3.3.1. Effect of EF-LSEs on Cherry Tomato Seeds

Germination of cherry tomato seeds was initiated on day 2 in seeds treated with 80% and 100% of EF-LSEs. The EF-LSEs were able to initiate germination of the seedlings in a dose-dependent manner (Figure 5). EF-LSE treated seeds emerged early when compared to the control (Figure 6). In addition, the EF-LSE treated seeds in doses 20%, 40%, 60%, 80% and 100% exhibited lower MGT (Figure 7) of 4, 3.4, 3, 2.6 and 2.2 days ($F_{6,28} = 3.23$; $p < 0.0001$), respectively, compared with seeds treated with SFS, 4 days ($F_{6,28} = 3.23$; $p < 0.0001$) as well as that of the control, 4.9 days ($F_{6,28} = 3.23$; $p < 0.0001$).

Figure 5. Effect of LSEs on growth of tomato seedlings.

The germination time course for control seeds was longer (Figure 6), taking 86.54, 93.2, 110.64, 116.4 and 125.80 hours ($F_{4,20} = 30.33$; $p < 0.0001$) for 1%, 10%, 25%, 50%, 75% and 100% of seeds to germinate. Seeds treated with SFS exhibited a time course almost similar to that of control, 85.54, 91.8, 109.34, 114.51 and 125.37 hours ($F_{4,20} = 30.33$; $p < 0.0001$). The germination time course decreased with increase in EF-LSE concentrations. The time taken for emergence of 1% of seeds decreased from 76.4

($F_{4,20}$ = 44.77; p < 0.0001) to 56.8 ($F_{4,20}$ = 36.81; p < 0.0001) and 50.8 ($F_{4,20}$ = 31.39; p < 0.0001) hours in seeds treated with 20%, 40% and 60% EF – LSEs, respectively.

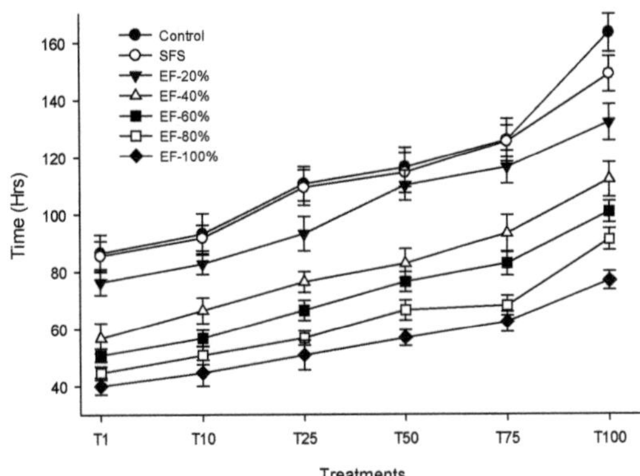

Figure 6. Effect of EF-LSEs on germination of tomato seeds—germination time course.

Similarly, the rate of emergence of seeds increased, resulting in a decreased germination time course, further to 44.6 ($F_{4,20}$ = 15.97; p < 0.0001) and 40 ($F_{4,20}$ = 13.66; p < 0.0001) hours in 80% and 100% EF-LSE treatments, respectively. As a series, the time taken for the emergence of 100% of seeds in treatment decreased from 163.2 ($F_{4,20}$ = 30.33; p < 0.0001) to 148.8 ($F_{4,20}$ = 38.16; p < 0.0001) hours in the respective control and SFS treated seeds. The EF-LSE treated seeds took 132 (($F_{4,20}$ = 44.77; p < 0.0001), 112.8 (($F_{4,20}$ = 36.81; p < 0.0001), 100.8 (($F_{4,20}$ = 31.39; p < 0.0001), 91.1 ($F_{4,20}$ = 15.97; p < 0.0001) and 76.8 (($F_{4,20}$ = 13.66; p < 0.0001) hours for 100% emergence at treatment concentrations of 20%, 40%, 60%, 80% and 100%, respectively.

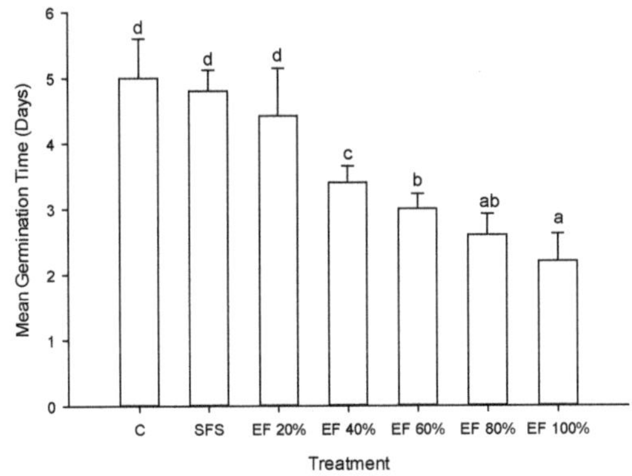

Figure 7. Effect of LSEs on germination of tomato seeds—mean germination time.

The seeds in the petri dishes treated with distilled water exhibited 84.8% GP ($F_{6,28}$ = 6.4; p < 0.0001) at the end of seven days. SFS treated seeds showed a GP of 86.6% GP ($F_{6,28}$ = 6.4; p < 0.0001). Seeds treated with 20%, 40%, 60%, 80% and 100% EF-LSE, displayed 93%, 94%, 95%, 96% and 97% ($F_{6,28}$ = 6.4; p < 0.0001) GP, respectively (Figure 8). The number of days taken for all the EF-LSE treated seeds to germinate also decreased to 5.5, 4.7, 4.2, 3.6 and 3.2 days ($F_{4,20}$ = 13.66; p < 0.0001).

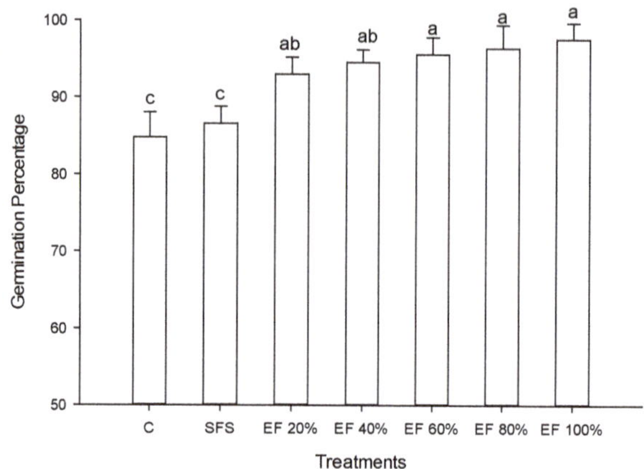

Figure 8. Effect of LSEs on germination percentage of cherry tomato seeds.

The germination energy (Figure 9) of the seeds treated with EF-LSEs were higher, in the range of 55.6, 85.4, 91.6, 94.6 and 97.8 ($F_{6,28}$ = 31.34; p < 0.0001) exposed to 20%, 40%, 60%, 80% and 100% EF–LSEs, respectively, which was very high in comparison with the control (10%) and that of SFS-treated (20%) ($F_{6,28}$ = 31.34; p < 0.0001) seeds.

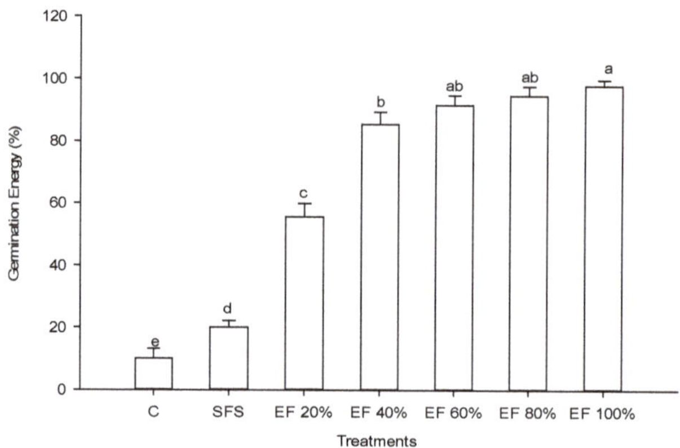

Figure 9. Effect of LSEs on germination energy of cherry tomato seeds.

The seedling vigour index of the control seeds and SFS treated seeds were 457.18 ($F_{6,28}$ = 32.86; p < 0.0001) and 743.75 ($F_{6,28}$ = 32.86; p < 0.0001), respectively. However, the SVI of seeds treated with EF-LSEs (Table 4) of dosages 20%, 40%, 60%, 80% and 100% were enhanced to 1003.59, 1047.04, 1060.24, 1236.96 and 1281.31 ($F_{6,28}$ = 32.86; p < 0.0001), respectively (Table 4).

The biomass of the cherry tomato seeds was determined after 48 hours of priming. Seeds that were not treated with any extracts unveiled respective dry and wet weights of 0.013 mg ($F_{6,28}$ = 32.08; p < 0.0001) and 0.113 mg ($F_{6,28}$ = 19.23; p < 0.0001). The dry weights of the tomato plants treated with SFS, EF-LSE extracts – 20%, 40%, 60%, 80% and 100% were found to be 0.0218 mg, 0.0262 mg, 0.0296 mg, 0.0316 mg, 0.042 mg and 0.0528 mg ($F_{6,28}$ = 32.08; p < 0.0001) respectively. The wet weights of the tomato plants treated with SFS, EF-LSE extracts—20%, 40%, 60%, 80% and 100% were found to be 0.1396 mg, 0.1434 mg, 0.144 mg, 0.152 mg, 0.2 mg and 0.308 mg ($F_{6,28}$ = 19.23; p < 0.0001), respectively (Table 4).

Table 4. Seedling vigour index and biomass (wet and dry weight) of tomato seeds treated with LSEs.

Treatments	Seedling Vigour Index (SVI)	Seed Weight (mg)	
		Wet Weight	Dry Weight
C	457.18 ± 2.76 [a]	0.113 ± 0.007 [a]	0.013 ± 0.007 [a]
SFS	743.75 ± 3.02 [b]	0.1396 ± 0.003 [b]	0.0218 ± 0.003 [b]
Ef 20%	1003.59 ± 1.98 [c]	0.1434 ± 0.009 [c]	0.0262 ± 0.009 [c]
Ef 40%	1047.04 ± 3.01 [d]	0.144 ± 0.005 [c]	0.0296 ± 0.001 [d]
Ef 60%	1060.24 ± 2.45 [e]	0.152 ± 0.003 [d]	0.0316 ± 0.003 [e]
Ef 80%	1236.96 ± 1.43 [f]	0.2 ± 0.003 [e]	0.042 ± 0.003 [f]
Ef 100%	1281.31 ± 2.65 [g]	0.308 ± 0.002 [f]	0.0528 ± 0.002 [g]

Columns denoted by a different letter are significantly different at $p \leq 0.05$.

The moisture content of the dry seeds was 7.8%, which increased after six hours in all SFS and EF-LSE treatments. The moisture content of control seeds increased to 8%, 9%, 9.4%, 9.7% and 9.9% at 6, 12, 24, 36 and 48 hours. A comparatively higher imbibition occurred in EF-LSE treated seeds, with a minimum imbibition of 8.5% and a maximum of 9.5 after six hours, in 20% and 100% primed seeds (Figure 10).

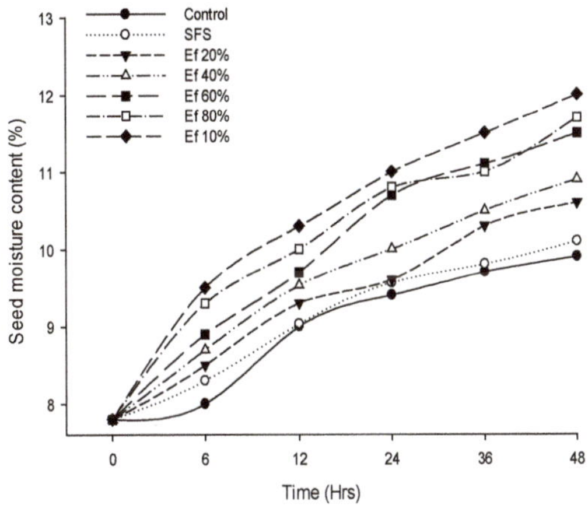

Figure 10. Effect of EF-LSEs on seed imbibition capabilities of tomato seeds.

3.3.2. Effect of EF-LSEs on Growth Parameters of Cherry Tomato

The effect of EF-LSEs on the growth of tomato seedlings tested exhibited significant differences in total seedling length (Figure 11), radicle, and plumule length (Figure 12).

The height of the tomato seedlings at the end of five days was 5.4 cm ($F_{6,28} = 30.98$; $p < 0.005$), with radicle and plumule lengths of 4 cm ($F_{6,28} = 42.16$; $p < 0.005$) and 3 cm ($F_{6,28} = 42.94$; $p < 0.005$), respectively (Figure 12), with 1.26 radicle: plumule ratio ($F_{5,24} = 32.20$, $p < 0.005$) (Figure 13).

The SFS was able to induce the seedling length to 8.6 cm ($F_{6,28} = 30.98$; $p < 0.005$) with a corresponding radicle and plumule lengths, ratio of 4.2 ($F_{6,28} = 42.16$; $p < 0.005$) and 3.72 cm ($F_{6,28} = 42.94$; $p < 0.005$), 1.36 ($F_{5,24} = 32.20$, $p < 0.005$), respectively. The EF-LSEs had a positive effect in stimulating the seedling height and their respective radicle and plumule lengths to 10.8 cm ($F_{6,28} = 30.98$; $p < 0.005$), 5.12 cm ($F_{6,28} = 42.16$; $p < 0.005$) and 4.12 cm ($F_{6,28} = 42.94$; $p < 0.005$), respectively, at 20% concentration, exhibiting radicle: plumule ratio of 1.36 ($F_{5,24} = 32.20$, $p < 0.005$).

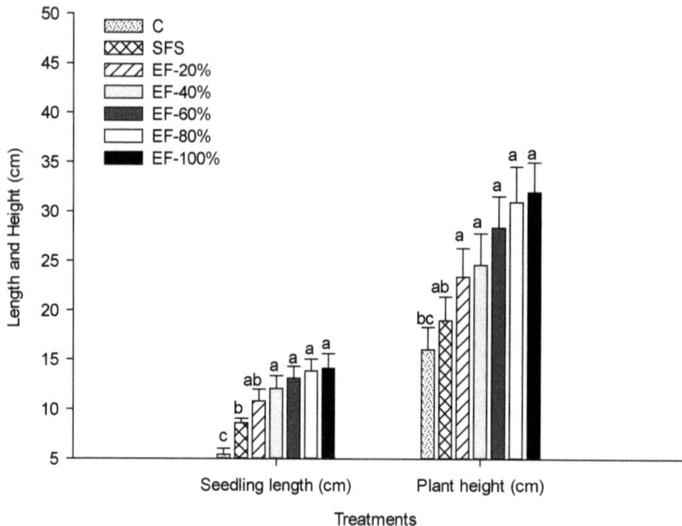

Figure 11. Effect of EF-LSEs on growth parameters of tomato seedling and plant height.

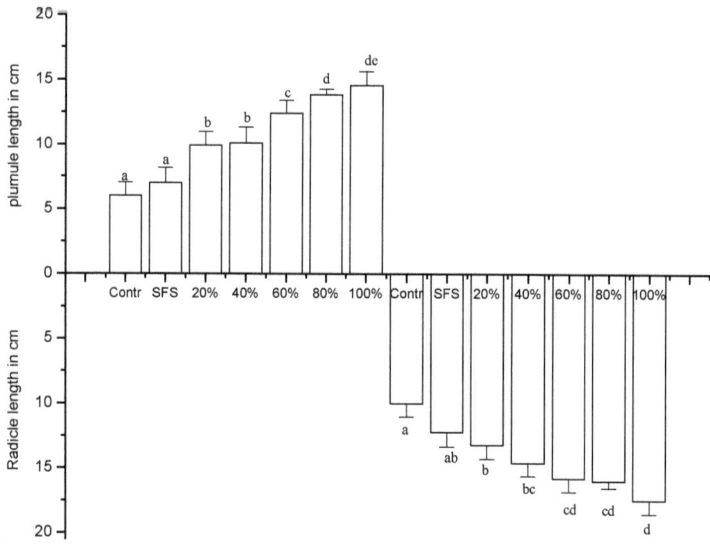

Figure 12. Effect of EF-LSEs on growth parameters of tomato seeds (plumule-radicle length).

The EF-LSEs had a positive effect in stimulating seedling height and their respective radicle and plumule lengths to 10.8 cm ($F_{6,28} = 30.98$; $p < 0.005$), 5.12 cm ($F_{6,28} = 42.16$; $p < 0.005$) and 4.12 cm ($F_{6,28} = 42.94$; $p < 0.005$), respectively, at 20% concentration, exhibiting radicle: plumule ratio of 1.36 cm ($F_{5,24} = 32.20$, $p < 0.005$). Similarly, the 40% EF-LSE treated seeds exhibited respective seedling height, radicle and plumule lengths, radicle: plumule ratio of 12.07 cm ($F_{6,28} = 30.98$; $p < 0.005$), 5.72 cm ($F_{6,28} = 42.16$; $p < 0.005$) and 5 cm ($F_{6,28} = 42.94$; $p < 0.005$) and 1.59 ($F_{5,24} = 32.20$, $p < 0.005$). Likewise 60% and 80% EF-LSE treated seeds were observed with seedling lengths of 13.09 cm and 13.84 cm ($F_{6,28} = 30.98$; $p < 0.005$), with corresponding radicle and plumule lengths of 7.6 cm ($F_{6,28} = 42.16$; $p < 0.005$) and 5.8 cm ($F_{6,28} = 42.94$; $p < 0.005$) as well as 8.2 cm ($F_{6,28} = 42.16$; $p < 0.005$) and 6.4 cm

($F_{6,28} = 42.94$; $p < 0.005$) besides the radicle: plumule ratios of 1.76 and 1.9 ($F_{5,24} = 32.20$, $P < 0.005$), respectively. The seeds treated with 100% EF-LSE exhibited the highest seedling length of 14.1 cm ($F_{6,28} = 30.98$; $p < 0.005$) with radicle and plumule lengths of 8.2 cm ($F_{6,28} = 42.16$; $p < 0.005$) and 6.4 cm ($F_{6,28} = 42.94$; $p < 0.005$), respectively. They revealed a radicle: plumule ratio of 2.04 ($F_{5,24} = 32.20$, $p < 0.005$).

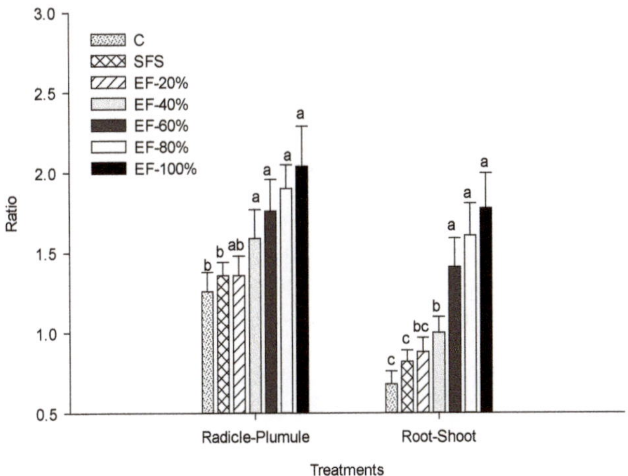

Figure 13. Radicle-plumule as well as root-shoot ratio of tomato seedlings.

The effect of EF-LSEs on the growth of tomato plants tested exhibited significant differences in total plant height (Figure 11), root, and shoot lengths (Figure 14).

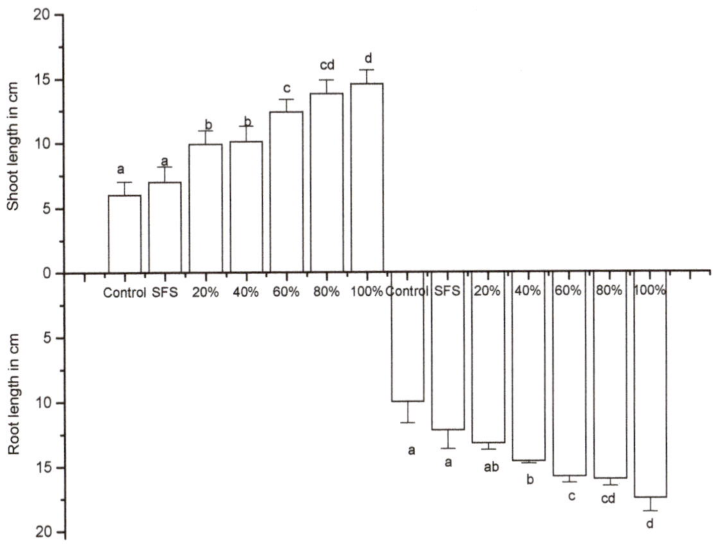

Figure 14. Effect of EF-LSE on root-shoot lengths of tomato plants.

The height of the tomato plant at the end of 20 days was 16 cm ($F_{6,28}$ = 40.45; p < 0.003), with root and shoot lengths of 6 cm ($F_{6,28}$ = 30.87; p < 0.005) and 10 cm ($F_{5,24}$ = 32.40; p < 0.005), respectively, with 0.68 root: shoot ratio ($F_{6,28}$ = 18.76; p < 0.000). The SFS was able to induce plant height to 19 cm ($F_{6,28}$ = 40.45; p < 0.003) with a corresponding root, shoot lengths and ratio of 7 cm ($F_{6,28}$ = 30.87; p < 0.005) and 12.2 cm ($F_{5,24}$ = 32.40; p < 0.005), and 0.82 ($F_{6,28}$ = 18.76; p < 0.00001), respectively. The EF-LSEs had a positive effect in stimulating tomato plant height and their respective root and shoot lengths to 23.4 cm ($F_{6,28}$ = 40.45; p < 0.003), 9.9 cm ($F_{6,28}$ = 30.87; p < 0.005) and 13.22 cm ($F_{5,24}$ = 32.40; p < 0.005), respectively, at 20% concentration, exhibiting root: shoot ratio 0.88 ($F_{6,28}$ = 18.76; p < 0.0001). Similarly, the 40% EF-LSE treated seeds exhibited respective plant height, root and shoot lengths, root: shoot of 24.6 cm ($F_{6,28}$ = 40.45; p < 0.003), 10.1 cm ($F_{6,28}$ = 30.87; p < 0.005) and 14.6 cm ($F_{5,24}$ = 32.40; p < 0.005) and 1 ($F_{6,28}$ = 18.76; p < 0.000). Likewise, 60% and 80% EF-LSE treated seeds were observed with total plant heights of 28.38 cm and 31cm ($F_{6,28}$ = 40.45; p < 0.003), with corresponding root and shoot lengths of 12.38 cm ($F_{6,28}$ = 30.87; p < 0.005) and 15.8 cm ($F_{5,24}$ = 32.40; p < 0.005) as well as 13.82 cm ($F_{6,28}$ = 30.87; p < 0.005) and 16 cm ($F_{5,24}$ = 32.40; p < 0.005) besides the root: shoot ratios of 1.41 and 1.606 ($F_{6,28}$ = 18.76; p < 0.0001), respectively (Figure 13). The seeds treated with 100% EF-LSE exhibited the highest plant height of 32 cm ($F_{6,28}$ = 40.45; p < 0.003) with root and shoot lengths of 14.54 cm ($F_{6,28}$ = 30.87; p < 0.005) and 17.48 cm ($F_{5,24}$ = 32.40; p < 0.005), respectively. They revealed a root: shoot ratio of 1.776 ($F_{6,28}$ = 18.76; p < 0.0001).

3.3.3. Effect of EF-LSEs on Fruit Quality Parameters of Cherry Tomato

The treated plants displayed a yield of 2.617 kg/plant compared with that of the control: 1.07 kg/plant ($F_{1,8}$ = 40.92; p < 0.0001). Further, the quality parameters analyses revealed the presence of increased amounts of 6.54 (Brix) TSS, 13.2 mg/100 g FM ascorbic acid, 88.45 µg g^{-1} FM lycopene and 7.26 mg g^{-1} DM phenol contents, compared with that of the control, which recorded 5.2 (Brix) TSS ($F_{1,8}$ = 57.1; p < 0.002), 10.48 mg/100 g FM ascorbic acid ($F_{1,8}$ = 53.52; p < 0.0001), 55.206 µg g^{-1} FM lycopene ($F_{1,8}$ = 61; p < 0.003) and 5.72 mg g^{-1} DM phenol ($F_{1,8}$ = 47. 8; p < 0.0001) contents (Table 5).

Table 5. Yield, TSS, Ascorbic acid, lycopene and phenol contents of treated cherry tomato fruits.

Parameters	C	EF-LSE Treated
Yield (kg/plant)	1.07 ± 0.96 [a]	2.617± 1.02 [b]
TSS (Brix)	5.2 ± 1.1 [a]	6.54 ± 1.07 [b]
Ascorbic acid (mg/100g FM)	10.48 ± 1.8 [a]	13.2 ± 1.5 [b]
Lycopene (µg g^{-1} FM)	55.206 ± 3.9 [a]	88.45 ± 4.3 [b]
Phenol (mg g^{-1} DM)	5.72 ± 1.47 [a]	7.26 ± 1.15 [b]

Rows denoted by a different letter are significantly different at $p \leq 0.05$.

4. Discussion

The indiscriminate application of fertilizers has not only intoxicated the environment, but also lost their efficiency. Alternative naturally benign bases of fertilizers, sourced from biological sources such as plants, animals and micro-organisms, have paved the way for the practice of "organic farming". Many eco-friendly bioactive compounds from seaweeds have been widely used in the agricultural field as plant growth promoters. Seaweeds are reported for their copious amounts of novel as well as assorted range of marine secondary metabolites [30]. Global population growth has seen leaps and bounds in the recent years, posing food insecurity [31]. With the foremost necessity of augmentation of crop production, farmers are in stress to improve yields of agriculturally important crops. As an imperative crop, the germination capability of tomato seed is valued to be around 70%. Seed emergence is mainly prejudiced by the equipoise, flanked by the growth skills of the embryo, in addition to the mechanical resistance of the endosperm, which should be debilitated for germination [32]. Seaweeds are being sought out as potential enhancers of crop growth and yield and are replacing chemical fertilizers owing to higher efficiencies, broader action range, eco-friendly nature, and cost-effective feature. Seaweeds, reported with outstanding plant growth promoting potentials, increased plant height, root as well as shoot lengths, consequently, are designated as plant growth biostimulants, as reviewed by Khan et al. [7] and Craigie [33] As a crucial and initial plant growth activity, the evaluation of a seed's germination and associated parameters can help in determining the rate of a crop success, in terms of yield and economy [34]. As contemplation, the current investigation was performed to determine the plant growth stimulant activities of green seaweed *U. flexuosa* (Chlorophyceae).

As the germination of a seed counts on various physical aspects, together with nutrient composition [12], preliminary tests of the extracts were performed by analyzing the pH and electrical conductivity (EC) of the extracts. The nutrient content of a solution, in terms of salts and electrolyte concentration can be determined by measuring their EC. The EF-LSE of *U. flexuosa* was found to possess a neutral pH and an optimum EC that indicates the presence of salts, for instance, boron, zinc, magnesium, calcium and other essential plant nutrients in a nutritive solution [35]. Higher rates of EC of nutrient or fertilizer solutions are proven with the stimulation of favourable agronomic traits, such as increase in nutritional quality, colour gradient and quality of tomato fruits [36,37]. However, solutions outside the optimum EC had an inhibitory effect on plant growth activities [38]. A nutrient solution within optimum EC was found optimal for the growth stimulation of lettuce in glasshouse conditions [39]. Henceforth, the EF-LSEs were designated as ideal to be tested for biostimulant potential by means of seed priming.

Additional experiments were carried out to determine the phytochemical as well as the elemental composition of the EF-LSEs. A preliminary phytochemical screening of the EF-LSE was done, which exposed the existence of more than a few compounds, such as carbohydrates, protein, phenols, flavonoids, saponins, tannins, and coumarins. Carbohydrates from different seaweeds were found to act as growth promoters of several crops such as tomato, soybean, duckweed and mung bean [40–42]. Proteins from seaweeds are recorded for their enhanced plant biostimulant activities in mung bean [43], and cherry tomato plants [44]. Proteins help plants to alleviate stress and increase their tolerance levels against abiotic stress like heat, cold, salt and even heavy metals [45].

Seaweeds are rich in phenolic compounds with varied bioactive properties [46,47]. Rajauria et al. [48] identified and characterized eight phenolic compounds from brown Irish seaweed *Himanthalia elongate*, which exhibited strong antioxidant activities. Chanthini et al. [6] correlated the levels of phenolic compound concentration with their antifungal potential. *U. flexuosa* had a considerate amount of phenols (1 mg/g of dry weight). Farasat et al. [49] detected higher phenolics as well as flavonoid levels from *U. flexuosa* and other edible green seaweeds. Besides, *Ascophyllun nodosum* extracts were able to increase the levels of phenols and flavonoids together, post application [50]. Saponins showcase a wide array of biological activities that play a pivotal role in plant growth as well as defense [51]. Coumarins also play a crucial part in plant development. These compounds have been proven with plant growth promotion capabilities alone and also in combination with phytohormones in faba bean [52]. Besides, the coumarin compounds were able to stimulate seed germination and seedling growth of wheat and sorghum seeds at optimum concentrations [53].

Tuhy et al. [54] testified that plant biomass surged by treatment with seaweed-derived micronutrients. The composition and functioning, together with the yield of all the plants, are reliant on their chlorophyll contents [55]. The chlorophyll content of *U. flexuosa* ranged up to 0.9 mg/g, which is relatively high among several other green seaweeds. This was also in agreement with the results published by Rathod [56]. The elemental composition performed revealed the presence of seven elements (O, Na, Mg, S, Cl, K and Ca) present on cell wall surface. The seaweed was 56.25% *w/v* of oxygen and along with high phenol content may be regarded as excellent candidates of antioxidizing agents [57]. Plant growth promoting elements such as sodium and potassium present in the cell wall of the seaweed makes them appropriate biofertilizers, besides a broad-spectrum of applications in the agricultural sector [58]. Several other mineral compounds such as chlorine, magnesium and calcium that are critical plant micronutrients are extant in the cell surface.

Tomato seeds treated with EF-LSEs displayed a positive response with respect to early germination, mean germination time, germination percentage, energy as well as better seed vigour index. Seed priming treatments achieved with quite a few plant derivatives such as plant hormones have been operational in the enhancement of seed germination of *Angelica glauca*, a threatened medicinal herb [59] as well as endive and chicory [60]. Seaweed extracts have been proven to show development enhancing properties on plant as well as seeds of various plants [61,62]. Also, priming seeds promoted early emergence of brinjal and tomato seeds compared with un-primed [63].

The seeds primed with EF-LSEs of *U. flexuosa* was analysed in different concentrations, comparing with the standard SFS solution and the control. The EF-LSE treated seeds displayed an increased germination percentage, exhibiting a lower mean germination time (MGT), taking only 2.2 days to emerge. *Codium tomentosum*, a green seaweed extract-treated aubergine seeds displayed lower MGT [64]. The potential of EF-LSEs to stimulate seed germination was reported long back in ornamental plants [65], green Chilies and Turnip [66]. Kavipriya et al. [67] also reported that priming of green gram seed with different seaweed extracts such as *Ulva lactuca* and *Caulerpa scalpelliformis* induced faster seed germination. Rapid seed emergence was recorded by priming the red gram seeds with the extracts of *Sargassum myriocystum* [68]. Furthermore, the positive effects of seaweed on the germination of green [69] and black gram [70] were also noted.

The comparative increase in seed emergence is correlated with seed eminence that is appeared to be augmented by the treatment of EF-LSEs. In addition, Amabika and Sujatha [68] proved that seed quality can be assessed by determining their seedling vigour index. Higher SVI of EF-LSE treated seeds implies an upliftment in seed quality. This was also evident from the results of other parameters of EF-LSE treated seeds, exhibiting higher seedling-plant height, radicle-root, as well as plumule-shoot lengths and dry-wet weight, in comparison with the control. Similar results of increased SVI from seed priming with EF-LSEs of *U. lactuca*, *U. reticulata*, *Padina pavonica*, *S. johnstonii* were correlated with that of increased seed germination and growth rates of brinjal and tomato, along with chilli [63].

As the primary developmental plant growth phase, the radicle and plumule are of prominent importance to determine the foundation of a plant. Seeds with an eminent radicle and plumule grow hastily, besides having an amplified competence [71]. Longer radicle lengths are also indicators of greater plant establishment efficiency. Seeds that produce shortened radicle-plumule might have issues in nutrient conduction to the embryo [72]. The EF-LSE primed seeds exhibited higher lengths of radicle and plumule, which improved with higher concentrations. EF-LSEs of seaweeds, *S. wightii* and *U. lactuca*, were demonstrated in their latent seed germination, besides plant growth promotion capabilities [73,74]. Similar increase of radicle-plumule lengths of tomato seeds was observed with the EF-LSEs of *C. sertularioides* and *S. liebmannii* [75].

The growth enhancement displayed by the EF-LSE primed seeds is owed to the occurrence of essential plant macro- and micro-nutrients, in addition to phytohormones. Di Filippo-Herrera et al. [76] reported the biostimulant activity of red seaweeds (*Acanthophora spicifera*, *Gelidium robustum*, and *Gracilaria parvispora*) and brown seaweeds (*Macrocystis pyrifera*, *Sargassum horridum* and *Ecklonia arborea*) primed on seeds of mung bean, which is primarily due to their nutritional and hormonal constituents. The fact that seaweeds stimulate plant growth has been documented by various researchers worldwide [77,78].

Similarly, EF-LSEs were earmarked for plant growth promotion by amplifying the growth of root and shoot of the tomato seeds, thereby displaying increased heights, compared to the control. Plant height was higher compared with seeds treated with SFS. Seaweed-treated seeds exhibiting higher root shoot lengths and ratio were previously reported in many studies [74,79,80]. Unlike the plant assessment results, comparatively developed shoot lengths remained. This suggests that the distribution of photosynthates and other compounds that aid in plant growth has shifted towards the shoot or increased in the above ground area. This could pave the way for the increase of plant yield [76].

Seed weight is considered an ecologically crucial character in plant progress, by way of influencing the establishment capacity of a seedling, as well as plant height and yield. This was proved by

Wuff [81], who found that the *Desmodium paniculatum* seeds of higher biomass produced a high yield. In addition, the EF-LSE treated seeds displayed increased wet and dry weights equated with control, besides improving with an EF-LSE concentration. This was in agreement with the results published by Karthikeyan and Shanmugam [82], who studied the effect of *Kappaphycus alvarezii* extract on peanuts. Vijayakumar et al. [83] also reported the increase in seed weight of *Capsicum annum* by treatment with *Codium decorticatum* EF-LSE. Increased seed weights produced plants with higher height, shoot mass, and yield [84]. Seed weight increase might be due to the production and accumulation of storage oils and several proteins that might promote plant growth abilities [85]. Seeds take in surrounding water, protoplasmic macromolecules by imbibition. The EF-LSEs are actively imbibed in the seed through capillary action, thus increasing biomass. Seeds with higher biomass have been reported to have better seedling growth. Sun et al. [86]. reported that maize seeds with increased biomass had high yields.

The biomass of EF-LSE treated seeds increased with time, thus revealing that the higher imbibing rate occurs at later hours of priming. This is attributed to the active process of enzymatic breakdown and mitosis essential for emergence. The increased imbibing rate of seeds exposed to EF-LSEs is primarily due to the abundance of plant essential nutrients and phytohormone composition of the EF-LSEs. *U. lactuca* and *P. gymnospora* primed tomato seeds exhibited higher imbibition rates during later stages of priming and hence were reported as more successful and better candidates for developing effective biostimulants to improve the growth of tomato plants (Hernández-Herrera [9]).

With respect to the biostimulant potentials of seaweed extract, the yield and quality parameters analyzed presented favorable results. The EF-LSE primed seeds observed an increase in yield, attributed to the presence of phyto-hormones and various plant growth promoting elements, as evident from the EDX analysis of the seaweed. Furthermore, the flavour of the cherry tomato fruit is directly proportional to the amount of TSS [87]. The nutritional value of these tomatoes is based on ascorbic acid content, which was also enhanced by the EF-LSE. Lycopene content, which is the reason for the ripening of fruits at an optimum stage, was also enhanced by EF-LSE treatment. Phenolic compounds, an indication of plant innate defense system, were augmented on treatment with seaweed extracts. These results are in agreement with that of Murtic et al. [88].

Since the cherry tomato gene pool, a wild relative of tomato, provides an opportunity to produce more nutritive and resilient tomato cultivar varieties, an attempt to preserve and conserve this inclusive gene pool in gene banks is critical [89]. Additionally, these wild relative crop types are nutritional repositories, whose cultivation shall be enhanced to meet the increasing global food security targets. The establishment of these species can be further stimulated by seed priming techniques. The application of seaweed extracts as a seed priming agent towards the improvement of agronomic traits of cherry tomato have resulted in positive responses to amplified seed germination capabilities, germination energy, and augmented seedling establishment. In addition, seed priming effects have induced a long-lasting priming effect by altering plant and fruit physiology in a favorable way, by increasing their biochemical constituents and fruit yield. Hence, this study proves the potential of *U. flexuosa* as a potential agricultural biostimulant that is both economic and effective.

Author Contributions: K.M.-P.C. and S.S.-N.; methodology, K.M.-P.C.; software, K.M.-P.C. and S.S.-N.; validation, V.S.-R., A.T., S.K., and R.S.; formal analysis, S.S.-N.; investigation, K.M.-P.C., resources, R.P., and A.T.; data curation, S.S.-N.; writing—original draft preparation, K.M.-P.C.; writing—review and editing, K.M.-P.C., and S.S.-N.; visualization, K.M.-P.C.,V.S.-R., N.S.S., H.S. and R.P.; supervision, S.S.-N.; project administration, R.S.; funding acquisition, S.S.-N.

Funding: This research was funded by the Department of Biotechnology, Ministry of Science and Technology grant number BT/IN/Indo-US/Foldscope//39/2015 dated 20 April 2018.

Abbreviations

LSE	Liquid seaweed extract
EF	*Ulva flexuosa*
SFS	Starter Fertilizer Solution
EDS	Energy Dispersive Spectroscopy
MGT	Mean Germination Time
SVI	Seedling Vigour Index
GE	Germination Energy
GP	Germination Percentage

References

1. Maarten, E.; Florian, S. Global demand for food is rising. Can we meet it? Available online: https://hbr.org/2016/04/global-demand-for-food-is-rising-can-we-meet-it (accessed on 9 September 2019).
2. Food and Agriculture Organization; International Fund for Agricultural Development; UNICEF; World Food Programme; WHO. The state of food security and nutrition in the world 2017: Building resilience for peace and food security. 2017. Available online: http://www.fao.org/3/a-i7695e.pdf (accessed on 6 April 2019).
3. Senthil-Nathan, S. A review of biopesticides and their mode of action against insect pests. In *Environmental Sustainability*; Springer India: New Delhi, India, 2015; pp. 49–63.
4. Hajjar, R.; Hodgkin, T. The use of wild relatives in crop improvement: A survey of developments over the last 20 years. *Euphytica* **2007**, *156*, 1–13. [CrossRef]
5. Senthil-Nathan, S. Natural pesticide research. *Physiol. Mol. Plant Pathol.* **2018**, *101*, 1–2. [CrossRef]
6. Chanthini, K.; Kumar, C.S.; Kingsley, S.J. Antifungal activity of seaweed extracts against phytopathogen Alternaria solani. *J. Acad. Indus. Res.* **2012**, *1*, 86–90.
7. Khan, W.; Rayirath, U.P.; Subramanian, S.; Jithesh, M.N.; Rayorath, P.; Hodges, D.M.; Critchley, A.T.; Craigie, J.S.; Norrie, J.; Prithiviraj, B. Seaweed Extracts as Biostimulants of Plant Growth and Development. *J. Plant Growth Regul.* **2009**, *28*, 386–399. [CrossRef]
8. Anisimov, M.M.; Chaikina, E.L.; Klykov, A.G.; Rasskazov, V.A. Effect of seaweeds extracts on the growth of seedling roots of buckwheat (*Fagopyrum esculentum* Moench) is depended on the season of algae collection. *Agric. Sci. Dev.* **2013**, *2*, 67–75.
9. HernÃ¡ndez-Herrera, R.M.; Santacruz-Ruvalcaba, F.; Ruiz-LÃ³pez, M.A.; Norrie, J.; HernÃ¡ndez-Carmona, G. Effect of liquid seaweed extracts on growth of tomato seedlings (*Solanum lycopersicum* L.). *J. Applied Phycol.* **2014**, *26*, 619–628. [CrossRef]
10. Hernández-Herrera, R.M.; Santacruz-Ruvalcaba, F.; Briceño-Domínguez, D.R.; Di Filippo-Herrera, D.A. Seaweed as potential plant growth stimulants for agriculture in Mexico. *Hidrobiológica* **2018**, *28*, 129–140. [CrossRef]
11. Rengasamy, K.R.; Kulkarni, M.G.; Papenfus, H.B.; Van Staden, J. Quantification of plant growth biostimulants, phloroglucinol and eckol, in four commercial seaweed liquid fertilizers and some by-products. *Algal. Res.* **2016**, *20*, 57–60. [CrossRef]
12. Lanka, S.K.; Senthil-Nathan, S.; Blouin, D.J.; Stout, M.J. Impact of thiamethoxam seed treatment on growth and yield of rice, Oryza sativa. *J. Econ. Entomol.* **2017**, *110*, 479–486. [CrossRef]
13. Baskin, C.C.; Baskin, J.M. *Ecologically Meaningful Germination Studies, Seeds: Ecology, Biogeography, and, Evolution of Dormancy and Germination*; Elseiver: New York, NY, USA, 1998.
14. Martinez, E.; Carbonell, M.V.; FlÃ¡rez, M.; Amaya, J.M.; Maqueda, R. Germination of tomato seeds (*Lycopersicon esculentum* L.) under magnetic field. *Int. Agrophys.* **2009**, *23*, 45–49.
15. Soltani, E.; Ghaderi-Far, F.; Baskin, C.C.; Baskin, J.M. Problems with using mean germination time to calculate rate of seed germination. *Aust. J. Bot.* **2016**, *63*, 631–635. [CrossRef]
16. Charlo, H.; Castoldi, R.; Ito, L.; Fernandes, C.; Braz, L. Productivity of cherry tomatoes under protected cultivation carried out with different types of pruning and spacing. In Proceedings of the XXVII International Horticultural Congress-IHC2006: International Symposium on Advances in Environmental Control, Automation, Seoul, Korea, 13–19 August 2006; Volume 761, pp. 323–326.

17. Lin, T.; Zhu, G.; Zhang, J.; Xu, X.; Yu, Q.; Zheng, Z.; Zhang, Z.; Lun, Y.; Li, S.; Wang, X.; et al. Genomic analyses provide insights into the history of tomato breeding. *Nat. Genet.* **2014**, *46*, 120. [CrossRef] [PubMed]

18. Wang, P.; Zhong, R.-B.; Yuan, M.; Gong, P.; Zhao, X.-M.; Zhang, F. Mercury (II) detection by water-soluble photoluminescent ultra-small carbon dots synthesized from cherry tomatoes. *Sci. Tech.* **2016**, *27*, 35. [CrossRef]

19. Sadasivam, S. *Biochemical Methods*, 2nd ed.; New Age International (p) Ltd. Publisher: New Delhi, India, 1996; p. 179.

20. Senthil-Nathan, S. Physiological and biochemical effect of neem and other Meliaceae plants secondary metabolites against Lepidopteran insects. *Front. Physiol.* **2013**, *4*, 359. [CrossRef] [PubMed]

21. Sadasivam, S.; Manickam, A. Chlorophylls. In *Biochemical Methods for Agricultural Sciences*; Wiley Eastern Limited: New Delhi, India, 1992; pp. 184–185.

22. Lowry, O.H.; Rosebrough, N.J.; Farr, A.L.; Randall, R.J. Protein measurement with the Folin phenol reagent. *J. Boil. Chem.* **1951**, *193*, 265–275.

23. Ellis, R.H.; Roberts, E.H. Improved Equations for the Prediction of Seed Longevity. *Ann. Bot.* **1980**, *45*, 13–30. [CrossRef]

24. Orchard, T.J. Estimating the parameters of plant seedling emergence. *Seed Sci. Technol.* **1977**, *5*, 61–69.

25. Larreta, M.M.; Upton, J.L.; HernÃ¡ndez, J.J.V.; Livera, A.H. Germination and Vigour of seeds in Pseudotsuga Menziesii of Mexico. *Ra. Ximhai* **2008**, *4*, 119–134. [CrossRef]

26. Crop production techniques of horticultural crops. Available online: http://agritech.tnau.ac.in/pdf/2013/cpg_horti_2013.pdf (accessed on 4 December 2019).

27. Hedge, J.E.; Hofreiter, B.T.; Whistler, R.L. *Carbohydrate Chemistry*; Academic Press: New York, NY, USA, 1962; Volume 17.

28. Horwitz, W.; Chichilo, P.; Reynolds, H. *Official Methods of Analysis of the Association of Official Analytical Chemists*; Association of Official Analytical Chemists: Washington, DC, USA, 1970.

29. Rangana, S. *Manual of Analysis of Fruit and Vegetable Products*; Tata McGraw-Hill: New Delhi, India, 1979.

30. Priyadharshini, S.; Arya, S.; Kiran, G.S.; Ninawe, A.S.; Selvin, J. Bioactive Marine Natural Products: Insights into Marine Microbes, Seaweeds, and Marine Sponges as Potential Sources of Drug Discovery. In *Marine Microorganisms*; CRC press: Boca Raton, FL, USA, 2016; pp. 17–30.

31. Godfray, H.C.J.; Garnett, T.; Godfray, H.C.J. Food security and sustainable intensification. *Philos. Trans. Soc. B Boil. Sci.* **2014**, *369*, 20120273. [CrossRef] [PubMed]

32. Seethalakshmi, S.; Umarani, R. Biochemical changes during imbibition stages of seed priming in Tomato. *Int. J. Cosmet. Sci.* **2018**, *6*, 454–456.

33. Craigie, J.S. Seaweed extract stimuli in plant science and agriculture. *J. Appl. Phycol.* **2011**, *23*, 371–393. [CrossRef]

34. Kalaivani, K.; Kalaiselvi, M.M.; Senthil-Nathan, S. Effect of methyl salicylate (MeSA), an elicitor on growth, physiology and pathology of resistant and susceptible rice varieties. *Sci. Rep.* **2016**, *6*, 34498. [CrossRef] [PubMed]

35. A Basic Guide to pH and EC and How They Affect Plants. available online: http://www.plant-magic.co.uk/post/2016/12/16/a-basic-guide-to-ph-and-ec-and-how-they-affect-plants/54/ (accessed on 4 October 2018).

36. Elia, A.; Serio, F.; Parente, A.; Santamaria, P.; Ruiz Rodriguez, G. Electrical conductivity of nutrient solution, plant growth and fruit quality of soilless grown tomato. In Proceedings of the V International Symposium on Protected Cultivation in Mild Winter Climates: Current Trends for Sustainable Technologies, Cartagena, Spain, 7 March 2000; Volume 559, pp. 503–508.

37. Krauss, S.; Schnitzler, W.H.; Grassmann, J.; Woitke, M. The Influence of Different Electrical Conductivity Values in a Simplified Recirculating Soilless System on Inner and Outer Fruit Quality Characteristics of Tomato. *J. Agric. Food. Chem.* **2006**, *54*, 441–448. [CrossRef]

38. Ding, X.; Jiang, Y.; Zhao, H.; Guo, D.; He, L.; Liu, F.; Zhou, Q.; Nandwani, D.; Hui, D.; Yu, J. Electrical conductivity of nutrient solution influenced photosynthesis, quality, and antioxidant enzyme activity of pakchoi (*Brassica campestris* L. ssp. chinensis) in a hydroponic system. *PLoS ONE* **2018**, *13*, 0202090. [CrossRef] [PubMed]

39. Samarakoon, U.C.; Weerasinghe, P.A.; Weerakkody, W.A.P. Effect of electrical conductivity (EC) of the nutrient solution on nutrient uptake, growth and yield of leaf lettuce (*Lactuca sativa* L.) in stationary culture. *Trop. Agric. Res.* **2006**, *18*.

40. Hernández-Herrera, R.M.; Santacruz-Ruvalcaba, F.; Zañudo-Hernandez, J.; Hernández-Carmona, G. Activity of seaweed extracts and polysaccharide-enriched extracts from Ulva lactuca and Padina gymnospora as growth promoters of tomato and mung bean plants. *J. Appl. Phycol.* **2016**, *28*, 2549–2560. [CrossRef]

41. Koh, Y. The Effect of Oligosaccharides in an Extract of the Brown Seaweed *Ascophyllum nodosum* on Plant Growth and Plant Immune Responses in Soybean (*Glycine max* L.) and Duckweed (*Lemna minor*). Ph.D. Thesis, The Saint Mary's University, Halifax, NS, Canada, April 2016.

42. Mzibra, A.; Aasfar, A.; El Arroussi, H.; Khouloud, M.; Dhiba, D.; Kadmiri, I.M.; Bamouh, A. Polysaccharides extracted from Moroccan seaweed: A promising source of tomato plant growth promoters. *J. Appl. Phycol.* **2018**, *30*, 2953–2962. [CrossRef]

43. Paul, J. Influence of seaweed liquid fertilizer of Gracilaria dura (AG.) J. A. G. (red seaweed) on Vigna radiata (l.) r. wilczek., in Thoothukudi, Tamil Nadu, India. *World J. Pharm Res.* **2014**, *3*, 968–978.

44. Polo, J.; Mata, P. Evaluation of a Biostimulant (Pepton) Based in Enzymatic Hydrolyzed Animal Protein in Comparison to Seaweed Extracts on Root Development, Vegetative Growth, Flowering, and Yield of Gold Cherry Tomatoes Grown under Low Stress Ambient Field Conditions. *Front. Plant Sci.* **2018**, *8*, 2261. [CrossRef] [PubMed]

45. Van Oosten, M.J.; Pepe, O.; De Pascale, S.; Silletti, S.; Maggio, A. The role of biostimulants and bioeffectors as alleviators of abiotic stress in crop plants. *Chem. Boil. Technol. Agric.* **2017**, *4*, 3. [CrossRef]

46. Corona, G.; Coman, M.M.; Guo, Y.; Hotchkiss, S.; Gill, C.; Yaqoob, P.; Spencer, J.P.E.; Rowland, I.; Spencer, J.P.J.P.E. Effect of simulated gastrointestinal digestion and fermentation on polyphenolic content and bioactivity of brown seaweed phlorotannin-rich extracts. *Mol. Nutr. Food Res.* **2017**, *61*, 1700223. [CrossRef] [PubMed]

47. Austin, C.; Stewart, D.; Allwood, J.W.; McDougall, G.J. Extracts from the edible seaweed, Ascophyllum nodosum, inhibit lipase activity in vitro: Contributions of phenolic and polysaccharide components. *Food Funct.* **2018**, *9*, 502–510. [CrossRef] [PubMed]

48. Rajauria, G.; Foley, B.; Abu-Ghannam, N. Identification and characterization of phenolic antioxidant compounds from brown Irish seaweed Himanthalia elongata using LC-DAD–ESI-MS/MS. *Innov. Food Sci. Emerg. Technol.* **2016**, *37*, 261–268. [CrossRef]

49. Farasat, M.; Khavari-Nejad, R.-A.; Nabavi, S.M.B.; Namjooyan, F. Antioxidant Activity, Total Phenolics and Flavonoid Contents of some Edible Green Seaweeds from Northern Coasts of the Persian Gulf. *Iran. J. Pharm. IJPR* **2014**, *13*, 163–170.

50. Lola-Luz, T.; Hennequart, F.; Gaffney, M. Effect on yield, total phenolic, total flavonoid and total isothiocyanate content of two broccoli cultivars (*Brassica oleraceae* var *italica*) following the application of a commercial brown seaweed extract (Ascophyllum nodosum). *Agric. Food Sci.* **2014**, *23*, 28–37. [CrossRef]

51. Faizal, A.; Geelen, D. Saponins and their role in biological processes in plants. *Phytochem. Rev.* **2013**, *12*, 877–893. [CrossRef]

52. Saleh, A.M.; Madany, M.M.Y.; GonzÃ¡lez, L. The effect of coumarin application on early growth and some physiological parameters in Faba Bean (*Vicia faba* L.). *Plant Growth Regul.* **2015**, *34*, 233–241. [CrossRef]

53. Chattha, F.A.; Munawar, M.A.; Ashraf, M.; Kousar, S.; Nisa, M.U. Plant growth regulating activities of coumarin-3-acetic acid derivatives. *Allelopathy J.* **2015**, *36*.

54. Tuhy, Ł.; Samoraj, M.; Basladynska, S.; Chojnacka, K. New Micronutrient Fertilizer Biocomponents Based on Seaweed Biomass. *Pol. J. Environ. Stud.* **2015**, *24*, 2213–2221. [CrossRef]

55. Palta, J.P. Leaf chlorophyll content. *Remote Sens. Rev.* **1990**, *5*, 207–213. [CrossRef]

56. Rathod, P.J. Metabolomics and morphological characterization of macroalgae. Ph.D. Thesis, Junagadh Agricultural University, Junagadh, India, 2018.

57. Farvin, K.S.; Jacobsen, C. Phenolic compounds and antioxidant activities of selected species of seaweeds from Danish coast. *Food Chem.* **2013**, *138*, 1670–1681. [CrossRef] [PubMed]

58. Ciepiela, G.Y.A.; Godlewska, A.; Jankowska, J. The effect of seaweed Ecklonia maxima extract and mineral nitrogen on fodder grass chemical composition. *Environ. Sci. Pollut. R.* **2016**, *23*, 2301–2307. [CrossRef] [PubMed]

59. Butola, J.S.; Badola, H.K. Effect of pre-sowing treatment on seed germination and seedling vigour in Angelica glauca, a threatened medicinal herb. *Curr. Sci.* **2004**, *87*, 796–799.

60. Tzortzakis, G.N. Effect of pre-sowing treatment on seed germination and seedling vigour in endive and chicory. *Hort. Sci.* **2009**, *36*, 117–125. [CrossRef]

61. Yildirim, E.; Dursun, A.; GÃvenc, I.; Kumlay, A.M. The effects of different salt, biostimulant and temperature levels on seed germination of some vegetable species. In Proceedings of the II Balkan Symposium on Vegetables and Potatoes, Thessaloniki, Greece, 11 October 2000; Volume 579, pp. 249–253.

62. Hassan, S.M.; Ghareib, H.R. Bioactivity of Ulva lactuca L. acetone extract on germination and growth of lettuce and tomato plants. *Afr. J. Biotechnol.* **2009**, *8*.

63. Patel, R.V.; Pandya, K.Y.; Jasrai, R.T.; Brahmbhatt, N. Significance of Green and Brown Seaweed Liquid Fertilizer on Seed Germination of *Solanum melongena*, *Solanum lycopersicum* and *Capsicum annum* by Paper Towel and Pot Method. *Int. J. Recent Sci. Res.* **2018**, *9*, 24065–24072.

64. Demir, N.; Dural, B.; Yildirim, K. Effect of seaweed suspensions on seed germination of tomato, pepper and aubergine. *J. Biol. Sci.* **2006**, *6*, 1130–1133.

65. Senn, T.L.; Aitken, J.B. Seaweed Products as a Fertilizer and Soil Conditioner for Horticultural Crops. *Bot. Mar.* **1965**, *8*, 144–147.

66. Dhargalkar, V.K.; Untawale, A.G. Some observations of the effect of SLF on higher plants. *Indian J. Mar. Sci.* **1983**, *12*, 210–214.

67. Kavipriya, R.; Dhanalakshmi, P.K.; Jayashree, S.; Thangaraju, N. Seaweed extract as a biostimulant for legume crop, green gram. *J. Ecobiotechnol.* **2011**, *3*, 16–19.

68. Amabika, S.; Sujatha, K. Effect of priming with seaweed extracts on germination and vigour under different water holding capacities. *Seaweed Res. Utiln* **2015**, *37*, 37–44.

69. Kumar, N.A.; Vanlalzarzova, B.; Sridhar, S.; Baluswami, M. Effect of liquid seaweed fertilizer of Sargassum wightii Grev. on the growth and biochemical content of green gram (*Vigna radiata* (L.) R. Wilczek). *Recent Res. Sci. Technol.* **2012**, *4*, 40–45.

70. Ganapathy Selvam, G.; Balamurugan, M.; Thinakaran, T.; Sivakumar, K. Developmental changes in the germination, growth and chlorophyllase activity of *Vigna mungo* L. using seaweed extract of Ulva reticulata Forsskal. *Int. Res. J. Pharm.* **2013**, *4*, 252–254.

71. Khatami, S.R.; Sedghi, M.; Garibani, H.M.; Ghahremani, S. The effect of priming on seed germination characteristics of maize under salt stress. In *Annales of West University of Timisoara. Series of Biology*; West University of Timisoara: Timișoara, Romania, 2017; Volume 20, p. 65.

72. Basra, S.; Farooq, M.; Tabassam, R.; Ahmad, N. Physiological and biochemical aspects of pre-sowing seed treatments in fine rice (*Oryza sativa* L.). *Seed Sci. Technol.* **2005**, *33*, 623–628. [CrossRef]

73. Kumar, G.; Sahoo, D. Effect of seaweed liquid extract on growth and yield of Triticum aestivum var. Pusa Gold. *J. Appl. Phycol.* **2011**, *23*, 251–255. [CrossRef]

74. Castellanos-Barriga, L.G.; Santacruz-Ruvalcaba, F.; Hernández-Carmona, G.; Ramírez-Briones, E.; Hernández-Herrera, R.M. Effect of seaweed liquid extracts from Ulva lactuca on seedling growth of mung bean (*Vigna radiata*). *J. Appl. Phycol.* **2017**, *29*, 2479–2488. [CrossRef]

75. Zodape, S.T.; Gupta, A.; Bhandari, S.C.; Rawat, U.S.; Chaudhary, D.R.; Eswaran, K.; Chikara, J. Foliar application of seaweed sap as biostimulant for enhancement of yield and quality of tomato (*Lycopersicon esculentum* Mill.). *J. Sci. Ind. Res.* **2011**, *70*, 215–219.

76. Di Filippo-Herrera, D.A.; Muñoz-Ochoa, M.; Hernández-Herrera, R.M.; Hernández-Carmona, G. Biostimulant activity of individual and blended seaweed extracts on the germination and growth of the mung bean. *J. Appl. Phycol.* **2018**, *33*, 1–13. [CrossRef]

77. Arioli, T.; Mattner, S.W.; Winberg, P.C. Applications of seaweed extracts in Australian agriculture: Past, present and future. *J. Appl. Phycol.* **2015**, *27*, 2007–2015. [CrossRef] [PubMed]

78. Battacharyya, D.; Babgohari, M.Z.; Rathor, P.; Prithiviraj, B. Seaweed extracts as biostimulants in horticulture. *Sci. Hortic.* **2015**, *196*, 39–48. [CrossRef]

79. Rayorath, P.; Jithesh, M.N.; Farid, A.; Khan, W.; Palanisamy, R.; Hankins, S.D.; Critchley, A.T.; Prithiviraj, B. Rapid bioassays to evaluate the plant growth promoting activity of *Ascophyllum nodosum* (L.) Le Jol. using a model plant, *Arabidopsis thaliana* (L.) Heynh. *J. Appl. Phycol.* **2008**, *20*, 423–429. [CrossRef]

80. Ghaderiardakani, F.; Collas, E.; Damiano, D.K.; Tagg, K.; Graham, N.S.; Coates, J. Effects of green seaweed extract on Arabidopsis early development suggest roles for hormone signalling in plant responses to algal fertilisers. *Sci. Rep.* **2019**, *9*, 1983. [CrossRef] [PubMed]

81. Wulff, R.D. Seed size variation in Desmodium paniculatum: III. Effects on reproductive yield and competitive ability. *J. Ecol.* **1986**, *74*, 115–121. [CrossRef]

82. Karthikeyan, K.; Shanmugam, M. Grain yield and functional properties of red gram applied with seaweed extract powder manufactured from Kappaphycus alvarezii. *Int. J. Rec. Adv. Mul. Dis. Res.* **2016**, *3*, 1353–1359.

83. Vijayakumar, S.; Durgadevi, S.; Arulmozhi, P.; Rajalakshmi, S.; Gopalakrishnan, T.; Parameswari, N. Effect of seaweed liquid fertilizer on yield and quality of *Capsicum annum* L. *Acta Ecol. Sin.* **2018**, in press. [CrossRef]

84. Poveda, K.; Steffan-Dewenter, I.; Scheu, S.; Tscharntke, T. Effects of below- and above-ground herbivores on plant growth, flower visitation and seed set. *Oecologia* **2003**, *135*, 601–605. [CrossRef]

85. Baud, S.; Boutin, J.-P.; Miquel, M.; Lepiniec, L.; Rochat, C. An integrated overview of seed development in Arabidopsis thaliana ecotype WS. *Plant Physiol. Biochem.* **2002**, *40*, 151–160. [CrossRef]

86. Sun, X.; Cahill, J.; Van Hautegem, T.; Feys, K.; Whipple, C.; Novák, O.; Delbare, S.; Versteele, C.; Demuynck, K.; De Block, J.; et al. Altered expression of maize PLASTOCHRON1 enhances biomass and seed yield by extending cell division duration. *Nat. Commun.* **2017**, *8*, 14752. [CrossRef] [PubMed]

87. Venkadeswaran, E.; Vethamoni, P.I.; Arumugam, T.; Manivannan, N.; Harish, S. Evaluating the yield and quality characters of cherry tomato [*Solanum lycopersicum* (L.) var. *cerasiforme* Mill.] genotypes. *Int. J. Chem. Stud.* **2018**, *6*, 858–863.

88. Murtic, S.; Oljaca, R.; Murtic, M.S.; Vranac, A.; Koleska, I.; Karic, L. Effects of seaweed extract on the growth, yield and quality of cherry tomato under different growth conditions. *Acta Agric. Slov.* **2018**, *111*, 315–325. [CrossRef]

89. Castañeda-Álvarez, N.P.; Khoury, C.K.; Achicanoy, H.A.; Bernau, V.; Dempewolf, H.; Eastwood, R.J.; Guarino, L.; Harker, R.H.; Jarvis, A.; Maxted, N.; et al. Global conservation priorities for crop wild relatives. *Nat. Plants* **2016**, *2*, 16022. [CrossRef] [PubMed]

Article

Wild *Vigna* Legumes: Farmers' Perceptions, Preferences, and Prospective Uses for Human Exploitation

Difo Voukang Harouna [1,3], Pavithravani B. Venkataramana [2,3], Athanasia O. Matemu [1,3] and Patrick Alois Ndakidemi [2,3,*]

[1] Department of Food Biotechnology and Nutritional Sciences, Nelson Mandela African Institution of Science and Technology (NM-AIST), P.O. Box 447 Arusha, Tanzania; harounad@nm-aist.ac.tz (D.V.H.); athanasia.matemu@nm-aist.ac.tz (A.O.M.)

[2] Department of Sustainable Agriculture, Biodiversity and Ecosystems Management, Nelson Mandela African Institution of Science and Technology, P.O. Box 447 Arusha, Tanzania; pavithravani.venkataramana@nm-aist.ac.tz

[3] Centre for Research, Agricultural Advancement, Teaching Excellence and Sustainability in Food and Nutrition Security (CREATES-FNS), Nelson Mandela African Institution of Science and Technology, P.O. Box 447 Arusha, Tanzania

* Correspondence: patrick.ndakidemi@nm-aist.ac.tz; Tel.: +255-757744772

Received: 17 April 2019; Accepted: 24 May 2019; Published: 31 May 2019

Abstract: The insufficient food supply due to low agricultural productivity and quality standards is one of the major modern challenges of global agricultural food production. Advances in conventional breeding and crop domestication have begun to mitigate this issue by increasing varieties and generation of stress-resistant traits. Yet, very few species of legumes have been domesticated and perceived as usable food/feed material, while various wild species remain unknown and underexploited despite the critical global food demand. Besides the existence of a few domesticated species, there is a bottleneck challenge of product acceptability by both farmers and consumers. Therefore, this paper explores farmers' perceptions, preferences, and the possible utilization of some wild *Vigna* species of legumes toward their domestication and exploitation. Quantitative and qualitative surveys were conducted in a mid-altitude agro-ecological zone (Arusha region) and a high altitude agro-ecological zone (Kilimanjaro region) in Tanzania to obtain the opinions of 150 farmers regarding wild legumes and their uses. The study showed that very few farmers in the Arusha (28%) and Kilimanjaro (26%) regions were aware of wild legumes and their uses. The study further revealed through binary logistic regression analysis that the prior knowledge of wild legumes depended mainly on farmers' location and not on their gender, age groups, education level, or farming experience. From the experimental plot with 160 accessions of wild *Vigna* legumes planted and grown up to near complete maturity, 74 accessions of wild *Vigna* legumes attracted the interest of farmers who proposed various uses for each wild accession. A X^2 test (likelihood ratio test) revealed that the selection of preferred accessions depended on the farmers' gender, location, and farming experience. Based on their morphological characteristics (leaves, pods, seeds, and general appearance), farmers perceived wild *Vigna* legumes as potentially useful resources that need the attention of researchers. Specifically, wild *Vigna* legumes were perceived as human food, animal feed, medicinal plants, soil enrichment material, and soil erosion-preventing materials. Therefore, it is necessary for the scientific community to consider these lines of farmers' suggestions before carrying out further research on agronomic and nutritional characteristics toward the domestication of these alien species for human exploitation and decision settings.

Keywords: non-domesticated legumes; *Vigna racemosa*; *Vigna ambacensis*; *Vigna reticulata*; *Vigna vexillata*; Tanzania; wild food legumes

---

1. Introduction

Legumes (family: Fabaceae) possess an undeniable vital nutritional value for both humans and animals due to their protein content. They are known to be the second most valuable plant source of nutrients for both humans and animals, and the third largest family among flowering plants, with about 650 genera and 20,000 species [1]. Some of the most commonly domesticated, grown, and commercialized legumes such as soybeans, cowpeas, common beans, and other forms have demonstrated considerable contribution to the global food security [2]. Yet, their production rate remains unsatisfying compared with their consumption rate due to biotic and abiotic challenges [3]. Therefore, there is a need to look for alternatives. A systematic screening of the hitherto wild non-domesticated and wild relatives of the domesticated species within the commonly known and the little-known genera of legumes might be a promising strategy.

The *Phaseolus* and *Vigna* genera comprises the most widely consumed legumes, namely common beans (*Phaseolus vulgaris*) and cowpea (*Vigna unguiculata*) [2,4,5]. Within each genus, there are fewer domesticated edible species as compared with the numerous non-domesticated wild species. Some domesticated or semi-domesticated species have been termed as neglected and underutilized species due to little attention being paid to them or the complete ignorance of their existence by agricultural researchers, plant breeders, and policymakers [6]. This study mainly focuses on the genus *Vigna*.

The genus *Vigna* is a huge and important set of legumes consisting of more than 200 species [7]. It comprises several species of agronomic, economic, and environmental importance. The most common domesticated ones include the mung bean [*V. radiata* (L.) Wilczek], urd bean [*V. mungo* (L.) Hepper], cowpea [*V. unguiculata* (L.) Walp.], azuki bean [*V. angularis* (Willd.) Ohwi & Ohashi], bambara groundnut [*V. subterranea* (L.) Verdc.], moth bean [*V. aconitifolia (Jacq.) Maréchal*], and rice bean [*V. umbellata* (Thunb.) Ohwi & Ohashi]. Many of these species are valued as forage, green manure, and cover crops, besides their value as high protein grains. The genus *Vigna* also comprises more than 100 wild species that do not possess common names apart from their scientific appellation yet [8]. They are given different denotations such as underexploited wild *Vigna* species, non-domesticated *Vigna* species, wild *Vigna*, or alien species, depending on the scientist [2,7,9].

The rapid evolution, distribution, and spreading of improved bred crop varieties due to breeding programs and domestication in order to respond to food security challenges have also impacted positively on the negligence and disappearance of wild crop relatives [10,11]. This is certainly a negative impact vis-à-vis the species' biodiversity conservation. From that perspective, one could imagine and question the awareness, beliefs, and preferences of some generations regarding the origin of the consumed modern crops. This may explain the stigma about the consumption and even the existence of these wild legumes, and therefore their rejection as food while they have been used as such in the past in some cases.

Food acceptability and food choices are usually influenced by many factors in which sensory preferences play an important role [12]. The nutritional composition is also a very essential characteristic to consider in food selection and consumption, as it is directly linked to consumers' health and well-being. Unfortunately, this parameter may only be seriously considered in parts of the world where food accessibility, availability, and affordability are not challenged. Hence, much is needed to be done in this line to investigate the nutritional composition of wild crop foods together with close understanding of their social acceptability.

Investigations on the chemical composition of wild *Vigna* legumes seem to be less attractive to the scientific community for reasons yet to be established. Research in that line has remained silent and undocumented for more than a decade [13]. The latest report shows that some of the wild *Vigna* accessions studied present nutrient levels comparable to those of some domesticated species with exceptionally higher levels of sulfur amino acids [13]. However, it is highly necessary at this point to think about the acceptability of these wild legumes by farmers and consumers before any further research is conducted in order to orient the improvement, adoption, and domestication for a proper exploitation to the benefit of mankind.

This study explores experienced legumes, cultivating farmers' awareness, perception, acceptability, and preferred uses for some accession of wild *Vigna* legumes (*Vigna racemosa*, *Vigna ambacensis*, *Vigna reticulata*, and *Vigna vexillata*). The study has been organized into two parts, considering farmers' awareness in the first part and preferences for wild legumes in the second.

2. Materials and Methods

2.1. Study I: Explorative Survey

The aim of this study was to ascertain farmers' awareness about the existence of wild non-domesticated legumes and their uses in addition to challenges and experiences related to the growth and domestication of wild legumes.

The study was conducted among legume farmers in a mid-altitude agro-ecological zone (Arusha Region) and a high altitude agro-ecological zone (Kilimanjaro Region) of Tanzania where legume cultivation is intensified, as shown in Figure 1A [14]. A purposive sampling from a crop-growing population of 0.13% (37,985) from Arusha [15] and 0.17% (56,710) from Kilimanjaro [16] were used to obtain a representative sample size. The total number of farmers involved in legume improvement programs included 50 from the Seliani Agricultural Institute (TARI), Arusha and 100 from the Tanzania Coffee Research Institute (TaCRI), Moshi, Kilimanjaro regions, respectively (Figure 1B). A systematic selection of farmers who had at least two years of trying locally improved legume varieties was performed. An individual face-to-face interview with the help of a semi-structured questionnaire prior to participant experimental plot visit was executed to obtain a broad range of individual opinions and explore their awareness of wild legumes. The questionnaire consisted of 24 items including sociodemographic characteristics. The items were categorized and analyzed to assess the sociodemographic characteristics of participants, their prior knowledge/awareness about wild legumes, and the uses of wild legumes as known by experienced farmers as well as some challenges faced by legume farmers.

2.2. Study II: Farmers' Preferences and Perceptions of Wild Vigna Legumes

The main aim of this study was to identify farmers' perceptions and prospective uses of preferred accessions of wild legumes based on morphological agronomic characteristics in order to direct the domestication process.

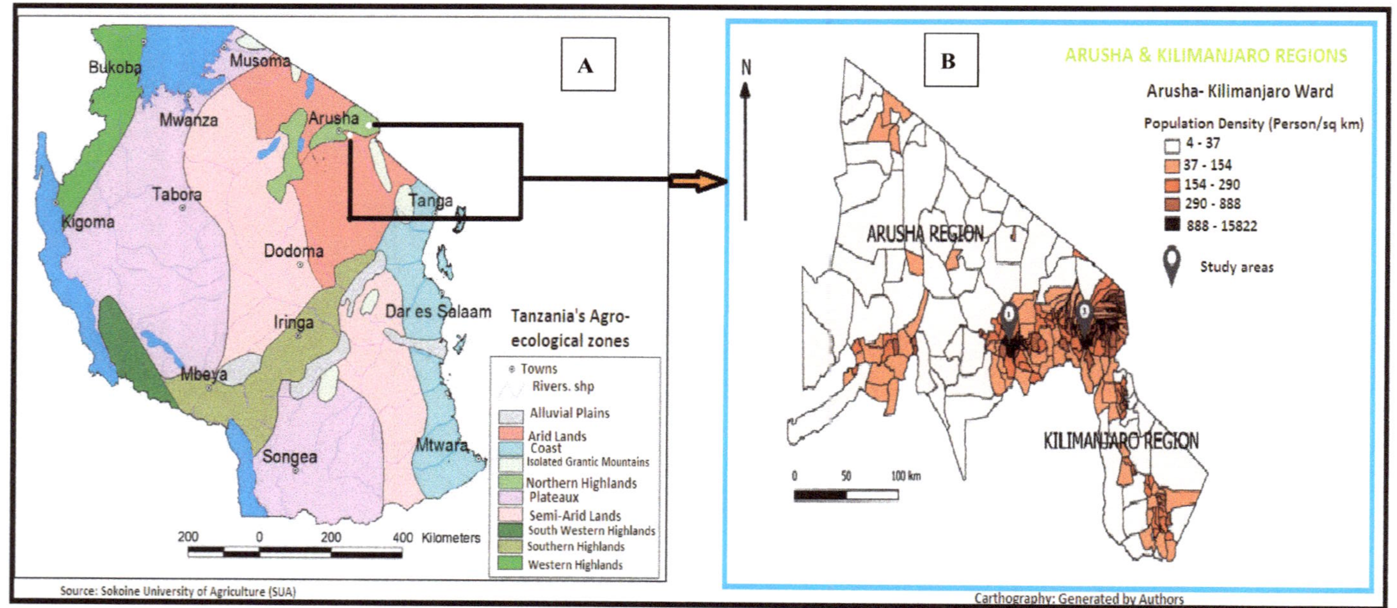

Figure 1. Tanzania map showing agro-ecological zones of Tanzania (**A**) [17] and the study sites (**B**): 1 = Arusha district (Arusha region) and 2 = Hai district (Kilimanjaro region).

Table 1. Wild *Vigna* species collected from the gene banks/self.

Vigna Species	Genebank/ Number of Accession			Total
	GRC, IITAIbadan, Nigeria	AGGHorsham, Victoria	Self-Collected	
Vigna racemosa	-	4	-	4
Vigna reticulata	48	3	-	51
Vigna vexillata	47	13	-	60
Vigna ambacensis	42	0	-	42
Unknown *V. racemosa* accession (Nigeria)	-	-	1	1
Unknown *V. reticulata* Accession (Nigeria)	-	-	1	1
Unknown *Vigna* (Tanzania)	-	-	1	1
Total	137	20	3	160

GRC, IITA: Genetic Resource Center, Germplasm Health Unit, International Institute of Tropical Agriculture (IITA), Headquaters, PMB 5320, Oyo Road, Idi-Oshe, Ibadan-Nigeria. AGG: Australian Grain Genebank, Department of Economic Development, Jobs, Transport and Resources, Private Bag 260, Horsham, Victoria 3401.

Figure 2. Microphotographs illustrating seed morphology of some wild *Vigna* species; Four (4) seeds per accession were pictured under the same conditions to give an image of the morphology and the relative size. Distances of lines in the background are 1 cm in the vertical and horizontal directions. Source: Images taken and compiled by the authors based on seeds requested from the Australian Grain Genebank (AGG) (**a–e,q–t**) and the Genetic Resources Center, International Institute of Tropical Agriculture, (IITA), Ibadan-Nigeria (**f–p**).

2.2.1. Sample Collection

One hundred and sixty (160) accessions of wild *Vigna* species of legume were obtained from gene banks as presented in Table 1 with their details in Appendix A. All the accessions were planted in an experimental plot following the augmented block design arrangement [18] and allowed to grow until near maturity before inviting the farmers to explore their opinions. Since the accessions did not show uniform growth patterns due to their genetic differences, farmers were invited when more than 50% of the accessions reached maturity. An illustration of the seeds of some of the samples is also shown in Figure 2. In addition, three domesticated *Vigna* legumes—that is, cowpea (*V. unguiculata*), rice bean

(*V. umbellata*), and a semi-domesticated landrace (*V. vexillata*)—were used as checks. The checks were also obtained from the Genetic Resource Center (GRC-IITA), Nigeria, the National Bureau of Plant Genetic Resources (NBPGR), India and the Australian Grain Genebank (AGG), Australia respectively.

2.2.2. Experimental Design and Study Site

The study was conducted in two agro-ecological zones located at two research stations in Tanzania during the main cropping season (March-September 2018). One was at the TaCRI, located at Hai district, Moshi, Kilimanjaro region (latitude 3°13′59.59″ S, longitude 37°14′54″ E). The site is at an elevation of 1681 m above sea level, with a mean annual rainfall of 1200 mm and mean maximum and minimum temperatures of 21.7 °C and 13.6 °C, respectively. The second site was at the Tanzania Agricultural Research Institute (TARI), Selian Arusha in the northern part of Tanzania. TARI-Selian lies at latitude 3°21′50.08″ N and longitude 36°38′06.29″ E at an elevation of 1390 m above sea level (a.s.l.) with mean annual rainfall of 870 mm. The mean maximum and minimum temperatures ranged from 22 °C to 28 °C and 12 °C to 15 °C, respectively.

The 160 accessions of wild *Vigna* legumes were planted in an augmented block design field layout following the randomization generated by the statistical tool on the website (http://www.iasri.res.in/design/Augmented%20Designs/home.htm) [19] for 160 treatments with three checks. The field was monitored and maintained in good conditions from germination to near maturity of 75% of all the accessions before inviting farmers to assess their opinions.

2.2.3. Participants and Data Collection

Participants in the previous study (Study 1) in the Arusha ($N_1 = 50$) and Kilimanjaro ($N_2 = 100$) regions also participated in this study. Field visits were done in groups of five participants. A trained research assistant was recruited to guide the participants around the experimental field from the first to the last block or vice versa. A semi-structured questionnaire was used to collect information on the most preferred accessions (at least 10), and reasons for each selection were given. Every accession was assigned a number to ease participant selection. The number of times each accession was selected was divided by the total number of selections and multiplied by 100 to give the percentage of selection of each accession.

2.2.4. Focus Group Discussion

Participants in their respective regions were further grouped into two groups based on their gender, men and women, giving a total of four group interviews. Each group was invited to participate in an animated video-recorded focus group interview to ascertain their opinion about wild *Vigna* legumes, as obtained in the previous studies. The recorded videos (04) were transcribed verbatim and translated from Swahili language to English. The transcripts were cross-checked with the recordings by the interviewers to align transcripts with notes on non-verbal responses. A coding framework was developed based on the interview objectives and the interview guide. The qualitative data analysis package NVivo 11 (QRS International, 2015) was used to code and organize the data systematically as described by other workers [12]. Key concepts and categories were identified.

2.3. Data Analysis

For study I, the collected information during the survey was grouped, coded, organized, and analyzed using the statistical package IBM SPSS Statistic 20.0 (New York, NY, USA). Analysis consisted of the descriptive statistics as well as the binary logistic regression to test for the relationship between the prior knowledge about the wild *Vigna* legumes and the farmers' sociodemographic characteristics.

In the case of study II, data were coded and entered in the statistical package IBM SPSS Statistic 20.0 and analyzed. Analysis included descriptive statistics and likelihood ratio test of X^2 to determine the relationship between the preferences and the farmers' gender, farming experience, and research location [20].

3. Results

3.1. Study I

3.1.1. Sociodemographic Characteristics of Participants

The results from the sociodemographic characteristics showed that 64% and 36% were female and male farmers, respectively (Figure 3a). Most of the participants were above 45 years old, with the highest level of education being primary (Kilimanjaro) and secondary (Arusha). Furthermore, most of the farmers had a reasonable number of years of experience farming legumes, varying from two to more than 35 years of farming (Figure 3d). The intervals of years of farming experience and the percentages of participants with the longest farming experience were 6–10 and 16–20%, respectively (Figure 3d).

3.1.2. Prior Knowledge/Awareness about Wild Legumes

Less than 30% (28% and 26% in both study sites) of the experienced participants involved in the study were aware of the existence of wild legumes (Figure 4). According to the binary logistic regression analysis (Table 2), the model including the farmers' sociodemographic characteristics as explanatory variables and prior knowledge of legumes as a dependent variable is a good fit with the data as $p = 0.633 > 0.05$ (*Hosmer and Lemeshow test*). This explains that the variance in the outcome is significant ($X^2 = 40.632$, df = 19, p.003) (Omnibus Tests of Model Coefficients). The results show that there is no significant association between the prior knowledge about wild legumes and the overall gender (Wald = 0.495, df = 1, $p > 0.05$) (Table 2). However, there is a slightly effect associated with being a female farmer and prior knowledge (B = 0.303, $p = 0.482$). No significant relationship existed between the overall farmers' age groups and their prior knowledge of wild legumes (Wald = 7.061, df = 6, $p = 0.315 > 0.05$), although there is a slight significance relationship with the youngest age group [15–20] (Wald = 4.113, df = 1, B = 2.982, $p = 0.043$), as shown in Table 2. In the same vein, the test shows that the education level (Wald = 3.962, df = 4, $p = 0.411$) as well as their farming experience (Wald = 5.462, df = 7, $p = 0.604$) do not have any influence to their prior knowledge about wild legumes. On the contrary, the location (research site) has a significant effect on their prior knowledge of wild legumes (Wald = 9.884, df = 1, B = 1.687, $p = 0.002$).

3.1.3. Prior Uses of Wild Legumes

A few participants who had prior knowledge of wild legumes mentioned several uses attributed to the wild legumes they had seen before. Some of the uses mentioned were livestock feed, human food, and soil fertility ingredients as well as botanical pesticides (Table 3).

3.1.4. Challenges Faced by Legume Farmers

Diseases and drought (or reduced rainfall) were the most challenges faced by the farmers in both mid and high altitude agro-ecological zones (Figure 5). Apart from diseases and reduced rainfall issues, other reported challenges were related to market, pest, and storage (Figure 5). Taste and cooking aspects were not of very serious concern to the farmers in the two zones, since most of them seemed to be comfortable with the taste and cooking aspects of their legumes.

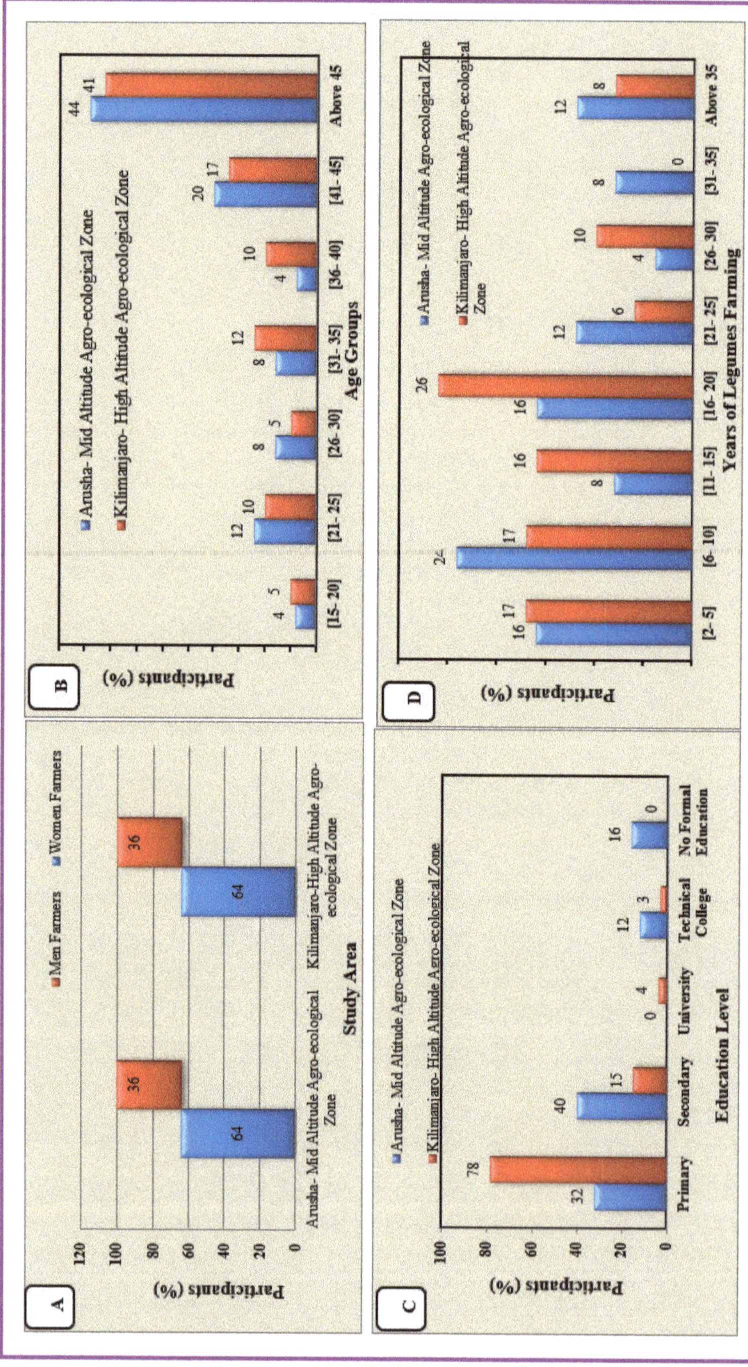

Figure 3. Sociodemographic characteristics of participants (**A**): participants' gender per study area (%); (**B**): participants' age groups; (**C**): participants' education level; and (**D**): participants' legumes farming experience.

3.2. Study II

3.2.1. Farmers' Preferred Accessions of Wild *Vigna* Legumes

The study shows that 74 accessions out of the 160 planted and grew to an appreciable level at the screening moment and were selected based on the participants' personal preferences (Figure 6). In the high-altitude zone (Kilimanjaro), only five (5) accessions (TVNu-293, TVNu-758, AGG308107WVIG 2, AGG308101WVIG 1, and TVNu-1546) were selected by the farmers more than half of the time, while in the mid-altitude zone (Arusha), none of the accessions had up to 50% selection (Figure 6). The five most selected accessions in the mid-altitude zone—TVNu-293 (36%), TVNu-758 (36%), AGG51603WVIG 1 (30%), AGG308099WVIG 2 (40%), and AGG53597WVIG 1 (34%)—were different from those selected in the high-altitude zone, except for TVNu-293 and TVNu-758.

The likelihood ratio test revealed that the wild *Vigna* selection (preferences) significantly depended on the farmers' gender ($G^2 = 130.813$, df = 73, $p < 0.000$), farming experience ($G^2 = 669.196$, df = 511, $p < 0.000$), and location ($G^2 = 1110.606$, df = 73, $p < 0.000$).

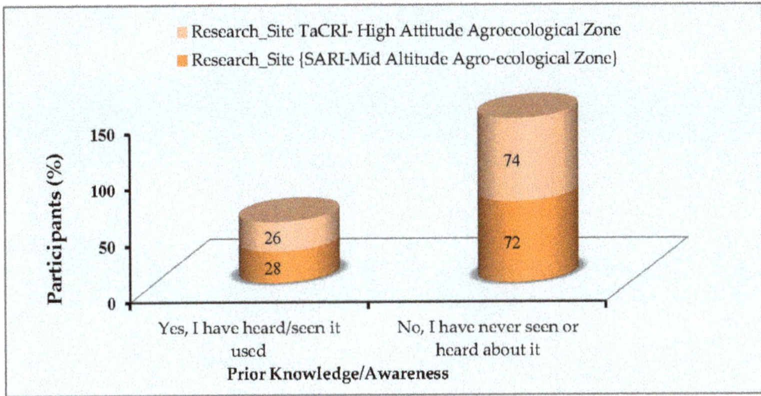

Figure 4. Participants' prior knowledge of wild legumes. TaCRI: Tanzania Coffee Research Institute.

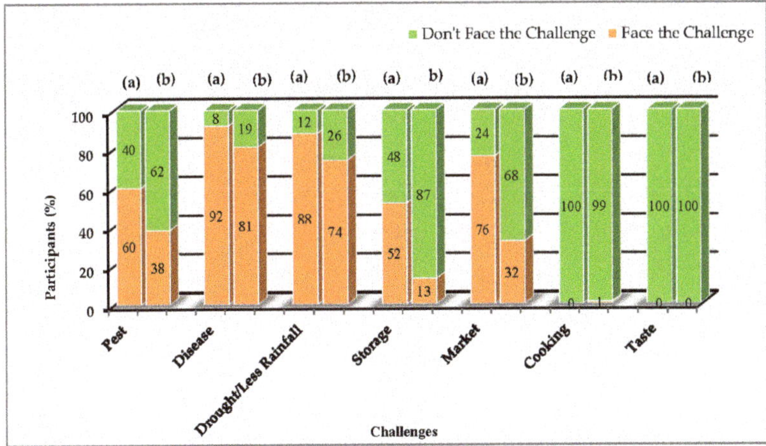

Figure 5. Participants' challenges faced during legumes cultivation in the two study areas: (**a**) Arusha and (**b**) Kilimanjaro.

Table 2. Binary logistic analysis result.

		Variables in the Equation							
		B	S.E.	Wald	df	Sig.	Exp(B)	95% C.I. for EXP(B)	
								Lower	Upper
Step 1 [a]	**Gender (1)**	0.303	0.431	0.495	1	0.482	1.354	0.582	3.153
	Age			7.061	6	0.315			
	Age (1)	2.982	1.471	4.113	1	0.043	19.732	1.105	352.281
	Age (2)	1.162	1.010	1.325	1	0.250	3.197	0.442	23.123
	Age (3)	1.755	1.124	2.440	1	0.118	5.786	0.639	52.342
	Age (4)	1.154	0.876	1.733	1	0.188	3.171	0.569	17.668
	Age (5)	1.010	0.798	1.601	1	0.206	2.745	0.575	13.111
	Age (6)	−0.255	0.622	0.168	1	0.681	0.775	0.229	2.620
	Education_Level			3.962	4	0.411			
	Level (1)	1.817	1.269	2.049	1	0.152	6.155	0.511	74.087
	Level (2)	2.334	1.285	3.299	1	0.069	10.316	0.831	127.995
	Level (3)	1.694	1.763	0.923	1	0.337	5.439	0.172	172.291
	Level (4)	1.407	1.504	0.876	1	0.349	4.084	0.214	77.805
	Research_Site (1)	1.687	0.537	9.884	1	0.002	5.402	1.887	15.460
	Farming_Experience			5.462	7	0.604			
	Experience (1)	−1.005	1.216	0.683	1	0.408	0.366	0.034	3.966
	Experience (2)	−1.245	1.118	1.242	1	0.265	0.288	0.032	2.573
	_Experience (3)	−1.222	1.022	1.430	1	0.232	0.295	0.040	2.183
	_Experience (4)	0.121	0.873	0.019	1	0.890	1.129	0.204	6.248
	_Experience (5)	0.409	1.025	0.159	1	0.690	1.505	0.202	11.216
	_Experience (6)	−0.559	0.998	0.313	1	0.576	0.572	0.081	4.046
	_Experience (7)	21.259	194,50.255	0.000	1	0.999	1,708,644,034.887	0.000	.
	Constant	−2.586	1.058	5.975	1	0.015	0.075		

[a]. Variable(s) entered on step 1: Gender, Age, Education_Level, Research_Site, Farming_Experience. B: represent the values for the logistic regression equation for predicting the dependent variables from the independent variables; S.E.: Standard errors associated with coefficients; Wald: Wald X^2 value; df: Degree of freedom for each of the tests of the coefficients; Sig.: Significance level (p-value); EXP(B): Exponentiation of the coefficients (odd ratios for the predictors); C.I.: Confidence Interval.

Table 3. Wild legumes uses as known by participants with prior knowledge of wild legumes. *

	Percentage (%)	Livestock Feed	Human Food	Soil Fertility Ingredient	Traditional Botanical Pesticides
Participants in a mid-altitude agro-ecological zone	28	12 Animal feed = 'Chakula cha mifugo', 'chakula cha ng'ombe'	16 Human food = 'Chakula cha binadamu', Vegetable = 'Mboga'	0	0
Participants in a high-altitude agro-ecological zone	26	4 Animal feed = 'Chakula cha mifugo', 'chakula cha ng'ombe'	4 Human food = 'chakula cha binadamu' Vegetable = 'Mboga'.	14 Rattlepod (Crotalaria ochroleuca) = 'Marejea' used as fertilizer = 'mbolea', Nourish the soil = 'Hurutubisha ardhi', Cover crop = 'Kutandaza shambani'	4 Pesticide = 'kunyunyuzia shambani', 'kutengeza dawa ya kunyunyuzia shambani'

* words in single quotation marks (' ') are exact expressions given by participants in Swahili, which has been translated.

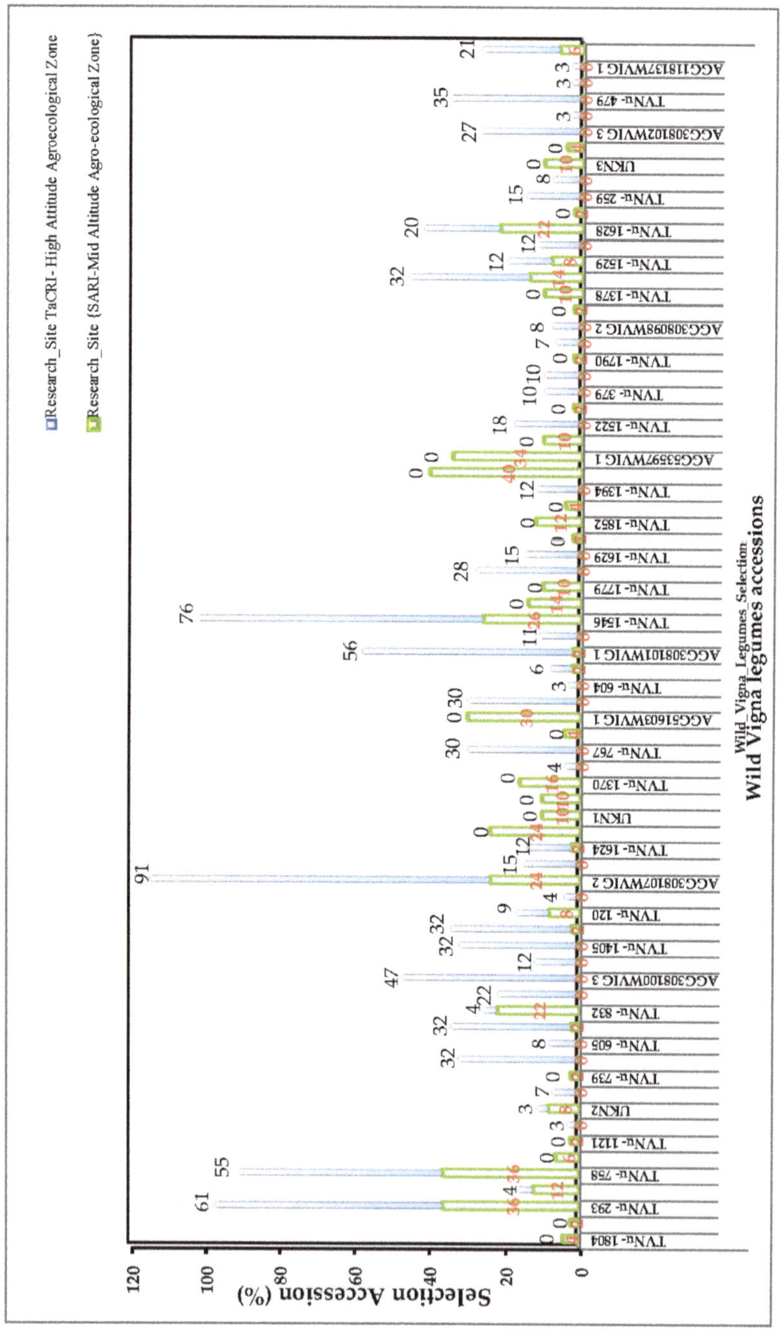

Figure 6. Wild *Vigna* legumes preferred (selected) by participants from the two agro-ecological zones.

3.2.2. Prospective Uses of Farmers' Preferred Accessions of Wild *Vigna* Legumes

The suggested uses of selected accessions were based on their personal assessment and preferences. Some accessions were selected for more than one use, and the number of selections for every accession is shown on Figure 7a–e. Other uses were proposed by farmers that better suited the accession of their choice. Four main uses (human food, animal feed, forage, and cover crop) were proposed as a result of farmer's preferences and perceptions. Therefore, a total of 31 accessions were preferred as human food (Figure 7a), 49 were preferred as animal feed (Figure 7b), 27 were preferred as forage (Figure 7c), 28 were preferred as cover crop (Figure 7d), and 44 were given specific personal uses (Figure 7e), respectively.

Four accessions were selected at least 30 times or more as human food, while 27 accessions were selected less than 30 times for the same purpose (Figure 7a). The four most selected accessions for this purpose were TVNu-1359 (36), AGG308099WVIG 2 (34), AGG53597WVIG 1 (32), and AGG51603WVIG 1 (30), respectively.

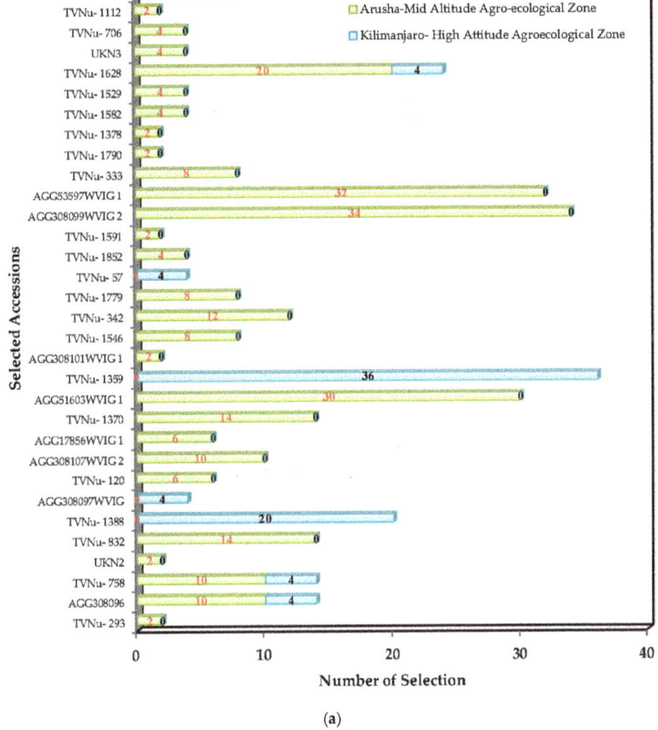

Figure 7. *Cont.*

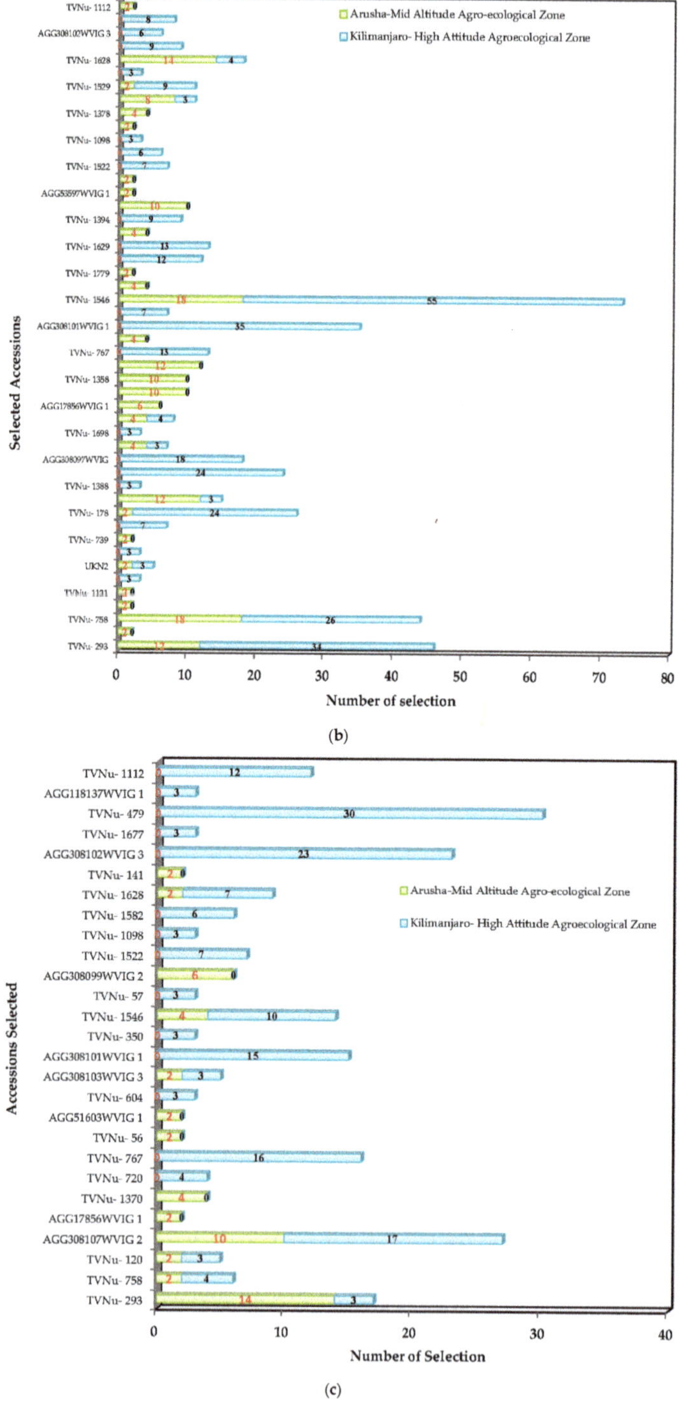

Figure 7. *Cont.*

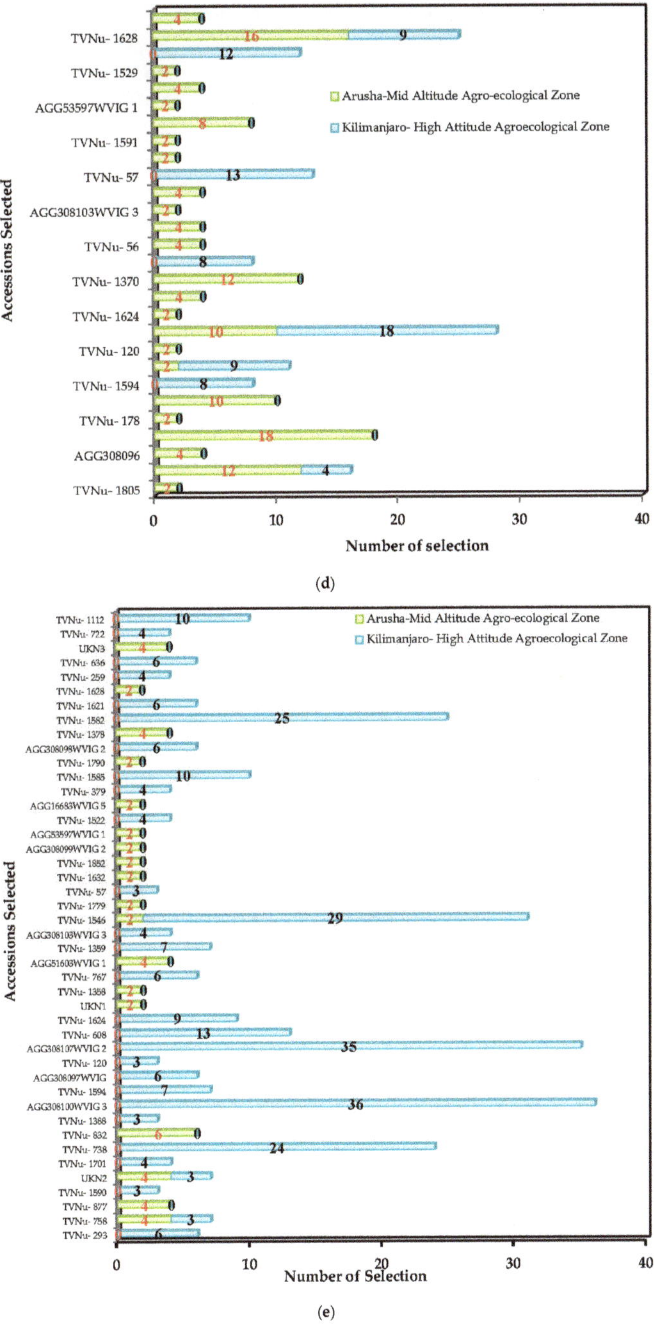

(d)

(e)

Figure 7. (**a**) Wild *Vigna* legumes suggested as a human food; (**b**) Wild *Vigna* legumes suggested as animal feed; (**c**) Wild *Vigna* legumes suggested as forage; (**d**) Wild *Vigna* legumes suggested as cover crop; and (**e**) Wild *Vigna* legumes given specified uses.

Four other accessions were also selected at least 30 times or more by participants as animal feed in the two study sites combined. The selected accessions were TVNu-1546 (18 + 55), TVNu-293 (12 + 34), TVNu-758 (18 + 26), and AGG308101WVIG 1 (35), respectively (Figure 7b).

Only one accession was selected up to 30 times to serve as forage (Figure 7c), while none of the preferred as cover crop accessions were chosen up to 30 times by the participants in both study sites (Figure 7d).

Out of the 44 selected accessions with specified uses, only two accessions—AGG308107WVIG 2 (35) and AGG308100WVIG 3 (36)—were selected more than 30 times (Figure 7e).

All of the non-domesticated wild *Vigna* legumes subjected to this study belonged to four species, *V. racemosa*, *V. reticulata*, *V. vexillata*, and *V. ambacensis*. In summary, it has been shown that the *V. vexillata* accessions were more preferred, followed by *V. reticulata* and *V. racemosa* (Figure 8). Despite the higher number of *V. ambacensis* accessions as compared with *V. racemosa*, it was less selected than *V. racemosa*.

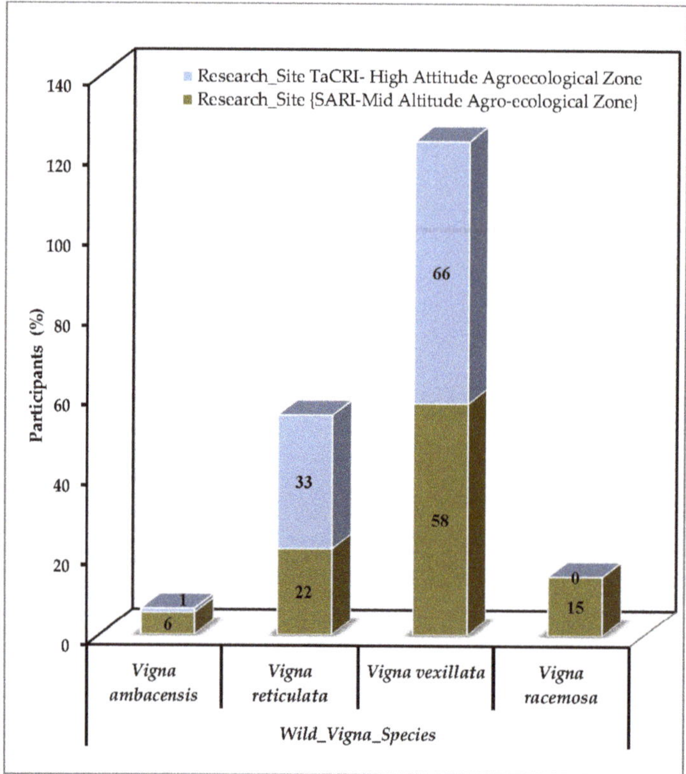

Figure 8. Wild *Vigna* legumes selected according to their species.

From their sight and appraisal of the wild *Vigna* legumes, other uses could be organic manure (locally known as '*Mbolea*'—fertilizer), business use, medicinal uses, preventers of soil erosion, and vegetable food for accessions with nice leaves (Figure 9). For personal uses, none of the accessions was selected up to 30 times or more. However, five accessions were selected more than 20 times at least for a specific use. The selected accessions were AGG308100WVIG 3 (24) and TVNu-738 (24) for soil erosion mitigation, and TVNu-1582 (22), TVNu-1546 (26), and AGG308107WVIG 2 (28) for soil fertility as an organic manure agent, respectively (Figure 9).

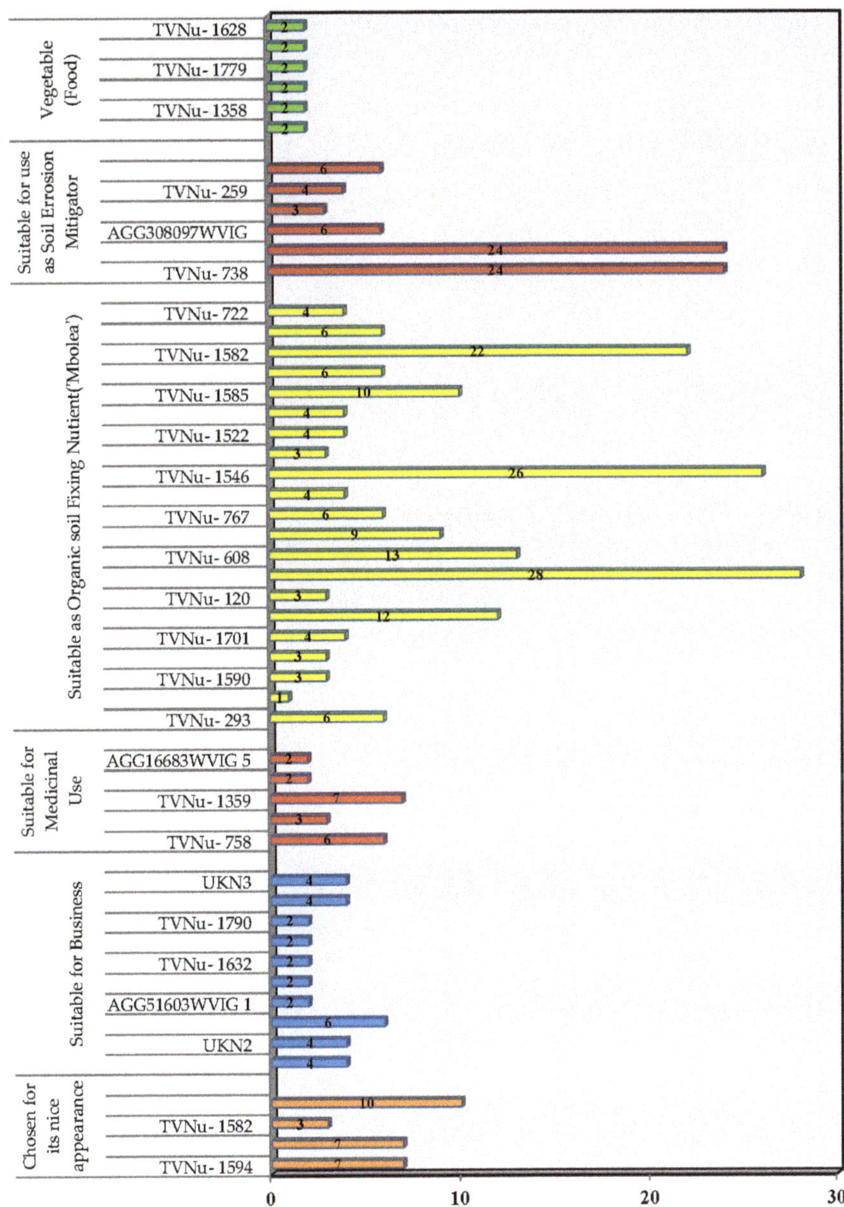

Figure 9. Specified uses of wild *Vigna* legumes as proposed by farmers in the two agro-ecological zones of Tanzania.

3.2.3. Farmers' Perception of Wild *Vigna* Legumes

From the focus group discussion, most farmers perceived some accessions of wild *Vigna* legumes as good material for future promising business in the field of agriculture due to their high seed production, resistance to drought conditions, and high production of leaves, which can benefit both humans and animals as forage. For example, a male farmer from the Arusha region during the group discussion enthusiastically responded when the interviewer asked whether they would be willing to adopt some of the wild *Vigna* legumes presented to them for the first time. He said: "Yes, I like some of these beans because many people don't know about them, they are found in the bush but people don't know that they can be useful, so if we discover their usefulness, this can be a great source of good business because they seem to have a higher productivity as compared with other known beans." To support the view, another voice rose in the hall and said: "Yes, I also like some, because after seeing these crops planted in the farm (referring to the wild legumes of study), I discovered that there are other new varieties of legumes, and this may be another source of food. I also realized that some of them have nice leaves that can be used as vegetables, and some can help us feed our cattle."

A smaller proportion of farmers (represented by 26% and 28% in study I, as shown in Figure 3), who curiously noticed the existence of wild legumes before the study, confirmed having seen some of the planted legumes of the study and having consumed them or used them as medicine for animals and even humans. One of the most interesting views that supported this point was from one of the old female farmers in the Arusha region, who said: "This variety with [a] large number of leaves lying on the ground (referring to one of the varieties of the study with a spreading growth habit), I have consumed them several times when I was a kid. Back then, our mothers used to go to the bush and harvest their leaves, and then go to town and buy maize and come back to cook them together. Myself, I have eaten them and we used to call that meal {*Ngolowo*}, which is very delicious and when we mix it with milk, it looks similar to another meal called {*Rojo*}. So for that one, it is not a poison, because I have eaten it before, it is a food, the leaves are eaten and the seeds are also eaten; it is called {*Ngolowo*}". All her mates in the hall during the group discussion listened to her speech with very attuned ears and clapped at the end. A similar view came from the group of men, which was articulated in these terms: "I have seen these beans before growing in the bush and we were using them as food and feed for animals; then, when I saw it here, I just confirmed that it is edible. Animals enjoy them so much. We used to take them from the bush and consume them and we had no health problems with them, and after I saw it here in the farm, I just realized that it is a normal food. It has never affected our people negatively after consuming them.

However, most of the participants in general proposed that more research and improvements were needed, especially in terms of the toxicity and nutritional benefits, as well as the seed color of the legumes to increase their acceptability for efficient exploitation and utilization. "One of the varieties I saw in the farm numbered 132 looks nice; it looks similar to (Choroko, Swahili word for Mung bean). So, I think that if it can be improved, it will be good for business because it has high productivity and nice leaves, but we don't know if it is not toxic or can negatively affect our health", said a participant who was supported by another one, who said: "similar to this one (participant showing some seeds harvested from the experimental fields), if the color can be improved, it will be very nice, because people in the market don't like buying black-colored beans. Their reason is that the black-colored seeds turn the cooking water black and that is not preferable for them. The black-colored seed beans are only preferred during hunger seasons; that is, seasons where less rainfall has affected the crop yield in the community."

4. Discussion

The explorative survey above shows that women were more engaged in legume farming in the two zones compared with men. Similarly, the contribution of women in agricultural activities is well-known in Africa [21]. In this study, no statistical significance was found between gender influence and prior knowledge about legumes. This means that being a woman or a man does not influence the probability of being aware of wild legumes.

Legume farming was mainly practiced by the older participants (Figure 3b). This indicates that the younger generations in the areas were not very interested in legume-farming activities or farming other crops. In general, belonging to any age group did not influence the prior knowledge about the legumes, due to the long period of disappearance of the wild genotypes, which led to the ignorance of many generations of people [2,8,11]. However, belonging to the 15 to 20-year-old age group showed a slight influence on the prior knowledge of wild legumes. This may suggest that farmers in this age range may possess some understanding of wild legumes.

The education level of farmers and their farming experience showed no significant influence on their prior knowledge of wild legumes, which meant that being educated or well experienced in farming legumes did not influence the knowledge of wild legumes. This showed that both experienced and non-experienced farmers as well as educated and non-educated farmers might have the same perception and prior background about wild legumes. In addition, it implied that both farming experience and level of education may not be necessary when making policy decisions about the implementation or adoption of a wild legume as a new crop. However, this is in contradiction with other studies carried out using other domesticated crops such as rice and maize [22,23]. Then, it is necessary for further research to try such experiences with other wild crops in other parts of the world to ascertain this fact.

From the results, the location (research site) has a significant effect on the prior knowledge of wild legumes, meaning that being in the Arusha region increased the chance of knowing wild legumes. Decision making regarding the adoption of wild *Vigna* legumes needs to take the location of farmers into consideration. This is in line with earlier reports [24]. This could be explained by the Arusha region being more populated by a certain ethnic group of people (called the Maasai) who are well-known in Tanzania for their indigenous ethno-medical knowledge of plants [25,26]. They are also found in the high-altitude agro-ecological zone (Kilimanjaro), but they are more concentrated in the mid-altitude agro-ecological zone of Arusha [25].

The ignorance of the wild legumes by the majority of participants in the two study sites may be due to the high and long-term distribution of bred, improved, and landrace varieties of legumes that led to the disappearance, rejection, and negligence of the original wild legumes [2]. However, the numerous challenges (biotic, abiotic, and policy) faced by the improved varieties have recently raised scientific concerns [3]. Therefore, it might be important to go back to the wild and investigate other legumes with good characteristics in relation to their acceptability in order to mitigate the global food insecurity challenge, as pointed out by earlier reports [27].

It is noted from this study that despite the high ignorance noted by the majority, the wild legumes are still used for various purposes, including human consumption by a minority. It has also been noted that ignorance or knowledge/awareness of wild legumes significantly depends on the location of the farmers rather than their gender, age group, or farming experience. This could be explained by some ethnic groups of people with significant traditional and indigenous knowledge of plants being concentrated in some parts of the world [25]. Then, it would be wise to carry out more investigation on such legumes in order to domesticate more varieties possessing resistance to the current legumes challenges. From this study, the main challenges experienced by legume farmers in the two study sites were diseases and low rainfall, which might definitely be due to climate change, as it is global challenge [28]. Therefore, alternatives varieties of legumes with resistance to climate variability and diseases would be of great benefit to such similar communities. The study also attempted to screen some accession of choice by the same farmers based on the general appearance, pods, and seeds of some of the wild legumes in order to select varieties for domestication.

Furthermore, it was observed that the prior knowledge about wild legumes is independent of gender, age, education level, and farming experience, but dependent on the farmers' location. However, it is curiously noted that after carefully sighting the wild *Vigna* legumes performing in the field by participants, it is revealed that there is a significant relationship between the farmers' preferences and their gender, farming experience, and location (likelihood ratio test). This could explain that the knowledge of wild legumes increases farmers' attraction and preferences of wild legumes depending on their gender, farming experience, and location. Less than 50% (74 out of 160) of the planted accessions were preferred by farmers in both research sites (Figure 6), showing that several accessions had common preferences depending on the locations. Although this could be influenced by the number of accessions that reached an appreciable growth level by the selection period, the selection should depend on other parameters such as farmers' gender (G^2 = 130.813, df = 73, p < 0.000), farming experience (G^2 = 669.196, df = 511, p < 0.000), and location (G^2 = 1110.606, df = 73, p < 0.000), as confirmed by the X^2 test. In a similar study, significant correlations between preferences of male and female farmers in an on-farm trial indicated that both groups have similar criteria for the selection of rice varieties in India [29]. Experiments investigating farmers' knowledge about unknown or wild food crops are lacking or almost non-existent in the literature [12]. The wild *Vigna* species are not well-known legumes, which could be the reason taxonomic characterizations have still been under investigation by scientists until recently [30].

The ignorance of wild legumes by the majority explains the few uses suggested by the farmers as compared with the uses suggested after field visits to farms with wild *Vigna* legumes (Figure 4 and Table 3). Several uses have been suggested by farmers after sighting the wild *Vigna* legumes in farms, showing their interest and motivation to adopt some of the wild crops for human benefit. This is in accordance with findings from earlier research studies carried out with domesticated legumes possessing characteristics that are not well-known [31,32]. It was observed that the farmers were willing to adopt some of the crops for several human exploitation purposes, although some need more improvement. It is also noted that some farmers even had experience consuming some of the wild *Vigna* legumes. Therefore, farmers generally perceived the wild *Vigna* legumes as exploitable resources for a variety of purposes that lack awareness and scientific attention. A recent report also demonstrated participant eagerness to adopt wild vegetables (duckweed) as human food upon first-time observations from a picture [12].

This study also shows that there is a high probability that any sample of farmers taken in Tanzania and any other region of the world would ignore the existence of wild legumes. Therefore, considering food insecurity levels in the developing world, the dependence on a few accessions of legumes, and the challenges faced by farmers and consumers regarding domesticated legumes, there is a need to further study these un-exploited legumes and orient their utilization. Very limited reports approaching the assessment of participants, farmers, or consumers' perception, appreciation, or adoption of wild plants as human food exist.

5. Conclusions

The existence of non-domesticated wild legumes is highly ignored by many farmers despite the presence of a large existing number in the gene banks and bushes around the world. The ignorance of wild legumes is generally not related to the farmers' gender, age group, or farming experience, while it is significantly related to their location. Besides, preferences in some accessions of wild legumes depend on the gender, farming experience, and location. In addition, the discovery of the wild *Vigna* legumes for the first time motivates the attraction of farmers to prefer them for various purposes. Farmers perceived wild *Vigna* legumes as human food, animal feed, medicinal plants, soil enrichment material, and soil erosion-preventing materials. Therefore, it is necessary for the scientific community to give better attention to these so-called alien species in order to improve their agronomic, nutritional, and physiological characteristics with prior consideration of farmers' and consumers' preferences and perception to orient their domestication, as it is the case here.

Author Contributions: P.A.N. conceived and designed the experiments; D.V.H. performed the experiments, collected data, analyzed the data, and wrote the first draft of the manuscript; P.B.V. and A.O.M. supervised the research and internally reviewed the manuscript; and P.A.N. made the final internal review and revised the final draft of the manuscript.

Funding: This research was partially funded by the Centre for Research, Agricultural Advancement, Teaching Excellence and Sustainability in Food and Nutrition Security (CREATES-FNS) through the Nelson Mandela African Institution of Science and Technology (NM-AIST) under the reference number P220/CAM.16. The research also received funding support from the International Foundation for Science (IFS) through the grant number I-3-B-6203-1.

Acknowledgments: This research was partially funded by the Centre for Research, Agricultural Advancement, Teaching Excellence and Sustainability in Food and Nutrition Security (CREATES-FNS) through the Nelson

Mandela African Institution of Science and Technology (NM-AIST). The authors acknowledge the additional funding support from the International Foundation for Science (IFS) through the grant number I-3-B-6203-1. The authors also acknowledge the Genetic Resources Center, International Institute of Tropical Agriculture (IITA), Ibadan-Nigeria as well as the Australian Grains Genebank (AGG) for providing detailed information on wild *Vigna* and seed materials for this research. The support this study received from the N2 Africa project based at the Nelson Mandela African Institution of Science and Technology (NM-AIST) is also well acknowledged. The authors are thankful to Frank E. Mmbando, Theresia L. Gregory, and Lameck Makoye of the Tanzania Agricultural Research Institute (TARI), Selian-Arusha, Tanzania for their technical support in recruiting farmers and reviewing the questionnaire. The authors also thank Djomo Raoul Fani (Department of Agricultural Economics, Federal University of Agriculture, Nigeria) and Elmugheira Mockarram Ibrahim Mohammed (Department of Forest Management Sciences, Faculty of Forest Sciences & Technology, University of Gezira, Sudan) for their technical advises regarding the data analysis.

Conflicts of Interest: The authors declare no conflict of interest.

Appendix A

Table A1. Wild *Vigna* legumes accessions used in the study.

S/N	Accession Number	Species Name	Genebank
1	TVNu-313	*Vigna ambacensis*	GRC, IITA
2	TVNu-557	*Vigna ambacensis*	GRC, IITA
3	TVNu-1186	*Vigna ambacensis*	GRC, IITA
4	TVNu-375	*Vigna ambacensis*	GRC, IITA
5	TVNu-1212	*Vigna ambacensis*	GRC, IITA
6	TVNu-1792	*Vigna ambacensis*	GRC, IITA
7	TVNu-947	*Vigna ambacensis*	GRC, IITA
8	TVNu-1679	*Vigna ambacensis*	GRC, IITA
9	TVNu-1840	*Vigna ambacensis*	GRC, IITA
10	TVNu-219	*Vigna ambacensis*	GRC, IITA
11	TVNu-720	*Vigna ambacensis*	GRC, IITA
12	TVNu-877	*Vigna ambacensis*	GRC, IITA
13	TVNu-706	*Vigna ambacensis*	GRC, IITA
14	TVNu-216	*Vigna ambacensis*	GRC, IITA
15	TVNu-722	*Vigna ambacensis*	GRC, IITA
16	TVNu-1631	*Vigna ambacensis*	GRC, IITA
17	TVNu-1677	*Vigna ambacensis*	GRC, IITA
18	TVNu-1791	*Vigna ambacensis*	GRC, IITA
19	TVNu-765	*Vigna ambacensis*	GRC, IITA
20	TVNu-1843	*Vigna ambacensis*	GRC, IITA
21	TVNu-629	*Vigna ambacensis*	GRC, IITA
22	TVNu-452	*Vigna ambacensis*	GRC, IITA
23	TVNu-1185	*Vigna ambacensis*	GRC, IITA
24	TVNu-342	*Vigna ambacensis*	GRC, IITA
25	TVNu-1125	*Vigna ambacensis*	GRC, IITA
26	TVNu-1678	*Vigna ambacensis*	GRC, IITA
27	TVNu-223	*Vigna ambacensis*	GRC, IITA
28	TVNu-1644	*Vigna ambacensis*	GRC, IITA
29	TVNu-1781	*Vigna ambacensis*	GRC, IITA
30	TVNu-1851	*Vigna ambacensis*	GRC, IITA
31	TVNu-1069	*Vigna ambacensis*	GRC, IITA
32	TVNu-456	*Vigna ambacensis*	GRC, IITA
33	TVNu-148	*Vigna ambacensis*	GRC, IITA
34	TVNu-3	*Vigna ambacensis*	GRC, IITA
35	TVNu-1827	*Vigna ambacensis*	GRC, IITA
36	TVNu-1691	*Vigna ambacensis*	GRC, IITA
37	TVNu-1804	*Vigna ambacensis*	GRC, IITA
38	TVNu-1699	*Vigna ambacensis*	GRC, IITA

Table A1. *Cont.*

S/N	Accession Number	Species Name	Genebank
39	TVNu-1184	*Vigna ambacensis*	GRC, IITA
40	TVNu-374	*Vigna ambacensis*	GRC, IITA
41	TVNu-1150	*Vigna ambacensis*	GRC, IITA
42	TVNu-1213	*Vigna ambacensis*	GRC, IITA
43	AGG52867WVIG 1	*Vigna racemosa*	AGG
44	AGG51603WVIG 1	*Vigna racemosa*	AGG
45	AGG53597WVIG 1	*Vigna racemosa*	AGG
46	AGG60436WVIG 1	*Vigna racemosa*	AGG
47	*Unknown Vigna racemosa*	*Vigna racemosa*	Self-collected
48	AGG60441WVIG 1	*Vigna reticulata*	AGG
49	AGG17856WVIG 1	*Vigna reticulata*	AGG
50	AGG118137WVIG 1	*Vigna reticulata*	AGG
51	TVNu-259	*Vigna reticulata*	GRC, IITA
52	TVNu-302	*Vigna reticulata*	GRC, IITA
53	TVNu-161	*Vigna reticulata*	GRC, IITA
54	TVNu-1790	*Vigna reticulata*	GRC, IITA
55	TVNu-138	*Vigna reticulata*	GRC, IITA
56	TVNu-604	*Vigna reticulata*	GRC, IITA
57	TVNu-1112	*Vigna reticulata*	GRC, IITA
58	TVNu-312	*Vigna reticulata*	GRC, IITA
59	TVNu-224	*Vigna reticulata*	GRC, IITA
60	TVNu-1394	*Vigna reticulata*	GRC, IITA
61	TVNu-995	*Vigna reticulata*	GRC, IITA
62	TVNu-1405	*Vigna reticulata*	GRC, IITA
63	TVNu-1522	*Vigna reticulata*	GRC, IITA
64	TVNu-379	*Vigna reticulata*	GRC, IITA
65	TVNu-609	*Vigna reticulata*	GRC, IITA
66	TVNu-1191	*Vigna reticulata*	GRC, IITA
67	TVNu-766	*Vigna reticulata*	GRC, IITA
68	TVNu-343	*Vigna reticulata*	GRC, IITA
69	TVNu-349	*Vigna reticulata*	GRC, IITA
70	TVNu-916	*Vigna reticulata*	GRC, IITA
71	TVNu-758	*Vigna reticulata*	GRC, IITA
72	TVNu-491	*Vigna reticulata*	GRC, IITA
73	TVNu-767	*Vigna reticulata*	GRC, IITA
74	TVNu-608	*Vigna reticulata*	GRC, IITA
75	TVNu-1808	*Vigna reticulata*	GRC, IITA
76	TVNu-1825	*Vigna reticulata*	GRC, IITA
77	TVNu-1852	*Vigna reticulata*	GRC, IITA
78	TVNu-1698	*Vigna reticulata*	GRC, IITA
79	TVNu-932	*Vigna reticulata*	GRC, IITA
80	TVNu-450	*Vigna reticulata*	GRC, IITA
81	TVNu-524	*Vigna reticulata*	GRC, IITA
82	TVNu-605	*Vigna reticulata*	GRC, IITA
83	TVNu-1156	*Vigna reticulata*	GRC, IITA
84	TVNu-607	*Vigna reticulata*	GRC, IITA
85	TVNu-1779	*Vigna reticulata*	GRC, IITA
86	TVNu-325	*Vigna reticulata*	GRC, IITA
87	TVNu-324	*Vigna reticulata*	GRC, IITA
88	TVNu-57	*Vigna reticulata*	GRC, IITA
89	TVNu-56	*Vigna reticulata*	GRC, IITA
90	TVNu-1520	*Vigna reticulata*	GRC, IITA

Table A1. *Cont.*

S/N	Accession Number	Species Name	Genebank
91	TVNu-602	*Vigna reticulata*	GRC, IITA
92	TVNu-1388	*Vigna reticulata*	GRC, IITA
93	TVNu-141	*Vigna reticulata*	GRC, IITA
94	TVNu-738	*Vigna reticulata*	GRC, IITA
95	TVNu-739	*Vigna reticulata*	GRC, IITA
96	TVNu-350	*Vigna reticulata*	GRC, IITA
97	TVNu-142	*Vigna reticulata*	GRC, IITA
98	TVNu-1805	*Vigna reticulata*	GRC, IITA
99	*Unknown Vigna reticulata*	*Vigna reticulata*	Self-collected
100	AGG308102WVIG 3	*Vigna vexillata*	AGG
101	AGG308105WVIG 2	*Vigna vexillata*	AGG
102	AGG308098WVIG 2	*Vigna vexillata*	AGG
103	AGG16683WVIG 5	*Vigna vexillata*	AGG
104	AGG308099WVIG 2	*Vigna vexillata*	AGG
105	AGG308097WVIG 1	*Vigna vexillata*	AGG
106	AGG308101WVIG 1	*Vigna vexillata*	AGG
107	AGG308100WVIG 3	*Vigna vexillata*	AGG
108	AGG58678WVIG 2	*Vigna vexillata*	AGG
109	AGG308103WVIG 3	*Vigna vexillata*	AGG
110	AGG308107WVIG 2	*Vigna vexillata*	AGG
111	AGG308096 WVIG 2	*Vigna vexillata*	AGG
112	AGG62154WVIG 1	*Vigna vexillata*	AGG
113	TVNu-1098	*Vigna vexillata*	GRC, IITA
114	TVNu-1629	*Vigna vexillata*	GRC, IITA
115	TVNu-1718	*Vigna vexillata*	GRC, IITA
116	TVNu-1590	*Vigna vexillata*	GRC, IITA
117	TVNu-1378	*Vigna vexillata*	GRC, IITA
118	TVNu-120	*Vigna vexillata*	GRC, IITA
119	TVNu-178	*Vigna vexillata*	GRC, IITA
120	TVNu-1796	*Vigna vexillata*	GRC, IITA
121	TVNu-1529	*Vigna vexillata*	GRC, IITA
122	TVNu-1092	*Vigna vexillata*	GRC, IITA
123	TVNu-1546	*Vigna vexillata*	GRC, IITA
124	TVNu-1370	*Vigna vexillata*	GRC, IITA
125	TVNu-1626	*Vigna vexillata*	GRC, IITA
126	TVNu-1358	*Vigna vexillata*	GRC, IITA
127	TVNu-1624	*Vigna vexillata*	GRC, IITA
128	TVNu-1585	*Vigna vexillata*	GRC, IITA
129	TVNu-1617	*Vigna vexillata*	GRC, IITA
130	TVNu-1621	*Vigna vexillata*	GRC, IITA
131	TVNu-479	*Vigna vexillata*	GRC, IITA
132	TVNu-1344	*Vigna vexillata*	GRC, IITA
133	TVNu-1628	*Vigna vexillata*	GRC, IITA
134	TVNu-381	*Vigna vexillata*	GRC, IITA
135	TVNu-792	*Vigna vexillata*	GRC, IITA
136	TVNu-1586	*Vigna vexillata*	GRC, IITA
137	TVNu-1582	*Vigna vexillata*	GRC, IITA
138	TVNu-293	*Vigna vexillata*	GRC, IITA
139	TVNu-1359	*Vigna vexillata*	GRC, IITA
140	TVNu-955	*Vigna vexillata*	GRC, IITA
141	TVNu-1591	*Vigna vexillata*	GRC, IITA
142	TVNu-1701	*Vigna vexillata*	GRC, IITA
143	TVNu-1443	*Vigna vexillata*	GRC, IITA

Table A1. *Cont.*

S/N	Accession Number	Species Name	Genebank
144	TVNu-832	*Vigna vexillata*	GRC, IITA
145	TVNu-1121	*Vigna vexillata*	GRC, IITA
146	TVNu-636	*Vigna vexillata*	GRC, IITA
147	TVNu-1476	*Vigna vexillata*	GRC, IITA
148	TVNu-1748	*Vigna vexillata*	GRC, IITA
149	TVNu-781	*Vigna vexillata*	GRC, IITA
150	TVNu-969	*Vigna vexillata*	GRC, IITA
151	TVNu-1592	*Vigna vexillata*	GRC, IITA
152	TVNu-1632	*Vigna vexillata*	GRC, IITA
153	TVNu-333	*Vigna vexillata*	GRC, IITA
154	TVNu-1360	*Vigna vexillata*	GRC, IITA
155	TVNu-1594	*Vigna vexillata*	GRC, IITA
156	TVNu-1369	*Vigna vexillata*	GRC, IITA
157	TVNu-593	*Vigna vexillata*	GRC, IITA
158	TVNu-1593	*Vigna vexillata*	GRC, IITA
159	TVNu-837	*Vigna vexillata*	GRC, IITA
160	*Unknown*	*Vigna*	Self-collected, NM-AIST, Tanzania

References

1. Bhat, R.; Karim, A.A. Exploring the nutritional potential of wild and underutilized legumes. *Compr. Rev. Food Sci. Food Saf.* **2009**, *8*, 305–331. [CrossRef]

2. Harouna, D.V.; Venkataramana, P.B.; Ndakidemi, P.A.; Matemu, A.O. Under-exploited wild *Vigna* species potentials in human and animal nutrition: A review. *Glob. Food Sec.* **2018**, *18*, 1–11. [CrossRef]

3. Ojiewo, C.; Saxena, R.K.; Monyo, E.; Boukar, O.; Mukankusi-mugisha, C.; Thudi, M.; Pandey, M.K.; Gaur, P.M.; Chaturvedi, S.; Varshney, R.K.; et al. Genomics, genetics and breeding of tropical legumes for better livelihoods of smallholder farmers. *Plant Breed.* **2018**, *00*, 1–13. [CrossRef]

4. Garcia, H.E.; Pena-Valdivia, B.C.; Rogelio, J.R.A.; Muruaga, S.M.J. Morphological and agronomic traits of a wild population and an improved cultivar of common bean (*Phaseolus vulgaris* L.). *Ann. Bot.* **1997**, *79*, 207–213. [CrossRef]

5. Gepts, P. Origins of plant agriculture and major crop plants. In *Our Fragile World, Forerunner Volumes to the Encyclopedia of Life-Supporting Systems*; Tolba, M.K., Ed.; Eolss Publishers: Oxford, UK, 2001; pp. 629–637.

6. Padulosi, S.; Thompson, J.; Rudebjer, P. *Fighting Poverty, Hunger and Malnutrition with Neglected and Underutilized Species (NUS): Needs, Challenges and the Way Forward*; Bioversity International: Rome, Italy, 2013; ISBN 9789290439417.

7. Pratap, A.; Malviya, N.; Tomar, R.; Gupta, D.S.; Jitendra, K. Vigna. In *Alien Gene Transfer in Crop Plants, Volume 2: Achievements and Impacts*; Pratap, A., Kumar, J., Eds.; Springer Science+Business Media, LLC: Berlin/Heidelberg, Germany, 2014; pp. 163–189.

8. Tomooka, N.; Naito, K.; Kaga, A.; Sakai, H.; Isemura, T.; Ogiso-Tanaka, E.; Iseki, K.; Takahashi, Y. Evolution, Domestication and Neo-Domestication of the Genus *Vigna*. *Plant Genet. Resour. Charact. Util.* **2014**, *12*, S168–S171. [CrossRef]

9. Tomooka, N.; Kaga, A.; Isemura, T.; Vaughan, D.; Srinives, P.; Somta, P.; Thadavong, S.; Bounphanousay, C.; Kanyavong, K.; Inthapanya, P. Vigna Genetic Resources. In Proceedings of the 14th NIAS International Workshop on Genetic Resources: Genetic Resources and Comparative Eenomics of Legumes (*Glycine* and *Vigna*), Tsukuba, Japan, 14 September 2009; pp. 11–21.

10. Mba, C.; Guimaraes, E.P.; Ghosh, K. Re-orienting crop improvement for the changing climatic conditions of the 21st century. *Agric. Food Secur.* **2012**. [CrossRef]

11. Briggs, H. IUCN Red List: Wild Crops Listed as Threatened—BBC News. Available online: http://www.bbc.com/news/science-environment-42204575 (accessed on 11 February 2018).

12. De Beukelaar, M.F.A.; Zeinstra, G.G.; Mes, J.J.; Fischer, A.R.H. Duckweed as human food. The influence of meal context and information on duckweed acceptability of Dutch consumers. *Food Qual. Prefer.* **2019**, *71*, 76–86. [CrossRef]

13. Macorni, E.; Ruggeri, S.; Carnovale, E. Chemical Evaluation of wild under-exploited *Vigna* spp. seeds. *Food Chem.* **1997**, *59*, 203–212.

14. Binagwa, P.; Kilango, M.; Williams, M.; Rubyogo, J.C. Bean research and development in Tanzania (Including Bean Biofortification). In Proceedings of the BNFB Inception Meeting, Arusha, Tanzania, 16–18 March 2016; SARI: Timur, Indonesia; NZARDI: Pabra, India; CIAT: London, UK, 2016.

15. MAFC; MLDF; MWI; Ministry of Agriculture, Livestock and Environment, Zanzibar; PMO; RALG; MITM; NBS; the Office of the Chief Government Statistician, Zanzibar. *National Sample Census of Agriculture 2007/2008 Regional Report: Arusha Region*; The Ministry of Agriculture Food Security and Cooperatives: Dar es Salaam, Tanzania, 2012; Volume Vb.

16. MAFC; MLDF; MWI; Ministry of Agriculture, Livestock and Environment, Zanzibar; PMO; RALG; MITM; NBS; the Office of the Chief Government Statistician, Zanzibar. *National Sample Census of Agriculture 2007/2008 Regional Report: Kilimanjaro Region*; The Ministry of Agriculture Food Security and Cooperatives: Dar es Salaam, Tanzania, 2012; Volume Vc.

17. MAFSC. *Agriculture Climate Resilience Plan*; Chiza, C., Kaduma, S.E., Eds.; The United Republic of Tanzania Ministry of Agriculture, Food Security and Cooperatives: Dar es Salaam, Tanzania, 2014.

18. Bisht, I.S.; Bhat, K.V.; Lakhanpaul, S.; Latha, M.; Jayan, P.K.; Biswas, B.K.; Singh, A.K. Diversity and genetic resources of wild *Vigna* species in India. *Genet. Resour. Crop Evol.* **2005**, *52*, 53–68. [CrossRef]

19. IASRI Augmented Block Designs. Available online: http://www.iasri.res.in/design/AugmentedDesigns/home.htm (accessed on 15 May 2019).

20. Mchugh, M.L. The X^2 of independence Lessons in biostatistics. *Biochem. Med.* **2013**, *23*, 143–149. [CrossRef]

21. Njuki, J.; Kruger, E.; Starr, L. *Increasing the Productivity and Empowerment of Women Smallholder Farmers Results of a Baseline Assessment from Six Countries in Africa and Asia*; Care: Atlanta, GA, USA, 2013.

22. Hussein, S.; Alhassan, A.; Salifu, K. Determinants of farmers adoption of improved maize varieties in the wa University for development studies. *Am. Int. J. Contemp. Res.* **2015**, *5*, 27–35.

23. Himire, R.G.; Wen-chi, H.U.; Hrestha, R.B.S. Factors affecting adoption of improved rice varieties among rural farm households in central Nepal. *Agrekon* **2015**, *22*, 35–43.

24. Mwangi, H.W.; Kihurani, A.W.; Wesonga, J.M.; Ariga, E.S.; Kanampiu, F. Factors influencing adoption of cover crops for weed management in Machakos and Makueni counties of Kenya. *Eur. J. Agron.* **2015**, *69*, 1–9. [CrossRef]

25. Ibrahim, F.; Ibrahim, B. The maasai herbalists in Arusha town, Tanzania. *GeoJournal* **1998**, *46*, 141–154. [CrossRef]

26. Ibrahim, H.A.; Halliru, S.N.; Usaini, S.; Abdullahi, I.I. Ethnobotanical survey of the wild edible food plants consumption among local communities in Kano state, north-western, Nigeria. *Int. J. Sci. Technol.* **2012**, *2*, 713–717.

27. Porch, T.; Beaver, J.; Debouck, D.; Jackson, S.; Kelly, J.; Dempewolf, H. Use of wild relatives and closely related species to adapt common bean to climate change. *Agronomy* **2013**, *3*, 433–461. [CrossRef]

28. Boukar, O.; Close, T.J.; Roberts, P.A.; Fatokun, C.A.; Huynh, B.-L. Genomic tools in Cowpea breeding programs: Status and perspectives. *Front. Plant Sci.* **2016**, *7*, 1–13. [CrossRef]

29. Burman, D.; Maji, B.; Singh, S.; Mandal, S.; Sarangi, S.K.; Bandyopadhyay, B.K.; Bal, A.R.; Sharma, D.K.; Krishnamurthy, S.L.; Singh, H.N.; et al. Participatory evaluation guides the development and selection of farmers' preferred rice varieties for salt- and flood-affected coastal deltas of South and Southeast Asia. *Field Crop. Res.* **2017**, *220*, 67–77. [CrossRef]

30. Gore, P.G.; Tripathi, K.; Pratap, A.; Bhat, K.V.; Umdale, S.D.; Gupta, V.; Pandey, A. Delineating taxonomic identity of two closely related *Vigna* species of section Aconitifoliae: V. trilobata (L.) Verdc. and V. stipulacea (Lam.) Kuntz in India. *Genet. Resour. Crop Evol.* **2019**, *66*, 1155–1165. [CrossRef]

31. Bruno, A.; Katungi, E.; Nkalubo, S.; Mukankusi, C.; Malinga, G.; Gibson, P.; Rubaihayo, P.; Edema, R. Participatory farmers' selection of common bean varieties (*Phaseolus vulgaris* L.) under different production constraints. *Plant Breed.* **2018**, *137*, 283–289. [CrossRef]
32. Bekele, W. Participatory variety selection of Faba bean for yield components and yield at highlands of West Hararghe, Eastern Ethiopia. *Int. J. Plant Breed. Crop Sci.* **2016**, *3*, 99–102.

 *agronomy*

Review

Local Solutions for Sustainable Food Systems: The Contribution of Orphan Crops and Wild Edible Species

Teresa Borelli [1,*], Danny Hunter [1], Stefano Padulosi [1], Nadezda Amaya [1], Gennifer Meldrum [1], Daniela Moura de Oliveira Beltrame [2], Gamini Samarasinghe [3], Victor W. Wasike [4], Birgül Güner [5], Ayfer Tan [6], Yara Koreissi Dembélé [7], Gaia Lochetti [1], Amadou Sidibé [7] and Florence Tartanac [8]

[1] Alliance of Bioversity International and CIAT, HQ, via dei Tre Denari 472/a, 00054 Rome, Italy; D.Hunter@cgiar.org (D.H.); S.Padulosi@cgiar.org (S.P.); N.Amaya@cgiar.org (N.A.); G.Meldrum@cgiar.org (G.M.); G.Lochetti@cgiar.org (G.L.)

[2] Biodiversity for Food and Nutrition Project, Ministry of the Environment, CEP 70068-900 Brasília-DF, Brazil; dani.moura.oliveira@gmail.com

[3] Horticultural Crops Research and Development Institute, B365, Peradeniya 20400, Sri Lanka; gaminisam@yahoo.com

[4] Kenya Agricultural and Livestock Research Organization, P.O. Box 57811-00200, Nairobi, Kenya; victor.wasike@kalro.org

[5] General Directorate of Agricultural Research and Policies, Ministry of Agriculture and Forestry, 06800 Ankara, Turkey; birgul.guner@tarimorman.gov.tr

[6] Plant Genetic Resources Department, Aegean Agricultural Research Institute, P.O. Box 9 Menemen, 35661 Izmir, Turkey; ayfer_tan@yahoo.com

[7] Institut d'Economie Rurale, BP 438 Bamako, Mali; ykoreissidemb@gmail.com (Y.K.D.); amadousidibe57@yahoo.fr (A.S.)

[8] Food and Agriculture Organization of the United Nations, 00153 Rome, Italy; Florence.Tartanac@fao.org

* Correspondence: t.borelli@cgiar.org

Received: 7 January 2020; Accepted: 1 February 2020; Published: 5 February 2020

Abstract: Calls for a global food system transformation and finding more sustainable ways of producing healthier, safe and nutritious food for all have spurred production approaches such as sustainable intensification and biofortification with limited consideration of the copious amounts of orphan crops, traditional varieties and wild edible species readily available in many countries, mostly in and around smallholder farmers' fields. This paper explores the potential role of locally available, affordable and climate-resilient orphan crops, traditional varieties and wild edible species to support local food system transformation. Evidence from Brazil, Kenya, Guatemala, India, Mali, Sri Lanka and Turkey is used to showcase a three-pronged approach that aims to: (i) increase evidence of the nutritional value and biocultural importance of these foods, (ii) better link research to policy to ensure these foods are considered in national food and nutrition security strategies and actions, and (iii) improve consumer awareness of the desirability of these alternative foods so that they may more easily be incorporated in diets, food systems and markets. In the seven countries, this approach has brought about positive changes around increasing community dietary diversity and increasing market opportunities for smallholder growers, as well as increased attention to biodiversity conservation.

Keywords: orphan crops; neglected and underutilized species; wild edibles; biodiversity; food composition; nutrition; policy

---

1. Introduction

Global food production is at a critical juncture. According to the Food and Agriculture Organization of the United Nations (FAO), the world produces food in sufficient quantity to satisfy global food demand [1]; yet, in 2017, an estimated 821 million people in the world still lacked sufficient food to lead an active and healthy life [2]. Globalization, market and trade forces, as well as years of research and agricultural subsidies targeting productivity growth, have skewed crop production towards a handful of crops that can be grown to scale in the world's main breadbasket regions: India, USA, Russian Federation, China and Brazil [1]. Wheat, rice and maize now dominate global markets, contributing 51% of the world's caloric intake and underpinning global diets that are becoming increasingly homogenous [3], much like the landscapes in which they grow [4]. Market incentives to grow these crops are such that in many parts of the world farmers have replaced traditional crops with these commodities. However, these crops alone are unable to support healthy and balanced diets. Research by Krishna Bahadur KC et al. [5] comparing modern food-based dietary guidelines with agricultural production statistics highlighted that our food systems are overproducing energy-dense foods, such as sugar, grains, fats and oils, while other studies show that there is insufficient production of protein-rich food and fruits and vegetables [6,7].

Growing commodity crops in sufficient quantity to meet rising demands for food has also come at the expense of agriculture's natural resource base and wider planetary systems [8] and is threatening its future production potential. Monocultures use large quantities of external inputs, such as pesticides and fertilizers, while conversion to agriculture is the main cause of deforestation, soil erosion, greenhouse gas emission and ecosystem pollution [8]. Food production systems are also major causes of biodiversity loss—including loss of agricultural biodiversity—as the recent State of the World's report on Biodiversity for Food and Agriculture [4] and the 2019 Global Assessment Report on Biodiversity and Ecosystem Services [9] remind us. Of the different earth ecosystem components assessed in the sixth Global Environment Outlook (GEO 6) report, biodiversity health is considered the most affected by environmental degradation, with negative repercussions on the resilience of ecosystems, including agricultural systems and food security [10]. Highlighted in these reports is the rapid decline and global disappearance of many local varieties of domesticated plants and crop wild relatives, many of which are underutilized and are maintained by custodian farmers exclusively for home consumption or for informal trade [11–13]. Those species and varieties, domesticated, semi-domesticated or wild, that are being cultivated, traded and maintained mainly by farmers, and which have been marginalized by specialized modern agricultural production systems and science, are referred to in this paper as "orphan crops". Efforts to conserve and use these plant genetic resources for sustainable agriculture, food systems and sustainable and healthy diets are mostly local and poorly connected to each other [4,14]. They nevertheless have a tremendous potential to redress key challenges in sustainable development, viz., vulnerability of production systems to climate change, disempowerment of vulnerable groups (women and indigenous peoples), widespread poverty, shrinking food biodiversity and pervasive malnutrition [14–16].

The quest to achieve as many of the sustainable development goals (SDG) as possible by the target date of 2030 [17] and, particularly, SDG2, has prompted scientists and practitioners to think of innovative ways of increasing the supply of sufficient quantities of safe and nutritious food for current and future generations without undermining the environment. One of these ways—which is slowly gaining traction in international agendas—is to make greater and better use of the abundance and diversity of orphan crops and wild edible plant species in agriculture, food systems and supply chains while exploring creative approaches that increase consumer demand and desirability for these plant genetic resources. Starting with the Rome Declaration on Nutrition [18] and Recommendation 10 of the Framework for Action [19] adopted in 2014 during the Second International Conference on Nutrition (ICN2), which reads "Promote the diversification of crops including underutilized traditional crops, more production of fruits and vegetables, and appropriate production of animal-source products as needed, applying sustainable food production and natural resource management practices", increasingly more UN agencies, along with Agriculture, Nutrition and Health think tanks, acknowledge the important role biodiversity plays in supporting sustainable food systems and balanced diets. In 2016, the General Assembly of the United Nations (UNGA) proclaimed the UN Decade of Action on Nutrition (2016–2025) [20], calling upon the FAO and the World Health Organization (WHO) to lead its implementation, in collaboration with the World Food Programme (WFP), the International Fund for Agricultural Development (IFAD) and the United Nations Children's Fund (UNICEF). Specific attention to biodiversity mainstreaming for food and nutrition was accorded by several UNGA Resolutions [21], as well as by the World Health Assembly (WHA) [22], which supports policies that promote the increased use of healthy local agricultural products and foods. In 2020, the Committee on World Food Security (CFS), which sits at the science-policy interface, is set to launch the *Voluntary Guidelines on Food Systems and Nutrition*, which are particularly supportive of local food systems and the need to sustainably utilize cultivated and non-cultivated diversity [23].

Priority actions being advocated for making these species available to consumers include: directing increased research investments to explore their climate resilience and nutritional characteristics; providing greater support for their mainstreaming in food security policies and programs (e.g., public procurement); encouraging their use to diversify farming systems and create more biodiverse landscapes and healthier ecosystems and upgrading their value chains and markets to ensure their continued and sustainable use [4,24,25].

Brazil, Kenya, Guatemala, India, Mali, Sri Lanka and Turkey are setting the pace in terms of exploring the potential role of locally available, affordable and climate-resilient orphan crops and wild edible species to support local food system transformation. With funding from the Global Environment Facility (GEF), the International Fund for Agricultural Development (IFAD) and the European Commission (EC), among other donors, Bioversity International—under its Healthy Diets from Sustainable Food Systems Initiative—has supported the seven countries in prioritizing a select range of nutritious orphan crops and wild edible species and placed them at the center of a holistic food system approach that considers both the production and consumption extremes of the food value chain. Within this initiative, the Local Agri-Foods Program sets out to deploy underused local agrobiodiversity to diversify the diets of vulnerable communities and provide a rich source of naturally available nutrients all year round. Following a brief overview of the methodology used, the paper will focus on the three-pronged approach that the countries used to: (i) increase evidence of the nutritional value and biocultural importance of these foods; (ii) better link research to policy to ensure these foods are considered in national food and nutrition security strategies and actions; and (iii) increase awareness of the desirability of these alternative food options among consumers so that they may more easily be incorporated in diets, food systems and markets (Figure 1).

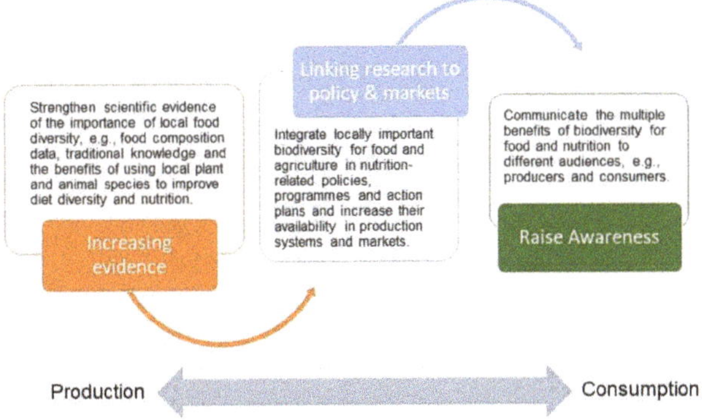

Figure 1. The three-pronged approach used by countries to establish enabling environments for mainstreaming biodiversity for food and nutrition into local food systems, from production to consumption (adapted from [26–28]).

Although the methodology and approaches have been described in greater detail in other publications for the GEF-supported Biodiversity for Food and Nutrition (BFN) Project [26–28] and for the IFAD Project [25,29,30], this is the first time the authors attempt to pull together the portfolio of work carried out by Bioversity International and its partner countries to diversify food systems using orphan crops, traditional varieties and wild edible species. Table 1 below provides an overview of the countries' focus by pillar, followed by the indicative length of implementation and budget investments. While it is understood that implementation of the approach may involve significant resources, countries with lower funding levels (e.g., Kenya) were able to successfully implement the full approach by reducing the scope of work, trimming down the number of target species and project sites and leveraging additional national and international resources for project activities.

Table 1. The table indicates the project timeframes and level of investment used by countries to implement the three-pronged approach to mainstream biodiversity for food and nutrition.

	Increasing Evidence	Linking Research to Policy and Markets	Raising Awareness	Time Frame (Years)	Project Budget (US$ Million)	Country Co-Financing (US$ Million)
Brazil	X	X	X	7	1.72	59.61
Kenya	X	X	X	7	0.22	0.24
Sri Lanka	X	X	X	7	0.83	1.03
Turkey	X	X	X	7	1.04	2.26
India	X	X	X	5	0.90	0.40 *
Guatemala	X	X	X	5	0.33	0.13 *
Mali	X		X	5	0.36	0.13 *

* Co-financing mobilized from country and other supporting agencies.

Over several years, the countries have generated scientific evidence of the nutritional value of 188 orphan crops and wild edibles species and achieved positive changes around improving community dietary diversity, increasing market opportunities for smallholder growers and gaining increased attention for biodiversity conservation and climate-smart agricultural practices. The paper concludes by identifying global opportunities to create momentum around the conservation and use of orphan crops and wild edible species for sustainable agriculture, diversified food systems and sustainable and healthy diets.

2. Background

The extraordinary diversity of ecosystems and food species that Brazil, Kenya, Guatemala, India, Sri Lanka and Turkey support made these countries ideal candidates for the interventions established under the different funding mechanisms outlined above. With the exception of Mali (Mali was chosen in response to a specific request made by IFAD to support one of its agricultural investment programs operating in the country, focusing on enhancing and leveraging local agrobiodiversity to support nutrition, climate change and income outcomes), the six other countries are among the highest in terms of species richness and endemism. Despite this natural abundance, they suffer alarming rates of one or more forms of malnutrition (Table 2). At the same time, the native edible biodiversity that they harbor, both wild and cultivated, is threatened by environmental pressures or lack of use.

Table 2. The table shows the countries' respective ranking according to the Convention on Biological Diversity's (CBD's) Biodiversity Index and the Global Hunger Index. It also captures the multidimensional nature of hunger, where overweight and stunting often occur in the same population, despite the high levels of biodiversity.

Country	Biodiversity Index Rank *	Global Hunger Index Rank **	Malnutrition ***		
			Prevalence of Overweight (in Women +18 Years)	Prevalence of Anemia in Women (15–49 Years)	Prevalence of Stunting (Children <5 Years)
Brazil	4	18	55.4%	27.2%	7%
Guatemala	17	72	59.9%	16.4%	46.7%
India	18	102	21.6%	51.4%	37.9%
Kenya	44	86	34.3%	27.2%	26.2%
Mali	140	83	35.1%	51.3%	30.4%
Sri Lanka	36	66	27.4%	32.6%	17.3%
Turkey	70	1–17	69.3%	30.9%	9.9%

* NBI—The National Biodiversity Index of the Convention on Biological Diversity is based on estimates of country richness and endemism in four terrestrial vertebrate classes and vascular plants. Vertebrates and plants are ranked equally. Index values range between 1.000 (maximum: Indonesia) and 0.000 (minimum: Greenland). The NBI includes adjustments allowing for country size. The values shown are expressed over 161 countries. Countries with land area less than 5000 km^2 are excluded [31]. ** 2019 Global Hunger Index. Based on 117 countries ranked on a 100-point GHI Severity Scale, where 0 is the best score (no hunger) and 100 is the worst [32]. *** Global Nutrition Report country nutrition profiles [33].

Brazil, for example, is one of the 17 most diverse countries in the world, hosting between 15%–20% of the world's biological diversity, with the greatest number of endemic species on a global scale [34,35]; yet, micronutrient deficiencies are prevalent, with 27.2% of women of reproductive age suffering from anemia and 54% of the adult population overweight [36].

At the other end of the spectrum, Mali carries the so-called "triple burden of malnutrition", in which hunger, overweight and micronutrient deficiencies coexist in the same population and often in the same individual across the life cycle. In 2015, 26% of children and adolescents in Mali were underweight, while an average 27.7% of adults were overweight and 51.3% of women of reproductive age suffered from anemia [37]. Although the abundance of agricultural biodiversity in the West African countries is significantly lower than Brazil, a number of nutrient-rich seasonal green leaves, semi-domesticated crops and wild foods—such as baobab leaves (*Adansonia digitata*), African locust beans (*Parkia biglobosa*), jute mallow (*Corchorus olitorius*) and amaranth (*Amaranth sp.*)—could potentially increase dietary quality, particularly in rural and resource-poor areas [38,39].

One of the immediate causes of malnutrition is either lack of food or the excessive consumption of calorie-dense, heavily processed foods and/or the insufficient consumption of key foods such as nutritious fruits, vegetables, nuts and seeds [40]. In the Sikasso region of Mali, rice, millet and sorghum constitute the bulk of the daily diet, with vegetables consumed in small quantities as a side dish [41]. Reportedly, less than 200g of vegetables per day per person are consumed, well below the recommended 400g/day recommended by the World Health Organization [37]. At the same time, a study by Siegel et al. [6] highlighted that the global and national fruit and vegetable supply is insufficient to meet the nutritional needs of current and growing populations, urging the global nutrition and agricultural communities, particularly in low-income countries, to find innovative ways of increasing fruit and vegetable production and consumption to meet population health needs.

One approach, which the seven countries embraced, is to better harness local agricultural biodiversity, including orphan crops and wild species, which are equally or often nutritionally superior to exotic crops [26,42,43], as well as being more resistant to biotic and abiotic stresses [44]. However, a number of barriers exist to the successful integration of orphan crops and wild edible species in food systems. These can be summarized as: (i) limited and fragmented data available on their nutritional value and potential benefits; (ii) limited capacity (technical, structural and financial) and research partnerships to fill this knowledge gap and generate more evidence; (iii) disabling policies and regulatory frameworks that fail to encourage the greater production and consumption of food biodiversity; (iv) unfavorable trade policies and poorly developed markets and infrastructure; (v) low consumer awareness and negative perceptions associated with their consumption and (vi) limited awareness and understanding of the nutrition, environmental and economic benefits that could arise by mainstreaming biodiversity for food and nutrition [45–47]. The sections below provide some insights on how these barriers were tackled through multi-stakeholder collaborative efforts realized by Bioversity International and its partners in Central America, Africa and South Asia over the last few years. As explained in greater detail in Section 3.2, although not easy to establish, multi-stakeholder policy platforms can make use of pre-existing frameworks, such as the ones established in Brazil under the Zero Hunger program [48], or be created from scratch, building on the momentum of national and international efforts to tackle malnutrition or biodiversity conservation, to develop biodiversity-sensitive school-feeding policies/programs [49] or a nutrition-sensitive biodiversity conservation plan (e.g., the Busia County Biodiversity Policy [50]).

3. Materials and Methods

3.1. Assessing Available Food Diversity

Trimming down the large amount of edible biodiversity that is present in each of the countries to a manageable sample size required an initial understanding of the breadth of this diversity, as well as national consensus on the desirable traits for selection. In the case of the BFN, in the project's preparatory phase, countries had undertaken a set of background studies and literature reviews revealing that information was present but mostly scattered across information sources and incomplete [28]. Only for Brazil had attempts been made to prioritize existing biodiversity for food and nutrition under the *Plants for the Future* (PFF) initiative spearheaded by the Ministry of the Environment. PFF had singled out 674 species of potential economic value across its five ecoregions (i.e., South, South–East, Central, North and North–East), of which 41 were edible. Kenya, Sri Lanka and Turkey had to carry out extensive literature reviews, ethnobotanical assessments and focus group discussions with key informants, as well as convening national experts, to prioritize the most appropriate species to include in the interventions [28].

Surveys were also used to capture traditional knowledge related to the species, as well as local names; habitats; parts used; specific uses (fruit, vegetable, staple, etc.) and users (by gender); seasonality; preparation and conservation and availability. Traditional knowledge, which is recognized by SDG Target 2.5 as underpinning the conservation and management of biodiversity and local and indigenous food production systems [14], is itself threatened by the loss of traditional lifestyles and food cultures, following rapid urbanization and the industrialization of agriculture and food processing [4]. In Turkey, market surveys targeting local plant collectors, sellers and consumers in three different ecogeographic regions helped gather information on 43 wild edible species, including mushrooms and landraces [26,28], while interviews with quilombola communities living in the Central–West and North–East regions of Brazil helped document traditional knowledge on the use of 16 native fruit species, such as the native dwarf cashew (*Anacardium humile*), which is eaten as a fruit, as well as a nut, by local communities. The continued cultivation of orphan crops, such as different cowpea varieties (*Vigna unguiculata*), in Turkey continues mostly thanks to women who use these ingredients in traditional dishes [28].

In Mali and Guatemala, seasonal agrobiodiversity assessments and consultations with local communities focused on existing food plant diversity (wild and cultivated), related uses and the assessment of climate resilience traits, as perceived by farmers. These investigations revealed a large diversity of cereals, legumes, fruits and vegetables that could be better leveraged to increase dietary quality and diversity while contributing to a greater climate adaptation of local production systems [39,51]. Turkey, whose focus was on wild edible plants, went one step further and developed a prioritization matrix that considered issues of environmental, economic and market sustainability to rank the species [28]. Of relevance to the environmental sustainability ranking was the existence of national domestication programs targeting the species. This is especially important when dealing with the increased commercialization of wild plants or fruits to ensure the plants are not overexploited and their growing habitats maintained. Although lengthy to set up, domestication programs ensure that farmers have access to good quality seed and can grow the species easily in their fields while reducing collection pressure on the wild populations. Many such programs are likely already taking place at the national level, as was the case for the golden thistle (*Scolymus hispanicus*) in Turkey and for African indigenous vegetables in Kenya. As interest for these forgotten resources grows, there may be a need to expand the programs to increase seed production, accompanied by sustainable collection guidelines where domestication programs cannot be set up.

3.2. Assessing the Quality of Diversity

Within the BFN Project [52], Brazil, Kenya, Sri Lanka and Turkey worked through national universities and research institutes to determine the nutrient composition of 188 orphan crops and wild edible species using methodologies developed by the FAO/INFOODS [53] while engaging with local communities to document traditional knowledge associated with these crops. For example, spearheaded by the Ministry of the Environment, Brazilian federal universities and research institutes compiled and generated food composition data for 70 native fruit species, many of which are wild-harvested, demonstrating their nutritional advantage over more commonly consumed exotic fruits. In Kenya, the Kenya Agricultural and Livestock Research Organization, working in partnership with the Ministry of Health and FAO, generated nutritional data on a select number of orphan crops and semi-domesticated species that are traditionally grown in home gardens or by family farmers. A full list of research partners and institutes is provided in Hunter et al. [26].

4. A holistic Approach to Diversifying Diets and Markets for Orphan Crops

4.1. Increasing Evidence of the Value of Orphan Crops

Evidence is the first necessary step to transform the enabling environment for biodiversity to improve nutrition (Figure 1) [47]. It is one of the three main pillars that underpin implementation of the FAO *Voluntary Guidelines for Mainstreaming Biodiversity into Policies, Programmes and National and Regional Plans of Action on Nutrition*, which aim at "addressing malnutrition in all its forms and, specifically, to promote knowledge, conservation, development and use of varieties, cultivars and breeds of plants and animals used as food, as well as wild, neglected and underutilized species contributing to health and nutrition" [54]. Stronger evidence of the nutritional value of biodiversity is also considered by thought leaders as one of the main areas for aid investment with the greatest untapped potential to support the reduction of undernutrition and increase dietary diversity, particularly among the poorest, who depend most on the sustainable use of natural resources [55]. However, limited information exists on the nutritional value of orphan crops and of the traditional knowledge associated with their collection, preparation and use. It is estimated that global databases, such as the International Network of Food Data Systems hosted by the FAO (FAO/INFOODS Food Composition Database for Biodiversity) [56] and Food Composition Tables (FCTs), capture a meager 1% of the nutritional information of foods consumed globally, let alone foods derived from local biodiversity [57]. Of the 10,156 foods recorded in the FAO/INFOODS database, only a modest 3118 (31%) are identified as wild plant and animal foods (belonging to a total of 1289 species). Of these, 1% are legumes, 4% nuts and seeds, 12% are starchy roots and tubers and 11% are vegetables [4]. In most cases, data are available for macronutrients and minerals, whereas vitamin and phytochemical compositions, which are particularly relevant to the promotion of wild and biodiverse foods, are seldom provided [4]. Food composition analysis carried out in Brazil as part of the BFN work confirmed that native species such as camu-camu (*Myrciaria dubia*), guabiroba (*Campomanesia xanthocarpa*), mangaba (*Hancornia speciosa*) and cagaita (*Eugenia dysenterica*) contain significantly higher levels of vitamin C compared to oranges, limes and tangerines [26], while the native fruits buriti (*Mauritia flexuosa*), tucumã (*Astrocaryum aculeatum*) and pitanga (*Eugenia uniflora*) are richer in provitamin A carotenoids than more commonly available and consumed fruits (Figure 2).

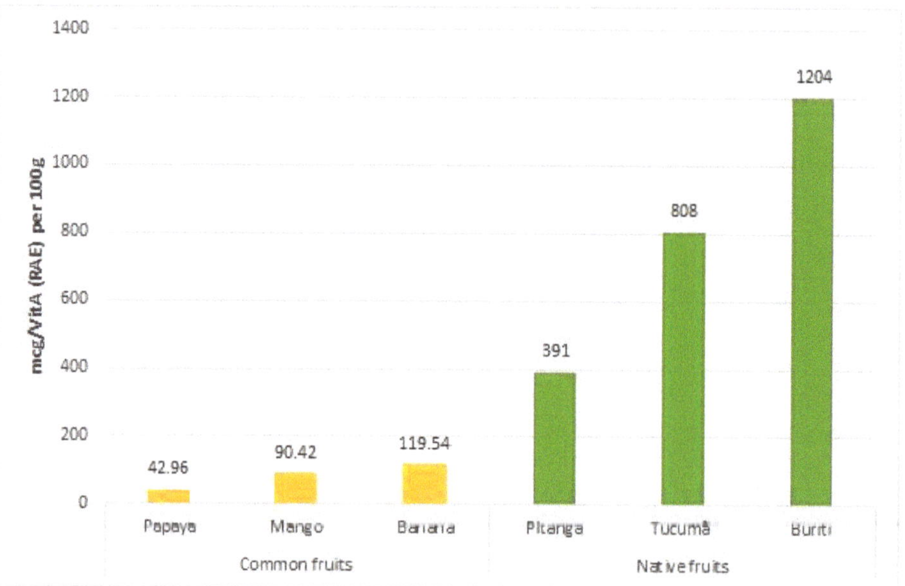

Figure 2. Vitamin A content (expressed in mcg RAE per 100g of fresh raw pulp/whole fruit, with or without peel) of the native Brazilian fruits pitanga (*Eugenia uniflora*), tucumã (*Astrocaryum aculeatum*) and buriti (*Mauritia flexuosa*) compared to more commonly available vitamin A-rich fruits. Source: Brazilian food composition tables [58] and the System on Brazilian Biodiversity (SiBBr) database [59].

Similar results were obtained in Kenya. In the case of African leafy vegetables, results showed that, compared to cabbage (*Brassica oleracea*), amaranth (*Amaranthus dubius*) contained almost 3.5 times as much beta-carotene equivalent—the primary plant-based source of dietary vitamin A— while Malabar spinach (*Basella alba*) contained 13.5 times as much iron [26]. In the same study, investigations into the nutrient content of different Sri Lankan rice varieties (*Oryza sativa*), comparisons of finger millet varieties (*Eleusine coracana*) across geographic locations and the nutritional analysis of wild edible plants in Turkey confirmed the high levels of intra- and inter-species diversity that have previously been reported in literature [60,61]. Table 3 provides a summary of the micronutrient content (important minerals and vitamins) of target underutilized species generated by food composition analyses in three of the four BFN countries. These results further highlight the importance of including a wide range of species and varieties in diets to improve diet quality [62].

Table 3. The table summarizes the micronutrient content of select underutilized fruits, vegetables and cereals obtained via food composition analyses in Brazil, Kenya and Sri Lanka. Values are expressed per 100g of raw parts used. Source: Biodiversity for Food and Nutrition (BFN) Species Database [63].

Country	Scientific Name	Common Name	Part Used	Ca (mg)	Fe (mg)	Mg (mg)	P (mg)	K (mg)	Na (mg)	Zn (mg)	VitA-RAE (mcg)	ß-Carotene (mcg)	Vit C (mg)
	Astrocaryum aculeatum	Tucumã	Pulp, peeled, raw	129	1.10	62	29	471	6	0.94	808	5191	24.22
	Campomanesia xanthocarpa	Guabiroba	Whole fruit, raw	26	1.76	24	30	215	1	0.92	484	2973	599.65
Brazil	*Eugenia dysenterica*	Cagaita	Pulp, unpeeled, raw	1	0.01	0.04	5	107	2	1.75	19	176	129.61
	Eugenia uniflora	Pitanga	Pulp, unpeeled, raw	17	0.97	11	39	147	4	0.22	73	438	5.75
	Hancornia speciosa	Mangaba	Pulp, unpeeled, raw	4	0.88	11	14	170	8	0.56	4	30	130.41

Table 3. *Cont.*

Country	Scientific Name	Common Name	Part Used	Ca (mg)	Fe (mg)	Mg (mg)	P (mg)	K (mg)	Na (mg)	Zn (mg)	VitA-RAE (mcg)	ß-Carotene (mcg)	Vit C (mg)	
	Mauritia flexuosa	Buriti	Pulp, peeled, raw	80	1.77	40		218	11	0.60	1204	13710	56.78	
	Myrciaria dubia	Camu-camu	Pulp, peeled, raw	10	0.51	3		83	1	0.48		3	2195.37	
	Amaranthus dubius	Amaranth	Fresh, raw	280	6.80	122	89	597	18	0.92	326	3913	77	
Kenya	*Basella alba*	Malabar spinach	Fresh, raw	267	10.9	40	56	446	8	0.50	184	2213	79.8	
		Finger millet, Bala Kurakkan local variety	Seed raw	13.98	144.28 ± 0.04		465.81 ± 0.03	4.45 ± 0.01	2.85 ± 0.0				454.86 ± 0.13	
Sri Lanka	*Eleusine coracana*	Finger millet, Wadimal Kurakkan local variety	Seed, raw	6.39	124.99 ± 0.04		536.63 ± 0.03	7.99 ± 0.01	2.64 ± 0.0				438.06 ± 0.21	

Collectively, the countries have made the single largest contribution to the FAO/INFOODS database and to broadening global knowledge of the nutritional value of orphan crops and wild edible species for potential integration into national and global food systems. The leafy vegetable chaya (*Cnidoscolus aconitifolius*), an evergreen, hardy shrub that was domesticated by the Mesoamericans in pre-Columbian times and is typically cultivated on a small scale in gardens and field margins for household use [64], contains high amounts of several macro– and micronutrients, including protein (60g per 1kg of leaves), vitamin A, niacin and vitamin C [65]. In Mali, the cereal fonio (*Digitaria exilis*) is rich in the amino acids methionine and cysteine [66], while Bambara groundnut (*Vigna subterranea*), with its chemical composition comparable to soy bean, is a good source of quality protein, fat and carbohydrates—enough to be considered a complete food [67,68]. Short growth cycles (for fonio) and the ability to grow well in resource-poor settings with limited water and minimal external inputs, make these crops particularly suited for growing in harsh environments, providing important contributions to food security, especially in times of food scarcity and extreme climate conditions [69]. Despite their importance, fonio and Bambara groundnut are presently cultivated only at a small scale and have been displaced by other crops that are better supported by the extension and marketing systems [70–72]. Through complementary investigations on markets and value chains of these climate-smart crops, farmers and other stakeholders also gained a better understanding of bottlenecks hindering the wider use of these local crops and acted to address these through multiple measures, as done effectively in the case of chaya in Guatemala [73].

4.2. Linking Research to Policy and Markets

Once obtained, reliable nutritional data on orphan crops and other nutritious biodiversity can be used as evidence to inform and shape global and national policies that currently incentivize the production of cheap, unhealthy food with a large environmental footprint, while minimizing diversity on farms and in agricultural landscapes [74,75]. This evidence has been recognized as crucial in the creation of enabling environments that are more conducive to enhancing nutrition and are aimed at making healthy, diversified and sustainable foods available to all and the easy default choice in local and global food markets [47].

The process, referred to as "biodiversity mainstreaming", indicates the "integration of actions that conserve and sustainably use biodiversity into strategies for production sectors such as agriculture, fisheries, and forestry" [27]. The approach can either be centralized (top–down), using country participation in international agreements to align national agendas with international priorities, or bottom–up, using a decentralized approach to guide national policies towards increasing the amount of attention, funding, technical and service support for orphan crops to make food systems more nutritious, while supporting sustainable agricultural practices and links with smallholder farmers. Both approaches would ideally be needed for the mainstreaming process to be effective and meaningful across key sectors of society (viz., agriculture, environment, nutrition, health, trade and education) and economically viable to farmers and all the other value chain actors, including consumers.

The Convention on Biological Diversity (CBD) and its Strategic Plan for Biodiversity 2011–2020 [76] offers a useful platform to mainstream the conservation and sustainable use of biodiversity for improved nutrition. The CBD has near-universal participation among countries and explicitly recognizes the potential of orphan crops and wild edibles to contribute to food systems' transformations and improve human health [77]. Sri Lanka is one of the 196 countries that are party to the CBD. It used evidence generated under the BFN Project on the nutritional importance of 28 traditional edible plant species and 58 varieties—including traditional rice varieties, cereals, pulses, root and tuber crops and vegetables—to update and revise its National Biodiversity Strategy and Action Plan (NBSAP) and better align its national targets with the global targets established by the Strategic Plan for Biodiversity—the so-called Aichi Biodiversity Targets (a set of 20 time-bound global targets under the Strategic Plan for Biodiversity 2011–2020) [76]. Data was used to justify and better integrate biodiversity concerns and nutrition-related objectives into NBSAP targets, so that National Target 2 now reads "Promote and mainstream underutilized, lesser-known or neglected food crops, livestock and food fishes, which provide nutrition" [78]. Strategic partnerships with multisectoral, key-line ministries established under the BFN Project led to similar results in Brazil, where Strategic Objectives and Targets of the most recent NBSAP [79] explicitly recognizes the nutritional value of native biodiversity (including orphan crops and wild edibles) and employs greater information and utilization of native nutrient-rich plant species as a successful measure of biodiversity conservation. The conservation of biodiversity for food and nutrition was included as a priority action for Target 13 (Conserve the gene pool) but recognized as having manifold benefits that directly impact Biodiversity Targets 1 (Understand values), 2 (Mainstream biodiversity), 3 (Address incentives), 4 (Sustainable production), 7 (Manage within limits), 14 (Restore ecosystems) and 18 (Respect and conserve traditional knowledge). Furthermore, many targets and/or initiatives targeting the conservation and sustainable use of native biodiversity were used to guide Brazil's Multi-Year Budget Plan (2016–2019). The continued integration of biodiversity research into national policies has helped establish long-term support for biodiversity as a link between agriculture and nutrition, translating into increased research funding for the nutritional profiling of orphan crops, incentives to grow these crops using agroecological practices and encouraging the diversification of the food supply by increasing their availability in markets.

One of the main policy instruments that has greatly enhanced market recognition for orphan crops and wild edibles using evidence generated by the BFN is Ordinance 284 (that supersedes ordinance 163/2016) on "Brazilian Sociobiodiversity Native Food Species of Nutritional Value", which identifies 101 marketable species, mostly wild edible fruits and nuts, and recognizes their key role for food and nutrition security [80]. The ordinance, jointly developed by the Ministries of Environment and Social Development, increases farmer incentives to continue growing and managing the species, which can be sold via two important public procurement programs: the National School Feeding Program (PNAE), which establishes that at least 30% of the food purchased with federal funds must be procured directly from family farmers, and the Food Acquisition Program (PAA), which pays an additional 30% for organic and agroecological produce grown by family farmers [81]. Complementary to PAA, the Minimum Price Guarantee Policy for Sociobiodiversity Products (PGPM-Bio) reduces farmer vulnerability to price volatility in food markets and steps in to compensate producers when their sociobiodiversity products do not attain the market value established by the National Supply Company [82]. Since the ordinance was launched in 2016, purchases for sociobiodiversity species by the PAA doubled from US $1M in 2014 to US $2.5M in 2016, while, between 2018 and 2019, minimum guarantee prices for wild edible fruits managed by extractivist communities have increased by 11.21% for buriti (*Mauritia flexuosa*) and 19.07% for juçara (*Euterpe edulis*) [83]. A structured demand for agrobiodiversity has been shown to engender other positive downstream effects, such as on farm diversification, improved management practices, increased land size devoted to horticultural production and increased availability of nutritious foods in local food systems [84]. However, the percentage of funds used by the PNAE and PAA to purchase native biodiversity products remains low (2%) relative to mainstream commodity crops [81].

In Kenya, where multisectoral platforms that tackle food security and nutrition issues are still emerging, and where the revised Constitution (2010) has devolved environmental management functions to the counties, a decentralized, bottom-up approach was used to mainstream biodiversity and nutrition-related concerns into the drafting of new policy documents and guidelines. Food composition data for traditional leafy greens such as malabar spinach (*Basella alba*), jute mallow (*Chorchorus olitorius*), spider plant (*Cleome gynandra*) and amaranth (*Amaranthus dubius*), as well as continuous engagement by the national BFN team with local policymakers, farmer groups, communities and schools, helped raise awareness of the value of traditional orphan crops and other nutrient-rich species and support the formulation and endorsement of a County Biodiversity Policy that recognizes the role of biodiversity in food security and nutrition, as well as livelihood and ecosystem resilience. The policy has also allocated resources to build the capacity of small-scale farmers to produce greater diversity and to conserve and promote regional biodiversity for food and nutrition, with specific provisions for designated conservation areas and the incorporation of biodiversity concerns into school curricula and nutrition education programs. First of its kind for any of Kenya's 47 counties, the policy has the potential to cause a ripple effect and positively influence neighboring counties and national policy-making. The policy is all the more important because it recognizes the need for greater investments in building farmer capacity to increase productivity of local biodiversity, to be able to apply food safety and quality standards and to access technology and markets that currently favor large-scale producers. At present, limited space exists for orphan crops and wild edibles in formal global food systems/markets, which apply rigorous quality standards that focus on product uniformity; limit production variability [85,86] and fail to recognize the social, cultural, economic and ecological benefits that diversified production systems engender. Demand for staple crops are common, while tenders for orphan crops, such as the broad range of African leafy vegetables (ALVs), are nonexistent. Even when market opportunities exist, smallholder farmers lack the capacity to pursue them. Few can write a business plan or a loan application to a local microfinance institution or a commercial bank. This means that smallholder farmers lack access to a steady market for orphan crops and are excluded from value addition opportunities that could generate extra income.

The seven countries have tried to fix some of the broken linkages along the value chain by creating a strong market for these crops, as well as exploring alternative, secondary markets. Model gardens and demonstration plots established in Aranayaka, Sri Lanka by the Department of Agriculture, in collaboration with the Community Development Centre (NGO), for example, have overseen the creation of a value chain (from production to consumption) for 36 traditional wild edible yams and tuber varieties belonging to the genus *Alocasia*, *Dioscorea* and *Colocasia*. Demonstration plots established in 18 villages provide training on sustainable land management practices, integrated pest management and the cultivation of these orphan crops. At the same time, training is provided on value addition and the creation of novel food products that are then marketed in agrobiodiversity outlets across Sri Lanka. The pilot is currently reaching 2000 individuals in Aranayaka, while around 100 families have directly engaged in the marketing of products made from traditional roots and tubers, reaching monthly profits of US $42 per family.

Public procurement policies also offer strategic entry points for the greater mainstreaming of agricultural biodiversity, offering a structured demand for this diversity and generating multiple benefits across all the elements of sustainability: economic, social and environmental [87,88]. In India, through the IFAD NUS Project [89]. Bioversity International and the M.S. Swaminathan Research Foundation (MSSRF) worked closely to provide scientific and development evidence of the comparative advantage of using minor millets rather than mainstream crops in the country's Public Distribution System (PDS) [90]. In 2013, sustained lobbying efforts led to the approval by Parliament of the National Food Security Act [91], which conceded the inclusion of a diverse range of food grains in the national food distribution schemes. These previously neglected grains include finger millet (*Eleusine coracana*), foxtail millet (*Setaria italica*), proso millet (*Panicum miliaceum*), kodo millet (*Paspalum scrobiculatum*), little millet (*Panicum sumatrense*) and barnyard millet (*Echinochloa colona*). The passing of this law represents an historical milestone in the pathway towards a more resilient and nutritious food system in India, a country which, until 2013, had been sourcing only wheat and rice for its food distribution schemes. Increased production of these highly resilient and nutritious cereals is also expected to enhance India's capacity to adapt to climate change and counteract food insecurity, especially in the more marginal areas where millets perform better than rice and wheat in terms of nutritional yields [92]. Other social benefits stemming from this supportive environment are foreseen in terms of biodiversity conservation, empowerment of women and maintenance of India's unique food traditions and cultural identity, which minor millets are an important part of [29,93].

Similarly, in Guatemala, collaboration with the Ministry of Education, the Ministry of Public Health and Social Assistance, the Ministry of Agriculture and local nongovernmental organizations (NGOs) is leading to the greater use of orphan crops in the country's school meal programs. The recently approved National School Feeding Law (Ley de Alimentación Escolar—Decree no. 16-2017) establishes that schoolchildren aged 6 to 12 are provided with culturally-relevant, healthy and nutritious meals that fulfill 25%–35% of a child's daily energy and protein requirements, and that includes all food groups (carbohydrates, proteins, fruits, vegetables and fat). It also emphasizes procuring fresh foods within the vicinity of the school, preferably from local producers who practice family farming. Using agronomic and nutritional data, Bioversity International, in partnership with the NGO Mancomunidad Copanch'orti', was able to promote the use of chaya (*Cnidoscolus aconitifolius*) in the school meals program of the Department of Chiquimula. Chaya's nutritional and drought-tolerant properties, accompanied by ease of production, affordability, acceptability and availability, ensured its inclusion as a key ingredient in two of the 20 dishes that composed Chiquimula's official school meal menu in 2019, benefiting more than 80,500 students. Chaya is also being considered as an alternative ingredient in four other dishes, to be used with other leafy vegetables, such as chipilín (*Crotalaria longirostrata*) and black nightshades (*Solanum americanum* and *Solanum nigrescens*), when suitable. Central to the greater use of chaya was the preparation of a recipe book that proposes new, creative and likeable recipes using chaya. Prepared in collaboration with local communities, the recipe book considers local practices and culinary specificities that are pleasing to local tastes. The recipes are easy-to-follow and incorporate colorful pictures while considering children's preferences and include foods like chaya popsicles, porridge and rice pudding. While schools can function as institutional markets for the procurement of orphan crops, as seen in Guatemala, or in the case of African leafy vegetables and other nonconventional vegetables and native fruits [82], they are also an important platform to improve nutrition through education and behavioral changes, as demonstrated in the next section.

4.3. Raising Awareness: The Power of Evidence to Drive Change

Information or guidance currently abound on how to change our food systems so that they nourish humanity in a healthy manner, do not harm the environment, reduce public health impacts and are equitable and just [94]. The EAT-Lancet Commission Report [95] and World Resources Report [96] are the latest, but there are many reports that suggest key opportunities and recommendations that can inform productive action. They also acknowledge that food systems and environments are not only driven by trade policies and the food industry but are equally underpinned by consumer demand, where consumers can drive supply via their food choices [8]. For a healthy food system transformation to occur [94], consumers need to be aware of the benefits of diversifying their diets and the impact of their food choices, not only on their own health and well-being, but also on the food system and the environment. The reality, however, is that most consumers are inadequately informed about what constitutes a balanced diet; they do not necessarily prioritize nutrition and health when choosing food, and they lack the capacity and resources to utilize the diversity that is available or tend to prefer convenience food that is often more easily available and accessible in today's food environments [85]. Furthermore, the relevance of biodiversity for food and nutrition may not be apparent to most consumers and is often an underappreciated and overlooked resource. In many cultures, the consumption of orphan crops and wild species occurs in times of food deprivation or illness—as many of the species have medicinal properties—thus evoking negative experiences and perceptions, as well as unfair labeling such as food "for the poor" or "food for the sick". However, in other parts of the world, orphan crops and wild edibles continue to be used in cultural and religious festivals [4,43] or in traditional cuisine and gastronomy. The seven countries in this paper used these positive connections to biocultural heritage as an entry point for increased appreciation and support for orphan crops and wild edible species, using unique, context-specific approaches to creating awareness. Nutritional data for these species and information about their occurrence across seasons and their uses and benefits can be included in educational activities and materials used by extension services as well as awareness campaigns encouraging consumers to make informed choices [27].

Nutrition education is the main conduit to support the adoption of food choices and other food-and-nutrition-related behaviors conducive to health and well-being in both youths and adults. Working with relevant government institutions, entry points can be identified for the inclusion of messaging around the importance of consuming varied diets and favoring local foods, as well as the importance of sustainable production practices. A useful tool for the greater appreciation of orphan crops is the use of school gardens. In Brazil, the BFN worked closely with the "Educating with School Gardens and Gastronomy" Project (PEHEG) implemented by the Ministry of Education and the University of Brasília to help children reconnect with their food system while promoting the inclusion of nonconventional vegetables and native fruit tree nurseries in schools. School gardens in Kenya were set up to complement school meal programs [97] but also doubled-up as educational tools with the potential to incorporate lessons from different disciplines, such as mathematics, biology, geography, agriculture and economics, into gardening activities. Similarly, in Turkey, school gardens [98] and nature walks were used to heighten the appreciation of traditional foods and wild edible species among gastronomy students of a vocational training college, contributing to "greening" and broadening their culinary horizons and accessing better job opportunities.

To further support the adoption of healthier food choices, countries such as Brazil have produced national public health resources that highlight the need for dietary diversification to meet essential nutritional needs. The Brazilian food-based dietary guidelines [99], for example, approach food consumption from a completely original perspective, promoting the consumption of a variety of whole foods rather than discussing separate nutrients and calories. The guidelines are used to guide government actions linked to the food and nutrition agenda—including patient care by doctors and nutritionists, school meals, nutrition education and food labeling. They focus on homemade, nutritious meals and encourage citizens to become more mindful of the social, emotional, cultural and political meaning of food and be critical of ultra-processed foods. They also include advice on cooking culturally important dishes; create appreciation for regional foods and traditional dietary habits and highlight the links between healthy eating patterns; the promotion of biodiversity; shorter value chains and physical, social and environmental health. Unsurprisingly, the adoption of healthier food choices will not happen overnight, but monitoring and evaluation exercises, such as household/population surveys, can be set up at the onset and at the end of the intervention to assess increased consumption of orphan crops. If possible, and in the case of long-term interventions, linkages can be sought with the agencies responsible for social and demographic surveys.

Seasonal availability calendars were developed using participatory community assessments in Guatemala, Mali and India to encourage the year-round consumption of locally available and diverse fruits and vegetables. The species available per month were recorded and identified using local names and easily recognizable pictures. The calendars and information booklets distributed to communities following these assessments included information on national food-based dietary guidelines, as well as results of local diet surveys that generally revealed low levels of fruit and vegetable intakes. As shown in Figure 3, fruits and vegetables were divided into food groups from the Minimum Dietary Diversity for Women indicator [100]—dark, green, leafy vegetables, vitamin A-rich fruits and vegetables, other fruits and other vegetables—and information was provided on the unique nutritional role of each group. Simple recommendations were included on how to improve diet quality using the seasonal calendar, including planning the home garden to ensure year-round availability of fruits and vegetables, adding seasonally available produce to recipes to give a touch of color and encouraging families to eat fruits and vegetables on a daily basis.

Food innovations and gastronomy are also playing an increasingly important role in reversing the negative perceptions associated with many nutritious orphan crops. In Mali, nutritionists and food technologists at the Institut d'Economie Rurale (IER) developed new culturally acceptable recipes and novel food products using fonio, Bambara groundnut, jute mallow and amaranth (Table 4). These foods were analyzed for their nutrient contents (e.g., energy, protein, carbohydrate, fat, fiber, iron and zinc); underwent sensory evaluation tests and were promoted within communities and at diversity fairs organized in the Ségou and Sikasso regions (Figure 4). Likewise, in Guatemala, local women were trained on a protein extraction process for chaya leaves. The protein can then be used to enrich traditional foods, such as tamales (steamed maize dough wrapped in maize husk), tortillas (thin flatbread made from ground maize cooked in limewater), lemonade and scrambled eggs. Follow-up actions with the communities who grow and maintain these crops included cooking and value addition training sessions with a focus on quality and food safety standards in processing, preservation and packaging.

Figure 3. Seasonal calendar of fruits and vegetables in the Segou region, Mali, in the local language. Source: Bioversity International. [41,101].

Table 4. Recipes and novel food products using orphan crops fonio (*Digitaria exilis*) and Bambara groundnut (*Vigna subterranea*) in nutrition interventions carried out by Bioversity and the Institut d'Economie Rurale (IER) in Mali.

Fonio	Bambara Groundnut
Fonio with vegetables	*Covafo*: a complementary food made with Bambara groundnut, fonio, wheat flour and dried fruits
Fonio with amaranth and jute mallow leaves	*Tô*: Bambara groundnut and fonio flours served with baobab leaf or okra sauce
	Couscous: Bambara groundnut served with a leafy vegetable or peanut sauce
	Accra: a paste of soaked and fried Bambara groundnut and spices
	Shô furu: fried Bambara groundnut flour and baobab leaf powder
	Fari: steamed Bambara groundnut flour served with vegetables
	Soup made from de-hulled and smoked grains of Bambara groundnuts
	Ragoût: made from de-hulled and soaked grains of Bambara groundnuts

Figure 4. Women with their processed products in Bolimasso, Mali promoting the use and commercialization of fonio and Bambara groundnuts. Credit: Bioversity International/G. Meldrum.

In all countries, engagement with the gastronomy sector and with celebrity chefs has also sparked new interest in these forgotten crops, particularly by urban consumers. In Brazil, the BFN collaboration with federal and state institutions, private sector actors, university professors, researchers and students resulted in the publication of a 900-page recipe book entitled Biodiversidade Brasileira. Sabores e Aromas (Brazilian Biodiversity. Flavors and Aromas) [102], which contains 335 recipes using 64 nutrient-rich underutilized plant species from the six Brazilian regions. This encyclopedic feat represents a huge contribution to the greater promotion of local species and the sustainable use of native biodiversity in general. By the same token, in Guatemala, engagement with Guatemalan chefs Paula Enriquez and Andrei Shrei and a prestigious gastronomy school (Escuela culinaria de alta cocina de Guatemala—ACAM) has led to the inclusion of chaya in the menus of two of the top ten restaurants in Guatemala and direct procurement linkages with the rural women's cooperative (the Integral Marketing Cooperative Chorti—COREDCHORTI) established in the framework of the IFAD-EU-supported project led by Bioversity International (2015–2019) [103].

Special events such as fairs, festivals or national celebrations for World Food Day (Figure 5a–d) that attract wide media coverage can also raise the profile of orphan crops and wild herbs. Notable examples include the Traditional Food Festival in Sri Lanka or the Alaçatı Wild Herb Festival in Turkey, a four-day event devoted entirely to the celebration of wild edibles that promotes the appreciation of local biodiversity—including orphan crops—and attracts tourists, food gourmets, TV food channels and chefs for cooking workshops, food tastings and scientific lectures about the importance of these foods in healthy and sustainable diets.

Much like linking with gastronomy, the festivals, which are becoming increasingly popular, have helped increase the market for orphan crops and wild edibles, an opportunity which has been mostly taken up by women. In Alaçatı, women producers have organized themselves into groups and transformed and sold traditional products/foods made with wild edibles. In Sri Lanka, 20 food outlets by the name of Hela Bojun (literally, True Sri Lankan Taste) sell freshly prepared traditional foods and provide employment to rural women trained by the Women Farmers' Extension Program of the Department of Agriculture on aspects of healthy nutrition, food safety and preparation. By working at the food outlets, women earn between US $600–$800 a month, sufficient to enroll their children in school and support the family household [104]. In addition to the above, awareness was also raised among local farmers on the double value of local crops in terms of nutrition and climate change adaptation through community workshops, radio programs and dissemination of dedicated brochures and booklets. These interventions were often carried out in combination with the dissemination to farmers of seeds of the best-performing varieties, which had been jointly evaluated by farmers and agronomists.

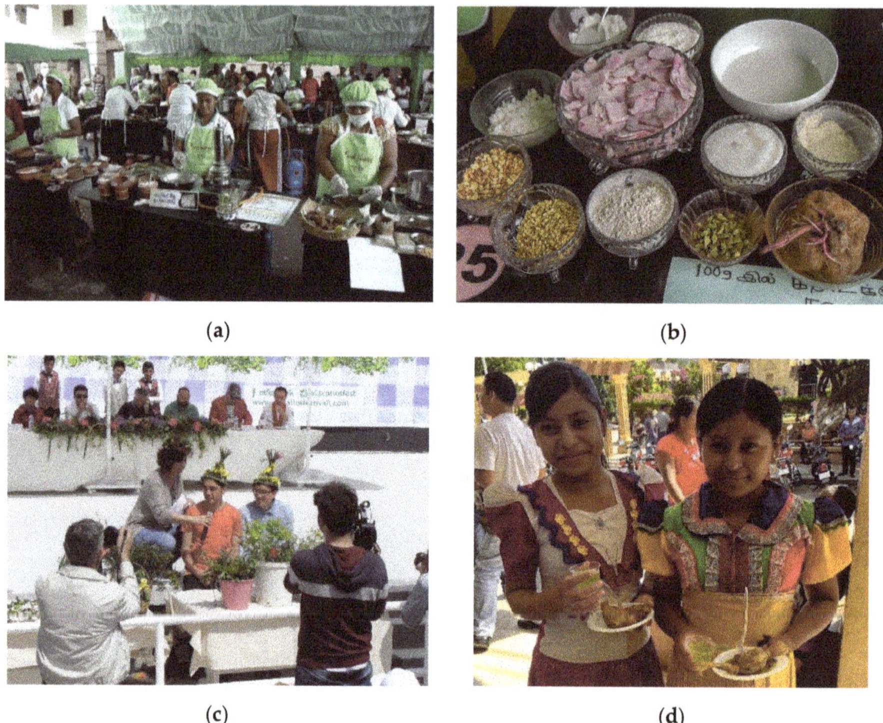

(a)

(b)

(c)

(d)

Figure 5. Festivals and gastronomy events help raise the profile of orphan crops in the intervention countries. (**a**) Food festival celebrating orphan crops and fruits carried out in Peradeniya, Sri Lanka for World Food Day. Credit: Bioversity International/D. Hunter. (**b**) Local yam is the main ingredient of a competition recipe at the Food Festival. Credit: Bioversity International/D. Hunter. (**c**) The young winners of the wild herb collection competition held during the Alaçatı Wild Herb Festival are interviewed by national television. Credit: Bioversity International/D. Hunter. (**d**) Chaya recipes promoted in Chiquimula Market, Guatemala. Credit: Bioversity International/ N. Amaya.

5. Conclusions and Recommendations

Recent global reports and scientific literature on food security and food systems [4,8,94–96], as well as those that are tracking global progress towards the achievement of SDG2 (Zero Hunger), are urging countries to make greater use of locally relevant biodiversity as a means of halting biodiversity loss and making food systems more sustainable. This includes adopting alternative agricultural models that focus on sustainable production and rural development, while empowering consumers to make better decisions around diets that are healthier for people and for the planet [74,105–108]. Thus, actions to achieve SDG2 are likely to fast-track progress towards other goals and targets, including poverty (SDG1); health (SDG3); economic growth (SDG8); sustainable production and consumption (SDG12); climate change (SDG13) and ecosystems, biodiversity and forests (SDG15), among the most relevant. The examples provided have demonstrated that many of these orphan crops and wild edible species provide valuable macro- and micronutrients, as well as beneficial bioactive non-nutrients that contribute to dietary health [26] and could play a key role in addressing local challenges linked to malnutrition, climate change, poverty and shrinking food biodiversity.

The case studies presented have also shown that, with an enabling environment in place, local nutrient-rich crops represent low-hanging fruits for improving dietary diversity and that rapid gains and progress can be made with relatively small effort, time and resources. Furthermore, reinforcing the knowledge and use of local plants provides holistic benefits, including strengthening local culinary traditions that are linked to culture and identity, while promoting food and nutrition sovereignty with foods that are available locally and suited to the local climate. Research and development efforts such as those presented in this paper play an important role in filling knowledge gaps and fostering connections between stakeholders in different sectors to promote a greater awareness, production, marketing and integration of orphan crops and wild edible species in public programs, including in sustainable intensification efforts, where orphan crops have been shown to contribute to increasing the output and quality (e.g., nutritional content) of food and other products while using less land, water and other inputs per unit output [4].

In the year 2020, countries will be called upon to critically assess successes and failures in the implementation of the Strategic Plan For Biodiversity (2011–2020) of the Convention on Biological Diversity, to reframe the new global biodiversity framework for the post-2020 era and to endorse the Voluntary Guidelines on Food Systems and Nutrition [23] at the next meeting of the Committee on World Food Security. These two important events represent a unique chance for the orphan crop community to use these tried and tested approaches and visionary thinking developed in these countries to create momentum around the conservation and use of orphan crops and wild edible species for sustainable agriculture, diversified food systems and sustainable and healthy diets.

Although the approach described in this paper is context-specific and likely to change depending on the geography, several broad recommendations are put forward to assist countries and regions wishing to step up efforts to promote the conservation and sustainable consumption of wild edible species and orphan crops. Divided by the three main pillars of the approach, these include:

5.1. Provide Evidence

- Encourage and fund research aimed at increasing knowledge and information on the value of orphan crops and wild edibles and of their impact on agriculture-nutrition pathways.
- Step up research investments for domestication programs, particularly of promising wild edible species.

5.2. Policies and Markets

- Develop coordinated, multisectoral policies that recognize the importance of orphan crops and wild edible species, and integrate explicit biodiversity for food and nutrition objectives and indicators and concerns in their investment-planning process.
- Develop policies that mainstream the conservation and sustainable use of orphan crops and wild edible species, including strengthening policies that protect unique wild edible collection sites and support ecotourism.
- Support public procurement mechanisms that favor the supply of orphan crops and wild edibles from family farmers/collectors.
- Support the value chain development for new biodiverse products to enhance local businesses and improve farmers' and collectors' livelihoods.

5.3. Raising Awareness

- Support nutrition education using culturally and socially appropriate nutrition messaging, which can create a demand for orphan crops and wild species.
- Create awareness campaigns promoting diet diversification and the nutritional, environmental and economic benefits of orphan crops and wild edible species.

Author Contributions: Conceptualization, T.B. and D.H.; investigation, D.M.d.O.B., G.S., V.W.W., G.L., B.G., A.T., Y.K.D. and A.S.; supervision, F.T.; writing—original draft, T.B.; writing—review and editing, D.H., S.P., N.A., G.M., D.M.d.O.B., G.S., V.W.W., G.L., B.G., A.T., Y.K.D., A.S. and F.T. All authors have read and agreed to the published version of the manuscript.

Funding: Overall support for the Biodiversity for Food and Nutrition (BFN) Project was provided by the Global Environment Facility (GEF Project ID 3808). Co-funding and implementation support were received from the UN Environment Programme; the Food and Agriculture Organization of the United Nations; Bioversity International and the governments of Brazil, Kenya, Sri Lanka and Turkey. Additional funding was received by the Australian Centre for International Agriculture Research for work in Kenya (HORT2014/100, GP2017/007 and GP2018/101), as well as the CGIAR Research Program on Agriculture for Nutrition and Health. For the work in India, Mali and Guatemala, the authors would like to warmly thank the International Fund for Agricultural Development (IFAD) and the European Commission for their support through the Grants 2000000526 and 978, which have contributed to promoting the wider use of neglected and underutilized species to improve the livelihood of people, especially the poor and vulnerable.

Acknowledgments: The authors would also like to gratefully acknowledge the many colleagues and institutions in participating countries that contributed to the work described in this paper. The activities and outcomes of the BFN Project contributes to the implementation of the Convention on Biological Diversity's Cross-Cutting Initiative on Biodiversity for Food and Nutrition.

Conflicts of Interest: The authors declare no conflicts of interest, and the funders had no role in the design of the study; in the collection, analyses or interpretation of data; in the writing of the manuscript or in the decision to publish the results.

References

1. FAO. *World Food and Agriculture—Statistical Pocketbook 2018*; Food and Agriculture Organization of the United Nations: Rome, Italy, 2018; 254p.

2. FAO; IFAD; UNICEF; WFP; WHO. *The State of Food Security and Nutrition in the World 2018. Building Climate Resilience for Food Security and Nutrition*; FAO: Rome, Italy, 2018.

3. Khoury, C.K.; Bjorkmanc, A.D.; Dempewolf, H.; Ramirez-Villegasa, J.; Guarino, L.; Jarvis, A.; Riesebergd, L.H.; Struik, P.C. Increasing homogeneity in global food supplies and the implications for food security. *Proc. Natl. Acad. Sci. USA* **2014**, *111*, 4001–4006. [CrossRef] [PubMed]

4. FAO. *The State of the World's Biodiversity for Food and Agriculture*, 1st ed.; Bélanger, J., Pilling, D., Eds.; FAO Commission on Genetic Resources for Food and Agriculture Assessments: Rome, Italy, 2019; 572p.

5. Kc, K.B.; Dias, G.M.; Veeramani, A.; Swanton, C.J.; Fraser, D.; Steinke, D.; Lee, E.; Wittman, H.; Farber, J.M.; Dunfield, K.; et al. When too much isn't enough: Does current food production meet global nutritional needs? *PLoS ONE* **2018**, *13*, e0205683.

6. Siegel, K.R.; Ali, M.K.; Srinivasiah, A.; Nugent, R.A.; Venkat Narayan, K.M. Do we produce enough fruits and vegetables to meet global health need? *PLoS ONE* **2014**, *9*, e104059.

7. Padulosi, S.; Sthapit, B.; Lamers, H.; Kennedy, G.; Hunter, D. Horticultural biodiversity to attain sustainable food and nutrition security. *Acta Hortic.* **2018**, *1205*, 21–34. [CrossRef]

8. Repeated 86 Willett, W.; Rockström, J.; Loken, B.; Springmann, M.; Lang, T.; Vermeulen, S.; Garnett, T.; Tilman, D.; DeClerck, F.; Wood, A.; et al. Food in the Anthropocene: The EAT–Lancet Commission on healthy diets from sustainable food systems. *Lancet* **2019**, *393*, 447–492.

9. IPBES. *Summary for policymakers of the global assessment report on biodiversity and ecosystem services of the Intergovernmental Science-Policy Platform on Biodiversity and Ecosystem Services*; Díaz, S., Settele, J., Brondizio, E.S., Ngo, H.T., Guèze, M., Agard, J., Arneth, A., Balvanera, P., Brauman, K.A., Butchart, S.H.M., et al., Eds.; IPBES Secretariat: Bonn, Germany, 2019.

10. UN Environment. *Global Environment Outlook—GEO-6: Healthy Planet, Healthy People*; Cambridge University Press: Cambridge, UK, 2019; 745p. [CrossRef]

11. Gruberg, H.; Meldrum, G.; Padulosi, S.; Rojas, W.; Pinto, M.; Crane, T. *Towards a Better Understanding of Custodian Farmers and Their Roles: Insights from a Case Study in Cachilaya, Bolivia*; Bioversity International, Rome and Fundación PROINPA: La Paz, Bolivia, 2013; 33p.

12. Sthapit, B.; Lamers, H.; Rao, R. (Eds.) *Custodian Farmers of Agricultural Biodiversity: Selected Profiles from South and South East Asia*; Bioversity International: Rome, Italy, 2013.

13. Sthapit, S.; Meldrum, G.; Padulosi, S.; Bergamini, N. (Eds.) Strengthening the role of custodian farmers in the national conservation programme of Nepal. In *Proceedings of the National Workshop, Pokhara, Nepal, 31 July–2 August 2013*; Bioversity International: Rome, Italy; LI-BIRD: Pokhara, Nepal, 2015.

14. Bioversity International. *Mainstreaming Agrobiodiversity in Sustainable Food Systems: Scientific Foundations for an Agrobiodiversity Index*; Bioversity International: Rome, Italy, 2017.

15. Egeland, G.M.; Harrison, G. Health disparities: Promoting indigenous peoples' health through traditional food systems and self-determination. In *Indigenous Peoples' Food Systems and Well-Being*; Kuhnlein, H.V., Erasmus, B., Spigelski, D., Burlingame, B., Eds.; FAO: Rome, Italy, 2013.

16. Turner, N.J.; Plotkin, M.; Kuhnlein, H.V. Global environmental challenges to the integrity of indigenous peoples' food systems. In *Indigenous Peoples' Food Systems and Well-Being*; Kuhnlein, H.V., Erasmus, B., Spigelski, D., Burlingame, B., Eds.; FAO: Rome, Italy, 2013.

17. United Nations. *Transforming our world: The 2030 Agenda for Sustainable Development*; 2015 A/RES/70/1; United Nations: San Francisco, CA, USA, 2015.

18. FAO and WHO. *Rome Declaration on Nutrition*; ICN2 2014/2; FAO: Rome, Italy, 2014.

19. FAO. Framework for Action. ICN2 2014/3 Corr.1. In Proceedings of the Conference Outcome Document from the Second International Conference on Nutrition, Rome, Italy, 19–21 November 2014. Available online: http://www.fao.org/3/a-mm215e.pdf (accessed on 3 February 2020).

20. United Nations. General Assembly (UNGA) Resolutions 70/259 and 72/306. Available online: https://undocs.org/en/A/RES/70/259 (accessed on 3 February 2020).

21. United Nations. General Assembly (UNGA). Resolution A/RES/73/2. Political Declaration of the Third High-Level Meeting of the General Assembly on the Prevention and Control of Non-Communicable Diseases. Available online: https://undocs.org/en/A/RES/73/2 (accessed on 3 February 2020).

22. WHO. Global Action Plan for the Prevention and Control of Noncommunicable Diseases 2013–2020. 2013. Available online: https://apps.who.int/iris/bitstream/handle/10665/94384/9789241506236_eng.pdf?sequence=1 (accessed on 3 February 2020).

23. CFS (Committee on World Food Security). Zero Draft Voluntary Guidelines on Food Systems and Nutrition. 2020. Available online: http://www.fao.org/fileadmin/templates/cfs/Docs1819/Nutrition/CFS_Zero_Draft_Voluntary_Guidelines_Food_Systems_and_Nutrition.pd (accessed on 3 February 2020).

24. United Nations. General Assembly (UNGA). A/RES/73/132. Global Health and Foreign Policy: A Healthier World through Better Nutrition. Available online: https://undocs.org/en/A/RES/73/132 (accessed on 3 February 2020).

25. United Nations. General Assembly (UNGA). A/RES/73/253 Agriculture Development, Food Security and Nutrition. Available online: https://undocs.org/en/A/RES/73/253 (accessed on 3 February 2020).

26. HLPE. Agroecological and other innovative approaches for sustainable agriculture and food systems that enhance food security and nutrition. In *A Report by the High-Level Panel of Experts on Food Security and Nutrition of the Committee on World Food Security*; HLPE: Rome, Italy, 2019.

27. Padulosi, S.; Roy, P.; Rosado-May, F.J. *Supporting Nutrition Sensitive Agriculture through Neglected and Underutilized Species—Operational Framework*; Bioversity International and IFAD: Rome, Italy, 2019; 39p, Available online: https://cgspace.cgiar.org/handle/10568/102462 (accessed on 3 February 2020).

28. Hunter, D.; Borelli, T.; Beltrame, D.M.O.; Oliveira, C.N.S.; Coradin, L.; Wasike, V.W.; Wasilwa, L.; Mwai, J.; Manjella, A.; Samarasinghe, G.W.L.; et al. Potential of orphan crops for improving diets and nutrition. Special Issue on Orphan Crops: Contributions to Current and Future Agriculture. *Planta* **2019**. [CrossRef]

29. Hunter, D.; Borelli, T.; Olsen Lauridsen, N.; Gee, E.; Rota Nodari, G.; Moura de Oliveira Beltrame, D.; Oliveira, C.; Wasike, V.W.; Samarasinghe, G.; Tan, A.; et al. *Biodiversity Mainstreaming for Healthy & Sustainable Food Systems: A Toolkit to Support Incorporating Biodiversity into Policies and Programmes*; Bioversity International: Rome, Italy, 2018; 52p.

30. Gee, E.; Borelli, T.; Beltrame, D.M.O.; Oliveira, C.N.S.; Coradin, L.; Wasike, V.; Manjella, A.; Samarasinghe, G.; Güner, B.; Tan, A.; et al. The ABC of mainstreaming biodiversity for food and nutrition: Concepts, theory and practice. In *Biodiversity, Food and Nutrition. A New Agenda for Sustainable Food Systems*; Hunter, D., Borelli, T., Gee, E., Eds.; Routledge: London, UK. (In Press)

31. CBD. National Biodiversity Index of the Convention on Biological Diversity. Extracted from the Global Biodiversity Outlook 1. Annex 1. Biodiversity Information by Country. Available online: https://www.cbd.int/gbo1/annex.shtml#note1 (accessed on 3 February 2020).

32. Convention on Biological Diversity. *Brazil—Country Profile*; Biodiversity Facts: Montreal, Canada, 2016. Available online: https://www.cbd.int/countries/profile/?country=br (accessed on 4 February 2020).

33. Global Nutrition Report. Country Nutrition Profiles. Available online: https://globalnutritionreport.org/resources/nutrition-profiles/ (accessed on 3 February 2020).

34. Padulosi, S.; Bhag Mal, O.; King, I.; Gotor, E. Minor Millets as a Central Element for Sustainably Enhanced Incomes, Empowerment, and Nutrition in Rural India. *Sustainability* **2015**, *7*, 8904–8933. [CrossRef]

35. Raneri, J.E.; Padulosi, S.; Meldrum, J.; King, O.I. Promoting neglected and underutilized species to boost nutrition in LMICs. In *UNSCN Nutrition 44—Food Environments: Where People Meet the Food System*; UNSCN: Rome, Italy, 2019; pp. 10–25. Available online: https://www.unscn.org/uploads/web/news/UNSCN-Nutrition44-WEB-version.pdf (accessed on 3 February 2020).

36. Mittermeier, R.A.; Gil, P.R.; Mittermeier, C.G. *Megadiversity: Earth's Biologically Wealthiest Nations*; Conservation International: Washington, DC, USA, 1997.

37. Development Initiatives. Brazil country nutrition profile. In *2018 Global Nutrition Report: Shining a Light to Spur Action on Nutrition*; Development Initiatives: Bristol, UK, 2018.

38. Von Grebmer, K.; Bernstein, J.; Patterson, F.; Wiemers, M.; Chéilleachair, R.N.; Foley, C.; Gitter, S.; Ekstrom, K.; Fritschel, H. *2019 Global Hunger Index. Climate Justice: A New Narrative for Action Concern Worldwide and Welthungerhilfe*; Helvetas: Zürich, Switzerland, 2019.

39. Development Initiatives. Mali country nutrition profile. In *2018 Global Nutrition Report: Shining a Light to Spur Action on Nutrition*; Development Initiatives: Bristol, UK, 2018.

40. Nordeide, M.B.; Hatløy, A.; Følling, M.; Lied, E.; Oshaug, A. Nutrient composition and nutritional importance of green leaves and wild food resources in an agricultural district, Koutiala, in Southern Mali. *Int. J. Food Sci. Nutr.* **1996**, *47*, 455–468. [CrossRef]

41. Bioversity International and Institut d'Economie Rurale. *Calendrier saisonnier des fruits et légumes pour une alimentation diversifiée dans la région de Sikasso*; Bioversity International and Institut d'Economie Rurale: Mali, Rome, Italy, 2018.

42. De la Peña, I.; Garrett, J.; Gelli, A. *Nutrition-Sensitive Value Chains from a Smallholder Perspective. A Framework for Project Design*; IFAD Research Series: Rome, Italy, 2018.

43. Diawara, F. *Characterization of Food Consumption Patterns in Southern Mali: Districts of Bougouni and Koutiala Sikasso*; IITA: Ibadan, Nigeria, 2013.

44. Kobori, N.C.; Rodriguez-Amaya, D.B. Uncultivated Brazilian green leaves are richer sources of carotenoids than are commercially produced leafy vegetables. *Food Nutr. Bull.* **2008**, *29*, 320–328. [CrossRef]

45. Bharucha, Z.; Pretty, J. The roles and values of wild foods in agricultural systems. *Philos. Trans. R. Soc. B* **2010**, *365*, 2913–2926. [CrossRef]

46. Padulosi, S.; Hunter, D.; Jarvis, A.; Heywood, V. Underutilized crops and climate change - current status and outlook. In *Crop Adaptation to Climate Change*; Yadav, S., Redden, B., Hatfield, J.L., Lotze-Campen, H., Eds.; Wiley-Blackwell: Ames, IA, USA, 2011; pp. 507–521.

47. Kahane, R.; Hodgkin, T.; Jaenicke, H.; Hoogendoorn, C.; Hermann, M.; Keatinge, J.D.H.; d'Arros Hughes, J.; Padulosi, S.; Looney, N. Agrobiodiversity for food security, health and income. *Agron. Sustain. Dev.* **2013**, *33*, 671–693. [CrossRef]

48. Padulosi, S.; Thompson, J.; Rudebjer, P. *Fighting Poverty, Hunger and Malnutrition with Neglected and Underutilized Species (NUS): Needs, Challenges and the Way Forward*; Bioversity International: Rome, Italy, 2013; 56p, ISBN 978-92-9043-941-7.

49. Hunter, D.; Özkan, I.; Beltrame, D.; Samarasinghe, W.L.G.; Wasike, V.W.; Charrondière, U.R.; Borelli, T.; Sokolow, J. Enabled or disabled: Is the environment right for using biodiversity to improve nutrition? *Front. Nutr.* **2016**, *3*, 14. [CrossRef] [PubMed]

50. Da Silva, J.G.; Del Grossi, M.E.; de França, C.G. (Eds.) *The Fome Zero (Zero Hunger) Program: The Brazilian Experience*; MDA: Brasília, Brazil, 2010. Available online: http://www.fao.org/3/a-i3023e.pdf (accessed on 28 January 2020).

51. UNSCN Discussion Paper. *Schools as a System to Improve Nutrition: A New Statement for School-Based Food and Nutrition Interventions*; United Nations System Standing Committee on Nutrition: Rome, Italy, 2017. Available online: https://www.unscn.org/uploads/web/news/document/School-Paper-EN-WEB.pdf (accessed on 30 October 2019).

52. Biodiversity for Food and Nutrition Project (BFN Project). Homepage. Available online: www.b4fn.org (accessed on 3 February 2020).

53. Government of the Republic of Kenya. *Busia County Biodiversity Policy. 2016–2023*; Busia County Government: Busia, Kenya, 2016. Available online: http://www.b4fn.org/fileadmin/templates/b4fn.org/upload/documents/Country_additional_resources/Kenya/Busia_County_Biodiversity_Policy_10_Oct_2017_Final.pdf (accessed on 3 February 2020).

54. Bioversity International and Universidad del Valle de Guatemala. *Calendario Estacional de Frutas y Vegetales para la Diversidad Dietética Jocotán y Camotán, Chiquimula, Guatemala*; Bioversity International and Universidad del Valle de Guatemala: Rome, Italy, 2018.

55. Greenfield, H.; Southgate, D.A.T. *Food Composition Data*; Production Management and Use, Food and Agriculture Organization of the United Nations: Rome, Italy, 2003.

56. FAO/INFOODS Food Composition Database for Biodiversity. Available online: http://www.fao.org/infoods/infoods/food-biodiversity/en/ (accessed on 3 February 2020).

57. FAO. *Voluntary Guidelines for Mainstreaming Biodiversity into Policies, Programmes and National and Regional Plans of Action on Nutrition*; FAO: Rome, Italy, 2016. Available online: http://www.fao.org/3/a-i5248e.pdf (accessed on 27 January 2020).

58. Haddad, L.; Isenman, P. Which aid spending categories have the greatest untapped potential to support the reduction of undernutrition? Some ideas on moving forward. *Food Nutr. Bull.* **2014**, *35*, 266–276. [CrossRef] [PubMed]

59. System on Brazilian Biodiversity (SiBBr) Database. Available online: https://ferramentas.sibbr.gov.br/ficha/bin/view/FN/?&firstIndex=180 (accessed on 3 February 2020).

60. Charrondière, U.R. *Personal Communication*; Food and Agriculture Organization of the United Nations: Rome, Italy, 2018.

61. *Tabela Brasileira de Composição de Alimentos—TBCA*; Versão 7.0.; Universidade de São Paulo (USP); Food Research Center (FoRC): São Paulo, Brazil, 2019. Available online: http://www.fcf.usp.br/tbca (accessed on 30 October 2019).

62. Burlingame, B.; Charrondière, U.R.; Mouille, B. Food composition is fundamental to the cross-cutting initiative on biodiversity for food and nutrition. *J. Food Compos. Anal.* **2009**, *22*, 361–365. [CrossRef]

63. Biodiversity for Food and Nutrition (BFN) Species Database. Available online: http://www.b4fn.org/resources/species-database/ (accessed on 3 February 2020).

64. FAO/INFOODS. Food Composition Database for Biodiversity (BioFoodComp 2.1). 2013. Available online: http://www.fao.org/docrep/019/i3560e/i3560e.pdf (accessed on 30 October 2019).

65. Lachat, C.; Raneri, J.E.; Walker Smith, K.; Kolsteren, P.; Van Damme, P.; Verzelen, K.; Penafiel, D.; Vanhove, W.; Kennedy, G.; Hunter, D.; et al. Dietary species richness as a measure of food biodiversity and nutritional quality of diets. *Proc. Natl. Acad. Sci. USA* **2018**, *115*, 127–132. [CrossRef]

66. Ross-Ibarra, J.; Molina-Cruz, A. The ethnobotany of chaya (*Cnidoscolus aconitifolius* ssp. aconitifolius Breckon): A nutritious maya vegetable. *Econ. Bot.* **2002**, *56*, 350–365. [CrossRef]

67. Molina-Cruz, A.; Curley, L.; Bressani, R. Redescubriendo el valor nutritivo de las hojas de chaya (*Cnidoscolus aconitifolius*; Euphorbiaceae). *Univ. Valle Guatem.* **1997**, *3*, 1–4.

68. Small, E. Teff & fonio—Africa's sustainable cereals. *Biodiversity* **2015**, *16*, 37–41.

69. Mubaiwa, J.; Fogliano, V.; Chidewe, C.; Linnemann, A.R. Hard-to-cook phenomenon in Bambara groundnut (*Vigna subterranea* (L.) Verdc.) processing: Options to improve its role in providing food security. *Food Rev. Int.* **2017**, *33*, 167–194. [CrossRef]

70. FAO/Government of Kenya. *Kenya Food Composition Tables*; FAO/Government of Kenya: Nairobi, Kenya, 2018; p. 254. Available online: http://www.fao.org/3/I9120EN/i9120en.pdf (accessed on 30 October 2019).

71. Vall, E.; Andrieu, N.; Beavogui, F.; Sogodogo, D. Les cultures de soudure comme stratégie de lutte contre l'insécurité alimentaire saisonnière en Afrique de l'Ouest: Le cas du fonio (*Digitaria exilis* Stapf). *Cahiers Agricoles* **2011**, *20*, 94–300.

72. Dewbre, J.; Borot de Battisti, A. Agricultural Progress in Cameroon, Ghana and Mali: Why It Happened and How to Sustain It. In *OECD Food, Agriculture and Fisheries Working Papers, No. 9*; OECD Publishing: Paris, France, 2008. [CrossRef]

73. Cruz, J.; Béavogui, F. *Fonio, an African Cereal*; Editions CIRAD: Montpellier, France, 2016; 153p, ISBN 978-2-87614-720-1.

74. Cooper, M.W.; West, C.T. Unravelling the Sikasso paradox: Agricultural change and malnutrition in Sikasso, Mali. *Ecol. Food Nutr.* **2017**, *56*, 101–123. [CrossRef] [PubMed]

75. Amaya, N.; Padulosi, S.; Meldrum, G. Value Chain Analysis of Chaya (Mayan Spinach) in Guatemala. *Econ. Bot.* **2019**, 1–15. [CrossRef]

76. CBD. Strategic Plan for Biodiversity 2011–2020 and the Aichi Targets. "Living in Harmony with Nature". 2011. Available online: https://www.cbd.int/doc/strategic-plan/2011-2020/Aichi-Targets-EN.pdf (accessed on 3 February 2020).

77. IPES-Food. *From Uniformity to Diversity: A Paradigm Shift from Industrial Agriculture to Diversified Agroecological Systems*; International Panel of Experts on Sustainable Food systems: Brussels, Belgium, 2016. Available online: http://www.ipes-food.org/_img/upload/files/UniformityToDiversity_FULL.pdf (accessed on 30 October 2019).

78. FOLU. *Growing Better: Ten Critical Transitions to Transform. Food and Land Use. The Global Consultation Report of the Food and Land Use Coalition*; The Food and Land Use Coalition: London, UK, 2019. Available online: https://www.foodandlandusecoalition.org/wp-content/uploads/2019/09/FOLU-GrowingBetter-GlobalReport.pdf (accessed on 30 October 2019).

79. WHO/CBD, World Health Organization and Secretariat of the Convention on Biological Diversity. *Connecting Global Priorities: Biodiversity and Human Health. A State of Knowledge Review*; WHO Press: Geneva, Switzerland, 2015. Available online: https://www.cbd.int/health/SOK-biodiversity-en.pdf (accessed on 30 October 2019).

80. MoMD&E, Ministry of Mahaweli Development and Environment. *National Biodiversity Strategic Action Plan. 2016–2022*; Biodiversity Secretariat, Ministry of Mahaweli Development and Environment: Colombo, Sri Lanka, 2016. Available online: https://www.cbd.int/doc/world/lk/lk-nbsap-v2-en.pdf (accessed on 30 October 2019).

81. MoE, Ministry of Environment. *National Biodiversity Strategy and Action Plan*; Ministry of Environment—MMA, Biodiversity Secretariat: Brasília, Brazil, 2017. Available online: https://www.cbd.int/doc/world/br/br-nbsap-v3-en.pdf (accessed on 30 October 2019).

82. MoE & MoSD, Ministry of Environment and Ministry of Social Development. *Diário oficial da união. Portaria Interministerial N° 284 de 30 de Maio de* **2018**, *131*, 92, Imprensa Nacional, Brazil. Available online: https://ciorganicos.com.br/wp-content/uploads/2018/07/PORTARIA-INTERMINISTERIAL-N%C2%BA-284.pdf (accessed on 30 October 2019).

83. Beltrame, D.M.; Oliveira, C.N.S.; Borelli, T.; de Santiago, A.C.; Monego, E.T.; Vera de Rosso, V.; Coradin, L.; Hunter, D. Diversifying institutional food procurement—Opportunities and barriers for integrating biodiversity for food and nutrition in Brazil. *Revista Raízes* **2016**, *36*, 55–69.

84. Santana, C.A.M.; Nascimento, J.R. *Public Policies and Agricultural Investment in Brazil*; Food and Agriculture Organization of the United Nations. Policy Assistance Support Service (TCSP): Brasília, Brazil, 2012.

85. MoAPA, Ministério da Agricultura, Pecuária e Abastecimento. Diário Oficial da União. *Portaria n° 141, de 8 de Janeiro de* **2019**, 7, 12, Gabinete do Ministro: Brasilia, Brazil. Available online: http://pesquisa.in.gov.br/imprensa/jsp/visualiza/index.jsp?data=10/01/2019&jornal=515&pagina=12 (accessed on 3 February 2020).

86. Valencia, V.; Wittman, H.; Blesh, J. Structuring Markets for Resilient Farming Systems. *Agron. Sustain. Dev.* **2019**, *39*, 25. [CrossRef]

87. Béné, C.; Prager, S.; Achicanoy, H.; Toro, P.; Lamotte, L.; Cedrez, C.; Mapes, B. Understanding food systems drivers: A critical review of the literature. *Glob. Food Secur.* **2019**, *23*, 149–159. [CrossRef]

88. FAO. *Sustainable Value Chains for Sustainable Food Systems: A Workshop of the FAO/UNEP Programme on Sustainable Food Systems*; FAO: Rome, Italy, 2016.

89. IFAD NUS Project. Homepage. Available online: http://www.nuscommunity.org (accessed on 3 February 2020).

90. Tartanac, F.; Swensson, L.F.J.; Polo Galante, A.; Hunter, D. Institutional food procurement for promoting sustainable diets. In *Sustainable Diets: The Transdisciplinary Imperative*; Burlingame, B., Dernini, S., Eds.; CABI: Wallingford, UK, 2018.

91. Mabhaudhi, T.; Chibarabada, T.P.; Chimonyo, V.G.P.; Murugani, V.G.; Pereira, L.M.; Sobratee, N.; Govender, L.; Slotow, R.; Modi, A.T. Mainstreaming Underutilized Indigenous and Traditional Crops into Food Systems: A South African Perspective. *Sustainability* **2019**, *11*, 172. [CrossRef]

92. Padulosi, S.; Bala Ravi, S.; Rojas, W.; Valdivia, R.; Jager, M.; Polar, V.; Gotor, E.; Mal, B. Experiences and lessons learned in the framework of a global UN effort in support of neglected and underutilized species. *ISHS Acta Hortic.* **2013**, *979*, 517–531. [CrossRef]

93. The Gazette of India. National Food Security Act. 10 September 2013. Available online: http://egazette.nic.in/WriteReadData/2013/E_29_2013_429.pdf (accessed on 30 October 2019).

94. Padulosi, S.; Mal, B.; Ravi, S.B.; Gowda, J.; Gowda, K.T.K.; Shanthakumar, G.; Yenagi, N.; Dutta, M. Food security and climate change: Role of plant genetic resources of minor millets. *Indian J. Plant Genet. Resour.* **2009**, *22*, 1–16.

95. King, O.I.; Padulosi, S. Agricultural biodiversity and women's empowerment: A successful story from Kolli Hills, India. In *Creating Mutual Benefits: Examples of Gender and Biodiversity Outcomes from Bioversity International's Research*; Del Castello, A., Bailey, A., Eds.; Bioversity International: Rome, Italy, 2017; 4p, Available online: https://www.bioversityinternational.org/fileadmin/user_upload/Creating_GSICP.pdf (accessed on 30 October 2019).

96. Caron, P.; de Loma-Osorio, G.F.; Nabarro, D.; Hainzelin, E.; Guillou, M.; Andersen, I.; Arnold, T.; Astralaga, M.; Beukeboom, M.; Bickersteth, S.; et al. Food systems for sustainable development: Proposals for a four-part transformation. *Agron. Sustain. Dev.* **2018**, *38*, 41. [CrossRef] [PubMed]

97. Grasso, A. *School Gardening in Busia County, Kenya*. A case study. BFN Project website. 2015. Available online: http://www.b4fn.org/from-the-field/stories/school-gardening-in-busia-county-kenya (accessed on 3 February 2020).

98. Tan, A.; Adanacioğlu, N.; Tuğrul Ay, S. *Students in Nature, in the Garden and in the Kitchen, Turkey*. A case Study. BFN Project Website. 2018. Available online: http://www.b4fn.org/case-studies/case-studies/students-in-nature-in-the-garden-and-in-the-kitchen (accessed on 3 February 2020).

99. Searchinger, T.; Waite, R.; Hanson, C.; Ranganathan, J.; Matthews, E. *World Resources Report Creating a Sustainable Food Future: A Menu of Solutions to Feed Nearly 10 Billion People by 2050*; Worlds Resources Institute: Washington, DC, USA, 2019.

100. Hawkes, C.; Blouin, C.; Henson, S.; Drager, N.; Dubé, L. *Trade, Food, Diet. and Health. Perspectives and Policy Options*; Wiley-Blackwell: Hoboken, NJ, USA, 2009.

101. Lochetti, G.; Robitaille, R. Adding colour to rural diets year-round with the Seasonal Food Availability Booklet. 2018. Available online: https://www.bioversityinternational.org/news/detail/adding-colour-to-rural-diets-year-round-with-the-seasonal-food-availability-booklet (accessed on 3 February 2020).

102. MoH, Ministry of Health. *Dietary Guidelines for the Brazilian Population*; Ministry of Health: Brasilia, Brazil, 2015. Available online: http://bvsms.saude.gov.br/bvs/publicacoes/dietary_guidelines_brazilian_population.pdf (accessed on 30 October 2019).

103. IFAD NUS Project. About the project. Available online: http://www.nuscommunity.org/initiatives/ifad-eu-ccafs-nus/ (accessed on 3 February 2020).

104. FAO and FHI 360. *Minimum Dietary Diversity for Women: A Guide for Measurement*; FAO: Rome, Italy, 2016.

105. De Andrade Cardoso Santiago, R.; Coradin, L. (Eds.) *Biodiversidade Brasileira: Sabores e aromas*; Ministério do Meio Ambiente: Brasília, Brazil, 2019. Available online: http://redesans.com.br/rede/wp-content/uploads/2019/08/Livro-de-Receitas-03-07-2019.pdf (accessed on 30 October 2019).

106. Rathnasinghe, D.; Samarasinghe, G.; Silva, R.; Hunter, D. 'True Sri Lanka Taste' food outlets: Promoting indigenous foods for healthier diets. *Nutr. Exch.* **2019**, *12*, 14–15.

107. HLPE. Nutrition and food systems. In *A Report by the High Level Panel of Experts on Food Security and Nutrition of the Committee on World Food Security*; HLPE Report 12; FAO: Rome, Italy, 2017. Available online: http://www.fao.org/3/a-i7846e.pdf (accessed on 30 October 2019).

108. Vermeulen, S.; Park, T.; Khoury, C.K.; Mockshell, J.; Béné, C.; Thi, H.T.; Heard, B.; Wilson, B. *Changing Diets and Transforming Food Systems*; CCAFS Working Paper no. 282; CGIAR Research Program on Climate Change, Agriculture and Food Security (CCAFS): Wageningen, The Netherlands, 2019. Available online: https://cgspace.cgiar.org/bitstream/handle/10568/103987/Changing%20diets%20and%20transforming%20food%20systems%20WP%20282_repaired.pdf (accessed on 30 October 2019).

Review

Understanding Molecular Mechanisms of Seed Dormancy for Improved Germination in Traditional Leafy Vegetables: An Overview

Fernand S. Sohindji, Dêêdi E. O. Sogbohossou, Herbaud P. F. Zohoungbogbo, Carlos A. Houdegbe and Enoch G. Achigan-Dako *

Laboratory of Genetics, Horticulture and Seed Science, Faculty of Agronomic Sciences, University of Abomey-Calavi, 01 BP 526 Tri Postal, Cotonou, Benin; sohindjisilverfer@gmail.com (F.S.S.); deedik@gmail.com (D.E.O.S.); phanuelherbaud@gmail.com (H.P.F.Z.); houdariscarl@gmail.com (C.A.H.)
* Correspondence: enoch.achigandako@uac.bj; Tel.: +229-95-39-32-83

Received: 28 October 2019; Accepted: 24 December 2019; Published: 1 January 2020

Abstract: Loss of seed viability, poor and delayed germination, and inaccessibility to high-quality seeds are key bottlenecks limiting all-year-round production of African traditional leafy vegetables (TLVs). Poor quality seeds are the result of several factors including harvest time, storage, and conservation conditions, and seed dormancy. While other factors can be easily controlled, breaking seed dormancy requires thorough knowledge of the seed intrinsic nature and physiology. Here, we synthesized the scattered knowledge on seed dormancy constraints in TLVs, highlighted seed dormancy regulation factors, and developed a conceptual approach for molecular genetic analysis of seed dormancy in TLVs. Several hormones, proteins, changes in chromatin structures, ribosomes, and quantitative trait loci (QTL) are involved in seed dormancy regulation. However, the bulk of knowledge was based on cereals and *Arabidopsis* and there is little awareness about seed dormancy facts and mechanisms in TLVs. To successfully decipher seed dormancy in TLVs, we used *Gynandropsis gynandra* to illustrate possible research avenues and highlighted the potential of this species as a model plant for seed dormancy analysis. This will serve as a guideline to provide prospective producers with high-quality seeds.

Keywords: seed dormancy; seed germination; molecular biology; genetics; traditional leafy vegetables; *Gynandropsis gynandra*

---

1. Introduction

More than 1000 species were recorded to be used by African rural communities for dietary diversity, medicine purpose, food traditions, and cultural identity [1–4]. Given the potential of African traditional leafy vegetables (TLVs) to cope with varying climate constraints and feed Africa, the production and consumption of the same have been promoted in the continent for the last two decades. For example, the antioxidant system of *Cleome spinosa* Jacq. and *Gynandropsis gynandra* L. (Briq) variously copes with reactive oxygen species (ROS) formation under drought conditions limiting damage to cell structures, lipids, proteins, carbohydrates, nucleic acids, and cell death [5]. Luoh et al. [6] reported also the lowest losses of nutrients in amaranth species and the African nightshade species when cultivating under deficient conditions. Few species such as *Amaranthus* spp., *Solanum scabrum* Mill., and *Solanum macrocarpon* L. have been well domesticated, whereas *Bidens pilosa* L., *Brassica carinata* A. Braun., *Gynandropsis gynandra*, *Corchorus* spp., *Launaea taraxacifolia* Willd., *Talinum triangulare* Willd., etc. are widely used over the continent, but still semi-domesticated species and can grow spontaneously and/or cultivated according to the sociolinguistic groups or regions [3]. The successful promotion of those crops requires a proper investigation of all aspects related to their life cycle. The first step toward such promotion is to focus on farmers' traits of interest and constraints during cultivation and marketing. Once farmers' priorities are clearly defined, a subsequent step is to investigate crop physiology in relation to the traits of interest for farmers. For several species including *Gynandropsis gynandra*, *Solanum nigrum* L., *Amaranthus* spp., and *Corchorus* spp. the lack of high-quality seeds is the main constraint for full domestication [7]. In fact, seed germination and seed dormancy processes are still unclear for many traditional leafy vegetables. Farmers are confronted with the loss of seed viability, poor and delayed germination of seeds, and inaccessibility to quality seeds limiting all-year-round production of those crops [8,9]. In addition, Adebooye et al. [10] and Sogbohossou et al. [11] brought to light several gaps of knowledge for the improvement of TLVs. They listed, for instance: (1) the lack of extensive germplasm collections; (2) the need to understand the genetic control of key traits; (3) the need to develop and evaluate new cultivars, assess end-users' preferences, and perform multi-environment experiments; (4) the demand for appropriate technical packaging; and (5) the call for sustained efforts for value chain development. However, to reach active domestication, it is important to ensure quality seed availability and find solutions for seed dormancy issues that limit TLVs adoption and production.

Seed dormancy is a state of a viable seed, expressed by the inhibition of germination under favourable environmental conditions required for adequate germination [12,13]. It is an adaptive trait that optimizes the distribution of germination over time in a population of seeds [14]. On the other hand, germination is usually related to radicle protrusion, which is normally the visible result of germination [15]. Before this visible aspect, there are many events that begin with the uptake of water by mature dry seed and imbibition, and end with the embryonic axis elongation [16]. Seed dormancy and the absence of favourable environmental conditions for germination result in the absence of germination [12]. Non-germination due to unfavourable conditions is referred to as "quiescence" and enables seed survival for further seedling development under adverse conditions [17]. Seed dormancy appears as a complex quantitative trait under the influence of several genetic, hormonal, physiological, and environmental factors [18]. The pre-harvest sprouting and the absence or Delay of Germination after-ripening are the two undesirable contrasting levels of seed dormancy. Consequently, the constraint about dormancy is twofold: either it is not present in seeds (zero level) leading to pre-harvest sprouting especially for cereals or it is present at high level leading to the absence or Delay of Germination at the desired time [19,20]. There are two types of seed dormancy based on its times of expression: "primary dormancy" developed on the mother plant [21], and "secondary dormancy" induced in previously non-dormant seeds or re-induced in seeds that have lost primary dormancy due to the unfavourable environment factors for germination after seed dispersal [22]. Baskin and Baskin [12] reported various classes of dormancy based on embryo growth potential, seed coat, and seed physiological responses to temperature. Some of the classes were: physical dormancy (PY), physiological dormancy (PD), morphological dormancy (MD), morpho-physiological dormancy (MPD), and combinational dormancy (physical and physiological dormancy).

As stated by Koornneef et al. [23], the challenge is to master the initiation and suppression of germination ability through gene identification based on changes in seed transcriptome, proteome, and hormones under different environmental conditions. So far, the majority of molecular genetic studies on seed dormancy were conducted in the model plant species *Arabidopsis thaliana* L. [24] and economically important crops such as *Oryza sativa* L. [25], *Triticum aestivum* L. [26], and *Lycopersicum esculentum* Mill. [27]. Such studies revealed that seed germinability and subsequent seedling development are controlled by two sets of factors: internal factors including proteins, plant hormones (Abscissic–Gibberellic acids balance), chromatin-related factors (methylation, acetylation, histone ubiquitination), related genes (maturating genes and hormonal and epigenetics-regulating genes), non-enzymatic processes, seed morphological and structural components (endosperm, pericarp, seed coat), and external factors, such as light, temperature, salinity, acidity, soil nitrate [28,29]. These biotic and abiotic factors interact and determine the presence or absence of dormancy during seed development, in imbibed mature seeds and in dry seeds. Despite the fact that efforts of plant biologists, crop geneticists, breeders, and food scientists to understand seed dormancy phenomenon have shed light on physiology, genetic and molecular aspects of seed dormancy, little is known about TLVs species, although seed dormancy is still a challenge in these species as above described.

In this paper, we provide a synthesis of the current state of knowledge about seed dormancy in TLVs and seed dormancy control in plants. We further discussed how to transfer such knowledge to leafy vegetables in tropical areas. We highlighted the case of spider plant (*Gynandropsis gynandra*), a traditional leafy vegetable closely related to the model species *A. thaliana*. This review addressed the following questions: what is the germination ability of the well consumed TLVs? How do we identify seed dormancy regulatory genes in TLVs? How do we study the effects of those genes on seed germination? What approach do we implement to develop new cultivars with higher seed quality and attributes?

2. Methods

Literature search was conducted in PubMed Central and Google Scholar databases. Well-studied crops such as *Arabidopsis thaliana, Oryza sativa, Zea mays* L., *Triticum aestivum, Hordeum vulgare* L., *Lycopersicum esculentum, Avena sativa* L., *Nicotiana plumbaginifolia* Viv., and African leafy vegetables such as *Gynandropsis gynandra, Solanum nigrum, Corchorus olitorius* L., *Talinum triangulare* (Jacq.) Willd., and *Amaranthus* spp. (Figure 1) were considered in the search. The keywords used to collect the documents included "seed dormancy", "molecular control", "hormonal control and genetic variation", "germination of African/indigenous/traditional leafy vegetables", and "seed constraints for African/indigenous/traditional leafy vegetables". No date coverage was specified during the search. The relevance of documents was assessed based on the molecular pathways, genetic control, and environmental factors reported about seed dormancy or seed germination. The process consisted of a preliminary screening of titles and abstracts of 243 papers. The reading of full texts helped select 194 relevant papers for further review.

Figure 1. Some of important African traditional leafy vegetables widely used across Africa: (**a**) *Gynandropsis gynandra*, (**b**) *Solanum nigrum*, (**c**) *Corchorus olitorius*, (**d**) *Talinum triangulare*, (**e**,**f**) *Amaranthus* spp. Pictures: courtesy of F.S. Sohindji, H.F.P. Zohoungbogbo and E.G. Achigan-Dako.

3. Seed Dormancy in Traditional Leafy Vegetables

Seed dormancy prevents seeds to germinate under unfavourable conditions for further growth and development of the plant. In many leafy vegetables, it appears as a challenge while rapid and uniform germination are expected after sowing. So far, physical and physiological dormancy are the main dormancy cases reported in TLVs species from mature freshly harvested seeds to dry seeds [30]. An overview of studies on seed dormancy in TLVs is presented in Table 1.

Table 1. Seed constraints and dormancy types reported for traditional leafy vegetables.

Common Name	Scientific Name (Family)	Seed Constraints	References
Spider plant	*Gynandropsis gynandra* (Cleomaceae)	Primary non-deep physiological dormancy Physical dormancy and secondary dormancy Oxygen barrier between embryo and tissue Low germination of freshly harvested seeds Delayed, poor, and absence of germination Inaccessibility of quality seed for seed analysts and gene bank curators Low vigor and reduced number of viable seeds harvested by farmers Physiological dormancy	[31,32]
Jute mallow	*Corchorus olitorius* (Malvaceae)	Loss of viability and poor germination of fresh and old seeds Impermeable seed coat	[8,33]
African Nightshade	*Solanum nigrum* (Solanaceae)	Poor germination of seeds Improper seed extraction Deeper level of primary dormancy	[8,34]
Waterleaf	*Talinum triangulare* (Portulacaceae)	Dormancy due to the nature of the seed testa Undetermined physiological factors	[35]
Amaranths species	*Amaranthus* spp. (Amaranthaceae)	Primary dormancy and secondary dormancy occur among amaranths species	[36]

Seeds of waterleaf (*Talinum triangulare*) are known to exhibit a kind of dormancy due to the impermeability of the seed testa and some undetermined physiological factors [35]. These authors reported that scarification, alternating temperatures (6–10 °C and 28–35 °C), and constant temperature (34 °C) should enhance germination of waterleaf seeds. Taab and Andersson [34] reported a deeper level of primary dormancy for nightshade (*Solanum nigrum*). Dormancy-breaking treatments with stratification, potassium nitrate, and gibberellic acid failed to show encouraging results, and are often not applicable at farmers' level [34,37]. A loss of viability and poor germination of fresh and old seeds in jute mallow (*Corchorus olitorius*) are associated with the impermeability of jute mallow seed coat [8,33]. In the case of spider plant (*Gynandropsis gynandra*), its cultivation is limited by the fact that its seeds can exhibit a high dormancy lasting for several months. Geneve [38] reported that a primary non-deep physiological dormancy occurs in spider plant while Ochuodho and Modi [39] suspected physical dormancy and secondary dormancy. Ekpong [40] clarified that spider plant seeds are permeable to water but this water is trapped in the tissue between the embryo and the seed coat creating an oxygen barrier. Baskin and Baskin [32] concluded that spider plant exhibits a physiological dormancy. Recently, Shilla et al. [31] reported that there were no dormancy cases on fresh seeds of spider plant according to World Vegetable Center preliminary results. Nevertheless, the seeds can stay in a dormant state for several months before germination is activated and improved with dry storage periods [41–43]. Various levels of seed dormancy such as primary and secondary dormancy occur among amaranths species. For instance, there are primary dormant *Amaranthus retroflexus* L., secondary dormant *Amaranthus paniculatus* L., and non-dormant *Amaranthus caudatus* L. seeds [36]. Seed treatments such as seed holding in low temperature, pre-chilling, and the application of ethylene induce dormancy-breaking and accelerate the germination process in *Amaranthus* seeds [44].

While other factors including harvest time, storage, and conservation conditions can be easily controlled by farmers, breaking seed dormancy requires a thorough knowledge of the seed intrinsic nature and physiology. Farmers' efforts to break seed dormancy are therefore still insufficient to assure 50% to 100% germination. Many seed pre-treatment techniques (Table 2) such as light/dark, cold/warm, tap/distilled water, and physical/chemical scarification have been tested by researchers to break the seed dormancy in crops [13]. Unfortunately, results obtained can be negatively affected by factors such as seed provenance, seasons of production, storage containers, storage period, and storage temperature, to list a few environmental conditions [43]. Therefore, traditional leafy vegetable seeds management is still traditional and farmers, seed companies, gene banks, and researchers require adequate methods for breaking seed dormancy of those species.

Table 2. Various strategies for seed dormancy-breaking in African leafy vegetables (*Gynandropsis gynandra, Amaranthus* spp., *Corchorus olitorius, Talinum triangulare, Solanum nigrum*).

TLVs Species	Strategies for Seed Dormancy-Breaking
Gynandropsis gynandra	- Stratification for two weeks at 5 °C - 12 h of seed soaking - Scarification with 1000 µM KNO_3, 1000 µM K_2SO_4, 1000 $(NH_4)_2SO_4$ - GA_3 application at a concentration of 500 ppm - 60 min of seed pre-washing in running water - 1–5 days of seed pre-heating at 40 °C - Seed dried to 5% moisture content and stored at −20 °C - 3–6 months of after-ripening - Seed puncturing at the radicle end - Darkness and either alternating 20–30 °C or continuously at 30 °C for germination - 1 day of moist chilling at 5 °C
Amaranthus species	- Holding seeds for 18 months at 6 °C - Pre-chilling treatment - Scarification with 1000 µM KNO_3, 1000 µM K_2SO_4, 1000 $(NH_4)_2SO_4$ - 100 µM GA_3 application - Immersion in 2% KNO_3 solution for 24 h - Cold seed stratification at 5 °C for 12 days - Germination conditions under light - Application of ethylene, ethephon or 1-aminocyclopropane-1-carboxylic acid
Corchorus olitorius	- Mechanical scarification - Leaching treatment - Soaking in boiling water for 5 min
Talinum triangulare	- Activated carbon - Scarification - 5% thiourea treatment - Constantly high temperature (34 °C) for germination - Immersion in 0.2% potassium nitrate solution (24 h) - Immersion in water (24 h)
Solanum nigrum	- Soaking in GA3 with concentration of 0, 25, 50, 100, 200, and 400 ppm - Wet and dry pre-chilling for 15, 30, and 45 days in 4 °C - Seed coat chemical scarification for 1–3 min - Exposing seed to UV-C radiation for 30 min

4. Seed Dormancy Regulation in Plants

Seed dormancy is regulated by genotypic (internal regulation) and environmental factors during three stages in the persistent soil seed bank such as seed development, after-ripening, and seed germination. [45]. During seed development, some reserves are accumulated in seeds (reserve accumulation). During after-ripening, seeds especially orthodox seeds have ability to survive desiccation (desiccation tolerance). The seed germination stands for mobilization of reserves under favourable condition (reserve mobilization). The viable, mature, and freshly harvested seed can be dormant (primary dormancy) or non-dormant (able to germinate). The favourable conditions during after-ripening lead to the release of primary dormancy seed which becomes non-dormant. When conditions are unfavourable, non-dormant seeds, even those for which primary dormancy was released, enter the quiescent state before entering into secondary dormancy when unfavourable conditions persist or could be able to germinate under favourable conditions.

The internal regulation of seed dormancy occurs in two main pathways acting in interaction with the environment. There is the hormone-level pathway (indirect pathway) and gene-level pathway (direct pathway). The different known regulators involved in seed dormancy control and their interactions are presented in Figure 2.

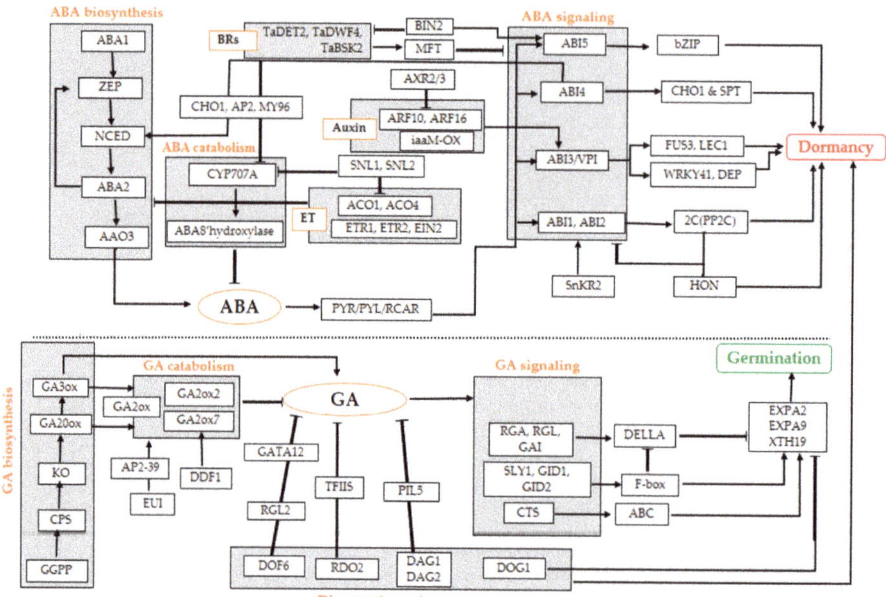

Figure 2. Mechanisms underlining seed dormancy and seed germination control in plants. This figure highlights abscisic and gibberellic acid metabolism and signalling pathways in plants and how ABA-GA balance is affected by the activity of other plant hormones and other genes enabling seed dormancy or seed germination. *ABA1*, abscisic acid deficient 1; *ABA2*, abscisic acid deficient 2; ZEP, zeaxanthin epoxidase; *NCED*, carotenoid cleavage dioxygenase; *AAO3*, abscisic aldehyde oxidase 3; ABA, abscisic acid; PYR/PYL/RCAR, pyrabactin resistance/pyrabactin-like/regulatory components of ABA receptors; *ABI*, abscisic acid insensitive; *VP1*, viviparous 1; bZIP, basic leucine zipper; *CHO1*, CHOTTO 1; *SPT*, SPATULA; *FUS3*, FUSCA3; *LEC1*, leafy cotyledon 1; WRKY41, WRKY DNA-binding protein 41; DEP, DESPIERTO; PP2C, protein phosphatase 2C; *HON*, HONSU; SnRK2, SNF1-related protein kinase 2; *CYP707A*, Cytochrome P450; ABA8'OH, ABA 8-hydroxylase; *AP2*, AP2-domain; *MYB96*, myeloblastosis 96; BRs, brassinosteroids; *TaDET2*, *Triticum aestivum* de-etiolated 2; *Ta DWF4*, *Triticum aestivum* DWF 4; *TaBSK2*, *Triticum aestivum* brassinosteroid signaling kinase 2; *BIN2*, brassinosteroid insensitive 2; *MFT*, mother of FT and TFL1; *ARF*, auxin-responsive factors; AXR2, auxin resistant 2; AXR3, auxin resistant 3; ACO, 1-aminocyclopropane-1-carboxylic acid oxidase; ET, Ethylene; *EIN2*, ethylene insensitive 2; *ETR1*, ethylene triple response 1; *ETR2*, ethylene triple response 2; *SNL1*, SIN3-like 1; *SNL2*, SIN3-like 2; GGDP, geranylgeranyl pyrophosphate; *CPS*, ent-copalyl diphosphate synthetase; *KO1*, ENT-kaurene oxidase 1; *GA3ox1*, gibberellins 3-oxidase 1; *GA20ox3*, gibberellins 20-oxidase 3; *GA2ox2*, gibberellins 2-oxidase 2; *GA2ox7*, gibberellins 2-oxidase 7; *AP2-39*, AP2 domain-containing transcription factor OsAP2-39; *EUI*, elongated uppermost internode; *DDF1*, delayed flowering 1; GA, gibberellins; GATA, GATA proteins; ABC, ATP-binding cassette; *DOF6*, DNA binding1 zinc finger 6; *DAG1*, DOF affecting germination1; *DAG2*, DOF affecting germination2; *DDF1*, delayed flowering 1; DELLA, DELLA proteins; F-box, F-box proteins;

DOG1, Delay of Germination 1; *RGA*, repressor of gibberellin acid; *RGL1*, repressor of gibberellin acid like 1; *RGL2*, repressor of gibberellin acid like 2; *RDO2*, reduced dormancy 2; *CTS*, comatose; *PIL5*, phytochrome interacting factor 3-like 5; *TFIIS*, transcription factor S-II; *SLY1*, sleepy1; *GAI*, gibberellin acid insensitive; *GID1*, gibberellin acid insensitive dwarf1; *GID2*, gibberellin acid insensitive dwarf2; *EXPA2*, expansin-A2; *EXPA9*, expansin-A9; *XTH19*, xyloglucan endo-transglycosylase 19.

4.1. Direct Pathway Regulation

Delay of Germination (*DOG*) genes are important determinants of seed dormancy within populations. The seed dormancy QTL called Delay of Germination 1 (*DOG1*) was identified as responsible for a strong dormancy and acting in interaction with *DOG3* [46]. *DOG1* was shown to be specifically expressed during seed development with detectable levels in dry seeds [47]. Global transcript analysis in *Arabidopsis* using microarrays indicated that the expression level of *DOG1* decreased during after-ripening [46]. The expression of *DOG1* was also reduced in the *hub1* mutant characterized by reduced dormancy in agreement with a role of *DOG1* in regulating dormancy levels [48]. Bentsink et al. [49] identified eleven *DOG* QTLs but there is the absence of strong epistasis interaction between different *DOG* loci suggesting that *DOG* loci affect dormancy via distinct genetic pathways. However, not all *DOG* genomic regions identified by Bentsink et al. [49] contain genes that have previously been associated with seed dormancy. For instance, *DOG3* collocates with *LEC2*, *DOG4* with *TT7* (Transparent Testa 7), *DOG5* with *ABI1*, *DOG19* with *GA2*, *DOG20* with *PIL5*, and *DOG22* with *RGL2*. *DOG1* is involved in the enhancement of dormancy by low temperatures during seed maturation and is a central factor of seed dormancy [29,50–53]. Homologs of Arabidopsis *DOG1* were characterized in other plants including *Brassica napus* L. [54,55], rice [54,56], *Lepidium sativum* L. [57], barley, and wheat [58]. Ashikawa et al. [58] reported that *DOG1*-like genes in wheat and barley were good candidate transgenes for reducing pre-harvest sprouting in wheat.

The DNA BINDING1 ZINC FINGER (DOF) proteins are a family of plant transcription factors in the plant kingdom and were identified as potential regulators of seed dormancy [59,60]. In *Arabidopsis* species, the genes encoding transcription regulators such as DOF affecting germination (*DAG*) have been identified and investigated for their efficiency in seed dormancy regulation [59,61]. Stamm et al. [62] reported 36 DOF proteins in *Arabidopsis*, many of which have been implicated in the regulation of germination. To support this, the DOF zinc finger proteins (DOF Affecting Germination 1 and 2) have been shown to possess opposing roles in the regulation of germination. For instance, *DAG1* inhibits germination by mediating *PIL5* activity and affecting gibberellin biosynthesis [63,64]. *DOF6* was shown to negatively regulate germination by affecting abscisic acid signalling in seeds [60]. Recently, Ravindran et al. [61] reported a novel crosstalk between DOF6 and *RGL2* that enables primary dormancy in *Arabidopsis* through GATA transcription factor regulation. This novel *RGL2*–DOF6 complex is required for activating *GATA12*, a gene encoding a GATA-type zinc finger transcription factor, as one of the downstream targets of *RGL2* expression in *Arabidopsis thaliana*, thus revealing a molecular mechanism to enforce primary seed dormancy by repressing GA signalling [59,61,62].

Other important genes are those related to reduced dormancy (*RDO*). A mutagenesis screen for seed dormancy in *Arabidopsis* yielded reduced dormancy (*RDO*) mutants, which appeared to be central for the dormancy mechanisms and an important target for seed dormancy research [65,66]. For example, *RDO2*, one of the genes identified from the screening, encoded *TFIIS* [67] and a mutation in *TFIIS* resulted in reduced seed dormancy [68]. Mutants *dog1* and *rdo4* presented a reduced seed longevity phenotype [55,69,70]. Then, the reduced dormancy mutants *rdo1* and *rdo2* are not affected in their response to ABA.

4.2. Hormonal Pathway Regulation

Abscisic acid (ABA) produced by embryo is fundamental for the promotion of seed dormancy unlike that produced by maternal tissues through ABA biosynthesis genes such as carotenoid cleavage dioxygenase (*NCED*), ABA-deficient (*ABA1*, *ABA2*), and abscisic aldehyde oxidase 3 (*AAO3*) [71–74]. In *Arabidopsis thaliana*, the induction of seed dormancy is due to *NCED5*, *NCED6*, and *NCED9* genes while *NCED1* and *NCED2* genes are reported regulating seed dormancy in cereals crops [17,75]. ABA is catabolized through ABA8′hydroxylase encoded by Cytochrome P450 (*CYP707A*) genes including *CYP707A1*, *CYP707A2* and *CYP707A3* [17]. ABA is signalled in *Arabidopsis* and monocot plants through ABA insensitive genes, such as *ABI1*, *ABI2*, *ABI3/VP1*, *ABI4*, *ABI5*, and *ABI8*. *ABI3/VP1* genes seem to be the most important in seed dormancy induction whereas *ABI4* regulates abiotic stress responses and different aspects of plant development and plays an important role in seed dormancy maintenance by binding itself to the promoter of *CYP707A1* and *CYP707A2* genes and repressing their expression [76]. Therefore, *NCED2* and *NCED3* activation is enhanced through *CHO1* encoding with APETELA 2 (*AP2*) domain, the ABA-responsive R3R2-type MYB transcription factor: myeloblastosis 96 (MYB96) [17,77]. These ABA signalling genes regulate seed dormancy through several transcription factors including Pyrabactin resistance proteins/PYR-like proteins/regulatory components of ABA receptor (PYR/PYL/RCAR), phosphatase 2C (PP2C), HONSU (HON), SNF1-related protein kinase 2 (SnRK2), and abscisic acid-responsive elements—binding factor (AREB) and basic leucine zipper (bZIP) [17,77–79].

Kucera et al. [80] identified an early and late GA biosynthesis within the embryo encoded respectively by ent-copalyl diphosphate synthetase (*CPS1*) gene and *GA3ox2* gene in the cortex and endodermis of the root. Over-expression of GA biosynthesis genes such as gibberellins 3-oxidase 1 (*GA3ox1*), gibberellins 20-oxidase 3 (*GA20ox3*), and ENT-kaurene oxidase 1 (*KO1*) results in seed germination [17,81,82]. The regulation of GA signalling gene comatose (*CTS*) is a key component to have dormancy or germination. Its activation results in germination through a peroxisomal protein of the ATP-binding cassette (ABC) transporter class [17,83]. DELLA proteins act as repressors of GA signalling and integrate environmental cues into GA signalling through the expression of the repressor of GA (*RGA*), RGA-like 1 (*RGL1*), RGA-like 2 (*RGL2*), gibberellic acid insensitive (*GAI*), and GA INSENSITIVE DWARF (*GID1*, *GID2*) genes [17,84]. The F-box protein is a receptor that mediates GA responses for the degradation of DELLA-type transcription repressors [80] through the activation of sleepy1 (*SLY1*) gene [17,85]. Contrarily, Delay of Germination 1 (*DOG1*) regulates the expression of GA biosynthesis genes by the inhibition of genes encoding cell wall remodelling enzymes and by regulating the appropriate time of germination according to ambient temperature [86]. Another GA catabolism genes include elongated uppermost internode (*EUI*) and delayed flowering 1 (*DDF1*). *ABI4* recruits an additional seed-specific transcription factor to repress the transcription of GA biogenesis gene or can directly bind itself to the promoter of *GA2ox3* as *ABI5*, activating its expression [76,87].

It is reported that auxin may suppress seed germination under high salinity [88], delay seed germination, and inhibit pre-harvest sprouting through indole-3-acetic acid: IAA [89]. Auxin synthesis decreases during after-ripening treatment enabling seed dormancy break [90] and helps ABA in dormancy induction and protection [91] by repressing, for example, the embryonic axis elongation during seed germination [92].

Ethylene (ET) participates in the seed dormancy regulation through its receptors such as ethylene triple response (*ETR1*, *ETR2*) and ethylene insensitive 2 (*EIN2*). ET may repress ABA accumulation and promote seeds dormancy release so that the high ET content in seeds is associated with dormancy loss through SIN3-like 1 (*SNL1*) and *SNL2* genes at the epigenetic level [17,77,93–95]. At the same time, *SNL1* and *SNL2* promote seed dormancy through another pathway [96].

Brassinosteroid (BR) action consists in suppressing the inhibitory effect of ABA during the germination process in wild types of *Arabidopsis* through an MFT (Mother of Flowering locus T (FT) and Terminal flower 1 (TFL1))-mediated pathway [97]. The lower content of BR during germination, due to *BIN2* (Brassinosteroid Insensitive 2) gene activation, stabilizes ABI5 protein to mediate ABA signaling unless there is BR treatment to repress the *BIN2*–*ABI5* interaction [17,98]. BR biosynthesis genes (TaDE-etiolated 2 (*TaDET2*) and *TaDWARF* 4) and BR signalling promotion genes (TaBR signalling kinase 2 (*TaBSK2*)) have been identified in wheat [17,99].

Jasmonic acid (JA) can inhibit the germination process promoting the effect of ABA biosynthesis genes regulatory action on *MFT* gene by 12-oxo-phytodienoic acid (OPDA) in dry seeds whereas during imbibition the transcription of those genes is repressed by JA [17,77,100–102]. Salicylic acid (SA) in a first way inhibits the expression of GA-induced a-amylase genes under normal growth conditions [77,103]. In the other way, it reduces oxidative damage under high salinity [77,104]. Cytokinin (CTKs) might effectively concentrate and direct cell division and elongation of the emerged root. In sorghum, cytokinin/ABA interaction controls germination by inducing ABI5 protein degradation [80,105].

Strigolactones (SLs) are involved in seed germination in *Arabidopsis*, in parasitic weeds, and in other species [77,106]. ABA is involved in SLs biosynthesis regulation in tomatoes [107]. Furthermore, some key components in the SL signalling pathway affect seed germination, including *SMAX1* (Suppressor of More Axillary Growth2 1) in *Arabidopsis* and its homolog *OsD53* in rice [77,108].

Numerous studies of the natural variation and various mutants have offered an opportunity to identify new seed dormancy regulators, their related genes or QTLs, and how they work. The description of seed dormancy regulators used in this review are presented in Tables 3 and 4.

Table 3. Description of mutant genes controlling seed dormancy reported by previous studies in *Arabidopsis thaliana*.

Mutants	Description/Action	References
nced6/nced9 and *nced5*	Promote germination	[75,109,110]
aao3, *aba1*, and *aba2*	Reduce dormancy	[73,111]
cyp707a	Enhance seed dormancy level	[17,77,112,113]
abi1	Reduce dormancy through chilling and dry storage, reduce ABA sensitivity for germination and no precocious germination	[80,114]
abi3	Leads to seed dormancy even in immature seeds	[115–118]
cts	Leads to the seed dormancy protection even after stratification and after-ripening	[80]
yuc1/yuc6 (Auxin)	Reduce seed dormancy	[89]
ein2 (Ethylene)	Leads to higher expression of *NCED3*	[17,119,120]
etr1	Induces lower activation of *CYP707A2* genes	
snl1 and *snl2*	Reduce seed dormancy together with the increased Ethylene content	[96]
hub1(*rdo4*)	Characterized by a reduced dormancy	[48]
tfiis	Reduces seed dormancy	[68]
dog1 and *rdo4*	Reduce seed longevity phenotype	[55,69,70]
rdo1 and *rdo2*	Not affected in their response to ABA	

Note: *yuc1/yuc6*, mutants of Yucca flavin monooxygenase genes; *hub1*, mutant of Homologous to UBIquitin; *nced*, mutant of NCED; *aao3*, AAO3 mutant; *aba1/aba2*, ABA1/ABA2 mutant; CYP707a, CYP707A mutant; *abi1/abi3*, ABI1/ABI3 mutant; *cts*, CTS mutant; *ein2*, EIN2 mutant; *snl1/snl2*, SNL1/SNL2 mutant; *tfiis*, TFIIS mutant, *dog1*, DOG1 mutant; *rdo1/rdo2/rdo4*, RDO1/RDO2/RDO4 mutant; *etr1*, ETR1 mutant.

Table 4. Processes and genes involved in seed dormancy regulation.

Process	Genes	Description	Related Species	References
ABA biosynthesis	NCED5, NCED6, NCED9	Induction of seed dormancy	Arabidopsis thaliana	[17,75,109,121,122]
	NCED1, NCED2	Induction of seed dormancy	Oryza sativa, Hordeum vulgare	[17]
	ABA1, ABA2	Encode for zeaxanthine poxidase	Arabidopsis thaliana; Zea mays; Nicotiana plumbaginifolia	[80]
	AAO3	Encodes final step of ABA biosynthesis	Arabidopsis thaliana	[80,123]
ABA catabolism	CYP707A1, CYP707A2, CYP707A3	Encode for ABA8'hydroxylase; loss of dormancy	Arabidopsis. thaliana, Hordeum vulgare	[17]
ABA signalling	ABI1, ABI2	Encode for Serine/threonine phosphatase 2C (PP2C) inducing seed dormancy	Arabidopsis thaliana and monocot	[80,114]
	ABI3/VP1	Regulation of chlorophyll, anthocyanin, and storage proteins accumulation with FUS3 and LEC1	Arabidopsis thaliana and monocot	[17,124,125]
		Regulated by WRKY41 and by DEP for primary seed dormancy establishment	Arabidopsis thaliana and monocot	[77,126,127]
	ABI4	Regulated by transcription factors CHO1 and SPT for dormancy establishment and maintenance through NCED2 and NCED3; Represses CYP707A1 and CYP707A2	Arabidopsis thaliana and monocot	[77,128–130]
	ABI5	Regulated by bZIP transcription factor for positive ABA signalling and repressing seed germination	Arabidopsis thaliana, Sorghum bicolor	[77,131]
GA biosynthesis	GA3ox1, GA20ox3, KO1	Inducing of hydrolytic enzymes that weaken the seed coat, inducing of mobilization of seed storage reserves, and stimulating of expansion of the embryo	Arabidopsis thaliana and monocot	[17,81,82]
	CPS	Catalyzed geranylgeranyl pyrophosphate (GGDP) cyclization reaction in the provascular tissue	Arabidopsis thaliana	[80]
GA signaling	CTS	Encodes a peroxisomal protein of the ATP-binding cassette (ABC) transporter class	Arabidopsis thaliana	[17,83]
	RGA, RGL1, RGL2, GAI	Encode DELLA proteins as a repressor of GA signalling	Arabidopsis thaliana	[17,79,80,84]
	SLY1	GA relieves DELLA repression of seed germination by F-box protein	Arabidopsis thaliana	[17,85]
	GID1	Induce release of seed dormancy by promoting interaction of DELLA with the F-box protein	Arabidopsis thaliana, Oryza. sativa	[81,132,133]
	GID2	Encodes for F-box subunits of an SCF E3 ubiquitin ligase that ubiquitinates DELLA proteins	Arabidopsis thaliana, Oryza sativa	[81,134–137]
GA catabolism (GA2ox2)	DOG1	Inhibition of genes encoding cell wall remodelling enzymes: EXPA2, EXPA9, XTH19 by regulates the expression of GA biosynthesis genes	Arabidopsis thaliana and monocot	[86]
	DDF1	Promotes transcription of the GA inactivation gene GA2ox7	Arabidopsis thaliana	[77,138]
	EUI	Promoted by AP2 domain-containing transcription factor OsAP2-39 for GA inactivation	Oryza sativa	[77,139]

Table 4. *Cont.*

Process	Genes	Description	Related Species	References
Auxin	*iaaM-OX*	Strong seed dormancy	*Triticum aestivum*	[89]
	ARF10 and *ARF16*	Activates *ABI3* by perceiving high level of IAA for dormancy maintenance	*Arabidopsis thaliana*	[17,77,91]
	AXR2/3	Repress *ARF10* and *ARF16*	*Arabidopsis thaliana*	
Ethylene	*ACO1, ACO4*	Ethylene biosynthesis genes	*Arabidopsis thaliana*	[17,140]
	ETR1, ETR2 EIN2	Contrasting roles for ABA biosynthesis during seed germination under salt-stress conditions	*Arabidopsis thaliana*	[141]
	SNL1 and *SNL2*	Reduce acetylation level of histone 3 lysine 9/18 and histone 3 lysine 14 repressing ABA accumulation at high level of ET	*Arabidopsis thaliana*	[96]
	SNL1 and *SNL2*	Promote seed dormancy through simultaneous modulation of *ACO1, ACO4* and *CYP707A1, CYP707A2*	*Arabidopsis thaliana*	[96]
Brassinosteroid biosynthesis	TaDE-etiolated 2 (*TaDET2*) and *TaDWARF 4*	Ensure BR production in plant	*Triticum aestivum*	[17,99]
Brassinosteroid signaling	TaBR signalling kinase 2 (*TaBSK2*)	Promote BR signalling	*Triticum aestivum*	[17,99]
	MFT	Forming a negative feedback loop to modulate ABA signalling	*Arabidopsis thaliana*	[142,143]
	BIN2	Key repressor of the BR signalling	*Arabidopsis thaliana*	[17,98]
Jasmonic acid	OPDA	Promote effect of *ABA1, ABI5,* and *RGL2* and its regulatory action on *MFT* gene for seed dormancy maintenance	*Arabidopsis thaliana*	[17,77,100,144].
Other genes	*DOG1*	Shows strong dormancy	*Arabidopsis thaliana, Hordeum vulgare, Triticum aestivum*	[29,50–53,55]
	DAG1 and *DAG2*	Inhibiting germination by mediating *PIL5* activity as well as directly affecting gibberellin biosynthesis	*Arabidopsis thaliana*	[63,64]
	DOF6	Negatively regulates germination by affecting abscisic acid signalling in seeds	*Arabidopsis thaliana*	[60]
	RDO2	Encodes *TFIIS* for strong dormancy	*Arabidopsis thaliana*	[67]
	GATA12	Encodes a GATA-type zinc finger transcription factor for novel *RGL2–DOF6* complex enforcing primary seed dormancy via GA signalling repression	*Arabidopsis thaliana*	[59,61,62]
	NR (Nitrate reductase)	Promotes dormancy release	*Arabidopsis thaliana*	[81,145]

5. Environmental Factors Influencing Seed Dormancy Regulation

After-ripening, stratification, light, and seed components influence seed germinability. After-ripening refers to the transition period from a dormant to a more readily germinable seed, where the seed is submitted to a set of environmental conditions after maturation and separation from the mother plant [146]. For instance, in *Gynandropsis gynandra*, this period is supposed to last about three to six months after harvest [31].

Temperature affects ABA-GA balance resulting in either germination or strong dormancy [51,77,147]. In wheat, seed germination was highly affected by genotype and temperature. Seed dormancy increased when seeds were developed under 15 °C whereas seed germination increased when seeds were sown at 20 °C [148–150]. For some African traditional leafy vegetables, the maximum germination occurs at 29 to 36 °C with 36 °C for *Vigna unguiculata* L., 35 °C for *C. olitorius*, and 30 °C for *G. gynandra* [33,39].

Light also plays a crucial role in seed germination by inducing a secondary dormancy in imbibed after-ripened seeds, increasing the expression of GA biosynthesis genes and repressing the expression of GA catabolism gene through the action of phytochrome [151,152]. For instance, red (R) light inhibits the expression of *NCED6*, fared (FR) light inhibits the expression of *CYP707A2*, and blue light enhances the transcription of ABA biosynthetic genes *NCED1* and *NCED2* [77,153,154]. While germination of *G. gynandra* increase under dark conditions [39], light is supposed to positively affect germination of TLVs such as *Amaranthus cruentus* L., *Brassica rapa* L. subsp. *chinensis*, *Corchorus olitorius*, *Citrullus lanatus* Thunbs., and *Solanum retroflexum* Dun. [33]. It is therefore important not only to assess the temperature and light requirements during seed development, seed storage, and seed germination for TLVs seed germination but also to explore the induced changes at the hormonal level. Ochuodho and Modi [155] reported negative photosensitivity at 20 °C in continuous white light during the germination of *G. gynandra*, probably controlled by changes in ABA and GA content. In addition, genetic and physiological studies of seed dormancy highlighted the effect of interactions between light and temperature on seed dormancy regulation in *A. thaliana* [156]. Some of the relations between environmental factors and seed dormancy regulators in plants reviewed by Skubacz and Daszkowska-Golec [17] are summarized in Table 5.

Table 5. Description of Environment factors and their role in seed dormancy regulation in plants reviewed by Skubacz and Daszkowska-Golec [17].

Environment Factors	Situations	Role in Seed Dormancy Regulation	Description	Species
After-ripening	Seed dry storage period at room temperature	Reduced dormancy	Positive relationship with CYP707A2 Induces GA insensitive dwarf1 *GID1b*	*Arabidopsis thaliana*
			Promotes expression of JA biosynthesis genes: Allene oxide synthase (*AOS*), 3-ketoacyl coenzyme A (*KAT3*) and Lipoxygenase 5 (*LOX5*); Induces *GA20ox1* and *GA3ox2*	*Triticum aestivum*
			Increases the expression of *ABA8'OH-1*	*Hordeum vulgare, Brachypodium distachyon*
Temperature	Low temperature	Reduced dormancy	Promotes *GA3ox1* expression; Represses *GA2ox2* gene	*Arabidopsis thaliana*
		Higher level of dormancy during seed development	Activates *MFT* gene	*Triticum aestivum*
	High temperature	Increased dormancy during seed imbibition	Represses *GA20ox1, GA20ox2, GA20ox3, GA3ox1*, and *GA3ox2* genes; Promotes the expression of ABA biosynthesis genes	*Arabidopsis thaliana*
Light	Red (R) light	Reduced dormancy	Inhibits the expression of *NCED6*	*Arabidopsis thaliana*
	Fared (FR) light	Increase dormancy	Inhibits the expression of *CYP707A2*	
	Blue light	Increased dormancy	Promotes *NCED1, NCED2, GA2ox3* and *GA2ox5* genes; Represses *GA3ox2*	*Hordeum vulgare*

6. Seed Coat Components

Seed coat components functions are related to flavonoids, which provide greater mechanical restraint and reduced permeability to water, gases, and hormones. Flavonoids can inhibit metabolic processes of after-ripening and germination. They provide protection from oxidative damage and the dehydration and desiccation tolerance of orthodox and recalcitrant seeds, which correlated with ABA and active oxygen species (AOS), respectively [81,157–159]. The seed coat or pod colour may be a good indication for seed physiological status and can provide insight on seed germination as reported in *G. gynandra* [41]. Adebo et al. [160] reported two colours for seed coat (black and brown) among *Corchorus olitorius* cultivars from Benin. In amaranth species, a *MYB*-like transcription factor gene controls the seed colour variation especially between the ancestor, which is black, and the domesticated species, which is white [161]. Most of the seed dormancy constraints observed in TLVs seeds are due to the seed coat structure. This may be probably due to the important content of flavonoids observed in TLVs species whether in leaf, shoot, and seed. Table 6 informs about flavonoid content in the seed and shoot of some African leafy vegetable species.

Table 6. Flavonoids content in some TLVs species. Values collected from Akubugwo et al. [162], Paśko et al. [163], Yang et al. [164].

Species	Part	Flavonoid (mg/100 g)
Gynandropsis gynandra	Shoot	64.3
Corchorus olitorius	Shoot	63.9
Solanum nigrum	Seed	1.01
Amaranthus cruentus	Seed	667
Chenopodium quinoa Willd.	Seed	2238

7. Pathway for Dormancy Studies in Traditional Leafy Vegetables

Genetic and molecular approaches help find out the mechanisms underlying each step in the life cycle of plants [165]. The quantitative nature of seed dormancy should help identify whether seed dormancy of TLVs is controlled by nuclear or maternal factors or both. It is also important to know if the seed dormancy is genetically dominant or recessive or an outcome of a mutation. Factors that regulate germinability of seeds and their relationship for seed development and germination need to be investigated at different levels. Apart from temperature and light requirement, ecological significance of seed dormancy mechanism for TLVs should be understood. The current knowledge of dormancy established in well-studied crops such as *A. thaliana* and cereals will serve as baseline information to provide guidance for dormancy studies in TLVs. Therefore, researchers should be able to retrieve and put forward appropriate methodologies including definition of objectives and hypotheses, type of plant material (shoot, seed, source), conditions around implementation (seed development, seed storage, seed pre-treatments, seed germination), data collection, and analysis to perform genetic, genomic, and physiological screening of seed dormancy in TLVs. Studies can be specifically directed toward identification of the main seed dormancy regulators such as genes (*DOG1, DAG, DOF, RDO, ABI3, GA3ox1, GA2ox3*), transcription factors and proteins (ABA8′hydroxylase, CHO1, DELLA, GATA), abscisic–gibberellin acids crosstalk, non-enzymatic processes, seed morphological and structural components (endosperm, pericarp, embryo, seed coat), and the external factors (light, temperature, salinity, acidity, soil nitrate).

Seed characterization for a given TLV species will help identify the dormant and non-dormant genotypes and develop genetic populations for further studies. Figure 3 illustrates the conceptual approach leading to complete successful seed dormancy study in TLVs. This approach includes six steps from germplasm collection to the development of new varieties with high-quality seeds.

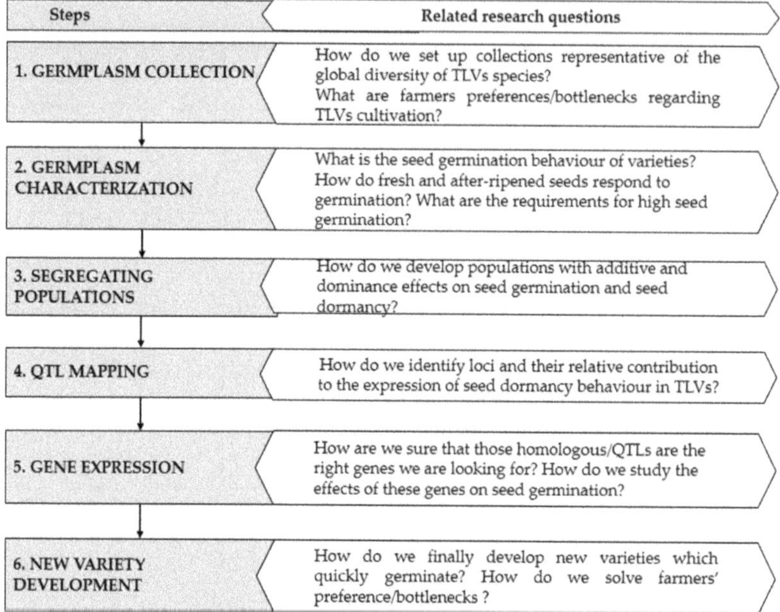

Figure 3. A conceptual approach for seed dormancy study in traditional leafy vegetables (TLVs) species. This proposed research avenue includes germplasm collection, germplasm characterization; development of mapping populations; quantitative trait loci (QTL) mapping; gene expression analysis; and new variety development. Germplasm collection is required in the range of distribution of the species for a developed core collection with cultivated and wild genotypes. Characterization of different factors influencing seed physiology and dormancy should help establish optimal germination conditions for fresh and after-ripened seeds and assess the natural variation of seed dormancy among genotypes. Development of mapping populations like biparental recombinant inbred lines (RILs), near-isogenic lines (NILs), F2, double haploid (HD), mutants, and multi-parental populations (multiparent advanced generation intercross (MAGIC)), multiparent recombinant inbred lines (AMPRIL)) will help access the genetic background existing among genotypes for seed dormancy. Development of molecular markers and use of modern genomic tools will lead to the successful implementation of mapping approaches such as in silico mapping, Bayesian mapping, linkage disequilibrium, multiple-QTL mapping, interval mapping, single marker-based, and marker regression for TLVs species. Tools such as expressed sequence tags (ESTs), serial analysis of gene expression (SAGE), massive parallel signature sequencing (MPSS), comparative genomics, microarray, and genome-wide association selection (GWAS) should be used for gene expression analysis. The development of improved varieties with high germination ability and many interest traits for farmers will be possible through Marker-assisted gene introgression, ultra-high-density bin map, and marker-assisted gene pyramiding.

7.1. Germplasm Collection

The use of wild and cultivated populations should enhance the possibility to identify new regulators with ecological relevance. The majority of TLV species are still neglected and underutilized and are poorly represented or even not present in gene banks. The first important task to achieve will therefore be to conduct germplasm collection missions to assemble as much as possible the natural diversity in these species by considering the geographical range of distribution of each given species. Seeds, roots, leaves, or any part of the reproductive material of plant should be collected in both wild and cultivated populations, and made available for scientists and gene bank curators. A good example is the *Gynandropsis gynandra* core collections developed by the World Vegetable Center, National

Plant Germplasm System of the United States Department of Agriculture, and the Laboratory of Genetics, Horticulture and Seed Science at University of Abomey-Calavi-Benin [11]. Such initiatives will contribute to the preservation of the genetic resources of TLV in the current context of increasing genetic erosion due to climate change, urbanization, and industrialization.

7.2. Seed Dormancy Characterization

Once germplasm collections are established or expanded, phenotypic characterization and genotypic characterization of seed dormancy related traits should be carried out. Phenotypic characterization includes variation in seed morphology and intrinsic germination capacity (radicle protrusion, seed, and seedling vigour). Seeds must be collected or developed at the same time for phenotypic characterization. Many studies of this type were conducted focusing on understanding the germination behaviour of various collections, comparisons of fresh dormant seed versus after-ripened non-dormant seeds, identifying light and temperature requirements, assessing the effects of various seed pre-treatments and seed origin [31,149,166]. These various experiments will lead to identifying populations with non-dormant seeds, populations with rather dormant seeds, and others with strong dormant seeds, in terms of germination rate, or time for dry storage to reach 50% germination or the number of days of seed dry storage required to reach 100% germination [49,167]. The genetic or molecular characterization of seed dormancy in mapping populations will be achieved using molecular markers as microsatellites or simple sequence repeats (SSR) and single nucleotide polymorphisms (SNPs). The combination of the data generated from the characterization will help predict the germination potential of the genotypes at an early stage of seed development. The observed variation about seed dormancy among populations could also be used to study gene expression profiles in different types of seeds. For instance, genetic and molecular dormancy investigations in *Arabidopsis* were performed using the reference accessions Landsberg erecta (Ler) and Colombia (Col) with low dormancy, the dormant accession Cape Verde Islands (Cvi), and various accessions with different levels of seed dormancy such as Antwerpen (An-1), St. Maria de Feira (Fei-0), Kondara (Kond), Shakdara (Sha), and Kashmir (Kas-2) [18,46,49,168]. Kępczynski et al. [36] have worked with primary dormant *Amaranthus retroflexus* seeds, secondary dormant *Amaranthus paniculatus* seeds, and non-dormant *Amaranthus caudatus* seeds.

7.3. Development of Mapping Populations for Identification of Candidate Genes Involved in Seed Dormancy

Seed dormancy is a quantitative trait with agronomic importance. Various segregating populations reviewed by Keurentjes et al. [169] have been used for QTLs identification, namely recombinant inbred lines (RILs), Near-Isogenic Lines (NILS), Multiparent Advanced Generation Intercross (MAGIC), and Multiparent Recombinant Inbred Line (AMPRIL). Thus far, RILs, NILs, or backcross inbred lines populations developed via crossing dormant and non-dormant accessions have been used in many seed dormancy studies [49,170]. Mutant lines were developed referring to transgenic techniques through overexpression or downregulation of genes and compared with wild types [54,58,171,172]. Doubled haploids, recombinant inbred lines, and near-isogenic lines are the most important types of population used in genetic mapping, gene discovery, and genomics-assisted breeding [173]. F_2 populations appear to be more powerful to help estimate both additive and dominance effects. Similar mapping populations should be developed for TLVs through mating systems such as diallel, nested, and factorial designs enabling to also assess the genetic variances and heritability of dormancy related trait.

The aim of developing mapping populations is to identify candidate loci and their relative contribution to the expression of seed dormancy behaviour. Several approaches have been developed for accuracy in QTLs detection such as single marker-based approaches, interval or LOD mapping, composite interval mapping, multiple interval mapping/multiple regression, marker regression, multiple-QTL mapping (MQM), Bayesian mapping, linkage disequilibrium mapping, meta-analysis, and in silico mapping [173,174]. For instance, the single marker regression and multiple-QTL mapping have been used in seed dormancy data analysis for *Arabidopsis thaliana* [18]. The heritability of major

QTLs through offspring will be important in the achievement of the breeding programs developed for traditional leafy vegetables.

The identified QTLs should help identify genes involved in cell organization and biogenesis, proteolysis, ribosomal proteins, hormones, elongation, and initiation factors that control seed dormancy in TLVs. Post-genomic analyses in previous studies about dormancy yielded interesting results through the use of proteomics, transcriptomics, metabolomics, and microarray analysis [175,176]. Their use should be focused on the comparison of transcription, post-transcription, gene expression either in imbibed seeds or fresh dormant seeds versus the non-dormant, after-ripened seeds. For example, a proteomic analysis of seed dormancy conducted in *Arabidopsis* with freshly harvested dormant seed and after-ripened non-dormant seed of accession Cvi revealed 71 proteins activities [177]; whereas 40 genes were reported by Li et al. [178] through the transcriptional profiling of imbibed *Brassica napus* L. seeds using *Arabidopsis* microarrays analysis. Various metabolic switches act during seed desiccation, vernalization, and early germination in *Arabidopsis* species [179] as well as in wheat [180]. In addition, genome-wide association analyses could provide a comprehensive understanding of gene activities in each of the three main regions of seed (coat, endosperm, embryo) during seed maturation, seed storage, and seed germination [181]. Comparative genomics will help construct strong phylogenies, identify changes in genome structure, annotate homologous genes, and understand novel traits [173,182]. Ayenan et al. [183] suggested the use of comparative genomics with rice to solve some of the constraints limiting the production of neglected African cereal crop fonio (*Digitaria* spp.). Comparative genomics is also relevant for TLVs genome screening as highlighted by Sogbohossou et al. [11] suggesting the use of available information on *Brassica* spp. and *Arabidopsis thaliana* genomes, *Solanum lycopersicum* and *Solanum tuberosum* L. genomes, and *Amaranthus hypochondriacus* L. and *Chenopodium quinoa*, respectively, for *Gynandropsis gynandra* and *Brassica carinata*, *Solanum* species (*Solanum nigrum*, *Solanum marcrocarpon*, *Solanum aethiopium* L., *Solanum scabrum*), and amaranths species (*Amaranthus* spp.).

7.4. Tapping into Comparative Genomics to Study Seed Dormancy in TLVs

Arabidopsis thaliana, as well as *Avena sativa*, *Solanum lycopersicum*, and *Nicotiana plumbaginifolia* have been used to understand many genetic and molecular determinants involved in germination and seed dormancy control for plants [175]. Recently, Sogbohossou et al. [11] presented spider plant as a good model plant for TLVs breeding studies. Spider plant is a promising vegetable species for food security and income generation for African communities in urban and in rural areas [184]. In addition, the genome size of spider plant is relatively small (~800 Mb) with the possibility to obtain three generations per year and offers many options in terms of breeding strategies [11]. Cleomaceae family and Brassicaceae family are closely related, sharing the At-β whole-genome duplication event within the Brassicales with fast molecular evolution rates [185]. Marshall et al. [186] argued the possibility to isolate orthologs of *Arabidopsis* genes from *Gynandropsis gynandra*, and van den Bergh et al. [187] reported that the genomes of *Gynandropsis gynandra* and *Arabidopsis thaliana* shared significant synteny, which will facilitate translational genomics between both species.

Seed dormancy is strongly reported in spider plant. Previous studies reviewed by Shilla et al. [31] support the assumption that spider plant accessions differ largely in their seed dormancy behaviour where the germination activation extends from zero (0) to more than six (6) months. Such a variation in the seed behaviour offers an opportunity to determine QTLs associated with germination in the species. *Gynandropsis gynandra* can be used as model plant for seed dormancy studies in plants the same way *Thellungiella salsuginea* has been used to investigate the physiological, metabolic, and molecular mechanisms of abiotic stress tolerance in plants [188]. Large physiological, genetic, and biochemical data should therefore be generated in spider plant for further investigations. As one of the priority species for the Mobility for Breeders in Africa project (in short "Mobreed") [189], some studies on *G. gynandra* are underway and will certainly provide us with more information about the potential of its genome. The use of comparative genomics, macrosynteny, and microsynteny is an alternative for the understanding of the seed dormancy determinants for *G. gynandra* and the generation of knowledge regarding molecular markers, QTLs, candidate genes and their functions [190,191]. The development of DNA molecular markers should accelerate *G. gynandra* plant breeding as diversity arrays technology (DArT) and single nucleotide polymorphism (SNP) markers can be used to locate seed dormancy genes across *G. gynandra* germplasm [192]. Once candidate genes are identified, the development of populations with different genetic backgrounds such as NILs, F2, or a backcross population are important for gene introgression and validation. Simple sequence repeats (SSRs) markers have been successfully used in this way for various crops. The multi-environment experiment is a factor related directly to the efficiency of the identified genes. The way of mutant development can also be explored for new allele identification compared to the wild phenotype.

The knowledge established with the model plant (*G. gynandra*) can be transferred to other TLVs species through the development of a platform integrating genomics, transcriptomics, phenomics, and ontology analyses [193]. For instance, Mutwil et al. [194] established whole-genome coexpression networks for Arabidopsis, barley, rice, Medicago, poplar, wheat, and soybean that may considerably improve the transfer of knowledge generated in *Arabidopsis* to crop species. The isolation and functional analysis of homologs/QTLs should clearly help characterize all the genes and understand their interaction with each other. The understanding of others genetic phenomena of economical traits such as late-flowering date, leaf yield, and disease resistance, should help develop cultivars combining several good characteristics for farmers. The construction of ultra-high-density genetic maps is therefore essential to perform gene transfer through marker-assisted gene introgression and marker-assisted gene pyramiding [195].

8. Conclusions

This review sheds light onto a better understanding of the regulation of seed dormancy and the main factors known to be involved in the control of seed dormancy at hormonal, transcriptomic, epigenetic, protein, and environmental levels. The abscisic-gibberellin acids (ABA-GA) balance is a key component involved in the indirect control of seed dormancy that determines whether seeds may germinate or not. Auxin and salicylic acid (SA) promote seed dormancy; jasmonic acid (JA), brassinosteroids (BRs) and ethylene (ET) play a dual role in seed dormancy regulation, and cytokinin (CTKs) promotes seed germination. Other dormancy regulators were shown to directly promote or repress the decision of seeds to germinate. Those include genes such as Delay of Germination (*DOG*), *DOF* affecting germination (*DAG*), and reduced dormancy (*RDO*). For *Gynandropsis gynandra* and other TLVs, research is needed to solve the seed dormancy constraints and to provide prospective producers with high-quality seeds. *Gynandropsis gynandra*, a closely related species to *Arabidopsis thaliana*, was considered as a model plant to propose a pathway into solving the dormancy constraints in TLVs. This proposed research avenue includes germplasm collection, germplasm characterization, development of mapping populations, QTL mapping; gene expression analysis, and a new of variety development. Therefore, research is needed to highlight the specific storage conditions and seed pre-treatment required to ensure the seed viability and germination of genotypes. The mechanisms occurring during after-ripening in *G. gynandra* should also be elucidated. Rapid progress for *G. gynandra* full domestication should be achieved starting with the analysis of the natural variation in seed dormancy in the available genotype collections. Further steps include the characterization of the different factors influencing seed physiology, dormancy, and dormancy-breaking approaches. The identification of major genetic and molecular factors underlying seed dormancy during seed development, seed storage, and germination needs to be elucidated for developing cultivars that farmers can effectively use. Multidisciplinary research teams including physiologists, geneticists, and bioinformaticians are therefore required to quickly and efficiently make significant progress toward breeding for non-dormancy in traditional leafy vegetables.

Author Contributions: E.G.A.-D. and F.S.S. developed the conceptual framework of the manuscript. F.S.S. wrote the manuscript. H.P.F.Z., C.A.H., D.E.O.S., and E.G.A.-D. reviewed and approved the final version. All authors have read and agreed to the published version of the manuscript.

Agronomy **2020**, *10*, 57

Funding: This work was supported by the Applied Research Fund of the Netherlands Organization for Science under the Project "Utilizing the genome of the vegetable species *Cleome gynandra* for the development of improved cultivars for the West and East African markets" (Project Number: W.08.270.350).

Acknowledgments: We thank the editors for the interactive improvement of the manuscript.

Conflicts of Interest: The authors declare no conflicts of interest.

References

1. Achigan-Dako, E.G.; N'Danikou, S.; Assogba-Komlan, F.; Ambrose-Oji, B.; Ahanchede, A.; Pasquini, M.W. Diversity, geographical, and consumption patterns of traditional vegetables in sociolinguistic communities in Benin: Implications for domestication and utilization. *Econ. Bot.* **2011**, *65*, 129–145. [CrossRef]

2. Kahane, R.; Temple, L.; Brat, P.; De Bon, H. Les légumes feuilles des pays tropicaux: Diversité, richesse économique et valeur santé dans un contexte très fragile. In Proceedings of the Colloque Les légumes: Un Patrimoine à Transmettre et à Valoriser, Angers, France, 7–9 September 2005.

3. Maundu, P.; Achigan-Dako, E.; Morimoto, Y. Biodiversity of African vegetables. In *African Indigenous Vegetables in Urban Agriculture*; Shackleton, C.M., Pasquini, M.W., Drescher, A.W., Eds.; Routledge: London, UK, 2009; pp. 97–136.

4. Towns, A.M.; Shackleton, C. Traditional, Indigenous, or Leafy? A Definition, Typology, and Way Forward for African Vegetables. *Econ. Bot.* **2018**, *72*, 461–477. [CrossRef]

5. Uzilday, B.; Turkan, I.; Sekmen, A.; Ozgur, R.; Karakaya, H. Comparison of ROS formation and antioxidant enzymes in *Cleome gynandra* (C4) and *Cleome spinosa* (C3) under drought stress. *Plant Sci.* **2012**, *182*, 59–70. [CrossRef] [PubMed]

6. Luoh, J.W.; Begg, C.B.; Symonds, R.C.; Ledesma, D.; Yang, R.-Y. Nutritional yield of African indigenous vegetables in water-deficient and water-sufficient conditions. *Food Nutr. Sci.* **2014**, *5*, 812. [CrossRef]

7. Kansiime, M.K.; Karanja, D.K.; Alokit, C.; Ochieng, J. Derived demand for African indigenous vegetable seed: Implications for farmer-seed entrepreneurship development. *Int. Food Agribus. Manag. Rev.* **2018**, *21*, 723–739. [CrossRef]

8. Abukutsa-Onyango, M. Seed production and support systems for African Leafy Vegetables in three communities In Western Kenya. *Afr. J. Food Agric. Nutr. Dev.* **2006**, *7*, 108–116.

9. Dube, P.; Struik, P.C.; Ngadze, E. Seed health tests of traditional leafy vegetables and pathogenicity in plants. *Afr. J. Agric. Res.* **2018**, *13*, 753–770.

10. Adebooye, O.; Ajayi, S.; Baidu-Forson, J.; Opabode, J. Seed constraint to cultivation and productivity of African indigenous leaf vegetables. *Afr. J. Biotechnol.* **2005**, *4*, 1480–1484.

11. Sogbohossou, E.D.; Achigan-Dako, E.G.; Maundu, P.; Solberg, S.; Deguenon, E.M.; Mumm, R.H.; Hale, I.; Van Deynze, A.; Schranz, M.E. A roadmap for breeding orphan leafy vegetable species: A case study of *Gynandropsis gynandra* (Cleomaceae). *Hortic. Res.* **2018**, *5*, 1–15. [CrossRef]

12. Baskin, J.M.; Baskin, C.C. A classification system for seed dormancy. *Seed Sci. Res.* **2004**, *14*, 1–16. [CrossRef]

13. Hilhorst, H.W. Standardizing seed dormancy research. In *Seed Dormancy*; Kermode, A.R., Ed.; Springer: Berlin/Heidelberg, Germany, 2011; Volume 773, pp. 43–52.

14. Fenner, M.; Thompson, K. *The Ecology of Seeds*; Cambridge University Press: Cambridge, UK, 2006; p. 260.

15. Bewley, J.D. Seed germination and dormancy. *Plant Cell* **1997**, *9*, 1055. [CrossRef] [PubMed]

16. Bewley, J.D.; Black, M. Seeds. In *Seeds Physiology of Development and Germination*, 3rd ed.; Springer: Berlin/Heidelberg, Germany; Plenum Press: New York, NY, USA, 1994; pp. 1–33.

17. Skubacz, A.; Daszkowska-Golec, A. Seed Dormancy: The Complex Process Regulated by Abscisic Acid, Gibberellins, and Other Phytohormones that Makes Seed Germination Work. In *Phytohormones-Signaling Mechanisms and Crosstalk in Plant Development and Stress Responses*; El-Esawi, M., Ed.; InTech: Vienna, Austria, 2017; pp. 77–100.

18. Van der Schaar, W.; Alonso-Blanco, C.; Léon-Kloosterziel, K.M.; Jansen, R.C.; Van Ooijen, J.W.; Koornneef, M. QTL analysis of seed dormancy in Arabidopsis using recombinant inbred lines and MQM mapping. *Heredity* **1997**, *79*, 190–200. [CrossRef] [PubMed]

19. Lin, M.; Liu, S.; Zhang, G.; Bai, G. Effects of TaPHS1 and TaMKK3-A Genes on Wheat Pre-Harvest Sprouting Resistance. *Agronomy* **2018**, *8*, 210. [CrossRef]

20. Li, C.; Ni, P.; Francki, M.; Hunter, A.; Zhang, Y.; Schibeci, D.; Li, H.; Tarr, A.; Wang, J.; Cakir, M. Genes controlling seed dormancy and pre-harvest sprouting in a rice-wheat-barley comparison. *Funct. Integr. enom.* **2004**, *4*, 84–93. [CrossRef]

21. Debieu, M.; Tang, C.; Stich, B.; Sikosek, T.; Effgen, S.; Josephs, E.; Schmitt, J.; Nordborg, M.; Koornneef, M.; de Meaux, J. Co-variation between seed dormancy, growth rate and flowering time changes with latitude in *Arabidopsis thaliana*. *PLoS ONE* **2013**, *8*, e61075. [CrossRef]

22. Hawkins, K.; Allen, P.; Meyer, S. Secondary dormancy induction and release in *Bromus tectorum* seeds: The role of temperature, water potential and hydrothermal time. *Seed Sci. Res.* **2017**, *27*, 12–25. [CrossRef]

23. Koornneef, M.; Alonso-Blanco, C.; Vreugdenhil, D. Naturally occurring genetic variation in *Arabidopsis thaliana*. *Annu. Rev. Plant Biol.* **2004**, *55*, 141–172. [CrossRef]

24. Zardilis, A.; Hume, A.; Millar, A.J. A multi-model framework for the Arabidopsis life cycle. *J. Exp. Bot.* **2019**, *70*, 2463–2477. [CrossRef]

25. Lu, Q.; Niu, X.; Zhang, M.; Wang, C.; Xu, Q.; Feng, Y.; Yang, Y.; Wang, S.; Yuan, X.; Yu, H. Genome-wide association study of seed dormancy and the genomic consequences of improvement footprints in rice (*Oryza sativa* L.). *Front. Plant Sci.* **2018**, *8*, 2213. [CrossRef]

26. Ali, A.; Cao, J.; Jiang, H.; Chang, C.; Zhang, H.-P.; Sheikh, S.W.; Shah, L.; Ma, C. Unraveling Molecular and Genetic Studies of Wheat (*Triticum aestivum* L.) Resistance against Factors Causing Pre-Harvest Sprouting. *Agronomy* **2019**, *9*, 117. [CrossRef]

27. Cota-Sánchez, J.H. Precocious Germination (Vivipary) in Tomato: A Link to Economic Loss? *Proc. Natl. Acad. Sci. India Sect. B Biol. Sci.* **2018**, *88*, 1443–1451. [CrossRef]

28. Miransari, M.; Smith, D. Plant hormones and seed germination. *Environ. Exp. Bot.* **2014**, *99*, 110–121. [CrossRef]

29. Née, G.; Xiang, Y.; Soppe, W.J. The release of dormancy, a wake-up call for seeds to germinate. *Curr. Opin. Plant Biol.* **2017**, *35*, 8–14. [CrossRef]

30. Baskin, J.M.; Baskin, C.C. New approaches to the study of the evolution of physical and physiological dormancy, the two most common classes of seed dormancy on earth. In *The Biology of Seeds: Recent Research Advances*; Nicolás, G., Bradford, K., Côme, D., Pritchard, H.W., Eds.; CAB International: Wallingford, UK, 2003; pp. 371–380.

31. Shilla, O.; Abukutsa-Onyango, M.O.; Dinssa, F.F.; Winkelmann, T. Seed dormancy, viability and germination of *Cleome gynandra* (L.) BRIQ. *Afr. J. Hortic. Sci.* **2017**, *10*, 45–52.

32. Baskin, C.; Baskin, J. *Seeds: Ecology, Biogeography, and Evolution of Dormancy and Germination*, 2nd ed.; Elsevier: Amsterdam, The Netherlands; Academic Press: San Diego, CA, USA, 2014.

33. Motsa, M.; Slabbert, M.; Van Averbeke, W.; Morey, L. Effect of light and temperature on seed germination of selected African leafy vegetables. *S. Afr. J. Bot.* **2015**, *99*, 29–35. [CrossRef]

34. Taab, A.; Andersson, L. Primary dormancy and seedling emergence of black nightshade (*Solanum nigrum*) and hairy nightshade (*Solanum physalifolium*). *Weed Sci.* **2009**, *57*, 526–532. [CrossRef]

35. Agble, F. Germination of seeds of *Talinum triangulare*. *Ghana J. Sci.* **1970**, *10*, 29–32.

36. Kępczynski, J.; Bihun, M.; Kępczynska, E. Ethylene Involvement in the Dormancy of Amaranthus Seeds. In *Biology and Biotechnology of the Plant Hormone Ethylene*; Kanellis, A.K., Chang, C., Kende, H., Grierson, D., Eds.; Springer: Dordrecht, The Netherlands, 1997; pp. 113–122.

37. Roberts, H.; Lockett, P.M. Seed dormancy and field emergence in *Solanum nigrum* L. *Weed Res.* **1978**, *18*, 231–241. [CrossRef]

38. Geneve, R.L. Seed dormancy in commercial vegetable and flower species. *Seed Technol.* **1998**, 236–250.

39. Ochuodho, J.; Modi, A. Temperature and light requirements for the germination of *Cleome gynandra* seeds. *S. Afr. J. Plant Soil* **2005**, *22*, 49–54. [CrossRef]

40. Ekpong, B. Effects of seed maturity, seed storage and pre-germination treatments on seed germination of cleome (*Cleome gynandra* L.). *Sci. Hortic.* **2009**, *119*, 236–240. [CrossRef]

41. Kamotho, G.; Mathenge, P.; Muasya, R.; Dullo, M. Effects of maturity stage, desiccation and storage period on seed quality of cleome (*Cleome gynandra* L.). *Res. Desk* **2014**, *3*, 419–433.

42. Yepes, J. Study of a weed *Cleome gynandra* L. *Rev. Comalfi* **1978**, *5*, 49–53.

43. Zharare, G. Differential requirements for breaking seed dormancy in biotypes of *Cleome gynandra* and two Amaranthus species. *Afr. J. Agric. Res.* **2012**, *7*, 5049–5059.

44. Enayati, V.; Esfandiari, E.; Pourmohammad, A.; Haj Mohammadnia Ghalibaf, K. Evaluation of different methods in seed dormancy breaking and germination of Redroot Pigweed (*Amaranthus retroflexus*). *Iran. J. Seed Res.* **2019**, *5*, 129–137. [CrossRef]

45. Foley, M.E. Seed dormancy: An update on terminology, physiological genetics, and quantitative trait loci regulating germinability. *Weed Sci.* **2001**, *49*, 305–317. [CrossRef]

46. Alonso-Blanco, C.; Bentsink, L.; Hanhart, C.J.; Blankestijn-de Vries, H.; Koornneef, M. Analysis of natural allelic variation at seed dormancy loci of *Arabidopsis thaliana*. *Genetics* **2003**, *164*, 711–729.

47. Kondou, Y.; Higuchi, M.; Takahashi, S.; Sakurai, T.; Ichikawa, T.; Kuroda, H.; Yoshizumi, T.; Tsumoto, Y.; Horii, Y.; Kawashima, M. Systematic approaches to using the FOX hunting system to identify useful rice genes. *Plant J.* **2009**, *57*, 883–894. [CrossRef]

48. Teng, S.; Rognoni, S.; Bentsink, L.; Smeekens, S. The Arabidopsis GSQ5/DOG1 Cvi allele is induced by the ABA-mediated sugar signalling pathway, and enhances sugar sensitivity by stimulating ABI4 expression. *Plant J.* **2008**, *55*, 372–381. [CrossRef]

49. Bentsink, L.; Hanson, J.; Hanhart, C.J.; Blankestijn-de Vries, H.; Coltrane, C.; Keizer, P.; El-Lithy, M.; Alonso-Blanco, C.; de Andrés, M.T.; Reymond, M. Natural variation for seed dormancy in Arabidopsis is regulated by additive genetic and molecular pathways. *Proc. Natl. Acad. Sci. USA* **2010**, *107*, 4264–4269. [CrossRef]

50. Chiang, G.C.; Bartsch, M.; Barua, D.; Nakabayashi, K.; Debieu, M.; Kronholm, I.; Koornneef, M.; Soppe, W.J.; Donohue, K.; de Meaux, J. DOG1 expression is predicted by the seed-maturation environment and contributes to geographical variation in germination in *Arabidopsis thaliana*. *Mol. Ecol.* **2011**, *20*, 3336–3349. [CrossRef] [PubMed]

51. Kendall, S.L.; Hellwege, A.; Marriot, P.; Whalley, C.; Graham, I.A.; Penfield, S. Induction of dormancy in Arabidopsis summer annuals requires parallel regulation of DOG1 and hormone metabolism by low temperature and CBF transcription factors. *Plant Cell* **2011**, *23*, 2568–2580. [CrossRef] [PubMed]

52. Nakabayashi, K.; Bartsch, M.; Xiang, Y.; Miatton, E.; Pellengahr, S.; Yano, R.; Seo, M.; Soppe, W.J. The time required for dormancy release in Arabidopsis is determined by DELAY OF GERMINATION1 protein levels in freshly harvested seeds. *Plant Cell* **2012**, *24*, 2826–2838. [CrossRef]

53. Nonogaki, H. Seed dormancy and germination—Emerging mechanisms and new hypotheses. *Front. Plant Sci.* **2014**, *5*, 233. [CrossRef] [PubMed]

54. Ashikawa, I.; Abe, F.; Nakamura, S. DOG1-like genes in cereals: Investigation of their function by means of ectopic expression in Arabidopsis. *Plant Sci.* **2013**, *208*, 1–9. [CrossRef] [PubMed]

55. Bentsink, L.; Jowett, J.; Hanhart, C.J.; Koornneef, M. Cloning of DOG1, a quantitative trait locus controlling seed dormancy in Arabidopsis. *Proc. Natl. Acad. Sci. USA* **2006**, *103*, 17042–17047. [CrossRef]

56. Sugimoto, K.; Takeuchi, Y.; Ebana, K.; Miyao, A.; Hirochika, H.; Hara, N.; Ishiyama, K.; Kobayashi, M.; Ban, Y.; Hattori, T. Molecular cloning of Sdr4, a regulator involved in seed dormancy and domestication of rice. *Proc. Natl. Acad. Sci. USA* **2010**, *107*, 5792–5797. [CrossRef]

57. Graeber, K.; Voegele, A.; Büttner-Mainik, A.; Sperber, K.; Mummenhoff, K.; Leubner-Metzger, G. Spatiotemporal seed development analysis provides insight into primary dormancy induction and evolution of the Lepidium DELAY OF GERMINATION1 genes. *Plant Physiol.* **2013**, *161*, 1903–1917. [CrossRef]

58. Ashikawa, I.; Abe, F.; Nakamura, S. Ectopic expression of wheat and barley DOG1-like genes promotes seed dormancy in Arabidopsis. *Plant Sci.* **2010**, *179*, 536–542. [CrossRef]

59. Boccaccini, A.; Santopolo, S.; Capauto, D.; Lorrai, R.; Minutello, E.; Serino, G.; Costantino, P.; Vittorioso, P. The DOF protein DAG1 and the DELLA protein GAI cooperate in negatively regulating the AtGA3ox1 gene. *BMC Plant Biol.* **2014**, *7*, 1486–1489. [CrossRef]

60. Rueda-Romero, P.; Barrero-Sicilia, C.; Gómez-Cadenas, A.; Carbonero, P.; Oñate-Sánchez, L. *Arabidopsis thaliana* DOF6 negatively affects germination in non-after-ripened seeds and interacts with TCP14. *J. Exp. Bot.* **2011**, *63*, 1937–1949. [CrossRef] [PubMed]

61. Ravindran, P.; Verma, V.; Stamm, P.; Kumar, P.P. A novel RGL2–DOF6 complex contributes to primary seed dormancy in *Arabidopsis thaliana* by regulating a GATA transcription factor. *Mol. Plant* **2017**, *10*, 1307–1320. [CrossRef] [PubMed]

62. Stamm, P.; Ravindran, P.; Mohanty, B.; Tan, E.L.; Yu, H.; Kumar, P.P. Insights into the molecular mechanism of RGL2-mediated inhibition of seed germination in *Arabidopsis thaliana*. *BMC Plant Biol.* **2012**, *12*, 179. [CrossRef] [PubMed]

63. Gabriele, S.; Rizza, A.; Martone, J.; Circelli, P.; Costantino, P.; Vittorioso, P. The Dof protein DAG1 mediates PIL5 activity on seed germination by negatively regulating GA biosynthetic gene AtGA3ox1. *Plant J.* **2010**, *61*, 312–323. [CrossRef]

64. Gualberti, G.; Papi, M.; Bellucci, L.; Ricci, I.; Bouchez, D.; Camilleri, C.; Costantino, P.; Vittorioso, P. Mutations in the Dof zinc finger genes DAG2 and DAG1 influence with opposite effects the germination of Arabidopsis seeds. *Plant Cell* **2002**, *14*, 1253–1263. [CrossRef]

65. Léon-Kloosterziel, K.M.; van de Bunt, G.A.; Zeevaart, J.A.; Koornneef, M. Arabidopsis mutants with a reduced seed dormancy. *Plant Physiol.* **1996**, *110*, 233–240. [CrossRef]

66. Peeters, A.J.; Blankestijn-de Vries, H.; Hanhart, C.J.; Léon-Kloosterziel, K.M.; Zeevaart, J.A.; Koornneef, M. Characterization of mutants with reduced seed dormancy at two novel rdo loci and a further characterization of rdo1 and rdo2 in Arabidopsis. *Physiol. Plant* **2002**, *115*, 604–612. [CrossRef]

67. Liu, Y.; Geyer, R.; Van Zanten, M.; Carles, A.; Li, Y.; Hörold, A.; van Nocker, S.; Soppe, W.J. Identification of the Arabidopsis REDUCED DORMANCY 2 gene uncovers a role for the polymerase associated factor 1 complex in seed dormancy. *PLoS ONE* **2011**, *6*, 22241. [CrossRef]

68. Grasser, M.; Kane, C.M.; Merkle, T.; Melzer, M.; Emmersen, J.; Grasser, K.D. Transcript elongation factor TFIIS is involved in Arabidopsis seed dormancy. *J. Mol. Biol.* **2009**, *386*, 598–611. [CrossRef]

69. Liu, Y.; Koornneef, M.; Soppe, W.J. The absence of histone H2B monoubiquitination in the Arabidopsis hub1 (rdo4) mutant reveals a role for chromatin remodeling in seed dormancy. *Plant Cell* **2007**, *19*, 433–444. [CrossRef]

70. Yazdanpanah, F.; Hanson, J.; Hilhorst, H.W.; Bentsink, L. Differentially expressed genes during the imbibition of dormant and after-ripened seeds—A reverse genetics approach. *BMC Plant Biol.* **2017**, *17*, 151–162. [CrossRef] [PubMed]

71. Finkelstein, R.R.; Gampala, S.S.; Rock, C.D. Abscisic acid signaling in seeds and seedlings. *Plant Cell Online* **2002**, *14*, S15–S45. [CrossRef] [PubMed]

72. Flematti, G.R.; Ghisalberti, E.L.; Dixon, K.W.; Trengove, R.D. A compound from smoke that promotes seed germination. *Science* **2004**, *305*, 977. [CrossRef] [PubMed]

73. Kanno, Y.; Jikumaru, Y.; Hanada, A.; Nambara, E.; Abrams, S.R.; Kamiya, Y.; Seo, M. Comprehensive hormone profiling in developing Arabidopsis seeds: Examination of the site of ABA biosynthesis, ABA transport and hormone interactions. *Plant Cell Physiol.* **2010**, *51*, 1988–2001. [CrossRef]

74. Nambara, E.; Marion-Poll, A. ABA action and interactions in seeds. *Trends Plant Sci.* **2003**, *8*, 213–217. [CrossRef]

75. Frey, A.; Effroy, D.; Lefebvre, V.; Seo, M.; Perreau, F.; Berger, A.; Sechet, J.; To, A.; North, H.M.; Marion-Poll, A. Epoxycarotenoid cleavage by NCED5 fine-tunes ABA accumulation and affects seed dormancy and drought tolerance with other NCED family members. *Plant J.* **2012**, *70*, 501–512. [CrossRef]

76. Shu, K.; Zhang, H.; Wang, S.; Chen, M.; Wu, Y.; Tang, S.; Liu, C.; Feng, Y.; Cao, X.; Xie, Q. ABI4 regulates primary seed dormancy by regulating the biogenesis of abscisic acid and gibberellins in Arabidopsis. *PLoS Genet.* **2013**, *9*, e1003577. [CrossRef]

77. Shu, K.; Liu, X.-D.; Xie, Q.; He, Z.-H. Two faces of one seed: Hormonal regulation of dormancy and germination. *Mol. Plant* **2016**, *9*, 34–45. [CrossRef]

78. Kim, W.; Lee, Y.; Park, J.; Lee, N.; Choi, G. HONSU, a protein phosphatase 2C, regulates seed dormancy by inhibiting ABA signaling in Arabidopsis. *Plant Cell Physiol.* **2013**, *54*, 555–572. [CrossRef]

79. Umezawa, T.; Sugiyama, N.; Mizoguchi, M.; Hayashi, S.; Myouga, F.; Yamaguchi-Shinozaki, K.; Ishihama, Y.; Hirayama, T.; Shinozaki, K. Type 2C protein phosphatases directly regulate abscisic acid-activated protein kinases in Arabidopsis. *Proc. Natl. Acad. Sci. USA* **2009**, *106*, 17588–17593. [CrossRef]

80. Kucera, B.; Cohn, M.A.; Leubner-Metzger, G. Plant hormone interactions during seed dormancy release and germination. *Seed Sci. Res.* **2005**, *15*, 281–307. [CrossRef]

81. Finkelstein, R.; Reeves, W.; Ariizumi, T.; Steber, C. Molecular aspects of seed dormancy. *Annu. Rev. Plant Biol.* **2008**, *59*, 387–415. [CrossRef] [PubMed]

82. Gómez-Cadenas, A.; Zentella, R.; Walker-Simmons, M.K.; Ho, T.-H.D. Gibberellin/abscisic acid antagonism in barley aleurone cells: Site of action of the protein kinase PKABA1 in relation to gibberellin signaling molecules. *Plant Cell* **2001**, *13*, 667–679. [CrossRef] [PubMed]

83. Footitt, S.; Slocombe, S.P.; Larner, V.; Kurup, S.; Wu, Y.; Larson, T.; Graham, I.; Baker, A.; Holdsworth, M. Control of germination and lipid mobilization by COMATOSE, the Arabidopsis homologue of human ALDP. *EMBO J.* **2002**, *21*, 2912–2922. [CrossRef]

84. Cao, D.; Hussain, A.; Cheng, H.; Peng, J. Loss of function of four DELLA genes leads to light-and gibberellin-independent seed germination in Arabidopsis. *Planta* **2005**, *223*, 105–113. [CrossRef]

85. Tyler, L.; Thomas, S.G.; Hu, J.; Dill, A.; Alonso, J.M.; Ecker, J.R.; Sun, T.-P. DELLA proteins and gibberellin-regulated seed germination and floral development in Arabidopsis. *Plant Physiol.* **2004**, *135*, 1008–1019. [CrossRef]

86. Graeber, K.; Linkies, A.; Steinbrecher, T.; Mummenhoff, K.; Tarkowská, D.; Turečková, V.; Ignatz, M.; Sperber, K.; Voegele, A.; de Jong, H. DELAY OF GERMINATION1 mediates a conserved coat-dormancy mechanism for the temperature-and gibberellin-dependent control of seed germination. *Proc. Natl. Acad. Sci. USA* **2014**, *111*, 3571–3580. [CrossRef]

87. Cantoro, R.; Crocco, C.D.; Benech-Arnold, R.L.; Rodríguez, M.V. In vitro binding of *Sorghum bicolor* transcription factors ABI4 and ABI5 to a conserved region of a GA 2-OXIDASE promoter: Possible role of this interaction in the expression of seed dormancy. *J. Exp. Bot.* **2013**, *64*, 5721–5735. [CrossRef]

88. Park, J.; Kim, Y.-S.; Kim, S.-G.; Jung, J.-H.; Woo, J.-C.; Park, C.-M. Integration of auxin and salt signals by the NAC transcription factor NTM2 during seed germination in Arabidopsis. *Plant Physiol.* **2011**, *156*, 537–549. [CrossRef]

89. Ramaih, S.; Guedira, M.; Paulsen, G.M. Relationship of indoleacetic acid and tryptophan to dormancy and preharvest sprouting of wheat. *Funct. Plant Biol.* **2003**, *30*, 939–945. [CrossRef]

90. Liu, A.; Gao, F.; Kanno, Y.; Jordan, M.C.; Kamiya, Y.; Seo, M.; Ayele, B.T. Regulation of wheat seed dormancy by after-ripening is mediated by specific transcriptional switches that induce changes in seed hormone metabolism and signaling. *PLoS ONE* **2013**, *8*, e56570. [CrossRef] [PubMed]

91. Liu, X.; Zhang, H.; Zhao, Y.; Feng, Z.; Li, Q.; Yang, H.-Q.; Luan, S.; Li, J.; He, Z.-H. Auxin controls seed dormancy through stimulation of abscisic acid signaling by inducing ARF-mediated ABI3 activation in Arabidopsis. *Proc. Natl. Acad. Sci. USA* **2013**, *110*, 15485–15490. [CrossRef]

92. Belin, C.; Megies, C.; Hauserová, E.; Lopez-Molina, L. Abscisic acid represses growth of the Arabidopsis embryonic axis after germination by enhancing auxin signaling. *Plant Cell* **2009**, *21*, 2253–2268. [CrossRef] [PubMed]

93. Arc, E.; Sechet, J.; Corbineau, F.; Rajjou, L.; Marion-Poll, A. ABA crosstalk with ethylene and nitric oxide in seed dormancy and germination. *Front. Plant Sci.* **2013**, *4*, 1–19. [CrossRef] [PubMed]

94. Corbineau, F.; Xia, Q.; Bailly, C.; El-Maarouf-Bouteau, H. Ethylene, a key factor in the regulation of seed dormancy. *Front. Plant Sci.* **2014**, *5*, 539. [CrossRef] [PubMed]

95. Subbiah, V.; Reddy, K.J. Interactions between ethylene, abscisic acid and cytokinin during germination and seedling establishment in Arabidopsis. *J. BioSci. (Bangalore)* **2010**, *35*, 451–458. [CrossRef] [PubMed]

96. Wang, Z.; Cao, H.; Sun, Y.; Li, X.; Chen, F.; Carles, A.; Li, Y.; Ding, M.; Zhang, C.; Deng, X. Arabidopsis paired amphipathic helix proteins SNL1 and SNL2 redundantly regulate primary seed dormancy via abscisic acid–ethylene antagonism mediated by histone deacetylation. *Plant Cell* **2013**, *25*, 149–166. [CrossRef]

97. Steber, C.M.; McCourt, P. A role for brassinosteroids in germination in Arabidopsis. *Plant Physiol.* **2001**, *125*, 763–769. [CrossRef]

98. Hu, Y.; Yu, D. BRASSINOSTEROID INSENSITIVE2 interacts with ABSCISIC ACID INSENSITIVE5 to mediate the antagonism of brassinosteroids to abscisic acid during seed germination in Arabidopsis. *Plant Cell* **2014**, *26*, 4394–4408. [CrossRef]

99. Chitnis, V.R.; Gao, F.; Yao, Z.; Jordan, M.C.; Park, S.; Ayele, B.T. After-ripening induced transcriptional changes of hormonal genes in wheat seeds: The cases of brassinosteroids, ethylene, cytokinin and salicylic acid. *PLoS ONE* **2014**, *9*, 87543. [CrossRef]

100. Nambara, E.; Okamoto, M.; Tatematsu, K.; Yano, R.; Seo, M.; Kamiya, Y. Abscisic acid and the control of seed dormancy and germination. *Seed Sci. Res.* **2010**, *20*, 55–67. [CrossRef]

101. Jacobsen, J.V.; Barrero, J.M.; Hughes, T.; Julkowska, M.; Taylor, J.M.; Xu, Q.; Gubler, F. Roles for blue light, jasmonate and nitric oxide in the regulation of dormancy and germination in wheat grain (*Triticum aestivum* L.). *Planta* **2013**, *238*, 121–138. [CrossRef] [PubMed]

102. Xu, Q.; Truong, T.T.; Barrero, J.M.; Jacobsen, J.V.; Hocart, C.H.; Gubler, F. A role for jasmonates in the release of dormancy by cold stratification in wheat. *J. Exp. Bot.* **2016**, *67*, 3497–3508. [CrossRef] [PubMed]

103. Xie, Z.; Zhang, Z.-L.; Hanzlik, S.; Cook, E.; Shen, Q.J. Salicylic acid inhibits gibberellin-induced alpha-amylase expression and seed germination via a pathway involving an abscisic-acid-inducible WRKY gene. *Plant Mol. Biol.* **2007**, *64*, 293–303. [CrossRef] [PubMed]

104. Lee, S.; Kim, S.G.; Park, C.M. Salicylic acid promotes seed germination under high salinity by modulating antioxidant activity in Arabidopsis. *New Phytol.* **2010**, *188*, 626–637. [CrossRef] [PubMed]

105. Dewar, J.; Taylor, J.; Berjak, P. Changes in selected plant growth regulators during germination in sorghum. *Seed Sci. Res.* **1998**, *8*, 1–8. [CrossRef]

106. Cook, C.; Whichard, L.P.; Turner, B.; Wall, M.E.; Egley, G.H. Germination of witchweed (*Striga lutea Lour.*): Isolation and properties of a potent stimulant. *Science* **1966**, *154*, 1189–1190. [CrossRef]

107. López-Ráez, J.A.; Charnikhova, T.; Fernández, I.; Bouwmeester, H.; Pozo, M.J. Arbuscular mycorrhizal symbiosis decreases strigolactone production in tomato. *J. Plant Physiol.* **2011**, *168*, 294–297. [CrossRef]

108. Stanga, J.P.; Smith, S.M.; Briggs, W.R.; Nelson, D.C. SUPPRESSOR OF MORE AXILLARY GROWTH2 1 (SMAX2) controls seed germination and seedling development in Arabidopsis. *Plant Physiol.* **2013**, *163*, 318–330. [CrossRef]

109. Lefebvre, V.; North, H.; Frey, A.; Sotta, B.; Seo, M.; Okamoto, M.; Nambara, E.; Marion-Poll, A. Functional analysis of Arabidopsis NCED6 and NCED9 genes indicates that ABA synthesized in the endosperm is involved in the induction of seed dormancy. *Plant J.* **2006**, *45*, 309–319. [CrossRef]

110. Martínez-Andújar, C.; Ordiz, M.I.; Huang, Z.; Nonogaki, M.; Beachy, R.N.; Nonogaki, H. Induction of 9-cis-epoxycarotenoid dioxygenase in *Arabidopsis thaliana* seeds enhances seed dormancy. *Proc. Natl. Acad. Sci. USA* **2011**, *108*, 17225–17229. [CrossRef] [PubMed]

111. Saez, A.; Apostolova, N.; Gonzalez-Guzman, M.; Gonzalez-Garcia, M.P.; Nicolas, C.; Lorenzo, O.; Rodriguez, P.L. Gain-of-function and loss-of-function phenotypes of the protein phosphatase 2C HAB1 reveal its role as a negative regulator of abscisic acid signalling. *Plant J.* **2004**, *37*, 354–369. [CrossRef] [PubMed]

112. Mehrotra, R.; Bhalothia, P.; Bansal, P.; Basantani, M.K.; Bharti, V.; Mehrotra, S. Abscisic acid and abiotic stress tolerance–Different tiers of regulation. *J. Plant Physiol.* **2014**, *171*, 486–496. [CrossRef] [PubMed]

113. Sah, S.K.; Reddy, K.R.; Li, J. Abscisic acid and abiotic stress tolerance in crop plants. *Front. Plant Sci.* **2016**, *7*, 571. [CrossRef] [PubMed]

114. Karssen, C.; Hilhorst, H.; Koornneef, M. The benefit of biosynthesis and response mutants to the study of the role of abscisic acid in plants. In *Plant Growth Substances 1988*; Springer: Berlin, Germany, 1990; pp. 23–31.

115. Bentsink, L.; Koornneef, M. Seed dormancy and germination. *Arab. Book* **2008**, *6*, 119. [CrossRef]

116. Jones, H.D.; Kurup, S.; Peters, N.C.; Holdsworth, M.J. Identification and analysis of proteins that interact with the *Avena fatua* homologue of the maize transcription factor VIVIPAROUS1. *Plant J.* **2000**, *21*, 133–142. [CrossRef]

117. Nambara, E.; Naito, S.; McCourt, P. A mutant of Arabidopsis which is defective in seed development and storage protein accumulation is a new abi3 allele. *Plant J.* **1992**, *2*, 435–441. [CrossRef]

118. Ng, D.W.; Chandrasekharan, M.B.; Hall, T.C. The 5′ UTR negatively regulates quantitative and spatial expression from the ABI3 promoter. *Plant Mol. Biol.* **2004**, *54*, 25–38. [CrossRef]

119. Beaudoin, N.; Serizet, C.; Gosti, F.; Giraudat, J. Interactions between abscisic acid and ethylene signaling cascades. *Plant Cell* **2000**, *12*, 1103–1115. [CrossRef]

120. Chiwocha, S.D.; Cutler, A.J.; Abrams, S.R.; Ambrose, S.J.; Yang, J.; Ross, A.R.; Kermode, A.R. The etr1-2 mutation in *Arabidopsis thaliana* affects the abscisic acid, auxin, cytokinin and gibberellin metabolic pathways during maintenance of seed dormancy, moist-chilling and germination. *Plant J.* **2005**, *42*, 35–48. [CrossRef]

121. Cadman, C.S.; Toorop, P.E.; Hilhorst, H.W.; Finch-Savage, W.E. Gene expression profiles of Arabidopsis Cvi seeds during dormancy cycling indicate a common underlying dormancy control mechanism. *Plant J.* **2006**, *46*, 805–822. [CrossRef] [PubMed]

122. Millar, A.A.; Jacobsen, J.V.; Ross, J.J.; Helliwell, C.A.; Poole, A.T.; Scofield, G.; Reid, J.B.; Gubler, F. Seed dormancy and ABA metabolism in Arabidopsis and barley: The role of ABA 8′-hydroxylase. *Plant J.* **2006**, *45*, 942–954. [CrossRef] [PubMed]

123. Xiong, L.; Gong, Z.; Rock, C.D.; Subramanian, S.; Guo, Y.; Xu, W.; Galbraith, D.; Zhu, J.-K. Modulation of abscisic acid signal transduction and biosynthesis by an Sm-like protein in Arabidopsis. *Dev. Cell* **2001**, *1*, 771–781. [CrossRef]

124. Parcy, F.; Valon, C.; Kohara, A.; Miséra, S.; Giraudat, J. The ABSCISIC ACID-INSENSITIVE3, FUSCA3, and LEAFY COTYLEDON1 loci act in concert to control multiple aspects of Arabidopsis seed development. *Plant Cell* **1997**, *9*, 1265–1277.

125. Parcy, F.; Valon, C.; Raynal, M.; Gaubier-Comella, P.; Delseny, M.; Giraudat, J. Regulation of gene expression programs during Arabidopsis seed development: Roles of the ABI3 locus and of endogenous abscisic acid. *Plant Cell* **1994**, *6*, 1567–1582.

126. Ding, Z.J.; Yan, J.Y.; Li, G.X.; Wu, Z.C.; Zhang, S.Q.; Zheng, S.J. WRKY41 controls Arabidopsis seed dormancy via direct regulation of ABI3 transcript levels not downstream of ABA. *Plant J.* **2014**, *79*, 810–823. [CrossRef]

127. Barrero, J.M.; Millar, A.A.; Griffiths, J.; Czechowski, T.; Scheible, W.R.; Udvardi, M.; Reid, J.B.; Ross, J.J.; Jacobsen, J.V.; Gubler, F. Gene expression profiling identifies two regulatory genes controlling dormancy and ABA sensitivity in Arabidopsis seeds. *Plant J.* **2010**, *61*, 611–622. [CrossRef]

128. Vaistij, F.E.; Gan, Y.; Penfield, S.; Gilday, A.D.; Dave, A.; He, Z.; Josse, E.-M.; Choi, G.; Halliday, K.J.; Graham, I.A. Differential control of seed primary dormancy in Arabidopsis ecotypes by the transcription factor SPATULA. *Proc. Natl. Acad. Sci. USA* **2013**, *110*, 10866–10871. [CrossRef]

129. Yamagishi, K.; Tatematsu, K.; Yano, R.; Preston, J.; Kitamura, S.; Takahashi, H.; McCourt, P.; Kamiya, Y.; Nambara, E. CHOTTO1, a double AP2 domain protein of *Arabidopsis thaliana*, regulates germination and seedling growth under excess supply of glucose and nitrate. *Plant Cell Physiol.* **2008**, *50*, 330–340. [CrossRef]

130. Yano, R.; Kanno, Y.; Jikumaru, Y.; Nakabayashi, K.; Kamiya, Y.; Nambara, E. CHOTTO1, a putative double APETALA2 repeat transcription factor, is involved in abscisic acid-mediated repression of gibberellin biosynthesis during seed germination in Arabidopsis. *Plant Physiol.* **2009**, *151*, 641–654. [CrossRef]

131. Skubacz, A.; Daszkowska-Golec, A.; Szarejko, I. The role and regulation of ABI5 (ABA-Insensitive 5) in plant development, abiotic stress responses and phytohormone crosstalk. *Front. Plant Sci.* **2016**, *7*, 1884. [CrossRef] [PubMed]

132. Griffiths, J.; Kohji, M.; Rieu, I.; Zentella, R.; Zhang, Z.-L.; Powers, S.J.; Gong, F.; Phillips, A.L.; Hedden, P.; Sun, T.-P.; et al. Genetic characterization and functional analysis of the GID1 gibberellin receptors in Arabidopsis. *Plant Cell* **2007**, *18*, 3399–3414. [CrossRef] [PubMed]

133. Ueguchi-Tanaka, M.; Nakajima, M.; Katoh, E.; Ohmiya, H.; Asano, K.; Saji, S. Molecular interactions of a soluble gibberellin receptor, GID1, with a rice DELLA protein, SLR1, and gibberellin. *Plant Cell* **2007**, *19*, 2140–2155. [CrossRef] [PubMed]

134. Dill, A.; Thomas, S.G.; Hu, J.; Steber, C.M.; Sun, T.-P. The Arabidopsis F-box protein SLEEPY1 targets gibberellin signaling repressors for gibberellin-induced degradation. *Plant Cell* **2004**, *16*, 1392–1405. [CrossRef] [PubMed]

135. Fu, X.; Richards, D.E.; Fleck, B.; Xie, D.; Burton, N.; Harberd, N.P. The Arabidopsis mutant sleepy1gar2-1 protein promotes plant growth by increasing the affinity of the SCFSLY1 E3 ubiquitin ligase for DELLA protein substrates. *Plant Cell* **2004**, *16*, 1406–1418. [CrossRef]

136. McGinnis, K.M.; Thomas, S.G.; Soule, J.D.; Strader, L.C.; Zale, J.M.; Sun, T.-P.; Steber, C.M. The Arabidopsis SLEEPY1 gene encodes a putative F-box subunit of an SCF E3 ubiquitin ligase. *Plant Cell* **2003**, *15*, 1120–1130. [CrossRef]

137. Sasaki, A.; Itoh, H.; Gomi, K.; Ueguchi-Tanaka, M.; Ishiyama, K.; Kobayashi, M.; Jeong, D.-H.; An, G.; Kitano, H.; Ashikari, M. Accumulation of phosphorylated repressor for gibberellin signaling in an F-box mutant. *Science* **2003**, *299*, 1896–1898. [CrossRef]

138. Magome, H.; Yamaguchi, S.; Hanada, A.; Kamiya, Y.; Oda, K. The DDF1 transcriptional activator upregulates expression of a gibberellin-deactivating gene, GA2ox7, under high-salinity stress in Arabidopsis. *Plant J.* **2008**, *56*, 613–626. [CrossRef]

139. Yaish, M.W.; El-kereamy, A.; Zhu, T.; Beatty, P.H.; Good, A.G.; Bi, Y.-M. The APETALA-2-like transcription factor OsAP2-39 controls key interactions between abscisic acid and gibberellin in rice. *PLoS Genet.* **2010**, *6*, 1001098. [CrossRef]

140. Narsai, R.; Law, S.R.; Carrie, C.; Xu, L.; Whelan, J.; Law, S.R.; Carrie, C.; Xu, L.; Whelan, J. In-depth temporal transcriptome profiling reveals a crucial developmental switch with roles for RNA processing and organelle metabolism that are essential for germination in Arabidopsis. *Plant Physiol.* **2011**, *157*, 1342–1362. [CrossRef]

141. Wilson, R.L.; Kim, H.; Bakshi, A.; Binder, B.M. The ethylene receptors ETHYLENE RESPONSE1 and ETHYLENE RESPONSE2 have contrasting roles in seed germination of Arabidopsis during salt stress. *Plant Physiol.* **2014**, *165*, 1353–1366. [CrossRef] [PubMed]

142. Xi, W.; Liu, C.; Hou, X.; Yu, H. MOTHER OF FT AND TFL1 regulates seed germination through a negative feedback loop modulating ABA signaling in Arabidopsis. *Plant Cell* **2010**, *22*, 1733–1748. [CrossRef] [PubMed]

143. Xi, W.; Yu, H. MOTHER OF FT AND TFL1 regulates seed germination and fertility relevant to the brassinosteroid signaling pathway. *Plant Signal. Behav.* **2010**, *5*, 1315–1317. [CrossRef] [PubMed]

144. Dave, A.; Vaistij, F.E.; Gilday, A.D.; Penfield, S.D.; Graham, I.A. Regulation of *Arabidopsis thaliana* seed dormancy and germination by 12-oxo-phytodienoic acid. *J. Exp. Bot.* **2016**, *67*, 2277–2284. [CrossRef]

145. Finch-Savage, W.E.; Cadman, C.S.; Toorop, P.E.; Lynn, J.R.; Hilhorst, W.H. Seed dormancy release in Arabidopsis Cvi by dry after-ripening, low temperature, nitrate and light shows common quantitative patterns of gene expression directed by environmentally specific sensing. *Plant J.* **2007**, *51*, 60–78. [CrossRef]

146. Simpson, G.M. *Seed Dormancy in Grasses*; Cambridge University Press: Cambridge, UK, 2007; pp. 197–206.

147. Footitt, S.; Douterelo-Soler, I.; Clay, H.; Finch-Savage, W.E. Dormancy cycling in Arabidopsis seeds is controlled by seasonally distinct hormone-signaling pathways. *Proc. Natl. Acad. Sci. USA* **2011**, *108*, 20236–20241. [CrossRef]

148. Buriro, M.; Oad, F.C.; Keerio, M.I.; Tunio, S.; Gandahi, A.W.; Hassan, S.W.U.; Oad, S.M. Wheat seed germination under the influence of temperature regimes. *Sarhad J. Agric.* **2011**, *27*, 539–543.

149. Nyachiro, J.; Clarke, F.; DePauw, R.; Knox, R.; Armstrong, K. Temperature effects on seed germination and expression of seed dormancy in wheat. *Euphytica* **2002**, *126*, 123–127. [CrossRef]

150. Reddy, L.; Metzger, R.; Ching, T. Effect of Temperature on Seed Dormancy of Wheat. *Crop Sci.* **1985**, *25*, 455–458. [CrossRef]

151. Lim, S.; Park, J.; Lee, N.; Jeong, J.; Toh, S.; Watanabe, A.; Kim, J.; Kang, H.; Kim, D.H.; Kawakami, N. ABA-INSENSITIVE3, ABA-INSENSITIVE5, and DELLAs interact to activate the expression of SOMNUS and other high-temperature-inducible genes in imbibed seeds in Arabidopsis. *Plant Cell* **2013**, *25*, 4863–4878. [CrossRef]

152. Contreras, S.; Bennett, M.A.; Metzger, J.D.; Tay, D. Maternal light environment during seed development affects lettuce seed weight, germinability, and storability. *HortScience* **2008**, *43*, 845–852. [CrossRef]

153. Barrero, J.M.; Downie, A.B.; Xu, Q.; Gubler, F. A role for barley CRYPTOCHROME1 in light regulation of grain dormancy and germination. *Plant Cell* **2014**, *26*, 1094–1104. [CrossRef] [PubMed]

154. Gubler, F.; Hughes, T.; Waterhouse, P.; Jacobsen, J. Regulation of dormancy in barley by blue light and after-ripening: Effects on abscisic acid and gibberellin metabolism. *Plant Physiol.* **2008**, *147*, 886–896. [CrossRef] [PubMed]

155. Ochuodho, J.O.; Modi, A.T. Light-induced transient dormancy in *Cleome gynandra* L. seeds. *Afr. J. Agric. Res.* **2007**, *2*, 587–591.

156. Derkx, M.; Karssen, C. Effects of light and temperature on seed dormancy and gibberellin-stimulated germination in *Arabidopsis thaliana*: Studies with gibberellin-deficient and-insensitive mutants. *Physiol. Plant* **1993**, *89*, 360–368. [CrossRef]

157. Debeaujon, I.; Lepiniec, L.; Pourcel, L.; Routaboul, J.-M. Seed coat development and dormancy. *Annu. Plant Rev.* **2007**, *27*, 25–49.

158. Groos, C.; Gay, G.; Perretant, M.-R.; Gervais, L.; Bernard, M.; Dedryver, F.; Charmet, G. Study of the relationship between pre-harvest sprouting and grain color by quantitative trait loci analysis in a white× red grain bread-wheat cross. *Theor. Appl. Genet.* **2002**, *104*, 39–47. [CrossRef]

159. Sweeney, M.T.; Thomson, M.J.; Pfeil, B.E.; McCouch, S. Caught red-handed: Rc encodes a basic helix-loop-helix protein conditioning red pericarp in rice. *Plant Cell* **2006**, *18*, 283–294. [CrossRef]

160. Adebo, H.O.; Ahoton, L.E.; Quenum, F.; Ezin, V. Agro-morphological characterization of *Corchorus olitorius* cultivars of Benin. *Annu. Res. Rev. Biol.* **2015**, *7*, 229–240. [CrossRef]

161. Stetter, M.G.; Vidal-Villarejo, M.; Schmid, K.J. Convergent seed color adaptation during repeated domestication of an ancient new world grain. *BioRxiv* **2019**. BioRxiv:547943.

162. Akubugwo, I.; Obasi, A.; Ginika, S. Nutritional potential of the leaves and seeds of black nightshade-*Solanum nigrum* L. Var *virginicum* from Afikpo-Nigeria. *Pak. J. Nutr.* **2007**, *6*, 323–326. [CrossRef]

163. Paśko, P.; Sajewicz, M.; Gorinstein, S.; Zachwieja, Z. Analysis of selected phenolic acids and flavonoids in *Amaranthus cruentus* and *Chenopodium quinoa* seeds and sprouts by HPLC. *Acta Chromatogr.* **2008**, *20*, 661–672. [CrossRef]

164. Yang, R.-Y.; Lin, S.; Kuo, G. Content and distribution of flavonoids among 91 edible plant species. *Asia Pac. J. Clin. Nutr.* **2008**, *17*, 275–279. [PubMed]

165. Foley, M.E.; Fennimore, S.A. Genetic basis for seed dormancy. *Seed Sci. Res.* **1998**, *8*, 173–182. [CrossRef]

166. Nguyen, T.-P.; Keizer, P.; van Eeuwijk, F.; Smeekens, S.; Bentsink, L. Natural variation for seed longevity and seed dormancy are negatively correlated in *Arabidopsis thaliana*. *Plant Physiol.* **2012**, *160*, 2083–2092. [CrossRef] [PubMed]

167. Huang, X.; Schmitt, J.; Dorn, L.; Griffith, C.; Effgen, S.; Takao, S.; Koornneef, M.; Donohue, K.J.M.E. The earliest stages of adaptation in an experimental plant population: Strong selection on QTLS for seed dormancy. *Mol. Ecol.* **2010**, *19*, 1335–1351. [CrossRef]

168. Clerkx, E.J.; El-Lithy, M.E.; Vierling, E.; Ruys, G.J.; Blankestijn-De Vries, H.; Groot, S.P.; Vreugdenhil, D.; Koornneef, M. Analysis of natural allelic variation of Arabidopsis seed germination and seed longevity traits between the accessions Landsberg erecta and Shakdara, using a new recombinant inbred line population. *Plant Physiol.* **2004**, *135*, 432–443. [CrossRef]

169. Keurentjes, J.J.; Willems, G.; van Eeuwijk, F.; Nordborg, M.; Koornneef, M. A comparison of population types used for QTL mapping in *Arabidopsis thaliana*. *Plant Genet. Resour.* **2011**, *9*, 185–188. [CrossRef]

170. Alonso-Blanco, C.; Peeters, A.J.; Koornneef, M.; Lister, C.; Dean, C.; van den Bosch, N.; Pot, J.; Kuiper, M.T. Development of an AFLP based linkage map of Ler, Col and Cvi *Arabidopsis thaliana* ecotypes and construction of a Ler/Cvi recombinant inbred line population. *Plant J.* **1998**, *14*, 259–271. [CrossRef]

171. Qin, X.; Zeevaart, J.A. Overexpression of a 9-cis-epoxycarotenoid dioxygenase gene in *Nicotiana plumbaginifolia* increases abscisic acid and phaseic acid levels and enhances drought tolerance. *Plant Physiol.* **2002**, *128*, 544–551. [CrossRef]

172. Thompson, A.J.; Jackson, A.C.; Symonds, R.C.; Mulholland, B.J.; Dadswell, A.R.; Blake, P.S.; Burbidge, A.; Taylor, I.B. Ectopic expression of a tomato 9-cis-epoxycarotenoid dioxygenase gene causes over-production of abscisic acid. *Plant J.* **2000**, *23*, 363–374. [CrossRef] [PubMed]

173. Xu, Y. *Molecular Plant Breeding*; CAB International: Oxfordshire, UK, 2010; pp. 195–247.

174. Kearsey, M.; Farquhar, A. QTL analysis in plants; where are we now? *Heredity* **1998**, *80*, 137. [CrossRef] [PubMed]

175. Bove, J.; Jullien, M.; Grappin, P. Functional genomics in the study of seed germination. *Genome Biol.* **2001**, *3*, 1002.1. [CrossRef] [PubMed]

176. Holdsworth, M.J.; Finch-Savage, W.E.; Grappin, P.; Job, D. Post-genomics dissection of seed dormancy and germination. *Trends Plant Sci.* **2008**, *13*, 7–13. [CrossRef] [PubMed]

177. Chibani, K.; Ali-Rachedi, S.; Job, C.; Job, D.; Jullien, M.; Grappin, P. Proteomic analysis of seed dormancy in Arabidopsis. *Plant Physiol.* **2006**, *142*, 1493–1510. [CrossRef]

178. Li, F.; Wu, X.; Tsang, E.; Cutler, A.J. Transcriptional profiling of imbibed *Brassica napus* seed. *Genomics* **2005**, *86*, 718–730. [CrossRef]

179. Fait, A.; Angelovici, R.; Less, H.; Ohad, I.; Urbanczyk-Wochniak, E.; Fernie, A.R.; Galili, G. Arabidopsis seed development and germination is associated with temporally distinct metabolic switches. *Plant Physiol.* **2006**, *142*, 839–854. [CrossRef]

180. Gao, F.; Ayele, B.T. Functional genomics of seed dormancy in wheat: Advances and prospects. *Front. Plant Sci.* **2014**, *5*, 458. [CrossRef]

181. Harada, J.J.; Pelletier, J. Genome-wide analyses of gene activity during seed development. *Seed Sci. Res.* **2012**, *22*, S15–S22. [CrossRef]

182. Schranz, M.E.; Song, B.-H.; Windsor, A.J.; Mitchell-Olds, T. Comparative genomics in the Brassicaceae: A family-wide perspective. *Curr. Opin. Plant Biol.* **2007**, *10*, 168–175. [CrossRef]

183. Ayenan, M.A.T.; Sodedji, K.A.F.; Nwankwo, C.I.; Olodo, K.F.; Alladassi, M.E.B. Harnessing genetic resources and progress in plant genomics for fonio (*Digitaria* spp.) improvement. *Genet. Resour. Crop Evol.* **2018**, *65*, 373–386. [CrossRef]

184. Ajaiyeoba, E. Phytochemical and antimicrobial studies of *Gynandropsis gynandra* and *Buchholzia coriaceae* extracts. *Afr. J. Biomed. Res.* **2000**, *3*, 161–165.

185. Barker, M.S.; Vogel, H.; Schranz, M.E. Paleopolyploidy in the Brassicales: Analyses of the Cleome transcriptome elucidate the history of genome duplications in Arabidopsis and other Brassicales. *Genome Biol. Evol.* **2009**, *1*, 391–399. [CrossRef] [PubMed]

186. Marshall, D.M.; Muhaidat, R.; Brown, N.J.; Liu, Z.; Stanley, S.; Griffiths, H.; Sage, R.F.; Hibberd, J.M. Cleome, a genus closely related to Arabidopsis, contains species spanning a developmental progression from C3 to C4 photosynthesis. *Plant J.* **2007**, *51*, 886–896. [CrossRef]

187. Van den Bergh, E.; Külahoglu, C.; Bräutigam, A.; Hibberd, J.M.; Weber, A.P.; Zhu, X.-G.; Schranz, M.E. Gene and genome duplications and the origin of C4 photosynthesis: Birth of a trait in the Cleomaceae. *Curr. Plant Biol.* **2014**, *1*, 2–9. [CrossRef]

188. Amtmann, A. Learning from evolution: Thellungiella generates new knowledge on essential and critical components of abiotic stress tolerance in plants. *Mol. Plant* **2009**, *2*, 3–12. [CrossRef]

189. Mobility for Breeders in Africa. Available online: https://mobreed.com (accessed on 15 October 2019).

190. Sorrells, M.E.; La Rota, M.; Bermudez-Kandianis, C.E.; Greene, R.A.; Kantety, R.; Munkvold, J.D.; Mahmoud, A.; Ma, X.; Gustafson, P.J.; Qi, L.L. Comparative DNA sequence analysis of wheat and rice genomes. *Genome Res.* **2003**, *13*, 1818–1827.

191. Zhu, H.; Choi, H.-K.; Cook, D.R.; Shoemaker, R.C. Bridging model and crop legumes through comparative genomics. *Plant Physiol.* **2005**, *137*, 1189–1196. [CrossRef]

192. Nadeem, M.A.; Nawaz, M.A.; Shahid, M.Q.; Doğan, Y.; Comertpay, G.; Yıldız, M.; Hatipoğlu, R.; Ahmad, F.; Alsaleh, A.; Labhane, N. DNA molecular markers in plant breeding: Current status and recent advancements in genomic selection and genome editing. *Biotechnol. Biotechnol. Equip.* **2018**, *32*, 261–285. [CrossRef]

193. Van den Bergh, E. Comparative Genomics and Trait Evolution in Cleomaceae, a Model Family for Ancient Polyploidy. Ph.D. Thesis, Wageningen University, Wageningen, The Netherlands, 2017.

194. Mutwil, M.; Klie, S.; Tohge, T.; Giorgi, F.M.; Wilkins, O.; Campbell, M.M.; Fernie, A.R.; Usadel, B.; Nikoloski, Z.; Persson, S. PlaNet: Combined sequence and expression comparisons across plant networks derived from seven species. *Plant Cell* **2011**, *23*, 895–910. [CrossRef]

195. Das, G.; Rao, G.J.; Varier, M.; Prakash, A.; Prasad, D. Improved Tapaswini having four BB resistance genes pyramided with six genes/QTLs, resistance/tolerance to biotic and abiotic stresses in rice. *Sci. Rep.* **2018**, *8*, 2413. [CrossRef] [PubMed]

 *agronomy*

Review

Pursuing the Potential of Heirloom Cultivars to Improve Adaptation, Nutritional, and Culinary Features of Food Crops

Sangam Dwivedi [1], Irwin Goldman [2] and Rodomiro Ortiz [3,*

1   Independent Researcher, Hyderabad, Telangana 500001, India
2   Department of Horticulture, College of Agricultural and Life Sciences, University of Wisconsin Madison, 1575 Linden Drive, Madison, WI 53706, USA
3   Department of Plant Breeding, Swedish University of Agricultural Sciences, Sundsvagen 10 Box 101, SE 23053 Alnarp, Sweden
*   Correspondence: rodomiro.ortiz@slu.se; Tel.: +46-4041-5527

Received: 19 July 2019; Accepted: 7 August 2019; Published: 9 August 2019

Abstract: The burdens of malnutrition, protein and micronutrient deficiency, and obesity cause enormous costs to society. Crop nutritional quality has been compromised by the emphasis on edible yield and through the loss of biodiversity due to the introduction of high-yielding, uniform cultivars. Heirloom crop cultivars are traditional cultivars that have been grown for a long time (>50 years), and that have a heritage that has been preserved by regional, ethnic, or family groups. Heirlooms are recognized for their unique appearance, names, uses, and historical significance. They are gaining in popularity because of their unique flavors and cultural significance to local cuisine, and their role in sustainable food production for small-scale farmers. As a contrast to modern cultivars, heirlooms may offer a welcome alternative in certain markets. Recently, market channels have emerged for heirloom cultivars in the form of farmer–breeder–chef collaborations and seed-saver organizations. There is therefore an urgent need to know more about the traits available in heirloom cultivars, particularly for productivity, stress tolerance, proximate composition, sensory quality, and flavor. This information is scattered, and the intention of this review is to document some of the unique characteristics of heirloom cultivars that may be channeled into breeding programs for developing locally adapted, high-value cultivars.

Keywords: consumer-oriented breeding; consumer-oriented germplasm conservation; culinary; farmer–breeder–chef–consumer nexus; genetic diversity; heritage seedbank; local food systems; seed-savers; stress tolerance

---

1. History and Survival of Heirloom or Folk Cultivars

Heirloom cultivars are characterized as traditional or older cultivars that are open-pollinated, passed down from gardener to gardener or handed down in families, and often not used in large-scale agricultural enterprises. The definition of heirloom varies, and the term itself does not carry precise scientific designations. One of the most typical concepts of an heirloom is its non-hybrid, or open-pollinated nature. Heirlooms can be from cross- or self-pollinated species, but if the crop species is cross-pollinated, then the heirloom is considered open-pollinated. Because of this fact, some heirloom cultivars may be quite variable, and it is therefore apparent why heirlooms may not fit well into modern agricultural systems that place great value on uniformity. Votava and Bosland [1] found that the pepper cultivar 'California Wonder' had substantial amounts of genetic variability compared to a standard modern cultivar. They recommended that this cultivar not be used for genetic analysis because of this variability. In many cases, heirloom cultivars are known to possess more inherent variation than modern cultivars. Some of this variation could be due to their open-pollinated nature, and some might be due to the fact that they are often associated with seed-saving. The appearance of F_1 hybrids during the 20th century was arguably a tremendous advance in terms of crop productivity and seed commercialization. Some advocates of heirloom cultivars may, however, view the advance of F_1 hybrids as an incursion on traditional cultivars, as they rapidly usurped the use of traditional, open-pollinated cultivars for crop production purposes. While open-pollinated cultivars became a source for extracting the inbred lines that became parents of F_1 hybrid cultivars [2], the populations themselves were often not maintained or advanced as they once had been.

As F_1 hybrid cultivars became more common, gardeners and farmers began to notice the disappearance of heirloom or traditional cultivars. Importantly, many plant breeders and seed companies have invested their time and resources in hybrid cultivars for decades, thereby leaving heirloom, open-pollinated cultivars with a lower commercial status. In some parts of Europe, laws were passed that made it illegal to sell cultivars that were not on government-approved lists. Some of these laws also required substantial time and effort in the testing of cultivars before they could be listed. Part of the rationale was to ensure that cultivars were distinct, uniform, and stable, but many heirloom cultivars did not fit well into these criteria [3]. Thus, heirloom cultivars were, in some cases, relegated to the agricultural fringe [4].

Kingsbury [2] commented that at times of technological and social change, it is common to look back on simpler times. He suggests that the interest in heirloom cultivars may be due in part to this phenomenon, which idealizes the past. Another aspect of this yearning for simplicity is the fact that gardens represent, for some, a retreat from the stress of modernity. In this way, cultivating heirlooms may offer an opportunity for the gardener or eater to reconnect with a different age.

Additional aspects of the heirloom cultivar experience are the unique flavors and culinary qualities that generations of people had come to know and appreciate. Such qualities may be absent in modern cultivars bred for modern cropping systems, creating a desire for the heirloom. An example of such a quality is the creamy mouthfeel caused by high levels of water-soluble polysaccharides (also described as phytoglycogen) in heirloom sweet corn cultivars. The creamy mouthfeel caused by the polysaccharides was an important attribute of the *sugary-1* allele, which reduced starch accumulation in the maize endosperm and increased the amount of sugars [5]. As newer, sweeter F_1 hybrid cultivars were developed with the *shrunken-2* allele, the *sugary-1* allele was replaced. Cultivars homozygous for *shrunken-2* are noticeably sweeter than, perhaps even twice as sweet as those carrying *sugary-1*. However, they lack the creamy mouthfeel and some of the aromatics associated with older sweet corn cultivars [5]. Thus, an heirloom may deliver flavors and culinary qualities associated with specific, older cultivars and may thus represent an important aspect of heritage.

Heirloom cultivars also are associated with seed-saving, and, in fact, the generation-to-generation transfer of heirloom seed is often one of the defining features of an heirloom. Navazio [6] commented that the post-World-War-II era in the USA saw a major transition in agriculture, where more farmers were willing to purchase seed every season from seed companies. Prior to that time, farmers were involved heavily with producing their own seed, and with selecting and maintaining good seed stock. As farmers began purchasing seed annually from seed companies, heirloom cultivars and their systems of maintenance through seed-saving gradually diminished. Fortunately, a resurgence of interest in heirlooms has taken place in the last few decades. Several examples may illustrate this point. Today, some seed companies feature and celebrate heirloom cultivars in their catalogues. Gardeners and farmers regularly seek out, save, and preserve heirloom seed to continue the traditions of these unique cultivars, and consumers are willing to pay substantially more at the market for their products. The development of "new heirlooms" also is discussed in plant breeding fora, where the breeding program may begin with heirloom populations that are subject to selection for current conditions, but under the careful guidance of a seed-saver or local plant breeder. The Slow Food movement (https://www.slowfoodusa.org/) maintains a catalogue of heritage foods, including heirloom cultivars, run by the Ark of Taste (https://www.slowfoodusa.org/ark-of-taste). These foods, which include crop cultivars, are, according to Ark of Taste, "culturally or historically linked to a specified region, locality, ethnicity, or traditional production practice."

Heirloom cultivars are a part of the farming system in many regions of the world. For example, southern Appalachia is an area of high crop biodiversity in the USA, where many heirloom cultivars remain. Veteto [7] documented 134 heirloom cultivars that were still being grown in the region in a recent survey. He found that even though one or two individuals in a community were usually involved in maintaining significant numbers of heirloom cultivars, many communities had lost their heirloom vegetable cultivars. He found that the decline of the farming population and the lack of cultural continuance in family seed-saving traditions were likely responsible for this loss.

Among the problems associated with heirloom cultivars is their susceptibility to pathogens. In many cases, modern breeding has helped to improve host-plant resistance; thus, it is not surprising that heirloom cultivars may lack resistance to important pests. Heirloom tomato production can be limited by soilborne diseases such as bacterial wilt and fusarium, caused by the pathogens *Ralstonia solanacearum* and *Fusarium oxysporum* f.sp. *lycopersici*, respectively. A creative approach to this problem was discussed by Rivard and Louws [8], who grafted heirloom scions onto resistant rootstock. In naturally infested soil, bacterial wilt incidence for nongrafted 'German Johnson' was 79% and 75% over two years, but it had no symptoms of bacterial wilt if grafted onto the resistant genotypes CRA 66 or Hawaii 7996. Fusarium wilt incidence was 46% and 50%, respectively, in nongrafted and self-grafted German Johnson controls, but no symptoms of fusarium wilt were seen if plants were grafted onto resistant genotypes. Thus, grafting may be an appropriate approach for heirloom production in infected soils.

A debate is occurring as to whether modern cultivars are less nutritious than their heirloom counterparts. Barker et al. [9] examined differences in mineral nutrient concentrations between modern F_1 hybrids and heirloom cultivars of cabbage, and also looked at fertilization practices with either organic fertilizer and compost or conventional fertilizers. Crop production increased with conventional or organic fertilizers compared to compost. Mineral nutrient composition did not vary between modern or heirloom cultivars or among different fertility practices; however, the authors did find cultivar differences for nutrient concentration. The fact that mineral nutrient content did not vary between modern and heirloom cultivars is important in that it runs counter to popular press articles that suggest the nutritional quality of our food supply is decreasing. Flores et al. [10] examined carotenoid levels of traditional tomato cultivars. They found substantial amounts of variation for many carotenoids in colored fruit from these traditional cultivars, suggesting new opportunities for breeding. There is little doubt that heirloom cultivars contain reservoirs of useful traits, including those that might be able to contribute to improved human nutrition.

Van der Knaap and Tanksley [11] examined the genetic basis of the unique phenotype of the heirloom tomato cultivar 'Yellow Stuffer,' which looks more like a bell pepper than a traditional tomato. Their analysis was based on a segregating population derived from a cross between Yellow Stuffer and a wild species of tomato. They found three quantitative trait loci (QTL) that influenced fruit shape and seven QTL that influenced fruit mass, many of which had already been identified in other tomato mapping research. They were able to pinpoint an allele at *fs8.1* causing the convex locule walls that were responsible for the extended, bumpy shape of the pepper-shaped fruit in Yellow Stuffer. They surmised that the evolution of bell-pepper-shaped tomato fruit may have proceeded through mutations of some of the same genes that led to bell-pepper-type fruit in garden pepper.

Heirloom potatoes offer unique flavors and qualities that are sought after by consumers. Production of these heirloom types is, however, not well understood in the context of modern farming systems. Fandika et al. [12] investigated how irrigation and nitrogen management might be best employed in heirloom potatoes. They found that modern cultivars were more responsive to irrigation and nitrogen than heirloom potatoes. Interestingly, they found that higher applications of nitrogen decreased the yield of heirloom potato cultivars, whereas yields of modern cultivars were increased. Heirloom cultivars were more drought-tolerant, but required larger water inputs because of their later maturities.

Overall conclusions of this study suggested that production of heirloom types could be more expensive than that of modern cultivars.

Interest in seed-saving continues to grow in many parts of the developed world, perhaps as a response to the increasing consolidation of the global seed industry and a desire for more local control of plant genetic resources. Interest in heirloom cultivars goes hand in hand with this expanded interest in seed-saving. For the most part, these efforts are celebrated by communities who wish to build closer relationships with their seed sources. However, not everyone is celebrating. In the last five years, efforts have been made to close certain seed-lending libraries, which allow gardeners to check out packets of seed and return the seeds that they save from the crop grown. The cultivars made available by lending libraries are typically open-pollinated and often heirloom. Efforts to close lending libraries were based on an interpretation of the US federal Seed Act, which would have required seed-lending libraries to test seeds for germination and purity. In 2016, the state department of agriculture in Pennsylvania determined that such seed libraries are noncommercial seed exchanges and therefore not subject to the Seed Act. Additionally, a number of efforts have been initiated to resume breeding and seed-saving of open-pollinated cultivars. The Open Source Seed Initiative (www.osseeds.org), which started operations in 2014, is a clearinghouse for primarily open-pollinated cultivars that have been released into a "protected" commons. The protected nature of this commons means that anyone can breed with or save seeds of these open source cultivars, but cannot restrict others' use of them. Some of the cultivars contained in this registry include open-pollinated types, such as the new carrot cultivar "Dulcinea," that were bred recently to fill gaps in commercial seed company catalogs. Such open-pollinated types might be considered a modern spin on an heirloom cultivar.

Navazio [6] stated that the seed "is a reflection of the farming system as it is grown, cultivated, selected, and fully incorporated into that system." In this way, heirloom cultivars represent a farming system that considers and prizes traits of long-term interest to the farmer and consumer, the incorporation of seed-saving, family and community traditions, and the transference of plant genetic resources from generation to generation. Heirloom cultivars long cultivated locally by people around the world may be a critically important tool in creating a more sustainable food supply, given the tremendous challenges of climate change, food production, and food security [13].

2. Heirlooms and Sustainable Agriculture

Local landraces—defined as traditional cultivars developed over time after adapting to natural and cultural environments—or heirloom cultivars may perform better than modern-bred cultivars, particularly in marginal and climatic vulnerable sites. Heirlooms are known for their great trait diversity, e.g., for color, shape, size, growth, height, phenology, yield, and flavor. This wide diversity—which is the main feature capturing the attention of consumers seeking unique, nutritious, local food sources—also plays a key role in the risk management strategy of farmers if modern-bred cultivars are unsuitable for the local context [14]. Crop heterogeneity and diversity should therefore be included in a national asset strategy by rural development policy-makers, because such a valuable genetic endowment will be useful for further breeding. In this regard, Newton et al. [15] noted that landraces are sources of host-plant resistance and abiotic stress tolerance genes, as well as of phytonutrients with desired micronutrient concentrations that alleviate human aging-related and chronic diseases, and nutrient-use efficiency traits—which are very important for sustaining agriculture. Indeed, landraces show variation in their response to diverse stress-prone environments, and this heritage may be used as a genetic resource for breeding future crops [16]. For example, sweet potato heirlooms exhibit moderate to high host-plant resistance to soilborne insects, which provides interesting sources for developing new cultivars [17]; Hopi farmers in Arizona in the United States still plant 'Hopi blue maize' because it is adapted to drought and the short growing season and because of cultural significance [18]; native black and yellow maize landraces from Los Tuxtlas, Mexico are efficient phosphorus (P) colonizers and thus adapted to low soil P conditions [19]; farmers from northeast India still cultivate and maintain traditional rice cultivars because of their adaptation to harsh growing conditions [20–22]; heritage durum wheat cultivars are more tolerant to drought than modern cultivars [23]; and Spanish and Italian farmers in some regions still cultivate tomato landraces that are highly adapted to drought and salinity [24,25].

3. Genetic Enhancement Using Heirlooms and Further Seed Supply of Bred-Cultivars for Organic Farming

Crossbreeding—based on controlled crossing, population improvement, and family selection—is pursued to develop cultivars that may deliver greater elasticity for adapting to climate shifts, energy limitations, and low inputs in organic farming. The main traits of a successful organic cultivar outside of industrial agricultural production systems are host-plant resistance to pathogens (e.g., bacteria, fungi, oomycetes, phytoplasma, viroid, or viruses) and pests (birds, insects, and nematodes), ability to outcompete weeds, resource-use efficiency, abiotic stress and pollution tolerance, adaptability to a range of soil quality factors, responsiveness to low plant density, improved nutritional content, and satisfactory yield under low-input conditions. Moreover, cultivars may be bred under organic conditions, thus providing suitable germplasm for such farming systems along with the desired quality traits for consumers. For example, dry bean consumers may like to buy heirlooms with unique color patterns, which may be also sold at premium prices [26]. Swegarden et al. [27] further indicated that both yield stability analysis and economic incentives suggest that heirloom dry bean cultivars—despite having a 44% lower average grain yield than commercial checks—may allow for diversifying production, differentiation in the market, and more attractive economic returns for their small-scale organic growers.

Seed production begins with the developer of the new cultivar, who often retains the original breeder stock that may be further used as the "golden standard" for such a cultivar and as source for the foundation stock [28]. The next step is producing registered seed for distribution to licensees, who produce certified seed, which is the last stage of large-scale seed production. Organic farming standards ask further for organically produced seed. Certified organic seed should be grown only in certified organic soil using the same inputs as in organic farming, and packaged in a certified facility. Crop husbandry includes protecting soil fertility, using manure, rotating crops, conserving biodiversity and natural resources, sound plant health management, and recycling, whereas any practices leading to accumulation of heavy metals and other pollutants are forbidden.

4. Promoting Conservation of Heirloom Germplasm

4.1. Seed-Savers and Heritage Seedbanks

Some studies have estimated that up to 75% of plant genetic diversity has been lost due to the rapid expansion of industrial agriculture and large-scale adoption of monoculture farming [29]. Many such studies have been based on a reduction in the number of cultivars of particular crops, and, as such, may be inaccurate estimates of the amount of actual plant genetic diversity present when both gene banks and cultivars are considered. However, it appears likely that the homogenization of agricultural environments around the world and the drive towards widely adapted uniform cultivars of a few major crop species have translated into reduced genetic diversity currently deployed in farmers' fields. In addition to a narrowing of diversity of crop germplasm, there are worrying reports of reductions in the global diversity of plant species. The Millennium Ecosystem Assessment Reports state that 60,000 to 100,000 species of plant are currently threatened with extinction [30]. It is apparent that some of these species play critical roles in agricultural and natural ecosystems, and serve as reservoirs of important genes that could play a role in crop breeding. Hence, this reduction in plant biodiversity is a significant global concern.

One traditional practice that may improve the current state of crop genetic diversity is seed-saving. Practiced since the beginnings of agriculture itself, the simple act of saving and replanting seed from a crop was an integral part of food production in many parts of the world until the 20th century.

The development of an efficient and wide-reaching global seed industry, the convenience and quality enhancement of purchasing seed each year, and the appearance of technologies such as F_1 hybrids and intellectual property rights have all played a role in reducing the practice of seed-saving. From the point of view of crop genetic diversity, revitalizing the practice of seed-saving may, however, be viewed as vital for the world's sustainable food production and nutritional security.

There are several seed-saving projects, in addition to 1700 ex situ gene banks worldwide, including the CGIAR gene banks and Svalbard Global Seed Vault [31], that are involved in collecting and maintaining heirloom cultivars across the globe to preserve agricultural biodiversity. Arche Noah in Austria (https://www.arche-noah.at) and ASEED Europe, Camino Verde, Hawai'i Public Seed Initiative, Irish Seed Savers Association, Louisiana Native Plant Initiative, Man and the Biosphere Programme, Millenium Seed Bank Project, Native Seed/SEARCH, Navdanya, National Laboratory for Genetic Resources Preservation, New York City Native Plant Conservation Initiative, Australian Plantbank, Seed Savers Exchange, Seesave.org, Slow Food International, USC Canada, Vavilov Research Institute, and the World Vegetable Center house many traditional and rare cultivars of fruit, vegetable, flower, and grain crops [32]. Native American seed-savers are dedicated to on-farm preservation of their agrobiodiversity (heirlooms), but unwilling to share their seed heritage for preservation in ex situ gene banks (for fear of loss of ownership and access), as they believe that community-based in situ conservation will maintain local control and seed viability better than ex situ gene banks [31]. One excellent example of a collection of heirloom cultivars is held at the Heritage Seed Library (HSL) of Garden Organic, UK. It curates and maintains a collection of 800 heirlooms of carrot (*Daucus carota* L.), cucumber (*Cucumis sativus* L.), *Brassica oleracea* L. var. *acephala* (DC.) Metzq, faba bean (*Vicia faba* L.), pea (*Pisum sutivum* L.), and lettuce (*Lactuca sativa* L.) [33]. Table 1 presents a list of a few websites providing further information on gene banks or seed-savers for heirloom cultivars.

Table 1. Some gene banks or seed-saver organizations providing heirloom cultivars or related information.

Name	Website
AVRDC—The World Vegetable Center	https://avrdc.org/seed/
Nature and Nurture Seeds	https://natureandnurtureseeds.com
Organic Seed Producer Directory	https://seedalliance.org/directory/
Seed Savers Exchanges	https://www.seedsavers.org/mission
Sustainable Seed Company	https://sustainableseedco.com/#
The Kerr Center for Sustainable Agriculture	http://kerrcenter.com/publication/heirloom-vegetables-genetic-diversity-and-the-pursuit-of-food-security/

4.2. South and Southeast Asia

Vrihi, the largest non-governmental in situ seed depository of traditional rice cultivars in eastern India, houses 610 rice landraces that withstand a much wider range of fluctuations in temperature and soil nutrient levels, as well as water stress, than modern rice cultivars. This collection includes numerous unique landraces, for example, the "Jugal," the doubled-grain rice, and "Sateen," the triple grain rice, long awn and erect flag leaf, or cultivars with distinct aroma, color, and taste. The Vrihi Seed Exchange Network has more than 6000 indigenous farmers who have received seeds from Vrihi and continue to cultivate and exchange seeds of these folk cultivars (synonymous to heirlooms) among themselves and to neighboring farmers [21]. At Basudha farm, located in the Rayagada district of Odisha, India, over 1400 folk rice cultivars and 30 other crops are grown every year, as a model of ecological agriculture and without any external inputs [www.cintdis.org/basudha]. The Philippines' Department of Agriculture, in collaboration with the International Rice Research Institute (IRRI, Philippines), has undertaken a project on raising productivity and enhancing the legacy of heirloom/traditional rice through empowering communities in unfavorable rice-based ecosystems in Philippines. This project has collected about 74 variants of 41 heirloom cultivars with distinct characteristics, with the sole aim to enhance productivity and livelihoods and conserve in situ, on-farm genetic resources [34].

4.3. USA

Native American seed-saving efforts are underway to preserve culturally significant seeds and knowledge to promote food sovereignty at the local or tribal level in the USA [31]. These consist of farmers and gardeners who share a common interest in keeping traditional and local crop diversity alive. They usually grow traditional crops and local cultivars of fruits and vegetables for cultural reasons, food preference, risk avoidance, local adaptation, and for niche market opportunities. Various groups, including the Indigenous Seed Keepers Network and Seedshed, have formed to foster the preservation and rematriation of heirloom strains of Native American crops, particularly of maize, bean, squash, tobacco, and sunflower. The Seed Savers Exchange (SSE), a non-profit heirloom seed gene bank based in Decorah, Iowa, saves and sells heirloom fruit, vegetable, and flower seeds in the USA. SSE has in its collection 24,000 rare fruit and vegetable cultivars [35]. This heirloom collection has immense variability, for example, its over 500 heirloom potato cultivars differ in tuber shape (round, oblong, and fingerling-shaped), skin (white, red, purple, and variegated), and flesh color (white, yellow, and purple), including varying skin and flesh color combinations, e.g., white skin–white flesh, white skin–yellow flesh, red skin–white flesh, red skin–yellow flesh, red skin–red or pink flesh [36]. Seed-saving efforts directed at heirloom cultivars may also be found in the southern/central Appalachia region, an area of relatively high crop biodiversity in the USA. In this region, a collection of 134 heirloom cultivars grown and saved by home-gardeners was documented, with beans being the most predominant and highly diverse group, followed by tomatoes, squash, corn, and potatoes. The decline of the farming population, combined with a lack of cultural continuance in family seed-saving traditions, however, threatens the ability of communities to maintain this crop biodiversity [7]. Culture and ethnicity also play an important role in preservation of heirloom cultivars [37–40].

4.4. Europe

There are numerous examples of seed-saver networks, for example, Réseau Semences Paysannes, Red de Semilla, and Rete Semi Rureli [41], involved in the organization of regional and local seed fairs, training workshops, participatory plant breeding activities, and transmission of farmers' knowledge about the selection and conservation of local cultivars [42]. The Réseau Semences Paysannes in France brings together all who are open to the development of peasant varieties. Peasant varieties, in this case, may be considered local cultivars originating from selection, identified as the collective heritage of the community and as heirlooms, and reproduced by all men and women who produce crops. The Red de Semillas is a technical, social, and political Spanish organization formed by the people that maintains agricultural biodiversity on peasant farms and on consumers' plates, whereas the Rete Semi Rureli has enacted regional laws to safeguard local agricultural biodiversity in Italy. The researchers associated with these networks are involved in the development of new cultivars shaped by history and participatorily bred with farmers. These cultivars are considered "peasant varieties" and are linked to the terroir of the plants [41].

4.5. Promoting Exchange of Heirloom Germplasm/Cultivars

Plant genetic resources are the basic raw materials for future genetic progress and an insurance against unforeseen threats to future agricultural production. Their use in crop improvement is one of the most sustainable ways to conserve valuable genetic resources. International and national policies, treaties, and agreements largely influence the use and distribution of plant genetic resources. Restrictive biodiversity laws in many countries prohibit or closely regulate the exchange of germplasm from other countries. This situation may, in the longer term, have a serious impact on the utilization of plant genetic diversity to cope with current and predicted challenges to agricultural production. Worldwide, over 7 million accessions (including some heirlooms) of plant genetic resources are preserved in over

1700 ex situ gene banks. However, obtaining seed samples from these gene banks is problematic, due to restrictions imposed by national governments. Ensuring conservation and sharing of such heritage germplasm and associated knowledge should be the priority. Adopting policies that foster consumer-oriented utilization and conservation of heritage germplasm could help ward off the dangers imposed by institutional and governmental restrictions on crop genetic resources [43].

5. Promoting Local Food Systems

Industrial agriculture, "Green Revolution" technologies, climate change, civil conflict, and change in market characteristics, including distance, have contributed to the erosion of farm-agrobiodiversity. The production and sale of heirloom cultivars are interdependent in local markets, suggesting that local food systems help preserve heritage diversity [44]. Modern food production systems based on Green Revolution technologies are neither sustainable nor nutritionally superior. Recent years have brought increased public awareness that the risk to onset of noncommunicable diseases (cancer, diabetes, heart diseases, and obesity) can be reduced by making changes in lifestyle and food habits [45,46]. Local food systems may be a critical component of lifestyle and of food habits that can contribute to improved public health.

Local food refers to food that is produced, sold, and consumed within a limited geographical area [47]. What drives consumers to buy locally produced food? People perceive that locally produced foods are fresher, taste better, are higher in minerals and vitamins, and safer than nonlocal food. The consumption of locally produced food is more environmentally sustainable than procuring food from global markets [44]. Attitudes towards supporting local agribusiness and preserving local heritage and tradition, keeping a connectedness with rural life, reducing one's carbon footprint (shorter transportation distance, minimizing the needs of packaging, processing, and refrigeration, among others), and protecting the environment for sustainable ecosystems also drive the public to consume locally produced foods [48–50]. Higher prices, accessibility, time constraints, and availability are, however, major barriers for consumption of local food [51,52]. Farmers' markets (FMs) refer to "markets where agricultural products are directly sold by producers to consumers through a common marketing channel" [53]. The "Buy local" initiative and "Buy local" movements in North America and western Europe have reinforced greater patronage at FMs and the belief of people that local food is fresher than food from farther away, and that buying local food benefits local farmers and the local economy as well as improving environmental sustainability [50,54].

6. Heirlooms' Nutritional, Sensory, and Culinary Characteristics

6.1. Pulses and Cereals

The literature suggests significant genetic variation for physico-chemical, sensory, and culinary characteristics among heirloom cultivars. Unique seed coat colors and patterns as well as seed size and shape differentiate heirlooms from modern bean (*Phaseolus vulgaris*) cultivars. Heirloom beans have been shown to have relatively higher protein, total fiber, and soluble fiber, and greater in vitro antioxidant capacity than modern cultivars, and took less time to hydrate (4.33 to 5.07 h compared to 8.1 h for the modern cultivars) but had a similar cooking time (~1 h). A few heirlooms had high flavonoids and differed in softness even after cooking. "Hutterite Soup" and "Jacob's Cattle" showed lower firmness than "Kornis Purple" and "Tiger's Eye," which had textures similar to the control [55,56]. A common bean landrace, "Ganxet," is appreciated greatly for its culinary value in northeast Spain. The inbred lines representing the variability within Ganxet germplasm showed greater protein, less total dietary fiber, more digestible dietary fiber, a higher proportion of seed coat, more glucose, and less starch than control cultivars ("White Kidney," "Navy," "Faba Asturiana," and "Tolosa") [57]. Twenty-four heirloom beans from western Washington in the USA have been identified, with distinct appearances and great culinary value [58]. The Italian heirloom "Monachine" is known famously as

Pellegrini in honor of Seattle's culinary icon Angelo Pellegrini, who enjoyed this bean for decades, and it is now featured by a high-end regional restaurant because of its great culinary quality [59].

There is growing demand for maize with blue kernels because of its significant health benefits and unique culinary applications. Blue-kernel maize typically produces anthocyanins in the aleurone layer of the endosperm, whereas in purple maize, the anthocyanin is produced predominantly in the pericarp of the kernel [60]. The variation in total anthocyanin content among maize heirlooms ranged from 17.6 to 65.1 mg 100 g seed^{-1}, with an average of 49.6 mg 100 g^{-1}. Cyanidin and pelargonidin were the major components, and peonidin and succinyl 3-glucoside were the minor components. Heirlooms with blue kernels had higher anthocyanin than those with purple or red kernels. "Navajo Blue" and "Ohio Blue" displayed highest anthocyanin values, whereas "Santa Clara Blue" and "Flor del Rio" had highest oil and protein contents [61]. The variants of a pericarp-pigmented heirloom, "Apache Red Purple," showed greater variation in anthocyanin concentrations (210–6183 µg g^{-1} pericarp), with some having greater proportions of either pelargonidin- or cyanidin/peonidin-derived anthocyanins [62].

Several Indian rice folk cultivars are either very nutritious or maintain their distinctive aroma and colors. "Kalanamak" (black-husked and short-grained rice) a non-basmati aromatic type famous for its distinct aroma, color, and taste, is a heritage rice from eastern India that has been in cultivation for the last 4000 years [63]. Unpolished brown rices such as "Bhadoi," "Kabiraj Sal," "Shatia," or "Agniban" are high in iron and antioxidants, whereas Shatia and "Kartiksal" are rich in fiber but low in carbohydrates [64]. Many of the traditional cultivars possess health-related properties. For example, "Pichha vari," "Karthigai samba," "Dudhsar," and "Bhejri" enhance milk production in lactating mothers; "Kelas" and "Bhutmoori" cure anemia; "Paramai-sal" improves child growth; "Nyavara" treats neurotic disorders; "Karhainy" alleviates paralysis; "Gudna" treats gastric ailments [22,65]; "Karanga" treats dysenteric complaints; "Bora" treats jaundice; "Pakheru," "Saraiphool," "Karia Gora," "Dani Gora," and "Punai Gora" are traditionally used as a tonic; and people consuming "Bhama," "Danigora," "Karhani," "Ramdi," "Muru," "Hindmauri," and "Punaigora" rices can work in their fields for a whole day without feeling hungry [65].

The white-seeded durum wheat landrace "Aybo" is highly preferred for use in holy communion and for making difo-dabo, and "Set-Akuri," "Arndeto," "Loko," "Kurkure," and "Mengesha" are known for their superior baking quality and used for making difo-dabo and injera. Difo-dabo is a traditional homemade bread prepared from flour by fermenting thick dough, and injera is a thin, flat, and spongy pancake-like product from fermented dough baked using traditional mitad [66]. Farmers in northeast Ethiopia maintain several landraces for specific end-uses; for example, "Nechita" is used for preparing thick porridge (genfo) and shorba for a semi-fluid drink from cracked grains. "Tikur gebs" is used for a beverage. "Temej" is used for kolo (roasted grain), and "Enat gebs," "Sene gebs," and "Meher gebs" are used for injera [67].

6.2. Vegetables

Cucurbita (pumpkins and squash), either young or mature, entered Italian kitchens by the middle of the 16th century. A greater proportion of colored exocarp and firm mesocarp in long and narrow cucurbits, compared to round-fruited, provided better eating quality if the young fruits were consumed whole. This culinary use of long-fruited cucurbits has been the driving force behind the several independent evolutions of long-fruitedness in *C. pepo* [68]. Some of the oldest Italian squash heirloom cultivars are "Costata Romanesca," which dates to 1590, "Cocozelle," dating back to the early 19th century, and "Tondo di Nizza," from the mid-19th century [68].

The Puglia region in Italy possesses a vast pool of local heirloom vegetable cultivars (https://drive.google.com/file/d/1uVemsY6zE4zt-ivPLEUaQjar7zgDb-ZI/view?usp=sharing) that are maintained and cultivated by farmers. Tomato landraces in the Puglia region of Italy are maintained by farmers by repeated selection generation after generation for desired organoleptic quality. Shelf-life after harvest is among the most sought-after quality traits in tomato. "Regina," a landrace adapted to the coastal saline soils of central Puglia, is known for its unique qualitative profile, characterized by high concentrations of tocopherols, lycopene, and ascorbic acid, and long shelf-life [69]. "Corbarino" and "Lucariello" tomatoes are prized highly by consumers in Italy for the superior quality of their fruits and shelf-life. Corbarino produces an intense red color and has high levels of soluble and total solids, whereas Lucariello possesses a heart-shaped fruit with a pronounced pointed apex of less intense red and a thick cuticle [25].

A large number of heirloom cultivars of melons and watermelons have been described in a book titled "Melons for the Passionate Grower" [70], squash and pumpkins in a book titled "The Compleat Squash" [71], and tomatoes in a book titled "The Heirloom Tomato" [72], which may be consulted for further details.

Heirloom tomatoes from North America show abundant diversity in fruit weight (5–150 g), shape (elongated, flattened, rounded, heart-shaped, pyriform), color (white ivory, yellow, orange, red, black), firmness, and brightness, as well as chemical quality characteristics TSS (total soluble solids), TA (titratable acidity), TSS:TA ratio, flavor intensity, and ascorbic acid [73]. "Criollo" tomatoes from the Andean valley in Argentina are valued highly for their flavor and taste [74]. A few populations of heirloom tomatoes from eastern Spain showed high mean values for ascorbic acid (308 mg kg^{-1} fruit weight, fw), lycopene (130 mg kg^{-1} fw), β-carotene (30 mg kg^{-1} fw), and total phenolics (89 mg caffeic acid 100 g^{-1} fw), and may therefore be best suited as sources of functional compounds [75].

A multicolored landrace of carrot, "yellow-purple Polignano," included in the "slow food" list of traditional products (https://www.fondazioneslowfood.com/en/what-we-do/slow-food-presidia/), has been in cultivation for decades by small farmers in Italy's Puglia region. This landrace is appreciated greatly for its multicolored roots (outer core ranges from yellow or deep orange to dark purple, with an inner core of pale yellow to light green), special taste, tenderness, crispness, flavor, and fragrance. On average, it has a 22% lower total glucose, fructose, and sucrose content, but has a similar sweetness to a commercial cultivar. Fructose is the major contributor to its distinctive flavor as well as to its glycemic index. The purple variants of this cultivar showed high levels of antioxidant activity, total phenols, carotenoids, and β-carotene [69,76]. Another colored race from the Apulia region of southern Italy, "Taggiano" (or Saint Ippazio), possesses an outer and inner core with purple and yellow-orange color, respectively. This heirloom is known for the popular cult of the Saint Ippazio; i.e., protection from Saint Ippazio against hernias or male impotency. It has high levels of bioactive compounds and antioxidant capacity compared to orange-rooted cultivars [77].

7. Assessing Diversity among Heirloom Germplasm

Assessment of genetic diversity and population structure of heirloom cultivars will be of great help in identification of diverse heirloom germplasm with beneficial traits for breeding or farming. For example, amplified fragment length polymorphism (AFLP) analysis of 171 heirlooms from the Heritage Seed Library (HSL) of Garden Organic revealed 1.5- to 2-fold differences in heterozygosity within carrot, cucumber, and *Brassica* spp. (*Brassica oleracea* var. acephala (DC.)) accessions, as well as 3.6- to 9-fold differences within faba bean, pea, and lettuce accessions [33]. Heirloom beans from southern Italy exhibited significant diversity in seed shape (cuboid, kidney, oval, round, truncate), seed coat pattern (absent, bicolor, pattern around hilum, speckled, spotted bicolor, stripped), seed color (black, brown, grey, vine, violet, white), seed weight (21–74 g 100 seed), and phaseolin (C, T, S) [78]. Phaseolin is a storage grain protein encoded by the *phas* gene in common bean. The three distinct forms of phaseolins in common bean are S-, T-, and C-types, identified respectively from "Sanilac" (S), "Tendergreen" (T), and "Contender" (C) cultivars [78]. "Badda," a round, large-seeded bean with a partially colored seed coat, has been in cultivation for more than two centuries by Sicilian farmers. Over the centuries of its cultivation, this landrace has diverged into two easily distinguished morphotypes, i.e., "Badda bianco" (white badda) and "Badda nero" (black badda). The Badda morphotypes can be grouped into three distinct clusters, namely, Badda bianco accessions in one cluster and those of Badda nero in two separate, well-distinguished clusters [79].

Microsatellite-based genetic diversity analysis of over 100 rice heirlooms from Northeast India detected three genetically distinct groups. Most *joha* rice accessions from Assam and *tai* rices from Mizoram and Sikkim were in cluster 1, whereas *chakhao* rices from Manipur were grouped in cluster 2, and aromatic accessions from Nagaland in cluster 3. Pairwise FST between three clusters varied from 0.223 to 0.453 [80]. FST is a fixation index, a measure of genetic distance, which ranges from 0 to 1, where 0 means complete sharing of genetic material and 1 means the populations are genetically distinct (i.e., no sharing of alleles).

"Candy Roaster" heirloom squash fruits differ in size (10–250+ lbs), shape (round, cylindrical, teardrop, blocky), and color (pink, tan, green, blue, gray, or orange), yet most have fine-textured orange flesh [81], whereas variants of "Cappello da prete," another squash heirloom of the 19th century, still grown in the Po Valley of northern Italy, differ significantly in fruit weight, pulp thickness, rind thickness, peduncle diameter, and seed weight [82]. Using highly polymorphic simple sequence repeats (SSRs) on a collection of 85 winter squash and pumpkin (*Cucurbita maxima* Duchesne) accessions, Kaźmińka et al. [83] noted two distinct clusters, with cluster 1 possessing modern breeding lines and cultivars characterized by small fruits (<5 kg) and bushy or vine growth habit, while cluster 2 contained old cultivars from central and eastern Europe, together with breeding lines and cultivars characterized by large fruits and vine growth habit. The Australian and New Zealand cultivars had longer fruit shelf-life. Thus, they concluded that old cultivars were an interesting source of genetic variation for breeding novel hybrids.

Italian tomato landraces Corbarino and Lucariello are known for adaptation to water deficit, prolonged fruit shelf-life, and good fruit quality. Whole-genome sequencing revealed 43,054 and 44,579 gene loci annotated in Corbarino and Lucariello. Both genomes exhibited novel regions with similarity to tomato wild species *Solanum pimpinellifolium* and *S. pennellii*, and single nucleotide polymorphisms (SNPs) or candidate genes associated with fruit quality, shelf-life, and stress tolerance [25]. Heirloom tomatoes, e.g., Criollo from an Andean valley in Argentina, are known for their excellent organoleptic quality, especially flavor and aroma. Aroma and sourness in Criollo fruit correlate with citrate and several volatile organic compounds, such as α-terpeneol, *p*-menth-1-en-9-al, linalool, and 3,6-dimethyl-2,3,3a,4,5,7a-hexahydrobenzofuran (DMHEX), which is a novel volatile compound discovered in this heirloom cultivar [74]. The fruit shape among Italian tomato landraces ranges from flattened or ribbed through pear or oxheart to round or elongated types. Although round or elongated types are rich in glycoalkaloids, the flattened types are rich in phenolic compounds [84]. North American heirloom tomatoes have been found to vary in fruit color (white, yellow, orange, red, or black), shape (elongated, flattened, heart-shaped, pyriform, or rounded), weight (4.2 to 265 g), firmness, total soluble solids (TSS) content, titrable acidity (TA), TSS:TA ratio, flavor intensity, and ascorbic acid concentration (Figure 1) [73]. Clearly, heirloom cultivars differ across crops, as noted above, maintain unique diversity, and can serve as a valuable resource for gardeners, farmers, and plant breeders.

The colocynth or desert watermelon (*Citrullus colocynthis*, CC) is a perennial watermelon adapted to desert soils throughout northern Africa, the Middle East, and southwestern Asia, and possesses genes for enhancing disease and pest resistance in cultivated watermelon (*Citrullus lanatus* var. *lanatus*, CLL) cultivars. A study on genetic diversity and relationships among CC, CLL, citron melon (*C. lanatus* ssp. *lanatus* var. *citroides*, CLC), and desert perennial (*C. ecirrhous*, CE) accessions revealed five groups corresponding to their geographic origins, and an additional group with admixture. Group 1 contained accessions from northern Africa that were distinguished by two subgroups. Group 2 and 3 included accessions mainly from the Middle East. The fourth and fifth groups were represented by single CC accessions, collected in Iran and Egypt, respectively. Each of these groups contained unique alleles, but also shared alleles with CLL, CLC, or CE, implying evolution from a common ancestor [85].

Figure 1. Diversity in fruit, root, and tuber size and color of heirloom cultivars of capsicum, carrot, potato, and tomato. Photo credit: University of Wisconsin—Madison College of Agricultural and Life Sciences.

8. Retaining Culinary and Nutritional Traits and Improving Heirloom Productivity

There is wide variation in productivity among heirloom cultivars when grown in organic farming systems. A few heirloom cultivars can compete with modern cultivars in organic systems, but almost no heirlooms can compete with modern cultivars under intensive production systems [26,27,86–88]. However, heirloom cultivars possess great diversity in fruit and seed size, shape, and appearance. They may often be superior in culinary and nutritional quality, and represent a rich source of many health-promoting compounds. Thus, improving the productivity while retaining the culinary, nutritional, and health-promoting compounds of heirloom cultivars is a great challenge for future plant breeding efforts [89]. For example, the heritage rice from eastern India Kalanamak is a tall, low-yielding, black-husked, short-grained rice superior to "Basmati" in aroma and taste. Its acreage has gone down (from 50,000 to almost zero) over years of cultivation due to its low grain yield. A systematic attempt was made to collect and evaluate variants of Kalanamak, collected from farmers, and to identify accessions as most aromatic and true to perceived Kalanamak quality. Using pure line selection, UPCAR-KN-1-5-1-1 was released as "Kalanamak 3" (KN3) for cultivation in eastern Uttar Pradesh, India. Subsequently, several semi-dwarf breeding lines, developed through hybridization or induced mutation, outyielded KN3 by 40% [63]. A first semi-dwarf, non-black husk cultivar, "Bauna Kalanamak 102," with comparable cooking quality and aroma to "KN3," has been released for cultivation in eastern Uttar Pradesh [90].

Spaniards highly appreciate the Ganxet bean, which is a landrace adopted on the Iberian Peninsula due to its culinary values, i.e., seed coat tenderness and buttery texture. Over the years of its cultivation, the original Ganxet bean became contaminated, possibly due to cross-hybridization with other beans. Several variants (inbred lines) derived from the Ganxet populations were evaluated against commercial cultivars. These lines showed greater variability in protein content, total dietary fiber, digestible dietary fiber, seed coat, glucose, and starch contents than commercial cultivars. An inbred line, "L67," had many favorable characteristics (greater protein, less total digestible fiber, more digestible dietary fiber, a higher proportion of seed coat, more glucose, and less starch than commercial cultivars) and resulted in commercialization as true to the original Ganxet [64]. "Caparrona" is another heirloom bean that was grown largely by farmers in Monzón, Italy. The increasing modernization of agriculture at the end of 20th century resulted in its replacement by modern cultivars. Today, only a few local growers continue producing Caparrona beans, and mainly for family use. A systematic effort was made to recover this landrace from extinction. Seed samples from growers were grown, and two progenies true to Caparrona types in morphology, chemical composition, and agronomic performance were identified for commercializing Caparrona beans as a gourmet product by a local producers' association [91].

A major seed company, Seminis Vegetable Seeds, has released a multiple-disease-resistant heirloom-type tomato hybrid, "Purple Boy," with great taste and unique appearance (purple), which are often the characteristics of heirloom cultivars [http://www.hortibiz.com/item/news/seminis--releases--first--purple--tomato--hybrid/].

Intensive breeding of crops with a focus on yield and stress tolerance has led indirectly to a reduction in nutrition and flavor in certain cases [92,93]. Environmental and agricultural practices may also impact the intensity of flavor and aroma [94–96]. Flavor improvement is most difficult to achieve because of the difficulty of assessing the phenotype, as well as a lack of basic knowledge about the chemicals driving consumer preferences, the pathways of their synthesis, and genes regulating the output of these pathways [97]. Analyses of consumer preferences, together with accurate phenotyping and the use of modern genomics and analytical chemistry tools in breeding, as evidenced in case of melon, strawberry, and tomato, and participatory farmer–breeder–chef–consumer collaborations may facilitate development of a next generation of crops to meet the growing demands of safe and nutritionally enhanced foods with good flavor, color, aroma, and texture [88,98].

9. Farmer–Breeder–Chef–Consumer Partnerships Preserving Heirlooms' Unique Cultural and Culinary Significance

Within the last decade, several organizations have begun collaborative breeding efforts among farmers, chefs, culinary professionals, and plant breeders. One of these is the Culinary Breeding Network (https://www.culinarybreedingnetwork.com/), which is based at Oregon State University in Corvallis, Oregon in the USA. The goal of this organization is to build communities of plant breeders, seed growers, farmers, produce buyers, chefs, and other stakeholders to improve quality in vegetables and grains. The Culinary Breeding Network is interested in facilitating communication, collaboration, and participation in selection, so that cultivars with superior performance, flavor, texture, and culinary attributes may be developed. They are also trying to promote and expand the awareness of cultivars developed by independent and public sector plant breeders, and particularly those that have been selected for organic systems. Similar efforts are being made by another collaborative endeavor formed at the University of Wisconsin in Madison under the direction of Professor Julie Dawson (https://dawson.horticulture.wisc.edu/). This effort is known as the Seed to Kitchen Collaborative (https://dawson.horticulture.wisc.edu/chef-farmer-plant-breeder-collaboration/). In addition, a seed company known as Row 7 (https://www.row7seeds.com/) has recently been established to foster close cooperation between chefs, farmers, and plant breeders to heighten the focus on culinary trait breeding and the sale of cultivars with special culinary characteristics.

These efforts are not directed solely at heirloom cultivars that possess unique culinary attributes. Instead, these efforts are an attempt to capitalize on the connection between the cultivar and the user of that cultivar, which is celebrated with many heirlooms, but with cultivars that have been selected for modern cropping systems. Thus, the goal is to combine the satisfaction of heirloom cultivars with the modern traits needed for high-intensity farming, often in organic systems.

Part of the impetus for these efforts is the realization that many plant breeders must necessarily focus their efforts on commodity markets. In so doing, they often need to place greater emphasis on traits such as postharvest storability, harvestable yield, and host-plant resistance to pathogens. There is little question that such traits are of great importance for crop production. Breeders also recognized that improving specific culinary qualities was possible, but often may not be as high on the priority list as those traits that deliver the productivity the market requires. Many plant breeders therefore continued to focus on commodity traits and relegated specific culinary objectives to minor parts of their work.

Within the last 10 years, what were once small projects by plant breeders to modify culinary traits have blossomed into full-blown breeding programs. One of the primary thought leaders in this area is Chef Dan Barber, of Blue Hill and Blue Hill at Stone Barns in New York (https://www.bluehillfarm.com/team/dan-barber). Dan was among the first to realize that partnerships among chefs, farmers, and plant breeders were necessary to bring a focus on culinary traits to cultivars that could be grown and enjoyed widely. His efforts with the seed company Row 7 are one of the best examples of how such partnerships have formed in recent years. The Culinary Breeding Network and Seed to Kitchen Collaborative are attempts to bring many professionals together to participate in the development and selection of new vegetable cultivars. As such, it represents a new form of participatory plant breeding that rapidly is gaining interest in many parts of the world.

The origin of the Culinary Breeding Network was in 2011, when Lane Selman, an agricultural researcher at Oregon State University (https://www.culinarybreedingnetwork.com/), observed chefs and plant breeders sharing knowledge during a taste test of nine different sweet pepper cultivars. At that time, Lane was involved in the management of vegetable trials for the Northern Organic Vegetable Improvement Collaborative (NOVIC; http://eorganic.info/novic/), a federally funded partnership that uses on-farm trials to identify cultivars of vegetables that will thrive in organic systems. Farmers were trying to find a replacement for a hybrid sweet pepper that was no longer being offered for sale in the seed industry. Among the sweet peppers trialed were several cultivars developed by Frank Morton, breeder and owner of Wild Garden Seed in Philomath, Oregon (https://www.wildgardenseed.com/). In the field, these plants stood out, and the participants in the trial began to discuss the various qualities of these peppers. This discussion was followed by more in-depth discussions among the various stakeholders who could be growing, cooking, eating, and producing seed of these peppers. A participatory network was thus born.

While plant breeders of horticultural crops often taste the breeding lines that they are developing, their evaluations typically lack the sort of analysis that may be common for chefs or other culinary professionals. For example, many plant breeders "bite test" their breeding materials in the field, rendering an opinion when comparing among lines, but often lacking in the sort of detailed observation that might be helpful in the culinary arts. When chefs and plant breeders finally started coming together in these efforts, the conversations that ensued were eye opening. Chefs often expressed interest in traits breeders had never considered, or typically would throw out. An expansion of the stakeholders involved in selection made for a much more robust selection process. Importantly, this sort of participatory breeding has a better chance of resulting in an adopted cultivar, given the involvement of many different stakeholders during cultivar development. Conversely, as culinary specialists come to appreciate the challenges and limitations of the plant breeding process, there is a much better overall understanding of what is possible for the future of our crops. Both the Culinary Breeding Network and the Seed to Kitchen Collaborative combine field trials with tasting events that allow for such conversations to occur.

The goal of the Seed to Kitchen Collaborative (SKC) is to connect farmers, plant breeders, and chefs to develop better cultivars for regional food systems in the Upper Midwest. The collaborative uses farmer focus groups and surveys to identify traits of interest, and flavor is consistently one of the top traits from these surveys. Currently, 80 farms are participating in this collaborative undertaking, and the majority of these farms use information gleaned from trials to find the cultivars that they use in their production systems. Another important aspect of the collaborative initiative is that it provides an opportunity for independent plant breeders and public sector plant breeders to trial their newest material. For many specialty crops, organized trailing programs do not exist; therefore, the collaborative provides a unique outlet for breeding programs.

Acorn squash is a very old cultivar group of *Cucurbita pepo* subsp. *texana*, characterized by its distinctive turbinate shape marked with ten alternating ridges and furrow. This cultivar group was developed by Native Americans of eastern North America by the late 15th and early 16th centuries [99]. It has fairly good quality. However, consumers often add sweeteners in its culinary preparations. Sensing this need, breeders in Israel developed a new hybrid cultivar of acorn squash called "Table Sugar," characterized by a bush growth habit and powdery mildew resistance, fruits with black-green exterior color, increased sweetness, chestnut flavor, and a solids content of 12–18%. The breeder teamed up with a seed producer, grower, supermarket chain, and chefs to introduce this new squash in Israel, and it was then proclaimed by the leading food magazine in that country to be "the finest-tasting pumpkin or winter squash ever developed" [100].

The Row 7 seed company began in 2017 as an effort to bring together breeders and chefs who want to make produce with better flavor and culinary characteristics. Founders of the Row 7 seed company, Chef Dan Barber, Professor Michael Mazourek of Cornell University, and Matthew Goldfarb of Fruition Seed Company, are filling a unique gap in the seed market. Their goal has been to promote and sell seeds for new vegetable and grain cultivars that might otherwise not have been sold by any seed company, because their primary characteristics could be their culinary attributes. Row 7's focus is to partner chefs and plant breeders, with the aim of creating new cultivars that will bring a focus on flavor to what may have otherwise been commodity crops [101].

Chefs have brought very useful and interesting perspectives to vegetable breeding efforts in Oregon and Wisconsin in recent years [102]. Chefs readily admit that the food that they prepare is only as good as the ingredients that they use, so cultivars with improved qualities will make their job easier and more satisfying. Sometimes, consumers and chefs can make an impact on cultivar development. Several examples described by Beans [102] include a new tomato cultivar developed by a cross between a modern breeding line, and an heirloom that was identified after a taste panel involving 100 consumers. A squash was developed by a plant breeder after extensive dialogue between himself and a chef, who trialed the product in restaurants during its development. Such collaborative breeding efforts widen the scope of traditional plant breeding, and likely forecast more consumer-friendly cultivars with enhanced culinary qualities.

10. Outlook

Heirlooms often embody particular shapes, colors, textures, flavors, and productivity traits for which they have come to be known and sought by farmers and consumers. They are recognized and prized for their specific qualities that have lent uniqueness to the cuisines of many of the world's cultures. Thus, the wealth of genetic variability encoded in heirloom cultivars of our crops is one of the treasures of our shared global food system; however, this treasure is not always freely shared, due to restrictions on the sharing and importation of germplasm.

As modern cultivars and the global seed industry have rapidly replaced heirlooms and the practice of seed-saving, this unique genetic heritage is, however, in danger of being lost. In recent decades, as our agricultural systems have become even more industrialized, many scientists have come to recognize that the genetic diversity of heirloom cultivars is of even greater importance for our shared food future. Modern crop breeding has improved agricultural productivity, but has simultaneously reduced genetic diversity in our major crops. As our farming systems become more industrialized and our climate becomes more erratic, enhancing, rather than shrinking, our crop genetic diversity will be critical for feeding the growing population of the world. Recent efforts to characterize, preserve, and enhance heirlooms abound, bringing these unique types to the forefront of many modern breeding efforts. Participatory breeding approaches with farmers, breeders, and chefs are but one example of modern approaches to expanding the diversity found in heirlooms into modern germplasm pools. Recently, molecular research has found markers associated with useful variants in heirloom cultivars that may be further used to introgress such traits into modern cultivars.

Heirloom cultivars also are closely associated with organic and sustainable farming systems, and typically do much better in such conditions than in modern, industrialized farming systems. As sustainable farming systems gain popularity in the developed world, the genetic diversity present in heirloom cultivars may become even more important. However, we realize that modern farming systems will continue to dominate many sectors of world agriculture. One of the most important breeding objectives going forward will therefore be to improve the productivity of heirloom types while retaining their unique qualities.

A number of groups have expanded their efforts to preserve traditional heirloom cultivars that have been important cultural touchstones for different ethnic groups around the world. One such example is the recent attempts to preserve, maintain, and repatriate germplasm of American crops domesticated and bred by Native American tribes. These efforts are closely tied to food sovereignty movements, which seek to bring particular heirloom types of staple food crops back under the control of the people who developed them. In such cases, heirloom cultivars represent a vital link to the past, as well as a critical bridge to the future. One of the most useful activities for the promotion of heirlooms would be documentation of these unique resources in text and photograph for the benefit of future generations.

One of the hallmarks of modern agriculture is the existence of high-performing cultivars that are bred for wide-area adaptation and productivity. These cultivars play a key role in feeding the world's growing population, but the narrowing genetic base of modern cultivars may make efforts to increase food production more difficult. Many decades ago, heirloom cultivars appeared to be relics of an earlier age. However, a renewed interest in heirloom cultivars has helped foster the sense that they may play an important role in future crop breeding efforts. New market channels have emerged for heirlooms and have re-energized this unique repository of crop germplasm. As they possess unique flavors, colors, texture, stress tolerances, and forms, heirlooms may represent an important collection of traits that can be of immediate value in crop production and as a source of breeding germplasm for future cultivars. The goal of this review has been to document some example of the unique characteristics of heirlooms, with the hope of further encouraging breeders to identify their useful exotic traits and introgress these traits into breeding programs aimed at developing high-quality, consumer-oriented cultivars.

Author Contributions: Conceptualization: S.D. and R.O.; writing—Original draft preparation: S.D., I.G. and R.O.; writing—Review and editing: S.D., I.G. and R.O.; Figure: I.G.

Funding: This research received no external funding.

Acknowledgments: Sangam Dwivedi acknowledges the contribution of Ramesh Kotana of Knowledge Sharing and Innovation Program of International Crops Research Institute for the Semi-Arid Tropics (ICRISAT) for arranging reprints on heirloom germplasm and cultivars as valuable literature resources that helped him draft his part of contribution to this manuscript.

Conflicts of Interest: The authors declare no conflict of interest.

References

1. Votava, E.J.; Bosland, P.W. Culivar by any other name: Genetic variability in heirloom bell pepper California Wonder. *HortScience* **2002**, *37*, 1100–1102. [CrossRef]
2. Kingsbury, N. *Hybrid: The History and Science of Plant Breeding*, 1st ed.; The University of Chicago Press: Chicago, IL, USA, 2009; pp. 1–493.
3. Wattnem, T. Seed laws, certification and standardization: Outlawing informal seed systems in the Global South. *J. Peasant Stud.* **2016**, *43*, 850–867. [CrossRef]
4. Jordan, J.A. The heirloom tomato as cultural object: Investigating taste and space. *Eur. Soc. Rural. Soc.* **2007**, *47*, 20–41. [CrossRef]
5. Tracy, W.F. Sweet corn. In *Specialty Corns*, 2nd ed.; Hallauer, A.R., Ed.; CRC Press: Boca Raton, FL, USA, 2001; pp. 155–197.
6. Navazio, J. *The Organic Seed Grower*; Chelsea Green Publishing: White River Junction, VT, USA, 2012; p. 388.

7. Veteto, J.R. The history and survival of traditional heirloom vegetable varieties in the southern Appalachian Mountains of western North Carolina. *Agric. Hum. Values* **2008**, *25*, 121–134. [CrossRef]

8. Rivard, C.L.; Louws, F.J. Grafting to manage soilborne diseases in heirloom tomato production. *HortScience* **2008**, *43*, 2104–2111. [CrossRef]

9. Barker, A.V.; Eaton, T.E.; Meagy, M.J.; Jahanzad, E.; Bryson, G.M. Variation of mineral nutrient contents of modern and heirloom cultivars of cabbage in different regimes of soil fertility. *J. Plant Nutr.* **2017**, *40*, 1–8. [CrossRef]

10. Flores, P.; Sánchez, E.; Fenoll, J.; Hellín, P. Genotypic variability of carotenoids in traditional tomato cultivars. *Food Res. Int.* **2017**, *100*, 510–516. [CrossRef] [PubMed]

11. Van Der Knaap, E.; Tanksley, S.D. The making of a bell pepper-shaped tomato fruit: Identification of loci controlling fruit morphology in Yellow Stuffer tomato. *Theor. Appl. Genet.* **2003**, *107*, 139–147. [CrossRef] [PubMed]

12. Fandika, I.R.; Kemp, P.D.; Millner, J.P.; Horne, D.; Roskruge, N. Irrigation and nitrogen effects on tuber yield and water use efficiency of heritage and modern potato cultivars. *Agric. Water Manag.* **2016**, *170*, 148–157. [CrossRef]

13. Petropoulos, S.A.; Barros, L.; Ferreira, I.C.F.R. Editorial: Rediscovering Local Landraces: Shaping Horticulture for the Future. *Front. Plant Sci.* **2019**, *10*. [CrossRef]

14. Coromaldi, M.; Pallante, G.; Savastano, S. Adoption of modern varieties, farmers' welfare and crop biodiversity: Evidence from Uganda. *Ecol. Econ.* **2015**, *119*, 346–358. [CrossRef]

15. Newton, A.C.; Akar, T.; Baresel, J.P.; Bebeli, P.J.; Bettencourt, E.; Bladenopoulos, K.V.; Czembor, J.H.; Fasoula, D.A.; Katsiotis, A.; Koutis, K.; et al. Cereal landraces for sustainable agriculture. A review. *Agron. Sust. Dev.* **2010**, *30*, 237–269. [CrossRef]

16. Dwivedi, S.L.; Cecarelli, S.; Blair, M.; Upadhyaya, H.D.; Are, A.K.; Ortiz, R. Landrace germplasm: A useful resource for improving yield and abiotic stress adaptation. *Trends Plant Sci.* **2016**, *21*, 31–42. [CrossRef] [PubMed]

17. Jackson, D.M.; Harrison, J.F. Insect resistance in traditional and heirloom sweetpotato varieties. *J. Econ. Entomol.* **2013**, *106*, 1456–1462. [CrossRef] [PubMed]

18. Cleveland, D.A.; Soleri, D.; Smith, S.E. Do folk crop varieties have a role in sustainable agriculture? *Bioscience* **2014**, *44*, 740–751. [CrossRef]

19. Sangabriel-Conde, W.; Negrete-Yankelevich, S.; Maldonado-Mendoza, I.E.; Trejo-Aguilar, D. Native maize landraces from Los Tuxtlas, Mexico show varying mycorrhizal dependency for P uptake. *Biol. Fertil. Soils* **2014**, *50*, 405–414. [CrossRef]

20. Das, T.; Das, A.K. Inventory of traditional rice varieties in farming systems of southern Assam: A case study. *Indian J. Tradit. Knowl.* **2014**, *13*, 157–163.

21. Deb, D. Valuing Folk Crop Varieties for Agroecology and Food Security. 2009. Available online: https://agrobiodiversityplatform.org/climatechange/2010/03/26/valuing-folk-crop-varieties-for-agroecology-and-food-security (accessed on 30 January 2019).

22. Deb, D. Folk Rice Varieties, Traditional Agricultural Knowledge and Food Security. 2012. Available online: https://www.researchgate.net/publication/233987520_Folk_Rice_Varieties_Traditional_Agricultural_Knowledge_and_Food_Security (accessed on 30 January 2019).

23. Slama, A.; Mallek-Maalej, E.; Ben Mohamed, H.; Rhim, T.; Radhouane, L. A return to the genetic heritage of durum wheat to cope with drought heightened by climate change. *PLoS ONE* **2018**, *13*, e0196873. [CrossRef] [PubMed]

24. Massaretto, I.L.; Albaladejo, I.; Purgatto, E.; Flores, F.B.; Plasencia, F.; Egea-Fernández, J.M.; Bolarin, M.C.; Egea, I. Recovering Tomato Landraces to Simultaneously Improve Fruit Yield and Nutritional Quality Against Salt Stress. *Front. Plant Sci.* **2018**, *9*, 1778. [CrossRef] [PubMed]

25. Tranchida-Lombardo, V.; Cigliano, R.A.; Anzae, A.; Landi, S.; Palombieri, S.; Colantuono, C.; Boston, H.; Termolino, P.; Aversano, R.; Batelli, G.; et al. Whole genome resequencing of two Italian tomato landraces reveals sequence variations in genes associated with stress tolerance, fruit quality and long shelf-life traits. *DNA Res.* **2018**, *25*, 149–160. [CrossRef]

26. Miles, C.; Atterberry, K.A.; Brouwer, B. Performance of heirloom dry bean varieties in organic production. *Agronomy* **2015**, *5*, 491–505. [CrossRef]

27. Swegarden, H.R.; Sheaffer, C.C.; Michaels, T.T. Yield stability of heirloom dry bean (*Phaseolus vulgaris* L.) cultivars in Midwest organic production. *HortScience* **2016**, *51*, 8–14. [CrossRef]

28. Adam, K.L. *Seed Production and Variety Development for Organic Systems*; ATTRA, the National Sustainable Agriculture Information Service; National Center for Appropriate Technology: Butte, MT, USA, 2005.

29. Jacques, P.J.; Jacques, J.R. Monocropping systems into ruin: The loss of fruit varieties and cultural diversity. *Sustainability* **2012**, *4*, 2970–2997. [CrossRef]

30. Millennium Ecosystem Assessment. *Ecosystems and Human Well-Being: Biodiversity Synthesis*; World Resources Institute: Washington, DC, USA, 2005.

31. Breen, S.D. Saving seeds: The Svalbard Global Seed Vault, Native American seed savers, and problems of property. *J. Agric. Food Syst. Commun. Dev.* **2014**, *5*, 39–52. [CrossRef]

32. Perroni, E. Twenty Initiatives Saving Seeds for Future Generations. 2017. Available online: https://foodtank.com/news/2017/07/seed-saving-initiatives/ (accessed on 30 January 2019).

33. Preston, J.M.; Ford-Lloyd, B.V.; Smith, L.M.J.; Sherman, R.; Munro, N.; Maxted, N. Genetic analysis of a heritage variety collection. *Plant Genet. Resour.* **2018**, *17*, 232–244. [CrossRef]

34. Manzanilla, D.; Lapitan, A.; Cope, A.; Vera Cruz, C. The heirloom rice story: A healthy fusion of science, culture and market. *CURE Matters* **2015**, *5*, 14–16.

35. Carolan, M.S. Saving seeds, saving culture: A case study of heritage seedbank. *Soc. Nat. Resour.* **2007**, *20*, 739–750. [CrossRef]

36. SSE. *Savers Potato Germplasm Collection: Using Phenotype and SSR Analysis to Characterize Heritage Solanum Spp Diversity Ex Situ*; Seed Savers Exchange: Decorah, IA, USA, 2014; Available online: https://pdfs.semanticscholar.org/f7de/6d9bdf9d4e91610e5a074544919856a40eea.pdf?_ga=2.242480258.1895016692.1550566917-687334884.1550566917 (accessed on 10 February 2019).

37. Hilgert, N.I.; Zamudio, F.; Furlan, V.; Cariola, L. The key role of cultural preservation in maize diversity conservation in the Argentine Yungas. *Evid. Based Complement. Altern. Med.* **2013**, *2013*, 1–10. [CrossRef]

38. Nazarea, V.D. Local Knowledge and Memory in Biodiversity Conservation. *Annu. Rev. Anthropol.* **2006**, *35*, 317–335. [CrossRef]

39. Perales, H.R.; Benz, B.F.; Brush, S.B. Maize diversity and ethnolinguistic diversity in Chipas, Mexico. *Proc. Natl. Acad. Sci. USA* **2005**, *102*, 949–954. [CrossRef]

40. Wang, Y.; Wang, Y.; Sun, X.; Caiji, Z.; Yang, J.; Cui, D.; Cao, G.; Ma, X.; Han, B.; Xue, D.; et al. Influence of ethnic traditional cultures on genetic diversity of rice landraces under on-farm conservation in southwest China. *J. Ethnobiol. Ethnomed.* **2016**, *12*, 51. [CrossRef] [PubMed]

41. Bocci, R.; Chable, V. Peasant seeds in Europe: Stakes and prospects. *J. Agric. Environ. Int. Dev.* **2009**, *103*, 81–93.

42. Da Via, E. Seed diversity, farmers' rights, and the politics of repeasantization. *Int. J. Sociol. Agric. Food* **2012**, *19*, 229–242.

43. Paris, H.S. Consumer-oriented exploitation and conservation of genetic resources of pumpkins and squash, Cucurbita. *Isr. J. Plant Sci.* **2018**, *65*, 202–221. [CrossRef]

44. Goland, C.; Bauer, S. When the apple falls close to the tree: Local food systems and the preservation of diversity. *Renew. Agric. Food Syst.* **2004**, *19*, 228–236. [CrossRef]

45. Arena, R.; Guazzi, M.; Lianov, L.; Whitsel, L.; Berra, K.; Lavie, C.J.; Kaminsky, L.; Williams, M.; Hivert, M.F.; Franklin, N.C.; et al. Healthy Lifestyle Interventions to Combat Noncommunicable Disease—A Novel Nonhierarchical Connectivity Model for Key Stakeholders: A Policy Statement from the American Heart Association, European Society of Cardiology, European Association for Cardiovascular Prevention and Rehabilitation, and American College of Preventive Medicine. *Eur. Heart J.* **2015**, *90*, 2097–2109.

46. Naicker, A.; Venter, C.S.; MacIntyre, U.E.; Ellis, S. Dietary quality and patterns and non-communicable disease risk of an Indian community in KwaZulu-Natal, South Africa. *J. Health Popul. Nutr.* **2015**, *33*, 12. [CrossRef] [PubMed]

47. Mirosa, M.; Lawson, R. Revealing the lifestyles of local food consumers. *Br. Food J.* **2012**, *114*, 816–825. [CrossRef]

48. Arsil, P.; Li, E.; Bruwer, J.; Lyons, G. Exploring consumer motivations towards buying local fresh food products: A mean-end approach. *Br. Food J.* **2014**, *116*, 1533–1549. [CrossRef]

49. Bianchi, C.; Mortimer, G. Drivers of local food consumption: A comparative study. *Br. Food J.* **2015**, *117*, 2282–2299. [CrossRef]

50. Dukeshire, S.; Garbes, R.; Kennedy, C.; Boudreau, A.; Osborne, T. Beliefs, attitudes, and propensity to buy locally produced food. *J. Agric. Food Syst. Commun. Dev.* **2011**, *1*, 19–29. [CrossRef]

51. Khan, F.; Prior, C. Evaluating the urban consumer with regard to sourcing local food: A heart or England study. *Int. J. Consum. Stud.* **2010**, *34*, 161–168. [CrossRef]

52. Tippins, M.J.; Rassuli, K.M.; Hollander, S.C. An assessment of direct farm-to-table food markets in the USA. *Int. J. Retail Distrib. Manag.* **2002**, *30*, 343–353. [CrossRef]

53. Ragland, E.; Tropp, D. *USDA National Farmer Market Manager Survey 2006*; Agricultural Marketing Service USDA: Washington, DC, USA, 2009. [CrossRef]

54. Giampietri, E.; Koemle, D.B.A.; Yu, X.; Finco, A. Consumers' Sense of Farmers' Markets: Tasting Sustainability or Just Purchasing Food? *Sustainability* **2016**, *8*, 1157. [CrossRef]

55. Garretson, L.; Marti, A. Pigmented Heirloom Beans: Nutritional and Cooking Quality Characteristics. *Cereal Chem. J.* **2017**, *94*, 363–368. [CrossRef]

56. Garretson, L.; Tyl, C.; Marti, A. Effect of Processing on Antioxidant Activity, Total Phenols, and Total Flavonoids of Pigmented Heirloom Beans. *J. Food Qual.* **2018**, *2018*, 1–6. [CrossRef]

57. Casañas, F.; Bosch, L.; Pujolá, M.; Sánchez, E.; Sorribas, X.; Baldi, M.; Nuez, F. Characteristics of a common bean landrace (Phaseolus vulgaris L.) of great culinary value and selection of a commercial breeding line. *J. Sci. Food Agric.* **1999**, *79*, 693–698.

58. Brouwer, B.; Winkler, L.; Atterberry, K.; Jones, S.; Miles, C. Exploring the role of local heirloom germplasm in expanding western Washington dry bean production. *Agroecol. Sustain. Food Syst.* **2016**, *40*, 319–332. [CrossRef]

59. Leson, N. The Monachine Bean is Truly a Pellegrini Family Heirloom, the Seattle Times. 2013. Available online: https://www.seattletimes.com/pacific-nw-magazine/the-monachine-bean-is-truly-a-pellegrini-family-heirloom/ (accessed on 30 October 2018).

60. Li, Q.; Somavat, P.; Singh, V.; Chatham, L.; Gonzalez de Mejia, E. A comparative study of anthocyanin distribution in purple and blue corn coproducts from three conventional fractionation processes. *Food Chem.* **2017**, *231*, 332–339. [CrossRef]

61. Nankar, A.; Holguin, F.M.; Scott, M.P.; Pratt, R.C. Grain and nutritional quality traits of southwestern U.S. blue maize landraces. *Cereal Chem. J.* **2017**, *94*, 950–955. [CrossRef]

62. Chatham, L.A.; West, L.; Berhow, M.A.; Vermillion, K.E.; Juvik, J.A. Unique flavanol-anthocyanin condensed forms in Apache Red purple corn. *J. Agric. Food Chem.* **2018**, *66*, 10844–10854. [CrossRef] [PubMed]

63. Chaudhary, R.C.; Mishra, S.B.; Yadav, S.K.; Ali, J. Extinction to distinction: Current status of Kalanamak, the heritage rice of eastern Uttar Pradesh and its likely role in farmers' prosperity. *LMA Conv. J.* **2012**, *8*, 7–14.

64. Basu, M. Reviving Traditional Rice Varieties in West Bengal. 2018. Available online: https://earthjournalism.net/stories/reviving-traditional-rice-varieties-in-west-bengal (accessed on 25 October 2018).

65. Rahman, S.; Sharma, M.P.; Sahai, S. Nutritional and medicinal values of some indigenous rice varieties. *Indian J. Tradit. Knowl.* **2006**, *5*, 454–458.

66. Tsegaye, B.; Berg, T. Utilization of durum wheat landraces in East Shewa, central Ethiopia: Are home uses as incentive for on-farm conservation? *Agric. Hum. Values* **2007**, *24*, 219–230. [CrossRef]

67. Shewayrga, H.; A Sopade, P. Ethnobotany, diverse food uses, claimed health benefits and implications on conservation of barley landraces in North Eastern Ethiopia highlands. *J. Ethnobiol. Ethnomed.* **2011**, *7*, 19. [CrossRef] [PubMed]

68. Lust, T.A.; Paris, H.S. Italian horticultural and culinary records of summer squash (Cucurbita pepo, Cucurbitaceae) and emergence of the zucchini in 19th-century Milan. *Ann. Bot.* **2016**, *118*, 53–69. [CrossRef] [PubMed]

69. Renna, M.; Serio, F.; Signore, A.; Santamaria, P. The yellow–purple Polignano carrot (Daucus carota L.): A multicoloured landrace from the Puglia region (Southern Italy) at risk of genetic erosion. *Genet. Resour. Crop Evol.* **2014**, *61*, 1611–1619. [CrossRef]

70. Goldman, A. *Melons: For the Passionate Grower*; Artison Books: New York, NY, USA, 2002.

71. Goldman, A. *The Compleat Squash: A Passionate Gowers's Guide to Pumpkins, Squash, and Gourds*; Artison Books: New York, NY, USA, 2004.

72. Goldman, A. *The Heirloom Tomato: From Garden to Table: Recipes, Portraits, and History of the World's Most Beautiful Fruit*; Bloomsbury: New York, NY, USA, 2008.

73. Rodríguez-Burruezo, S.; Prohens, J.; Roselló, J.; Nuez, F. "Heirloom" varieties as sources of variation for the improvement of fruit quality in greenhouse-grown tomatoes. *J. Hortic. Sci. Biotechnol.* **2005**, *80*, 453–460. [CrossRef]

74. D'Angelo, M.; Sance, M.; Asprelli, P.; Carrari, F.; Asis, R.; D'Sangelo, M.; Zanor, M.I.; Cortina, P.R.; Boggio, S.B.; Santiago, A.N.; et al. Contrasting metabolic profiles of tasty Andean varieties of tomato fruit in comparison with commercial ones. *J. Sci. Food Agric.* **2018**, *98*, 4128–4134. [CrossRef]

75. Cortés-Olmos, C.; Leiva-Brondo, M.; Roselló, J.; Raigón, M.D.; Cebolla-Cornejo, J. The role of traditional varieties of tomato as sources of functional compounds. *J. Sci. Food Agric.* **2014**, *94*, 2888–2904.

76. Cefola, M.; Pace, B.; Santamaria, P.; Signore, A.; Serio, F. Compositional analysis and antioxidant profile of yellow, orange and purple Polignano carrots. *Ital. J. Food Sci.* **2012**, *24*, 284–291.

77. Scarano, A.; Gerardi, C.; D'Amico, L.; Accogli, R.; Santino, A. Phytochemical analysis and antioxidant properties in colored Tiggiano carrots. *Agriculture* **2018**, *8*, 102. [CrossRef]

78. Piergiovanni, A.R.; Cerbino, D.; Brandi, M. The common bean populations from Basilicata (Southern Italy): An evaluation of their variation. *Genet. Resour. Crop Evol.* **2000**, *47*, 489–495. [CrossRef]

79. Loi, L.; Piergiovanni, A.R. Genetic diversity and seed quality of the 'Badda' common bean from Sicily (Italy). *Diversity* **2013**, *5*, 843–855. [CrossRef]

80. Roy, S.; Banerjee, A.; Mawkhleing, B.; Misra, A.K.; Pattanayak, A.; Harish, G.D.; Singh, S.K.; Ngachan, S.V.; Bansal, K.C. Genetic diversity and population structure in aromatic and quality rice (*Oryza sativa* L.) landraces from North-East India. *PLoS ONE* **2015**, *10*, e0129607. [CrossRef]

81. Neelamma, G.; Swamy, B.D.; Damodran, P. Phytochemical and pharmacological overview of Cucurbita maxima and future perspective as potential phytotherapeutic agent. *Eur. J. Pharm. Med. Res.* **2016**, *3*, 277–287.

82. Orsenigo, S.; Abeli, T.; Schiavi, M.; Cauzzi, P.; Guzzon, F.; Ardenghi, N.M.G.; Rossi, G.; Vagge, I. Morphological characterization of *Cucurbita maxima* Duchesne (*Cucurbitaceae*) landraces from the Po Valley (northern Italy). *Ital. J. Agron.* **2018**, *13*, 963. [CrossRef]

83. Kaźmińka, K.; Sobieszek, K.; Targońska-Karasek, M.; Korzeniewska, A.; Niemirowicz-Szczytt, K.; Bartoszewski, G. Genetic diversity assessment of a winter squash and pumpkin (*Cucurbita maxima* Duchesne) germplasm collection based on genomic Cucurbita-conserved SSR markers. *Sci. Hortic.* **2017**, *219*, 37–44. [CrossRef]

84. Baldina, S.; Picarella, M.E.; Troise, A.D.; Pucci, A.; Ruggieri, V.; Ferracane, R.; Barone, A.; Fogliano, V.; Mazzucato, A. Metabolite Profiling of Italian Tomato Landraces with Different Fruit Types. *Front. Plant Sci.* **2016**, *7*, 664. [CrossRef]

85. Levi, A.; Simmons, A.M.; Massey, L.; Coffey, J.; Wechter, W.P.; Jarret, R.L.; Tadmor, Y.; Nimmakayala, P.; Reddy, U.K. Genetic Diversity in the Desert Watermelon Citrullus colocynthis and its Relationship with Citrullus Species as Determined by High-frequency Oligonucleotides-targeting Active Gene Markers. *J. Am. Soc. Hortic. Sci.* **2017**, *142*, 47–56. [CrossRef]

86. Chapagain, T.; Riseman, A. Evaluation of heirloom and commercial cultivars of small grains under low input organic systems. *Am. J. Plant Sci.* **2012**, *3*, 655–669. [CrossRef]

87. Jankielsohn, A.; Miles, C. How do older wheat cultivars compare to modern wheat cultivars currently on the market in South Africa? *J. Hortic. Sci. Res.* **2017**, *1*, 42–47.

88. Nankar, A.; Grant, L.; Scott, P.; Pratt, R.C. Agronomic and kernel compositional traits of blue maize landraces from the southwestern United States. *Crop Sci.* **2016**, *56*, 2663–2674. [CrossRef]

89. Brouwer, B.O.; Murphy, K.M.; Jones, S.S. Plant breeding for local food systems: A contextual review of end-use selection for small grains and dry beans in western Washington. *Renew. Agric. Food Syst.* **2016**, *31*, 172–184. [CrossRef]

90. Kumar, S.; Mishra, S.B.; Chaudhary, R.C. Breeding Bauna Kalanamak 102 as new aromatic variety of heritage rice from Uttar Pradesh. *Int. J. Sci. Environ. Technol.* **2018**, *7*, 1690–1699.

91. Mallor, C.; Barberán, M.; Aibar, J. Recovery of a common bean landrace (*Phaseolus vulgaris* L.) for commercial purposes. *Front. Plant Sci.* **2018**, *9*, 1440. [CrossRef] [PubMed]

92. Marles, R.J. Mineral nutrient composition of vegetables, fruits and grains: The context of reports of apparent historical declines. *J. Food Compos. Anal.* **2017**, *56*, 93–103. [CrossRef]

93. White, P.J.; Broadley, M.R. Historical variation in the mineral composition of edible horticultural products. *J. Hortic. Sci. Biotechnol.* **2005**, *80*, 660–667. [CrossRef]

94. Dumas, Y.; Dadomo, M.; di Luca, G.; Grolier, P. Effects of environmental factors and agricultural techniques on antioxidant content of tomatoes. *J. Sci. Food Agric.* **2003**, *83*, 369–382. [CrossRef]

95. Cebolla-Cornejo, J.; Roselló, S.; Valcárcel, M.; Serrano, E.; Beltrán, J.; Nuez, F. Evaluation of genotype and environment effects on taste and aroma flavor components of Spanish fresh tomato varieties. *J. Agric. Food Chem.* **2011**, *59*, 2440–2450. [CrossRef]

96. Yanjie, X.; Yining, Y.; Shuhong, O.; Xiaolong, D.; Hui, S.; Shukun, J.; Shichen, S.; Jinsong, B. Factors affecting sensory quality of cooked japonica rice. *Rice Sci.* **2018**, *25*, 330–339. [CrossRef]

97. Klee, H.J.; Tieman, D.M. Genetic challenges of flavor improvement in tomato. *Trends Genet.* **2013**, *29*, 257–262. [CrossRef] [PubMed]

98. Folta, K.M.; Klee, H.J. Sensory sacrifices when we mass-produce mass produce. *Hortic. Res.* **2016**, *3*, 16032. [CrossRef] [PubMed]

99. Paris, H.S.; Nerson, H. Genes for intense pigmentation of squash. *J. Hered.* **1986**, *77*, 403–409. [CrossRef]

100. Paris, H.S.; Godinger, D. Sweet acorn squash, a new vegetable on the Israeli market. *Acta Hortic.* **2017**, *1127*, 451–456. [CrossRef]

101. Rao, T. Seeds Only a Plant Breeder Could Love, Until Now. *New York Times*. 27 February 2018. Available online: https://www.nytimes.com/2018/02/27/dining/row-7-seed-company-dan-barber.html (accessed on 20 January 2019).

102. Beans, C. Vegetable breeders turn to chefs for flavor boost. *Proc. Natl. Acad. Sci. USA* **2017**. [CrossRef] [PubMed]

Essay

Domesticating the Undomesticated for Global Food and Nutritional Security: Four Steps

Ajeet Singh, Pradeep Kumar Dubey, Rajan Chaurasia, Rama Kant Dubey, Krishna Kumar Pandey, Gopal Shankar Singh and Purushothaman Chirakkuzhyil Abhilash *

Institute of Environment & Sustainable Development, Banaras Hindu University, Varanasi 221005, India
* Correspondence: pca.iesd@bhu.ac.in; Tel.: +91-94156-44280

Received: 8 July 2019; Accepted: 27 August 2019; Published: 28 August 2019

Abstract: Ensuring the food and nutritional demand of the ever-growing human population is a major sustainability challenge for humanity in this Anthropocene. The cultivation of climate resilient, adaptive and underutilized wild crops along with modern crop varieties is proposed as an innovative strategy for managing future agricultural production under the changing environmental conditions. Such underutilized and neglected wild crops have been recently projected by the Food and Agricultural Organization of the United Nations as 'future smart crops' as they are not only hardy, and resilient to changing climatic conditions, but also rich in nutrients. They need only minimal care and input, and therefore, they can be easily grown in degraded and nutrient-poor soil also. Moreover, they can be used for improving the adaptive traits of modern crops. The contribution of such neglected, and underutilized crops and their wild relatives to global food production is estimated to be around 115–120 billion US$ per annum. Therefore, the exploitation of such lesser utilized and yet to be used wild crops is highly significant for climate resilient agriculture and thereby providing a good quality of life to one and all. Here we provide four steps, namely: (i) exploring the unexplored, (ii) refining the unrefined traits, (iii) cultivating the uncultivated, and (iv) popularizing the unpopular for the sustainable utilization of such wild crops as a resilient strategy for ensuring food and nutritional security and also urge the timely adoption of suitable frameworks for the large-scale exploitation of such wild species for achieving the UN Sustainable Development Goals.

Keywords: anthropocene; climate resilient; food and nutritional security; resource conservation; underutilized crops; Sustainable Development Goals

---

1. Producing More with Less Resources: The Need of the Hour

The sustainable utilization of our limited natural resources for maximizing the food production [1] within the planetary boundaries [2] is a serious challenge for attaining the United Nations Sustainable Development Goals (UN-SDGs), especially the first, second and third goals namely (i) no poverty, (ii) zero hunger and (iii) good health and wellbeing. Since the indiscriminate usage of a critical resource governing agricultural production, i.e., N, has already crossed planetary boundaries [2], there is a growing concern regarding the cultivation of high-input demanding modern crop varieties under resource-poor conditions. Moreover, the quality and availability of two other inputs vital for agricultural production, i.e., water and land, are already in a thinning state [3,4]. The changing climatic condition is another impediment for food production as it negatively affects the quality and availability of the critical resources as well as the quality and quantity of the agricultural production itself [5]. Since we have to enhance the food production by 70% for meeting the demand of the growing population in 2050, the cultivation of resilient, nutritionally rich, and low-resource intensive crops are of paramount importance for human wellbeing and environmental sustainability [6]. In this context, [7,8] the domestication of undomesticated, wild and neglected crops and exploiting their natural traits to efficiently use critical natural resources such as N, P, water, and land offers huge promise in attaining future food security as they are bestowed with high nutritional value [9,10] and adaptive traits [11]. Importantly, they need only minimal input and care so they can be easily cultivated in marginal and other nutrient-poor soil and even under changing climatic conditions [11]. Here we propose four important steps, i.e., (i) exploring the unexplored, (ii) refining the unrefined traits, (iii) cultivating the uncultivated, and (iv) popularizing the unpopular for the large-scale exploitation of such important but still underutilized and neglected crops for global food and nutritional security (Figure 1).

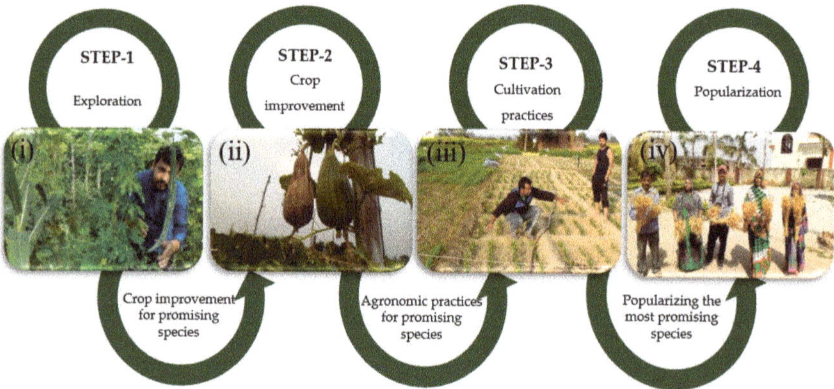

Figure 1. A casual loop diagram showing various steps involved in the sustainable utilization of wild and neglected crops for global food and nutritional security. (**i**) Step-1: exploration of various kind of wild and neglected crops; (**ii**) Step-2: improving the desirable traits in promising species by conventional as well as modern biotechnological approaches; (**iii**) Step-3: standardization and optimization of various agronomic practices for their large-scale exploitation and (**iv**) Step-4: the popularization of unpopular crops among farmers, policy makers, and other stakeholders.

2. Exploring the Unexplored: First Step

A list of nutritionally relevant, neglected and wild crops for global food security [10–14] is given in Table 1. While it has been reported that about 5538 crops have been used for food by humans throughout the history, only 12 crops contribute the lion share of the current global food production [12,15]. Among this, three crops namely rice, wheat and maize account for >50% the world's calories [15]. Though there are few estimates regarding the number of underutilized species (for example, Arora (2014) reported 992 species across the world), still a majority of them are unknown to various stakeholders. Therefore, the detailed exploration of such species from various agro-climatic regions of the world is important for identifying the promising species (Figure 2) like cereals/pseudo-cereals, roots and tubers, pulses, fruits and vegetables, nuts, seeds and spices etc. and their successful utilization in a dietary diversification program [9,15–17]. For this, a well-coordinated wild crop exploration program at various scales (i.e., national, regional, and global) are imperative. For example, Food and Agricultural Organization (FAO) has recently started an initiative to identify the Future Smart Food (FSF) crops on a regional basis and have identified 39 nutrition-sensitive and climate resilient crops from South and South East Asia as FSF [13] with the consultation of national experts from Bangladesh, Bhutan, Cambodia, India, Lao PDR, Myanmar, Nepal and Vietnam. While they proposed nine species of cereals/pseudo-cereals (buckwheat, tartary buckwheat, foxtail millet, porso millet, sorghum, amaranth, grain amaranth, quinoa, and specialty rice), six species of roots and tubers (taro, swamp taro, purple yam, fancy yam, elephants foot yam, and sweet potato), nine species of pulses (grasspea, fababean, cowpea, mungbean, blackbean, ricebean, lentil, horsebean, and soybean), 9 species of fruits and vegetables (drumstick, chayote, fenugreek, snake gourd, pumpkin, roselle, Indian gooseberry, jackfruit and wood apple), and five species of nuts, seeds and spices (linseed, walnut, Nepali butter tree, perilla and Nepali pepper) as FSFs for the South and South East Asia region [13] for further exploitation, there are many more neglected and wild species to be included and projected as FSFs (Figure 3A,B). For an instance, a detailed survey conducted by authors during the last few years in India (Figure 3) has revealed that there are many potential FSF such as finger millet, kodo millet, little millet (cereals/pseudo-cereals), air potato, turnip, kohlrabi, kudzu (roots and tubers), winged bean, sword bean, pigeon pea, chick pea (pulses), spine gourd, clove bean, phalsa, custard apple, kadamba, chenopodium, brown mustard, water spinach, jujube, ground cherry (fruits and vegetables) [10] etc. to be included in the regional list and thereby encouraged to be exploited them for future food security (for more such species, please see Figure 3A and Table 1). The cultivation of such species are not only important for global food security but also for attaining many other UN-SDGs (Figure 3B).

Figure 2. Exploration and germplasm collection of wild crops from diverse habitat ((**a**–**e**) collection from farmer's field, backyard and kitchen garden; (**f**) pond) is essential for identifying most promising species.

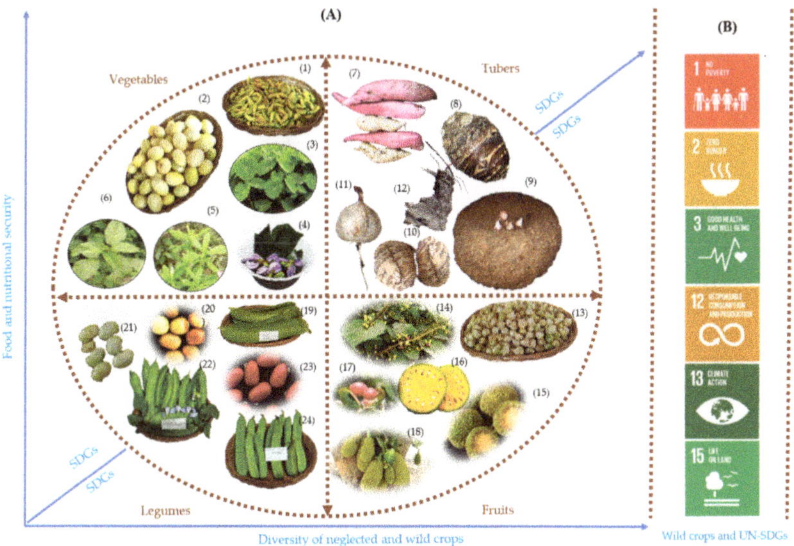

Figure 3. Schematic representation of the inter-relationship between diversity of wild and neglected crops, food and nutritional security and UN-SDGs. (**A**) food and nutritional security increases with the

large-scale exploitation of neglected and wild crops. Circle shows various neglected wild species: (1) Indian hemp (*Crotalaria juncea*); (2) Tarali (*Melothria heterophylla*); (3) Taro (*Colocasia esculenta*); (4) Flower and leaves of Winged bean (*Psophocarpus tetragonolobus*); (5) Water leaf (*Talinum fruticosum*); (6) Indian jute (*Corchorus olitorius*); (7) Wild sweet potato (*Ipomoea batatas*); (8) Elephant's ear (*Colocasia gigantea*); (9) Elephant foot yam (*Amorphophallus paeoniifolius*); (10) Taro (*Colocasia esculenta*); (11) Indian Kudzu (*Pueraria tuberosa*); (12) Tuber of purple yam (*Dioscorea alata*);(13) Winter cherry (*Physalis angulata*); (14) Phalsa (*Grewia asiatica*); (15) Bur flower tree (*Neolamarckia cadamba*); (16) Wood apple (*Aegle marmelos*); (17) Karanda (*Carissa carandas*); (18) Jackfruit (*Artocarpus heterophyllus*); (19) Pods of large sword bean (*Canavalia gladiata*); (20) Seeds of winged bean (*Psophocarpus tetragonolobus*); (21) Seeds of vegetable hummingbird (*Sesbania grandiflora*); (22) Pods of winged bean (*Psophocarpus tetragonolobus*); (23) Seeds of large sword bean (*Canavalia gladiata*); and (24) Sword bean (*Canavalia virosa*) (**B**) Sustainable utilization of wild and neglected crops are essential for attaining UN-SDGs.

Table 1. An indicative list of promising neglected and underutilized species such as roots and tubers, cereals and pseudo-cereals, fruits and nuts, vegetables, legumes, spices, condiments, and food-dye agents for global food and nutritional security [10–14].

SL No.	Common Name	Scientific Name
	(A) Roots and tubers	
1	Yams	*Dioscorea spp.*
2	Yacon	*Smallanthus sonchifolius*
3	Ulluco	*Ullucus tuberosus*
4	Taro	*Colocasia esculenta*
5	Arracacha	*Arracacia xanthorriza*
6	American yam bean	*Pachyrhizus spp*
7	Maca	*Lepidium meyenii*
8	Oca	*Oxalis tuberosa*
9	Parsnip	*Pastinaca sativa*
10	Cocoyam	*Xanthosoma sagittifolium*
11	Elephant foot yam	*Amorphophallus paeoniifolius*
12	Kohlrabi	*Brassica oleracea var. gongylodes L*
13	Wild turnip	*Brassica rapa var. rapa*
14	Eddoe	*Colocasia antiquorum*
15	Sweet potato	*Ipomea batatas*
16	Indian lotus	*Nelumbo nucifera Gaertn.*
17	Country potato	*Plectranthus rotundifolius*
18	Wild cassava	*Manihot spp.*
19	Indian Kudzu	*Pueraria tuberosa*
20	Edible Chlorophytum	*Chlorophytum tuberosum*
21	Asparagus	*Asparagus racemosus*
	(B) Cereals and pseudo-cereals	
22	Einkorn	*Triticum monococcum*
23	Emmer	*T. dicoccon*
24	Spelt	*T. spelta*
25	Tef	*Eragrostis tef*

Table 1. *Cont.*

SL No.	Common Name	Scientific Name
26	Fonio	*Digitaria exilis*
27	Cañihua	*Chenopodium pallidicaule*
28	Finger millet	*Eleusine coracana*
29	Kodo millet	*Paspalum scrobiculatum*
30	Foxtail millet	*Setaria italic*
31	Little millet	*Panicum sumatrense*
32	Proso millet	*Panicum miliaceum*
33	Amaranth	*Amaranthus caudatus*
34	Buckwheat	*Fagopyrum spp.*
35	Job's tears	*Coix lacryma-jobi*
36	Red amaranth	*Amaranthus cruentus L*
37	Pearl Millet	*Pennisetum glaucum (L.) R.Br.*
(C) Fruits and nuts		
38	Maya nut	*Brosimum alicastrum*
39	Breadfruit	*Artocarpus altilis*
40	Jackfruit	*Artocarpus heterophyllus*
41	Wild jackfruit	*Artocarpus hirsutus*
42	Fox nut	*Euryale ferox*
43	Baobab	*Adansonia digitate*
44	Jujube	*Ziziphus mauritiana*
45	Cherimoya	*Annona cherimola*
46	Cape gooseberry	*Physalis peruviana*
47	Naranjilla	*Solanum quitoense*
48	Pomegranate	*Punica granatum*
49	Noni	*Morinda citrifolia*
50	Marula	*Sclerocarya birrea*
51	Tamarind	*Tamarindus indica*
52	Annona	*Annona spp.*
53	Safou	*Dacryodes edulis*
54	Mangosteen	*Garcinia mangostana*
55	Salak	*Salacca spp.*
56	Nipa palm	*Nypa fruticans*
57	Monkey orange	*Strychnos cocculoides*
58	Duku	*Lansium domesticum*
59	Boscia	*Boscia spp.*
60	Carissa	*Carissa edulis*
61	Coccinia	*Coccinia trilobata*
62	Acacia	*Acacia toritilis*
63	Kei apple	*Dovyalis caffra*
64	Tree grapes	*Lamnea spp.*

Table 1. *Cont.*

SL No.	Common Name	Scientific Name
65	Medlars	*Vanguera spp.*
66	Pitanga	*Eugenia uniflora*
67	Malabar chestnut	*Pachira aquatica*
68	Camu camu	*Myrciaria dubia*
69	Dragon fruit	*Hylocereus spp.*
70	Brazil nut	*Bertholletia excels*
71	Egg nut	*Couepia longipendula*
72	Quince	*Cydonia oblonga*
73	Yara Yara	*Duguetia lepidota*
74	Araza	*Eugenia stipitate*
75	Lúcuma	*Lucuma obovate*
76	Miracle fruit	*Synsepalum dulcificum*
77	Water chestnut	*Trapa natans*
78	Indian bael	*Aegle marmelos*
79	Chilean wineberry	*Aristotelia chilensis (Molina)*
80	Lakoocha	*Artocarpus lacucha*
81	Karanda	*Carissa carandas*
82	Assyrian plum	*Cordia myxa*
83	Cluster Fig	*Ficus racemosa*
84	Phalsa	*Grewia asiatica*
85	Wood-apple	*Limonia acidissima L.*
86	Mulberries	*Morus alba*
87	Burflower-tree	*Neolamarckia cadamba*
88	Indian Gooseberry	*Phyllanthus emblica*
89	Angular winter cherry	*Physalis angulata*
90	Manila tamarind	*Pithecellobium dulce (Roxb.) Benth.*
91	Black nightshade	*Solanum nigrum L.*
92	Indian almond	*Terminalia catappa L.*
93	Jujube	*Ziziphus jujube*
94	Mahua	*Madhuca longifolia*
95	Hog Plum	*Spondias dulcis*
96	Starfruit	*Averrhoa carambola*
97	Bilimbi	*Averrhoa bilimbi*
98	Indian coffee plum	*Flacourtia jangomas*
99	Common guava	*Psidium guajava*
100	Soursop	*Annona muricata*
101	Spring Asparagus	*Asparagus officinalis*
102	Wild pear	*Pyrus communis*
103	Hill lemon	*Citrus psedolimon*

Table 1. *Cont.*

SL No.	Common Name	Scientific Name
(D) Vegetables		
104	Moringa	*Moringa oleifera*
105	African eggplant	*Solanum aethiopicum*
106	Thorny amaranth	*Amaranthus spinosa*
107	Wild amaranth	*Amranthus viridis*
108	Brassica	*Brassica rapa varieties*
109	Locust bean	*Parkia biglobosa*
110	Chayote	*Sechium edule*
111	Chrysanthemum	*Chrysanthemum oronarium*
112	Bitter gourd	*Momordica charantia*
113	Angle gourd	*Luffa acutangular*
114	Snake gourd	*Thrichosantes cucumerina var. anguina*
115	Indian spinach	*Basella rubra, Basella alba*
116	Spider plant	*Cleome gynandra*
117	Jute	*Corchorus olitorius*
118	Black nightshade	*Solanum nigrum*
119	Ivy gourd	*Coccinia grandis*
120	Gourd	*Lagenaria siceraria*
121	Celosia	*Celosia argentea*
122	Dika	*Irvingia spp.*
123	Egusi	*Citrullus lanatus*
124	Marama	*Tylosema esculentum*
125	Shea butter	*Vitellaria paradoxa*
126	Giant swamp taro	*Cyrtosperma merkusii*
127	Akoub	*Gundelia tournefortii*
128	Crambe	*Crambe spp.*
129	Cardoon	*Cynara cardunculus*
130	Eru	*Gnetum africanum*
131	Purslane	*Portulaca oleracea*
132	Golden thistle	*Scolymus hispanicus*
133	Bitter leaf	*Vernonia amygdalina*
134	Cabbage Leaf Mustard	*Brassica juncea var. rugosa*
135	Pigweed	*Chenopodium album*
136	Asian spiderflower	*Cleome viscosa*
137	False Amaranth	*Digera muricate (L.) Mart.*
138	Water spinach	*Ipomoea aquatica*
139	Thumbai	*Leucas aspera (Willd.) Linn*
140	Sweet neem	*Murraya koenigii*
141	Sickle Senna	*Senna tora*

Table 1. *Cont.*

SL No.	Common Name	Scientific Name
142	Waterleaf	*Talinum fruticosum (L.) Juss*
143	Fenugreek	*Trigonella foenum graecum*
144	Spiny gourd	*Momordica dioica Roxb. Ex Willd.*
145	Pointed gourd	*Trichosanthes dioica Roxb.*
146	Sunn hemp	*Crotalaria juncea L.*
147	Khejri Tree	*Prosopis cineraria*
148	Sweet leaf	*Sauropus androgynus*
149	Water spinach	*Ipomea aquatica*
150	Tarali	*Melothria heterophylla*
151	Kuda	*Holarrhena pubescens*
152	Korla	*Bauhinia malabarica*
153	Kawla	*Smithia hirsuta*
154	Dragon stalk yam	*Amorphophallus commutatus*
155	Bamboo	*Dendrocalamus strictus*
156	Wild senna	*Senna tora*
157	Dinda	*Leea indica*
158	Bharangi	*Rotheca serrata*
159	Edible fern	*Diplazium esculentum*
(E) Legumes		
160	Mungbean	*Vigna radiata*
161	Adzuki bean	*V. angularis*
162	Ricebean	*V. umbellata*
163	Lupin	*Lupinus mutabilis*
164	Bambara groundnut	*Vigna subterranean*
165	Jack bean	*Canavalia ensiformis*
166	Grasspea	*Lathyrus sativus*
167	Lablab	*Lablab purpureus*
168	Pigeon pea	*Cajanus cajan*
169	African yam bean	*Sphenostylis stenocarpa*
170	Kersting's groundnut	*Macrotyloma geocarpum*
171	Sword bean	*Canavalia gladiata*
172	Jack bean	*Canavalia virosa*
173	Winged bean	*Psophocarpus tetragonolobus*
174	Cluster bean	*Cyamopsis tetragonoloba (L.) Taub*
175	Agati	*Sesbania grandiflora (L.) Pers.*
176	Broad bean	*Vicia faba*
177	Chickpea	*Cicer arietinum*
178	Peanut	*Arachis hypogea*
179	Black gram	*Vigna mungo*
180	Black lentil	*Lens culinaris*

Table 1. *Cont.*

SL No.	Common Name	Scientific Name
(F) Spices, condiments, food-dye agents		
181	Makoni	*Fadogia ancylantha*
182	Annatto	*Bixa orellana*
183	Mustard seed	*Brassica juncea*
184	Fenugreek	*Trigonella foenumgraecum*
185	Pandan	*Pandanus amaryllifolius*
186	Polygonum	*Poligonum odoratum*
187	Antidesma	*Antidesma venosum*
188	Uer	*Lippia carviodora*
189	Rocket	*Diplotaxis spp*
190	Caper	*Capparis spinosa*
191	Monkey cola	*Cola lateritia*
192	Sea buckthorn	*Hippophae rhamnoides*
193	Nigella	*Nigella sativa*
194	Culantro	*Eryngium foetidum*
195	Coriander	*Coriandrum sativum*
196	Wild chilies	*Capsicum* spp.

The application of biotechnological advances for exploring crop diversity like genotyping by sequencing, genotyping arrays, and pangenomics approaches [18] is another important strategy for assessing genetic diversity. Crop genome sequencing can provide insight into genetic variation through the re-sequencing [19]. Jiao et al. (2012) re-sequenced 278 elite maize lines (USA and Chinese maize lines) and identified >27 million single-nucleotide polymorphisms (SNPs). Similarly, the 3000-rice genome project resulted in the identification of >18.9 million SNPs [20,21] and the methylation-sensitive digestion of genomic DNA coupled with next- generation sequencing in wheat resulted in ~23,500 SNPs [22]. Furthermore, whole-genome re-sequencing was also done for barley [23], soybean [24], and lupin [25].

However, whole-genome sequencing is yet to be done in many promising but neglected crops and there are many wild species yet to be explored for their food and nutritional significance. For example, exploring the diversity in root and tuber crops are essential for the food security, especially in tropical regions as they are nutritionally rich and widely found in tropical regions [26]. *Vigna* is another important legume genus having more than a 100 species of high nutritional value for human and animal consumption [27]. Among this, only 10 species have been domesticated so far and the potential of the rest of the remaining 90 species is yet to be explored. Prickly pear (*Opuntia* spp.) is another valuable fruit crop for the semi-arid regions [28]. Similarly, Indian spinach or Malabar spinach (*Basella* species), is a resilient, nutritionally significant, perennial leafy vegetable native to tropical region [11]. However, as in the case of many other tropical leafy vegetables such as wild amaranth [10], water spinach, and waterleaf etc., Basella species are yet to be exploited for the dietary diversification program and food security [11]. Therefore, the large-scale exploration of underutilized and neglected crop species from various agro-climatic regions of the world, with the active involvements of the Consultative Group on International Agricultural Research (CGIAR) institutions, Crop Trust, local and national institutions, and with the financial support from Food and Agricultural Organization of the United Nations (UN-FAO), corporate bodies (through corporate social responsibility), Non-Governmental

Organizations (NGOs), and also from concerned national governments, are imperative for exploring the most promising species as well as identifying the current diversity of such wild species.

Applications of geoinformatics tools such as remote sensing and geographical information system (RS & GISs) can be used for mapping the distribution of neglected and wild crops [29] in various agro-climatic regions of the world, as well as modelling their habitat suitability for identifying the most suitable areas for large-scale cultivation [17,30]. Apart from the large-scale exploration of neglected and underutilized species from diverse habitats and agroecological zones of the world, the conservation of promising species is also equally important for facilitating their future utilization [6,14]. The exploitation of indigenous and local knowledge (ILK) along with the successful incorporation of the biocultural practices of the local people are very much essential for applying the traditional mode of conservation (Figure 4), whereas the exploitation of modern techniques for germplasm storage is also essential for maintaining such vital resources for the future use [13,14]. Creating community gene banks at farmer's fields is an excellent strategy for the long-term conservation of such underutilized species (Figure 5).

Figure 4. Apart from the exploration of wild and neglected crops, conservation of the germplasms of most promising species are also equally important for their sustainable utilization. For this, a blend of traditional (**a–c**) as well as modern (**d,e**) germplasm storage technologies should be adopted.

Figure 5. Field gene bank for conservation of neglected and underutilized crops. Maintaining germplasms of promising species at farmer's field itself is a promising strategy for conservation. Figures shows the cultivation of wheat in the farmer's field of Rajgarh, Mirzapur district, UP, India.

3. Refining the Unrefined Traits: Second Step

While modern agriculture aiming for higher productions prefers new cultivars that are developed through selection and fixation of favorable alleles providing most desirable traits, unscientific practices coupled with due negligence results in loss of wild or traditional crops and crop varieties and thereby causing genetic erosion and decline in landraces and ecotypes [31]. Therefore, once the promising wild and neglected species have been identified, developing a suitable crop improvement program for traiting or featuring desirable traits is another major strategy. Though most of the wild species are endowed with many positive traits such as stress tolerance (drought, flood, salinity, heat, pest and diseases, etc.) and able to grow even in resource-poor conditions [5,7,9], these attributes can be further refined by crop improvement by using next generation sequencing (NGS) for genomic selection [32], the identification of candidate loci using quantitative trait locus (QTL) analyses, targeted gene replacement using zinc-finger nucleases and transcription activator-like effector nucleases and also by using clustered regularly interspaced short palindromic repeats (CRISPR/Cas system) etc. [18,33–36].

The application of genomic technologies [37] was found to increase the productivity of several orphan crops such as cassava, sweet potato, coconut, sorghum, yam, groundnut, common bean, chickpea, cowpea, cacao etc. [18], especially, exploring plant microRNAs (miRNAs) for developing climate resilient crops [38]. Khan et al. (2019) proposed the application of CRISPR-Cas9 for engineering the domestication traits in wild tomato for higher nutritive value and better adaptation to multiple stresses [39]. Pigeon pea (*Cajanus cajan*) is another versatile, stress tolerant and nutritious grain legume for enhancing the sustainability of dry tropical and sub-tropical system [40] and therefore, the crop breeding programs in pigeon pea has resulted in enhancing its tolerance to heat, cold, drought and waterlogging [40]. Muthamilarasan et al. (2016) explored the application of omics technologies for exploring suitable millet models for developing nutrient rich graminaceous crops [41]. De novo sequencing and comparative leaf transcriptome analysis of underutilized winged bean (*Psophocarpus tetragonolobus* (L.) DC) has been done by Singh et al. (2017) for removing the anti-nutritional factors in winged bean like condensed tannin (CT) and proanthocyanidin (PA) [42]. Such transcriptomic insights can be used for removing the antinutritional factors (high tannin and lignin content) in sword bean also. Lemmon et al. (2018) enhanced the domestication traits in an orphan crop of Solanaceae (groundcherry) by gene editing [36]. While biotechnological advances have revolutionized the crop breeding domain, there are many other potential species yet to be considered for crop breeding for improving their nutritional and domestication traits [10].

Apart from modern breeding programs, the exploitation of conventional breeding techniques [43,44] is also important for improving the traits as well as economic benefits of such underutilized species [43–50]. For instances, the traditional crop improvement techniques like mass selection, pure-line selection, pedigree method, bulk method, backcross method, single seed descent (SSD) method [43,49], mutation breeding [50] etc. are also essential for (i) improving the yield and quality attributing traits of the wild species, (ii) conferring resistance to biotic and abiotic stresses, (iii) enabling better adaptation by altering the crop duration, maturity/earliness etc., (iv) reducing the anti-nutritional factors, (v) enhancing the non-shattering characteristics, and (vi) improving the adaptation range as well as the photo and thermo sensitivity [43].

Targeted and participatory crop breeding programs [46,47] are necessary for some of the underutilized and neglected species such as Indian hemp (*Crotularia juncea*), tarali (*Melothria heterophylla*), water leaf (*Talinum fruticosum*), Indian jute (*Corchorus olitorius*), Indian kudzu (*Pueraria tuberosa*), winter cherry (*Physalis angulata*), phalsa (*Grewia asiatica*), bur flower tree (*Neolamarckia cadamba*), wood apple (*Aegle marmelos*), jack bean (*Canavalia ensiformis*) etc. For this, a globally coordinated wild crop breeding program (including both conventional as well as modern breeding) under the leadership of CGIAR institutes and various national agricultural organizations is essential for improving the traits of potential lesser utilized species and to bring them to mainstream agriculture.

4. Cultivating the Uncultivated: Third Step

As opined by Massawe et al. (2016), neglected and underutilized wild species are a treasure trove for food security [51]. However, majority of such lesser known species are undomesticated, and therefore, farmers are not aware of their cultivation and propagation methods. Therefore, standardizing the agronomic practices (Figure 6) for such uncultivated edible crops is another major step for optimizing their resource use and also for improving the yield and nutritional quality [7].

Figure 6. Optimization of agronomic practices and mass multiplication strategies are essential for the large-scale exploitation of underutilized crops. Photographs (**a–e**) shows the various stages of the mass propagation and cultivation of Indian spinach (*Basella alba* and *Basella rubra*).

Though the cultivation practices have been optimized for some of the wild species (for example, pigeon pea [40]; wild rocket [52]; terracrepolo [53]; Physalis sp. [54] etc.), such practices are yet to be standardized for a majority of the neglected species. For example, *Basella alba* and *Basella rubra* (Malabar spinach) are two of the highly nutritive, perennial leafy vegetable, capable of growing in diverse kind of habitat (including degraded and disturbed lands), there is no standardized protocols available for the large-scale cultivation of Basella sp. through stem cutting [11]. Similarly, there are many underutilized species yet to be introduced into farmers' fields (Figure 3A) and many optimal agronomic practices for improving their domestication traits. Moreover, the standardization of inventive agronomic practices is important for emission reduction [55] while increasing the soil's carbon pool [56,57].

Apart from validating the cultivation practices, standardization of suitable cropping models is also essential for successful crop diversification program. Since a majority of the wild crops are hardy and resilient to diverse climatic conditions, they can be used as border crops, as crop breaks, intercrops and also for multi-cropping and integrated farming practices [58]. Importantly, the cultivation of such wild crops will result in species diversification though multiple cropping and niche compartmentalization and thereby diversify the ecosystem functions and services [57]. Though monocropping system possess only limited traits to perform critical ecosystem functions (e.g., biological pest control, managing GHG emissions etc.), diversified cropping system have multiple traits to perform mosaic of functions (e.g., nutrient recycling, nitrogen fixation etc.) [56,59]. Moreover, crop diversification is also essential to ensure pollinator diversity such as honey bees, flies, beetles, moths, butterflies, wasps, ants, birds, and bats etc. as they are crucial for agrobiodiversity and other ecosystem services [60]. Therefore, internationally coordinated efforts are imperative to standardizing the cultivation practices of wild crops as well as for conducting their multi-locational trials in the various agro-climatic regions of the world, and such efforts must be linked to the concerned SDG targets of the appropriate national, regional, and international agencies for global sustainable development.

5. Popularizing the Unpopular: Final Step

Last, but not least, popularizing lesser known species as a climate resilient strategy for food security and resource conservation as well as promoting their role in maintaining ecosystem complexity to common people is another important step for their large-scale cultivation. For this, well-coordinated popularization efforts at various levels and scales are essential with the involvement of multiple stakeholders such as NGOs, State Governments, Local Help Groups, farming communities, National Science Academies, the FAO, CGIAR and other UN organizations. (Figures 7 and 8). Moreover, the standardization of cost-effective and suitable mass multiplication strategies (both micro and macro propagation) [61,62] are important for reducing the cost of the plating materials and also for the large-scale popularization of such neglected and underutilized species because such mass multiplication strategies will make the cultivation of novel crops profitable to farmers.

Figure 7. Popularization of neglected and wild edibles among diverse stakeholders especially women is important for dietary diversification program and thereby attaining food and nutritional security. Photographs shows the various interactive sessions of an awareness workshop on sustainable utilization of wild edibles for the health and well-being of the tribal women conducted by authors group in association with the National Academy of Sciences, India (NASI) and also with the support of Krishak, P.G College, Rajgarh, Mirzapur.

Figure 8. Periodic meetings and community involvements are essential for the awareness creation among indigenous and local community regarding the use of such neglected and underutilized crops for attaining global goals and also for encouraging them to use such species as an intermittent crop for intercropping and crop diversification in family and kitchen garden.

Harnessing indigenous and local knowledge and associated biocultural practices of the local people is an innovative strategy for conservation as well as popularization programs. Since the cultivation of such underutilized and neglected species will empower the indigenous community [5–7], suitable polices must be framed at the local and regional level to harness the ILK and biocultural practices of the local people for the exploration, large-scale exploitation, and also for developing the participatory crop improvement programs for the extensive utilization of such traditional and underutilized wild varieties for dietary diversification program. Therefore, conducting periodic awareness program for farmers especially women and other marginalized people is important for encouraging the introduction of various underutilized crops in their family and kitchen garden and also for developing suitable models for crop diversification based on such neglected and wild crops.

Developing a comprehensive, integrated and multi-utility database having the complete details of wild, neglected and underutilized crops in multiple languages (such as their origin, distribution, nutritional value, cultivation practices, crop genomics, crop improvement strategies, ecosystem services etc.) is of the utmost importance. Though there are websites and databases for crop wild relatives and wild crops, the content and coverage of them are limited. For example, the website of crop wild relatives only showcases 29 priority crops, listed in the Annex 1 of the International Treaty on Plant Genetic Resources for Food and Agriculture (ITPGRFA) (www.cwrdiversity.org) Similarly, the global distribution database of crop wild relatives published by the International Centre for Tropical Agriculture (CIAT) is another venture having the distribution database of crop gene pools of 80 species (www.gbif.org). The crop Wild Relatives Global Portal (www.cropwildrelatives.org) is another joint venture by Bioversity International, United Nations Environment Program (UNEP) and Global Environmental Facility (GEF). The International Legume Database and Information Services (ILDIS) is a promising database for legume species across the world (www.ildis.org). However, as we mentioned earlier, the species coverage in those databases must be expanded beyond the Annexure-1 of the ITPGRFA, and most promising species for global food security, especially from the lesser and underutilized leafy vegetables, wild tubers, wild fruits etc. must be included in the database.

The creation of a national or regional inventory of crop wild relatives is another promising strategy [63,64]. Cultivating wild crops in school gardens and using them for the mid-day meal programs of the developing countries, especially in South East Asia and Africa is another way for popularizing the lesser and under-utilized neglected species [17]. State governments can also play a vital role in encouraging the use of nutritionally rich wild crops. For example, the government of India has proposed nutritionally rich millets as 'nutrimillets' and urged for global solidarity in popularizing them by celebrating an international year for millets. As a result, the FAO council approves India's proposal to celebrate the year 2023 as the International Year of Millets (www.pib.nic.in). Furthermore, the existing 11 gene banks under CGIAR must also go beyond their proposed mandate (i.e., confined to the germplasm collection of a particular crop) and would start collecting germplasms of all uncollected species including wild leafy vegetables, wild vegetables, wild edible flowers, wild fruits, wild tubers etc. for ensuring the current and future food security.

6. Concluding Remarks

Unlocking the real potential of undomesticated and wild crops is imperative for global food and nutritional security and also for attaining many other UN-SDGs. However, targeted crop improvement, domestication as well as cost-effective mass cultivation programs are essential for exploiting their real potential. Therefore, international collaborations under the joint leaderships of FAO and CGIAR institutions, national agriculture institutions and State Governments are essential for the exploration, popularization, as well as the large-scale exploitation of undomesticated crops for ensuring the food and nutritional demand of the global population, and also for communicating their multipurpose benefits to various stakeholders, and framing suitable policies accordingly.

Author Contributions: Conceptualization, P.C.A.; initial draft, A.S. and P.C.A; revision, A.S., P.K.D., R.C., R.K.D., K.K.P., G.S.S., P.C.A.; editing, P.C.A., final approval, A.S., P.K.D., R.C., R.K.D., K.K.P., G.S.S., P.C.A.; supervision, P.C.A., co-supervision, G.S.S.

Funding: Authors have not received any specific funding for this work.

Acknowledgments: Authors are grateful to Head, DESD and Director, IESD for support and encouragement. A.S. is thankful to Jawaharlal Nehru Trust for Jawaharlal Nehru Scholarship; P.K.D. is thankful to UGC for Senior Research Fellowship; R.C. is grateful to CSIR for Junior Research Fellowship. We have the support of Nitin, Gangesh Kumar, Sujeet Bharati, Amit Bundela, and Avinash Kushwaha for field exploration. Thanks, are also due to Principal and Managers of Krishak PG College for their whole-hearted support for this study. Special thanks go to Paras Nath Singh for his kind help and support for this research study. Mahadav Karki, Vice-Chair, IUCN-CEM and Angela Andrade, Chair, IUCN-CEM for encouragements.

Conflicts of Interest: The authors declare no conflict of interest.

References

1. Godfray, H.C. Food security: The challenge of feeding 9 billion people. *Science* **2010**, *327*, 812–818. [CrossRef] [PubMed]

2. Rockstrom, J.; Steffen, W.; Noone, K.; Persson, A.; Chapin, F.S., III; Lambin, E.F.; Lenton, T.M.; Scheffer, M.; Folke, C.; Schellnhuber, H.J.; et al. A safe operating space for humanity. *Nature* **2009**, *461*, 472–475. [CrossRef] [PubMed]

3. Dubey, P.K.; Singh, G.S.; Abhilash, P.C. Agriculture in a changing climate. *J. Clean. Prod.* **2016**, *113*, 1046–1047. [CrossRef]

4. Tripathi, V.; Edrisi, S.A.; Chen, B.; Vilu, R.; Gathergood, N.; Abhilash, P.C. Biotechnological advances for restoring degraded lands for sustainable development. *Trends Biotechnol.* **2017**, *35*, 847–859. [CrossRef] [PubMed]

5. Parodi, A.; Leip, A.; De Boer, I.J.M.; Slegers, P.M.; Ziegler, F.; Temme, E.H.; Herrero, M.; Tuomisto, H.; Valin, H.; Van Middelaar, C.E.; et al. The potential of future foods for sustainable and healthy diets. *Nat. Sustain.* **2018**, *1*, 782. [CrossRef]

6. Nair, K.P. Chapter Four—Utilizing Crop Wild Relatives to Combat Global Warming. *Adv. Agron.* **2019**, *153*, 175–258.

7. Ewel, J.J.; Schreeg, L.A.; Sinclair, T.R. Resources for Crop Production: Accessing the Unavailable. *Trends Plant Sci.* **2018**, *24*, 121–129. [CrossRef] [PubMed]

8. Singh, A.; Abhilash, P.C. Agricultural Biodiversity for Sustainable Food Production. *J. Clean. Prod.* **2018**, *172*, 1368–1369. [CrossRef]

9. Gruber, K. The living library. *Nature* **2017**, *544*, S8–S10. [CrossRef]

10. Singh, A.; Dubey, P.K.; Abhilash, P.C. Food for thought: Putting wild edibles back on the table for combating hidden hunger in developing countries. *Curr. Sci.* **2018**, *115*, 611–613. [CrossRef]

11. Singh, A.; Dubey, P.K.; Chaurasiya, R.; Mathur, N.; Kumar, G.; Bharati, S.; Abhilash, P.C. Indian spinach: An underutilized perennial leafy vegetable for nutritional security in developing world. *Energy Ecol. Environ.* **2018**, *3*, 195–205. [CrossRef]

12. Padulosi, S.; Thompson, J.; Rudebjer, P. *Fighting Poverty, Hunger and Malnutrition with Neglected and Underutilized Species (NUS): Needs, Challenges and the Way Forward*; Bioversity International: Rome, Italy, 2013.

13. Singh, A.; Abhilash, P.C. Varietal dataset of nutritionally important *Lablab purpureus* (L.) sweet from Eastern Uttar Pradesh, India. *Data Brief* **2019**, *24*, 103935. [CrossRef] [PubMed]

14. Singh, A.; Singh, G.S.; El-Keblawy, A.; Abhilash, P.C. Neglected and Underutilized Crops: Towards UN-Sustainable Development Goals. In *Springer Brief in Environmental Sciences*; 2019; in press.

15. Kew, R.B.G. *The State of the World's Plants Report*; Royal Botanic Gardens: Kew, London, UK, 2016.

16. Arora, A.K. *Diversity in Underutilized Plant Species-An Asia Pacific Perspectives*; Bioversity International: New Delhi, India, 2014; 203p.

17. FAO. *Future Smart Food*; Unlocking Hidden Treasures in Asia and the Pacific; Regional Initiative on Zero Hunger Challenge Policy Brief Agricultural Diversification for a Healthy Diet; Food and Agricultural Organization: Rome, Italy, 2017; Available online: http://www.fao.org/3/a-i7717e.pdf (accessed on 10 May 2019).

18. Varshney, R.K.; Ribaut, J.-M.; Buckler, E.S.; Tuberosa, R.; Rafalski, J.A.; Langridge, P. Can genomics boost productivity of orphan crops? *Nat. Biotechnol.* **2012**, *30*, 1172. [CrossRef] [PubMed]

19. Leng, P.F.; Lübberstedt, T.; Xu, M.L. Genomics-assisted breeding—A revolutionary strategy for crop improvement. *J. Integr. Agric.* **2017**, *16*, 2674–2685. [CrossRef]

20. Jiao, Y.; Zhao, H.; Ren, L.; Song, W.; Zeng, B.; Guo, J.; Wang, B.; Liu, Z.; Chen, J.; Li, W.; et al. Genome-wide genetic changes during modern breeding of maize. *Nat. Genet.* **2012**, *44*, 812–815. [CrossRef] [PubMed]

21. Li, J.Y.; Wang, J.; Zeigler, R.S. The 3,000 rice genomes project: New opportunities and challenges for future rice research. *GigaScience* **2014**, *3*, 8. [CrossRef] [PubMed]

22. De Leeuw, M.; Martinant, J.P.; Duborjal, H.; Laffaire, J.B.; Beugnot, R. High throughput SNP discovery in wheat using methylation-sensitive digestion and next-generation sequencing. In *Proceedings of the 19th International Triticeae Mapping Initiative Meeting, Clermont-Ferrand, France, 31 August–4 September 2009*; INRA (Institut National de la Recherche Agronomique): Paris, France, 2009.

23. Takahagi, K.; Yamaguchi, Y.U.; Yoshida, T.; Sakurai, T.; Shinozaki, K.; Mochida, K.; Saisho, D. Analysis of single nucleotide polymorphisms based on RNA sequencing data of diverse bio-geographical accessions in barley. *Sci. Rep.* **2016**, *6*, 33199. [CrossRef]

24. Valliyodan, B.; Ye, H.; Song, L.; Murphy, M.; Shannon, J.G.; Nguyen, H.T. Genetic diversity and genomic strategies for improving drought and waterlogging tolerance in soybeans. *J. Exp. Bot.* **2016**, *68*, 1835–1849. [CrossRef]

25. Yang, H.; Jian, J.; Li, X.; Renshaw, D.; Clements, J.; Sweetingham, M.W.; Li, C. Application of whole genome re-sequencing data in the development of diagnostic DNA markers tightly linked to a disease-resistance locus for marker-assisted selection in lupin (*Lupinus angustifolius*). *BMC Genom.* **2015**, *16*, 660. [CrossRef]

26. Liu, Q.; Liu, J.; Zhang, P.; He, S. Root and tuber crops. In *Encyclopaedia of Agriculture and Food Systems*; Elsevier: London, UK, 2014; pp. 46–61.

27. Harouna, D.V.; Venkataramana, P.B.; Ndakidemi, P.A.; Matemu, A.O. Under-exploited wild *Vigna* species potentials in human and animal nutrition: A review. *Glob. Food. Secur.* **2018**, *18*, 1–11. [CrossRef]

28. Pimienta-Barrios, E. Prickly pear (*Opuntia* spp.): A valuable fruit crop for the semi-arid lands of Mexico. *J. Arid. Environ.* **1994**, *28*, 1–11. [CrossRef]

29. Greene, S.L.; Hart, T.C.; Afonin, A. Using geographic information to acquire wild crop germplasm for ex situ collections: II. Post-collection analysis. *Crop Sci.* **1999**, *39*, 843–849. [CrossRef]

30. Broegaard, R.B.; Rasmussen, L.V.; Dawson, N.; Mertz, O.; Vongvisouk, T.; Grogan, K. Wild food collection and nutrition under commercial agriculture expansion in agriculture-forest landscapes. *For. Policy Econ.* **2017**, *84*, 92–101. [CrossRef]

31. Reynolds, M.P.; Quilligan, E.; Aggarwal, P.K.; Bansal, K.C.; Cavalieri, A.J.; Chapman, S.C.; Chapotin, S.M.; Datta, S.K.; Duveiller, E.; Gill, K.S.; et al. An integrated approach to maintaining cereal productivity under climate change. *Glob. Food Secur.* **2016**, *8*, 9–18. [CrossRef]

32. Rasheed, A.; Hao, Y.; Xia, X.; Khan, A.; Xu, Y.; Varshney, R.K.; He, Z. Crop Breeding chips and genotyping platforms: Progress, challenges, and perspectives. *Mol. Plant* **2017**, *10*, 1047–1064. [CrossRef] [PubMed]

33. Varshney, R.K.; Bansal, K.C.; Aggarwal, P.K.; Datta, S.K.; Craufurd, P.Q. Agricultural biotechnology for crop improvement in a variable climate: Hope or hype? *Trends Plant Sci.* **2011**, *16*, 363–371. [CrossRef] [PubMed]

34. Huang, S.; Weigel, D.; Beachy, N.R.; Li, J. A proposed regulatory framework for genome-edited crops. *Nat. Genet.* **2016**, *48*, 109–111. [CrossRef] [PubMed]

35. Østerberg, J.T.; Xiang, W.; Olsen, L.I.; Edenbrandt, A.K.; Vedel, S.E.; Christiansen, A.; Landes, X.; Andersen, M.M.; Pagh, P.; Sandøe, P.; et al. Accelerating the domestication of new crops: Feasibility and approaches. *Trends Plant Sci.* **2017**, *22*, 373–384. [CrossRef] [PubMed]

36. Lemmon, Z.H.; Reem, N.T.; Dalrymple, J.; Soyk, S.; Swartwood, K.E.; Rodriguez-Leal, D.; Van Eck, J.; Lippman, Z.B. Rapid improvement of domestication traits in an orphan crop by genome editing. *Nat. Plants* **2018**, *4*, 766. [CrossRef]

37. D'Amelia, V.; Villano, C.; Aversano, R. Emerging Genetic Technologies to Improve Crop Productivity. *Encycl. Food Secur. Sustain.* **2019**, *3*, 152–158.

38. Xu, J.; Hou, Q.M., Khare, T.; Verma, S.K.; Kumar, V. Exploring miRNAs for developing climate-resilient crops: A perspective review. *Sci. Total Environ.* **2019**, *653*, 91–104. [CrossRef] [PubMed]

39. Khan, M.Z.; Zaidi, S.S.A.; Amin, I.; Mansoor, S. A CRISPR way for fast-forward crop domestication. *Trends Plant Sci.* **2019**, *24*, 293–296. [CrossRef] [PubMed]

40. Khoury, C.K.; Castanñeda-Alvarez, N.P.; Achicanoy, H.A.; Sosa, C.C.; Bernau, V.; Kassa, M.T.; Norton, S.L.; van der Maesen, L.J.S.; Upadhyaya, H.D.; Ramiírez-Villegas, J.; et al. Crop wild relatives of pigeon pea [*Cajanus cajan* (L.) Millsp.]: Distributions, ex situ conservation status, and potential genetic resources for abiotic stress tolerance. *Biol. Conserv.* **2015**, *184*, 259–270. [CrossRef]

41. Muthamilarasan, M.; Dhaka, A.; Yadav, R.; Prasad, M. Exploration of millet models for developing nutrient rich graminaceous crops. *Plant Sci.* **2016**, *242*, 89–97. [CrossRef] [PubMed]

42. Singh, V.; Goel, R.; Pande, V.; Asif, M.H.; Mohanty, C.S. *De novo* sequencing and comparative analysis of leaf transcriptomes of diverse condensed tannin-containing lines of underutilized *Psophocarpus tetragonolobus* (L.) DC. *Sci. Rep.* **2017**, *7*, 44733. [CrossRef]

43. Rakshit, S.; Bellundagi, A. Chapter 5—Conventional breeding techniques in Sorghum. In *Breeding Sorghum for Diverse End Uses*; Woodhead Publishing: Duxford, UK, 2019. [CrossRef]

44. Weltzein, E.; Christinck, A. Chapter 8—Participatory breeding: Developing improved and relevant crop varieties with farmers. In *Agricultural Systems*; Elsevier: London, UK, 2017; pp. 259–301. [CrossRef]

45. Balázs, E.; Divéki, Z. Boosting super-domestication: From crop domestication to genome editing. *S. Afr. J. Bot.* **2019**. [CrossRef]

46. Waldman, K.B.; Kerr, J.M.; Isaacs, K.B. Combining participatory crop trials and experimental auctions to estimate farmer preferences for improved common bean in Rwanda. *Food Policy* **2014**, *46*, 183–192. [CrossRef]

47. Annicchiarico, P.; Russi, L.; Romani, M.; Pecetti, L.; Nazzicari, N. Farmer-participatroy vs. conventional market-oriented breeding of inbred crops using phenotypic and genome-enabled approaches: A pea case study. *Field Crops Res.* **2019**, *232*, 30–39. [CrossRef]

48. Allard, R.W. *Principles of Plant Breeding*; John Wiley and Sons Inc.: New York, NY, USA, 1960.

49. Rakshit, S.; Gomashe, S. Basics of plant breeding with reference to sorghum. In *Basics of Sorghum Breeding and AICSIP Data Management*; Rakshit, S., Patil, J.V., Eds.; Directorate of Sorghum Research: Hyderabad, India, 2013; pp. 9–16.

50. Soeranto, H.; Trikoesoemaningtyas, T.; Sihono, S.; Sungkono, S. Development of sorghum tolerant to acid soil using induced mutation with gamma irradiation. *At. Indones.* **2010**, *36*, 11–15.

51. Massawe, F.; Mayes, S.; Cheng, M. Crop diversity: An unexploited treasure trove for food security. *Trends Plant Sci.* **2016**, *5*, 365–368. [CrossRef]

52. Schiattone, M.I.; Candido, V.; Cantore, V.; Montesano, F.F.; Boari, F. Water use and crop performance of two wild rocket genotypes under salinity conditions. *Agric. Water Manag.* **2017**, *194*, 214–221. [CrossRef]

53. Maggini, R.; Benvenuti, S.; Leoni, F.; Pardossi, A. Terracrepolo (*Reichardia picroides* (L.) Roth.): Wild food or new horticultural crop? *Sci. Hortic.* **2018**, *20*, 224–231. [CrossRef]

54. Saavedra, J.C.M.; Zaragoza, F.A.R.; Toledo, D.C.; Hernández, C.V.S.; Vargas-Ponce, O. Agromorphological characterization of wild and weedy populations of *Physalis angulata* in Mexico. *Sci. Hortic.* **2019**, *246*, 86–94. [CrossRef]

55. Wollenberg, E.; Richards, M.; Smith, P.; Havlik, P.; Obersteiner, M.; Tubiello, F.N.; Herold, M.; Gerber, P.; Carter, S.; Reisinger, A.; et al. Reducing emissions from agriculture to meet the 2 °C Target. *Glob. Chang. Biol.* **2016**, *22*, 3859–3864. [CrossRef] [PubMed]

56. Power, A.G. Ecosystem services and agriculture: Trade-offs and synergies. *Phils. Trans. R. Soc. B* **2010**, *365*, 2959–2971. [CrossRef]

57. Poppy, G.M.; Chiotha, S.; Eigenbrod, F.; Harvey, C.A.; Honzak, M.; Hudson, M.D.; Jarvis, A.; Madise, N.J.; Schreckenberg, K.; Shackleton, C.M.; et al. Food security in a perfect storm: Using the ecosystem services framework to increase understanding. *Phils. Trans. R. Soc. Lond. B* **2014**, *369*, 20120288. [CrossRef]

58. Dubey, P.K.; Singh, G.S.; Abhilash, P.C. Adaptive agricultural practices: Building resilience in a changing climate. In *SpringerBriefs in Environmental Science*; Springer: New York, NY, USA, 2019; ISBN 978-3-030-15518-6. [CrossRef]

59. Dubey, R.K.; Tripathi, V.; Dubey, P.K.; Singh, H.B.; Abhilash, P.C. Exploring rhizospheric interactions for agricultural sustainability: The need of integrative research on multi-trophic interactions. *J. Clean. Prod.* **2016**, *115*, 362–365. [CrossRef]

60. Potts, S.G.; Imperatriz-Fonseca, V.; Ngo, H.T.; Aizen, M.A.; Biesmeijer, J.C.; Breeze, T.D.; Dicks, L.V.; Garibaldi, L.A.; Hill, R.; Settele, J.; et al. Safeguarding pollinators and their values to human well-being. *Nature* **2016**, *540*, 220–229. [CrossRef]

61. ul-Haq, Z.; Rashid, A.; Khan, S.M.; Razzaq, A.; Al-Yahyai, R.A.; Kamran, S.; Ali, S.G.; Ali, S.; Saifullah; Abdullah; et al. In vitro and in vivo propagation of *Monotheca buxifolia* (Falc.) A. DC. An economical medicinal plant. *Acta Ecol. Sin.* **2019**, in press. [CrossRef]

62. Jalali, N.; Naderi, R.; Shahi-Gharahlar, A.; da Silva, J.A.T. Tissue culture of *Cyclamen* spp. *Sci. Hortic.* **2012**, *137*, 11–19. [CrossRef]

63. Maxted, N.; Scholten, M.; Codd, R.; Ford-Lloyd, B. Creation and use of a national inventory of crop wild relatives. *Biol. Conserv.* **2007**, *140*, 142–159. [CrossRef]

64. Khoury, C.K.; Amariles, D.; Soto, J.S.; Victoria Diaz, M.; Sotelo, S.; Sosa, C.C.; Ramírez-Villegas, J.; Achicanoy, H.A.; Velásquez-Tibatá, J.; Guarino, L.; et al. Comprehensiveness of conservation of useful wild plants: An operational indicator for biodiversity and sustainable development targets. *Ecol. Indic.* **2019**, *98*, 420–429. [CrossRef]

MDPI
St. Alban-Anlage 66
4052 Basel
Switzerland
Tel. +41 61 683 77 34
Fax +41 61 302 89 18
www.mdpi.com

Agronomy Editorial Office
E-mail: agronomy@mdpi.com
www.mdpi.com/journal/agronomy